This is an advanced text on electromagnetic theory, presenting a systematic discussion of electromagnetic waves and radiation processes in a wide variety of media. The treatment, taken from the field of plasma physics, is based on the dielectric tensor, and this permits the discussion of media outside the scope of the usual approach adopted in most textbooks on electromagnetism. The authors have thus unified the approaches used in plasma physics and astrophysics on the one hand, and in optics on the other.

The authors have written clearly and pedagogically for the student, and all the necessary mathematical tools such as tensor algebra, Fourier transforms and distributions, Greens functions etc, are set out in Part One. Parts Two and Three cover the properties of electromagnetic waves in various media, whilst Part Four treats the general theory of emission processes, such as multipole emission, bremsstrahlung and cyclotron emission. The fifth and final part is set at a more advanced level, and covers various specific emisssion processes in greater detail. The approach taken by the authors has notable advantages when applied to the conventional emission processes of electromagnetic theory.

The book will be of value to senior undergraduates, graduate students, lecturers and researchers in the fields of electromagnetism, plasma physics, astrophysics, optics and electrical engineering. Of particular usefulness to the student will be the exercises provided at the end of each chapter.

T0269098

Electromagnetic Processes in Dispersive Media

A Treatment Based on the Dielectric Tensor

Electromagnetic Processes in Dispersive Media

A Treatment Based on the Dielectric Tensor

D.B. Melrose and R.C. McPhedran
School of Physics
University of Sydney

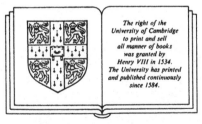

The right of the
University of Cambridge
to print and sell
all manner of books
was granted by
Henry VIII in 1534.
The University has printed
and published continuously
since 1584.

CAMBRIDGE UNIVERSITY PRESS
Cambridge
New York Port Chester
Melbourne Sydney

CAMBRIDGE UNIVERSITY PRESS
Cambridge, New York, Melbourne, Madrid, Cape Town, Singapore, São Paulo

Cambridge University Press
The Edinburgh Building, Cambridge CB2 2RU, UK

Published in the United States of America by Cambridge University Press, New York

www.cambridge.org
Information on this title: www.cambridge.org/9780521410250

First published 1991
This digitally printed first paperback version 2005

A catalogue record for this publication is available from the British Library

ISBN-13 978-0-521-41025-0 hardback
ISBN-10 0-521-41025-8 hardback

ISBN-13 978-0-521-01848-7 paperback
ISBN-10 0-521-01848-X paperback

Contents

Contents

Chapter 4 : Fourier Transforms
4.1	Fourier Integral Theorem	38
4.2	Properties of Fourier Transforms	40
4.3	The Dirac δ-function	42
4.4	Truncations	43
4.5	Relations Involving Generalized Functions	45
	Exercise Set 4	48

Chapter 5 : Greens Functions
5.1	Fourier Transformed Fields	52
5.2	Solution of Inhomogeneous Differential Equations	54
5.3	The Greens Function for Poisson's Equation	55
5.4	The Greens Function for d'Alembert's Equation	56
	Exercise Set 5	60

Part Two: Electromagnetic Responses of Media

Chapter 6 : The Response of a Medium
6.1	Static Responses	64
6.2	Properties of Media	68
6.3	Two Alternative Descriptions of the Linear Response	71
6.4	Non-linear Response Tensors	74
	Exercise Set 6	77

Chapter 7 : General Properties of Response Tensors
7.1	Positive and Negative Frequencies	79
7.2	Separation into Dissipative and Non-dissipative Parts	80
7.3	The Kramers–Kronig Relations	81
7.4	The Onsager Relations	83
7.5	The Principal Axes of Anisotropic Crystals	86
	Exercise Set 7	88

Chapter 8 : Analytic Properties of Response Functions
8.1	Complex Frequencies and Analytic Continuation	90
8.2	Laplace Transforms	92
8.3	Contour Integration	93
8.4	Poles in the Upper Half ω Plane	96
	Exercise Set 8	100

Chapter 9 : Response Tensors for Some Idealized Media
9.1	The Polarizability of Atoms and Molecules	102
9.2	The Lorenz–Lorentz Equation	103
9.3	A System of Forced Classical Oscillators	107
9.4	Quantum Calculation of Polarizability	110
	Exercise Set 9	113

Contents

Preface

In an undergraduate physics course it is common practice to introduce electromagnetic theory in two stages with a third stage at senior undergraduate level or higher. In a first course the integral forms of Maxwell's equations are introduced and used to treat a variety of problems relating electric and magnetic fields to their sources. The essential part of a second course is the introduction of the differential forms of Maxwell's equations, with a major ingredient being the development of the necessary mathematical tools of the differential vector calculus and the integral theorems of Gauss and Stokes. The physical content of this part of the second course differs little from that of the first course, and usually some additional chapters such as electromagnetic responses of media, propagation of electromagnetic waves in waveguides, Lorentz transformation of electromagnetic fields and so on, are included to add some new physical content. A third course in electromagnetic theory starts with the differential forms of Maxwell's equations, and is usually at senior undergraduate or first year graduate level. The present book is intended to be at the level of such a third course in electromagnetic theory.

There is an approach to the teaching of electromagnetic theory at this level that has become almost traditional due to the availability of some excellent textbooks that present a similar approach. Notable examples are Stratton (1941), Landau and Lifshitz, (1951) Jackson (1975) and Panofsky and Phillips (1962), as cited in the Bibliographic Notes. The content of these and of similar texts now essentially defines the content of advanced courses in "electromagnetic theory". Important formal parts of such a course include the properties of the electromagnetic field in material media and the emission of electromagnetic radiation. A conventional approach to the treatment of the responses of material media is based on induced dipole and multipole moments in the medium. This conventional approach is inadequate for at least one important application, specifically to plasmas. The description of the static response of a dielectric or of a magnetizable medium involves the introduction of the polarization **P** and the magnetization **M** and the associated fields **D** and **H**. Although this description may be applied to simple plasmas, more generally (specifically, for spatially dispersive media) it is ill defined and unsatisfactory.

There is a similar situation in the theory of the emission of radiation. The conventional treatment of the emission of electromagnetic radiation is based on the solution of d'Alembert's equation to relate time-varying potentials to their sources (in the case of emission by particles, the Lienard–Wiechert potentials), the construction of the fields from the potentials and the calculation of the Poynting flux. This procedure applies only to emission *in vacuo*. It can be extended relatively simply to treat the emission of radiation in a dispersive, isotropic dielectric, but an alternative approach needs to be used for emission in an arbitrary medium, specifically, in an arbitrary plasma. As a consequence of these weaknesses in the traditional treatment of electromagnetic theory, the important application to plasmas is either omitted or treated in a cursory manner, and plasma physics has become regarded as separate from electromagnetic theory rather than a natural application of electromagnetic theory.

Although the methods used to describe the response of a plasma and to treat the emission of radiation in plasmas were developed over three decades ago, there is still no wide consensus on the specific notation and terminology to be used. The essential difference, in terms of methodology, between the approach used in plasma physics and the conventional approach to electromagnetic theory outlined above is that the electromagnetic fields are described in terms of an expansion in plane waves. Thus, the electromagnetic fields are described in terms of their Fourier transforms in both space and time, so that they are functions of frequency ω and wave vector **k** rather than of time t and position **x**. One refers to the description in *Fourier space* rather than in *coordinate space*. This alternative approach is needed because the response of a plasma (other than in simple limiting cases) can be described simply only in

Fourier space, for example, as a linear relation between the induced current $\mathbf{J}(\omega, \mathbf{k})$ and the electric field $\mathbf{E}(\omega, \mathbf{k})$.

In this book we develop electromagnetic theory using the Fourier-space description from the plasma physics literature. Although the main advantage is in the application to processes in plasmas and plasma-like media, the method has notable advantages when applied to treat the conventional emission processes of electromagnetic theory. An example where the Fourier-space description has a major advantage is in the treatment of Cerenkov emission, which is the simplest of all emission processes in the approach adopted here, cf. Chapter 20. In contrast, in the more conventional approach, Cerenkov emission appears exceptional in two ways. First, in an approach based on the use of the Larmor formula, emission is attributed to the acceleration of a charge, whereas Cerenkov emission is due to a charge in constant rectilinear motion. Second, as a consequence, a special method is needed to treat Cerenkov emission, and this often involves artificially attributing the emission to the medium rather than to the particle. An advantage of the Fourier-space description is that it leads directly to a description of the emission in terms of its distribution in Fourier space, and hence in terms of frequency and angle of propagation, and this is what is of direct physical relevance. Thus even for emission *in vacuo* the Fourier-space description has an intrinsic advantage in this sense.

As the Fourier-space description is less familiar than the conventional approach, let us comment further on its advantages and its disadvantages. The advantages include a more general framework for the description of the response of an arbitrary medium, and a more general basis for the description of emission processes in dispersive media, of relevance to both plasma physics and to non-linear optics. There are two apparent disadvantages, one of which is real and the other is somewhat artificial. The latter apparent disadvantage is a loss of physical intuition when replacing the description in coordinate space by one in Fourier space. In reality, for a radiation field the Fourier-space description is of more direct physical relevance than coordinate-space description, but this point tends to be obscured by the traditional teaching of electromagnetic theory emphasizing the coordinate-space approach. Nevertheless, it is a complaint of students when they first encounter the Fourier-space description that it is difficult to build up a physical feeling for the description in Fourier space. The other disadvantage is an actual one: the Fourier-space description requires the use of mathematical tools that are unfamiliar to most students, and which need to be taught before much progress can be made. These tools include (a) the Fourier integral transform in both time and space, (b) tensor algebra, (c) generalized functions, and (d) some matrix algebra. This leads to the obvious disadvantage of the early part of the course being devoted to preliminary developments of the necessary mathematical tools.

To illustrate the application of this Fourier-space description to a conventional problem in electromagnetic theory, let us compare the treatments of emission of radiation *in vacuo* using the two approaches. The steps in the conventional approach involve: (1) reducing Maxwell's equations to d'Alembert's equation, (2) solving d'Alembert's equation for the potentials, (3) constructing the fields, and (4) constructing the Poynting vector. Formally, the most difficult step is solving d'Alembert's equation to find the retarded potentials in terms of their sources. In the Fourier-space description step (1) is replaced by deriving the inhomogeneous wave equation, and step (2) by solving this equation. The wave equation is an algebraic equation rather than a differential equation and hence its solution is trivial. The solution is complicated somewhat in the presence of an anisotropic medium in that one needs to solve a set of simultaneous algebraic equations using a matrix or equivalent approach. Steps (3) and (4) are replaced by identifying the energy radiated in terms of the solution of the wave equation. An important ingredient is the identification of the radiation field, which arises from a singular part of the solution, and the correct treatment of this singular term requires either the use of generalized functions or of contour integration to impose the causal condition. The subtlety involved is hidden in the more conventional approach by choosing the retarded solution of d'Alembert's equation.

The book is set out so that most of the preliminary mathematical tools are covered in Part One. Although we have attempted to relate the mathematical developments to specific problems in electromagnetic theory, it seems that a heavy initial emphasis on mathematical developments is unavoidable. From the viewpoint of the structure of the course, the essence of Part One is the reduction of Maxwell's equations to the "wave equation". The response of a medium is discussed in Part Two, with the emphasis on the distinction between the plasma-physics approach adopted here and the older more conventional approach. In Part Three the homogeneous wave equation is solved to find the properties of waves in various media. The inhomogeneous wave equation is solved in Part Four and a first treatment of emission processes is presented. A more formal and extended discussion of various emission and absorption processes is given in Part Five.

The present book developed from lecture notes for a course to Honours Year students at the University of Sydney. The objective of the course is to give the students an adequate background to embark on more specific chapters in astrophysical and laboratory plasma physics, in radioastronomy and, to a lesser extent, in physical optics. In our one-semester course we attempt to cover most (but not all) the material in Parts One–Four, with at most a brief introduction to the material in Part Five, which is more appropriate for graduate students.

University of Sydney   D.B. Melrose & R.C. McPhedran
January 1991

List of Symbols

PART ONE

Electromagnetic Fields in Vacuo

In order to relate the electromagnetic field to its sources one needs to solve Maxwell's equations. One method of solving Maxwell's equations is to introduce potentials (the scalar and vector potentials) which effectively reduces the number of independent equations. For static sources and fields one may reduce Maxwell's equations to Poisson's equation for the electromagnetic potentials, and for time-dependent sources and fields one may reduce Maxwell's equations to d'Alembert's equation for the electromagnetic potentials. An alternative approach is to Fourier transform. This approach applies only to fluctuating fields, and hence it is necessary to distinguish between fluctuating fields and any static field, which is regarded as an ambient field when describing the response to fluctuating fields using Fourier transforms. The use of Fourier transforms allows one to reduce Maxwell's equations for time-dependent sources and fields to a single algebraic equation, called the wave equation. A specific advantage of this approach is that it allows one to include the effect of an ambient medium in a simple but general way.

Maxwell's equations are written down and the electromagnetic potentials are introduced in Chapter 1. There are two mathematical tools that are required in the treatment of electromagnetic theory adopted here. One of these is tensor algebra, which is introduced in Chapter 2. Some electromagnetic applications of tensor algebra are described in Chapter 3 with particular emphasis on multipole moments. The other mathematical tool is the Fourier transformation, which is introduced in Chapter 4. The solution of Poisson's equation or d'Alembert's equation involves the use of Greens functions, and such functions are discussed in Chapter 5.

1

Electromagnetic Fields

Preamble

The electromagnetic field may always be described in terms of two basic fields: the electric field strength $\mathbf{E}$ and the magnetic induction $\mathbf{B}$. The sources for these fields are the charge density ρ and the current density $\mathbf{J}$. The sources and the fields are related by Maxwell's equations. Maxwell's equations are four simultaneous first order differential equations and one cannot solve them directly to find the fields given the source terms. One way of solving Maxwell's equations is to introduce an alternative description of an electromagnetic field in terms of the scalar potential ϕ and vector potential $\mathbf{A}$. These potentials are not uniquely defined and a specific choice for them satisfies a gauge condition.

1.1 Maxwell's Equations

The *electric field* $\mathbf{E}$ is defined as the force per unit charge acting on a test charge. Thus, if q is an arbitrarily small test charge (so that the electric field that it generates is negligible) located at a point a distance $\mathbf{x}$ from the origin at time t, then the electric force on it is $q\mathbf{E}(t, \mathbf{x})$. (For simplicity in notation, and where no confusion is likely to arise, the dependences on t and $\mathbf{x}$ are not shown explicitly.) The units of electric field strength are thus those of a force (kg m s^{-2}) per unit charge; the unit of charge is the Coulomb (C) so that the magnitude of $\mathbf{E}$ has units $\text{kg m s}^{-2}\text{C}^{-1}$. A more convenient unit follows by noting that the energy of a charge in an electric field is given by the charge times the electric potential; the unit of electric potential is the volt (V) so that $\mathbf{E}$ has the units of V m^{-1}.

The *magnetic induction* **B** cannot be defined in the same way as the electric field strength because there is no magnetic counterpart of a charge. Instead **B** is defined in terms of the force acting on a current **I**. The force per unit length is given by **I** × **B**. The unit of magnetic induction is the tesla (T). As the unit of current is the ampere (1 A = $1\,\mathrm{C\,s^{-1}}$), the tesla is related to the electric units by $1\,\mathrm{T} = 1\,\mathrm{kg\,s^{-1}C^{-1}}$. An older unit now replaced by the tesla involves the unit of magnetic flux, which is the weber (Wb); magnetic flux is defined by integrating the magnetic induction over a surface so that one has is $1\,\mathrm{T} = 1\,\mathrm{Wb\,m^{-2}}$. The cgs unit for magnetic induction is still widely used; it is the gauss (G) with $1\,\mathrm{T} = 10^4\,\mathrm{G}$.

The source terms are the charge density ρ and the current density **J**. The charge density is defined by considering an arbitrarily small volume about the point **x** at time t. Then ρ times the volume gives the total charge inside the volume. The units of ρ are $\mathrm{C\,m^{-3}}$. The current density **J** is defined by considering an arbitrarily small circular area centered on the point **x** at time t and oriented such that the normal to the surface of the circle is along the direction of **J**. Then the magnitude of **J** is equal to the charge per unit area crossing the surface per unit time. The units of current density are $\mathrm{A\,m^{-2}}$.

The fields **E** and **B** are related to each other and to ρ and **J** by Maxwell's equations. In SI units these are

$$\mathrm{curl}\,\mathbf{E} = -\partial\mathbf{B}/\partial t, \tag{1.1}$$

$$\mathrm{curl}\,\mathbf{B} = \mu_0\mathbf{J} + (1/c^2)\,\partial\mathbf{E}/\partial t, \tag{1.2}$$

$$\mathrm{div}\,\mathbf{E} = \rho/\varepsilon_0, \tag{1.3}$$

$$\mathrm{div}\,\mathbf{B} = 0, \tag{1.4}$$

where c is the speed of light. The first three equations are called are Faraday's law, Ampère's law and Gauss' law respectively. The parameters ε_0 and μ_0 are the permittivity and the permeability of free space, respectively, with $\mu_0\varepsilon_0 = 1/c^2$. The numerical values of these quantities are $\varepsilon_0 = 8.854 \times 10^{-12}\,\mathrm{F\,m^{-1}}$ and $\mu_0 = 4\pi \times 10^{-7}\,\mathrm{H\,m^{-1}}$, where the farad ($1\,\mathrm{F} = 1\,\mathrm{C\,V^{-1}}$) is the unit of capacitance, and the henry ($1\,\mathrm{H} = 1\,\mathrm{V\,s\,A^{-1}}$) is the unit of inductance.

Equations (1.1)–(1.4) are completely general forms of Maxwell's equations, and apply when a medium is present as well as in free space. In the presence of a medium the charge and current densities are separated into induced and extraneous parts by writing $\rho = \rho_{\mathrm{ind}} + \rho_{\mathrm{ext}}$ and $\mathbf{J} = \mathbf{J}_{\mathrm{ind}} + \mathbf{J}_{\mathrm{ext}}$, and the induced parts are identified as the response of the medium. In one description of the response of a certain class of media, the responses are incorporated into two new fields: the *electric induction* $\mathbf{D}$ and the *magnetic field strength* $\mathbf{H}$. Then (1.2) and (1.3) are replaced by $\mathrm{curl}\,\mathbf{H} = \mathbf{J}_{\mathrm{ext}} + \partial\mathbf{D}/\partial t$ and $\mathrm{div}\,\mathbf{D} = \rho_{\mathrm{ext}}$, respectively. However, the separation into induced and extraneous parts requires a model for the medium, and hence the introduction of $\mathbf{D}$ and $\mathbf{H}$ is model dependent. Thus the validity of this alternative form of Maxwell's equations relies on assumptions about the properties of the medium, and so this alternative form is less general than the form (1.1)–(1.4) in that it requires supplementary information to give it meaning.

In a vacuum $\mathbf{D}$ and $\mathbf{H}$ are related to $\mathbf{E}$ and $\mathbf{B}$ by $\mathbf{D} = \varepsilon_0\mathbf{E}$ and $\mathbf{H} = \mathbf{B}/\mu_0$. It is important to recognize that $\mathbf{E}$ and $\mathbf{B}$ are the basic fields, and that, apart from the case of a vacuum, the fields $\mathbf{D}$ and $\mathbf{H}$ are defined only in terms of a specific model of the medium. The description of the response of a medium is discussed in detail in Chapter 6.

1.2 Continuity Equations

Three continuity equations are implied directly by Maxwell's equations. A *continuity equation* expresses the fact that some quantity is conserved. The form of the continuity equation for a quantity Q is $\partial W/\partial t + \mathrm{div}\,\mathbf{F} = 0$, where W is the density of Q, that is, $Q = \int \mathrm{d}^3\mathbf{x}\,W$, and $\mathbf{F}$ is the flux of Q. On integrating the continuity equation over a volume V with a surface S, the first term gives the rate of change of Q with time, $\partial Q/\partial t$. The second term may be rewritten as a surface integral using the divergence theorem:

$$\int_V \mathrm{d}^3\mathbf{x}\,\mathrm{div}\,\mathbf{F} = \int_S \mathrm{d}^2 S\,\mathbf{n}\cdot\mathbf{F}, \tag{1.5}$$

where $\mathbf{n}$ is the outward normal to the surface S and where $\mathrm{d}^2 S$ denotes an element of area on the surface. The integral over the second term in the continuity equation is then interpreted as the rate of loss of Q from the volume V due to the outward flux of Q across the surface S. On moving the second term to the right hand side and interpreting the associated change in sign in terms of an inward rather than an outward flux, the integral of the continuity equation over a volume implies that the time rate of change of the quantity Q inside the volume is balanced by the flux of Q crossing the surface into the volume. In this way the continuity equation implies that the quantity Q is conserved. If there is a source (or sink, which is a negative source) of Q, then there is a *source term* on the right hand side of the continuity equation representing the rate per unit time and per unit volume at which the quantity Q is generated by the source.

The divergence of (1.2) and the time-derivative of (1.3) imply the *charge continuity equation*

$$\partial\rho/\partial t + \text{div } \mathbf{J} = 0. \tag{1.6}$$

The integral of the charge density over volume is the charge within that volume, and hence (1.6) expresses the fact that charge is conserved. That is, charges can neither be created nor destroyed.

An equation interpreted as expressing continuity of energy is derived by evaluating div $(\mathbf{E} \times \mathbf{B})$, and then using Maxwell's equations to rewrite the result. The details of the derivation are included as an exercise in the use of tensor notation in Chapter 2. In vector notation, one finds

$$\frac{\partial}{\partial t} \left(\tfrac{1}{2}\varepsilon_0 |\mathbf{E}|^2 + \tfrac{1}{2}|\mathbf{B}|^2/\mu_0 \right) + \text{div} \left(\mathbf{E} \times \mathbf{B}/\mu_0 \right) = -\mathbf{J} \cdot \mathbf{E}. \tag{1.7}$$

The interpretation of (1.7) is based on the fact that the source term on the right is interpreted in terms of mechanical work. For example, consider a system composed of classical charged particles. The equation of motion for a particle with charge q is $d\mathbf{p}/dt = q(\mathbf{E} + \mathbf{v} \times \mathbf{B})$. The rate of increase in the energy E of the particle is given by $dE/dt = \mathbf{v} \cdot d\mathbf{p}/dt = q\mathbf{v} \cdot \mathbf{E}$, which gives the rate at which the electric field does work on the particle. (Note that the magnetic fields do no work.) If there are n such particles per unit volume, then the rate per unit volume at which the electromagnetic field does work on these particles is $ndE/dt = \mathbf{J} \cdot \mathbf{E}$, where $\mathbf{J} = qn\mathbf{v}$ is the current density associated with the particles. The energy gained by the particles must be supplied by the electromagnetic field, and hence the rate work is done by the particles on the electromagnetic field is $-\mathbf{J} \cdot \mathbf{E}$, which is just the source term in (1.7). Thus (1.7) must be a continuity equation for energy in the electromagnetic field.

The interpretation of (1.7) as expressing continuity of electromagnetic energy implies that $\tfrac{1}{2}\varepsilon_0|\mathbf{E}|^2$ is the electric energy density, $|\mathbf{B}|^2/2\mu_0$ is the magnetic energy density, and $\mathbf{E} \times \mathbf{B}/\mu_0$, called the *Poynting vector*, is the electromagnetic energy flux. It should be emphasized that the interpretation of these terms on the left hand side of (1.7) is straight-forward only *in vacuo*. In the presence of a medium the induced part of the current is identified as the response of the medium, and the associated "source" term $-\mathbf{J}_{\text{ind}} \cdot \mathbf{E}$ should be moved to the left hand side and re-interpreted appropriately. In practice, this procedure is useful only for simple types of media, and an alternative procedure is more convenient for deriving the continuity equation for wave energy in dispersive media.

A third continuity equation that follows directly from Maxwell's equations is for the electromagnetic momentum. In this case the quantity that is conserved is a vector, and the associated "flux" is a tensor (the electromagnetic stress tensor). Thus even to write down the appropriate continuity equation one needs to use either tensor notation or some equivalent notation (such as dyadics). Discussion of this continuity equation is deferred to Chapter 3 after tensor notation is introduced in Chapter 2.

1.3 Scalar and Vector Potentials

In attempting to solve Maxwell's equations for the fields in terms of their sources, an obvious difficulty arises in solving four coupled partial differential equations. One first replaces them by an equivalent set of equations that may be solved in a standard way. There are several ways of proceeding, and different procedures are used in different contexts.

As a first step in solving Maxwell's equations it is often helpful to introduce the scalar potential ϕ and the vector potential $\mathbf{A}$. Although the primary motivation for introducing these fields is mathematical convenience, the fields ϕ and $\mathbf{A}$ are regarded as constituting an alternative description of an arbitrary electromagnetic field in place of $\mathbf{E}$ and $\mathbf{B}$. In simple cases, such as an electrostatic field, the potential ϕ is identified as the usual electric potential of the field, but in general the physical interpretation of the fields ϕ and $\mathbf{A}$ is less obvious than the physical interpretation of the fields $\mathbf{E}$ and $\mathbf{B}$. An important point is that an arbitrary electromagnetic field may be described in a variety of mathematically equivalent ways, and the only feature special about the description in terms of $\mathbf{E}$ and $\mathbf{B}$ is the straightforward physical interpretation of these fields.

The mathematical motivation for introducing the potential is to satisfy two of Maxwell's equations identically. Specifically, (1.1) and (1.4) are satisfied identically by writing

$$\mathbf{E} = -\operatorname{grad}\phi - \partial\mathbf{A}/\partial t, \tag{1.8}$$

$$\mathbf{B} = \operatorname{curl}\mathbf{A}. \tag{1.9}$$

The introduction of ϕ and $\mathbf{A}$ through (1.8) and (1.9) is a sufficient condition for the two Maxwell's equations (1.1) and (1.4) to be satisfied identically; it may be shown that the existence of ϕ and $\mathbf{A}$ satisfying (1.8) and (1.9) is a necessary condition of (1.1) and (1.4). With (1.1) and (1.4) satisfied identically, one rewrites $\mathbf{E}$ and $\mathbf{B}$ in terms of ϕ and $\mathbf{A}$ in the remaining two Maxwell equations, which then become two coupled first order partial differential equations for ϕ and $\mathbf{A}$ in terms of the sources. Thus the four Maxwell equations are reduced to two equations.

On reflection it should be obvious that one cannot reduce four independent equations to two independent equations without losing generality. In fact there is no loss of generality here, because the total number of equations has not changed. What is achieved by introducing the fields ϕ and $\mathbf{A}$ is the replacement of two of Maxwell's equations by the two equations (1.8) and (1.9), which are regarded as definitions, or alternatively as subsidiary equations. In this way two of Maxwell's equations are effectively relegated to a subsidiary status and do not need to be solved explicitly.

It is still impracticable to solve two coupled partial differential equations and it is desirable to reduce the set of equations to a single equation. One step in achieving this is to regard the equation of charge continuity (1.6) as another subsidiary equation in place of one of the two remaining Maxwell equations. This requires taking the time-derivative of (1.3) and then using (1.6) to eliminate ρ entirely. Then one has replaced the four Maxwell equations by three subsidiary equations and one basic equation, which is

$$\operatorname{curl}\operatorname{curl}\mathbf{A} = \mu_0 \mathbf{J} - \frac{1}{c^2}\frac{\partial}{\partial t}\left(\operatorname{grad}\phi + \frac{\partial}{\partial t}\mathbf{A}\right). \qquad (1.10)$$

One rewrites (1.10) in the form

$$\left(\frac{1}{c^2}\frac{\partial^2}{\partial t^2} - \nabla^2\right)\mathbf{A} + \frac{1}{c^2}\frac{\partial}{\partial t}\operatorname{grad}\phi + \operatorname{grad}\operatorname{div}\mathbf{A} = \mu_0\mathbf{J}. \qquad (1.11)$$

A related equation follows by taking the divergence of (1.11) and using the charge continuity condition (1.6), or more directly by substituting (1.8) into (1.3):

$$\nabla^2\phi + \operatorname{div}\partial\mathbf{A}/\partial t = -\rho/\varepsilon_0. \qquad (1.12)$$

Although Maxwell's equations are reduced to a single equation plus three subsidiary equations, it is still not possible to solve (1.11) or (1.12) directly because they involve both potentials ϕ and $\mathbf{A}$. However, one uses an arbitrariness in the definitions of ϕ and $\mathbf{A}$ to reduce (1.11) and (1.12) to equations that involve only one of these potentials.

1.4 Gauge Conditions

The relations (1.8) and (1.9) do not define $\mathbf{A}$ and ϕ uniquely. Any other pair of potentials $\mathbf{A}'$ and ϕ' related to $\mathbf{A}$ and ϕ by

$$\phi' = \phi + \partial\psi/\partial t, \tag{1.13}$$

$$\mathbf{A}' = \mathbf{A} - \operatorname{grad}\psi, \tag{1.14}$$

where ψ is an arbitrary differentiable function, also satisfy (1.8) and (1.9). The relations (1.13) and (1.14) are referred to as a *gauge transformation* from ϕ and $\mathbf{A}$ to ϕ' and $\mathbf{A}'$.

The freedom to make gauge transformations allows one to impose a *gauge condition* on the choice of potentials. Specific gauge conditions correspond to particular choices of the arbitrary function ψ. Three specific choices are:

Coulomb gauge	$\operatorname{div}\mathbf{A} = 0$,	(1.15)
Lorentz gauge	$(1/c^2)\,\partial\phi/\partial t + \operatorname{div}\mathbf{A} = 0$,	(1.16)
temporal gauge	$\phi = 0$.	(1.17)

Each of these choices of gauge is convenient for different purposes.

The *Coulomb gauge* is most convenient for treating static fields. If the time-derivatives in (1.11) are zero, as is the case for static fields, then (1.11) and (1.12) reduce to

$$\nabla^2\phi = -\rho/\varepsilon_0, \tag{1.18}$$

$$\nabla^2\mathbf{A} = -\mu_0\mathbf{J}, \tag{1.19}$$

so that the static fields are given in terms of their sources by Poisson's equation. In this case the only source for electric potential is the charge density and the only source of the vector potential is the current density. Moreover, in this time-independent case (1.13) and (1.14) imply that the electric field is independent of $\mathbf{A}$ and the magnetic field is independent of ϕ. Thus in the static limit there is a clear separation into electric and magnetic fields generated by charges and currents, respectively. This separation is clear physically only in the static case, and in this case it is obvious mathematically only if one uses the Coulomb gauge.

For time-varying fields it is sometimes useful to use the *Lorentz gauge*. In this case (1.13) and (1.14) reduce to

$$\left(\frac{1}{c^2}\frac{\partial^2}{\partial t^2} - \nabla^2\right)\phi = \rho/\varepsilon_0, \tag{1.20}$$

$$\left(\frac{1}{c^2}\frac{\partial^2}{\partial t^2} - \nabla^2\right)\mathbf{A} = \mu_0\mathbf{J}. \tag{1.21}$$

It follows that in the Lorentz gauge the potentials are related to their sources by d'Alembert's equation. In the Lorentz gauge ϕ is generated by ρ and $\mathbf{A}$ is generated by $\mathbf{J}$. However, there is not the clear separation between electric and magnetic effects that there is in the static case in the Coulomb gauge: in the Lorentz gauge both ϕ and $\mathbf{A}$ contribute to $\mathbf{E}$.

In the *temporal gauge* there is no scalar field, and (1.11) reduces to

$$\frac{1}{c^2}\frac{\partial^2 \mathbf{A}}{\partial t^2} + \text{curl curl}\,\mathbf{A} = \mu_0 \mathbf{J}. \tag{1.22}$$

This form is convenient for solving Maxwell's equations using Fourier transforms.

Note that any electromagnetic field may be described using the temporal gauge. This has the interesting implication that an arbitrary electromagnetic field, usually described in terms of $\mathbf{E}$ and $\mathbf{B}$, may be described by the single vector $\mathbf{A}$ in the temporal gauge. Thus one needs only one vector and not two vectors to describe an arbitrary electromagnetic field. The relations (1.8) and (1.9), with $\phi = 0$ in the temporal gauge, are to be used to find the fields $\mathbf{E}$ and $\mathbf{B}$ in terms of $\mathbf{A}$. Of course it is not always convenient to use the temporal gauge; for example, for static uniform fields $\mathbf{E}$ and $\mathbf{B}$ the vector potential in the temporal gauge depends on both t and $\mathbf{x}$, and it is rarely useful to introduce a time-dependent potential to describe a static field.

1.5 Solution of Poisson's and d'Alembert's Equations

The solutions of Poisson's equation and of d'Alembert's equation are relatively well known. They are written down here and, as in most introductions to electromagnetic theory, it is argued that the given solutions are indeed the desired solutions. A formal derivation of the solutions is given in Chapter 5 based on the use of Fourier transforms.

The solutions of Poisson's equation (1.18) and (1.19) for the electric and vector potentials are, respectively,

$$\phi(\mathbf{x}) = \frac{1}{4\pi\varepsilon_0} \int d^3\mathbf{x}'\,\frac{\rho(\mathbf{x}')}{|\mathbf{x} - \mathbf{x}'|}, \tag{1.23}$$

$$\mathbf{A}(\mathbf{x}) = \frac{\mu_0}{4\pi} \int d^3\mathbf{x}'\,\frac{\mathbf{J}(\mathbf{x}')}{|\mathbf{x} - \mathbf{x}'|}, \tag{1.24}$$

where the integrals over $\mathbf{x}'$ are over all space. These solutions apply to static fields.

To show that (1.23) is indeed a solution of Poisson's equation (1.18) one inserts (1.23) back into (1.18) and uses the identity $\nabla^2(1/r) =$

$-4\pi\delta^3(\mathbf{x})$, in the form

$$\nabla^2\left(\frac{1}{|\mathbf{x}-\mathbf{x}'|}\right) = -4\pi\delta^3(\mathbf{x}-\mathbf{x}').\tag{1.25}$$

The properties of the Dirac δ-function, cf. §4.3, then imply that (1.18) is satisfied. One shows that (1.24) is a solution of (1.19) in the same way.

The solutions of d'Alembert's equation (1.20) and (1.21) are

$$\phi(t,\mathbf{x}) = \frac{1}{4\pi\varepsilon_0}\int d^3x' \frac{\rho(t',\mathbf{x}')}{|\mathbf{x}-\mathbf{x}'|},\tag{1.26}$$

$$\mathbf{A}(t,\mathbf{x}) = \frac{\mu_0}{4\pi}\int d^3x' \frac{\mathbf{J}(t',\mathbf{x}')}{|\mathbf{x}-\mathbf{x}'|},\tag{1.27}$$

respectively, where $t' = t_r$ is the *retarded time*

$$t' = t_r := t - \frac{|\mathbf{x}-\mathbf{x}'|}{c}.\tag{1.28}$$

The retarded time is interpreted as the time t at which the field is observed minus the light propagation time from the point $\mathbf{x}'$ where the source was at time t' to the point $\mathbf{x}$ where the field is observed at time t. This is interpreted in terms of the electromagnetic field propagating from the source to the point of observation at the speed of light. This is the case only *in vacuo* however. In the presence of a medium, (1.26) and (1.27) do not apply.

Exercise Set 1

1.1 Consider a region of space R with a surface S in which $\mathbf{E}$ and $\mathbf{B}$ are static and non-parallel. Suppose that $\mathbf{J}$ is zero within R. Comment on the interpretation of the conservation of energy equation (1.7) in R, *viz.*,

$$\frac{\partial}{\partial t}\left(\tfrac{1}{2}\varepsilon_0|\mathbf{E}|^2 + \tfrac{1}{2}|\mathbf{B}|^2/\mu_0\right) + \operatorname{div}\left(\mathbf{E}\times\mathbf{B}/\mu_0\right) = -\mathbf{J}\cdot\mathbf{E}.$$

In particular explain the significance of the Poynting vector $\mathbf{E}\times\mathbf{B}/\mu_0$ in this case.

1.2 The electric induction $\mathbf{D}$ and the magnetic field strength $\mathbf{H}$ are defined in terms of the polarization $\mathbf{P}$ and magnetization $\mathbf{M}$ by writing

$$\mathbf{J} = \mathbf{J}_{\text{ind}} + \mathbf{J}_{\text{ext}}, \qquad \rho = \rho_{\text{ind}} + \rho_{\text{ext}}, \qquad (E1.1)$$

$$\mathbf{J}_{\text{ind}} = \partial\mathbf{P}/\partial t + \operatorname{curl}\mathbf{M}, \qquad \rho_{\text{ind}} = -\operatorname{div}\mathbf{P}, \qquad (E1.2)$$

$$\mathbf{D} = \varepsilon_0\mathbf{E} + \mathbf{P}, \qquad \mathbf{H} = \mathbf{B}/\mu_0 - \mathbf{M}. \qquad (E1.3)$$

(a) Show that (1.2) and (1.3) are then replaced by

$$\operatorname{curl}\mathbf{H} = \mathbf{J}_{\text{ext}} + \partial\mathbf{D}/\partial t, \qquad (E1.4)$$

$$\operatorname{div}\mathbf{D} = \rho_{\text{ext}}. \qquad (E1.5)$$

(b) Show that the equation of continuity of energy (1.7) is replaced by

$$\mathbf{E}\cdot\frac{\partial\mathbf{D}}{\partial t} + \mathbf{H}\cdot\frac{\partial\mathbf{B}}{\partial t} + \operatorname{div}(\mathbf{E}\times\mathbf{H}) = -\mathbf{J}_{\text{ext}}\cdot\mathbf{E}. \qquad (E1.6)$$

Remarks: (1) The separation $(E1.2)$ is not unique; a further assumption is required to make it unique, and one is free to assume $\mathbf{M} = 0$, as is done in plasma physics. (2) One can justify interpreting $\tfrac{1}{2}\mathbf{E}\cdot\mathbf{D}$ and $\tfrac{1}{2}\mathbf{H}\cdot\mathbf{B}$ as energy densities only if $\mathbf{D}$ and $\mathbf{E}$ vary in the same way in time, and if $\mathbf{H}$ and $\mathbf{B}$ vary in the same way in time. It follows that this interpretation is valid only in a non-dispersive medium.

1.3 Find an expression for $\mathbf{A}(t,\mathbf{x})$ in the temporal gauge for a static uniform electromagnetic field $\mathbf{E}_0$ and $\mathbf{B}_0$.

1.4 A magnetic field $\mathbf{B}$ is represented in terms of *Euler potentials* α and β:

$$\mathbf{B} = \operatorname{grad}\alpha \times \operatorname{grad}\beta. \qquad (E1.7)$$

(a) Show that

$$\mathbf{A} = \alpha\operatorname{grad}\beta, \quad \mathbf{A}' = -\beta\operatorname{grad}\alpha, \qquad (E1.8)$$

are two possible choices of vector potential.

(b) Show that the choices in ($E1.8$) are connected by a gauge trans-
     formation. Specifically, identify ψ in

$$\mathbf{A}' = \mathbf{A} - \text{grad}\,\psi.$$

1.5 One adds arbitrary constants to ϕ and $\mathbf{A}$ without affecting the
resulting values of $\mathbf{E}$ and $\mathbf{B}$. Thus even in a specific gauge ϕ and $\mathbf{A}$ are
not uniquely defined. Consider ϕ and $\mathbf{A}$ and another pair ϕ' and $\mathbf{A}'$,
both in the Lorentz gauge. Show that if these are related by

$$\mathbf{A}' = \mathbf{A} - \text{grad}\,\psi, \quad \phi' = \phi + \partial\psi/\partial t,$$

then ψ must satisfy

$$\left(\frac{1}{c^2}\frac{\partial^2}{\partial t^2} - \nabla^2\right)\psi = 0$$

in order that ϕ, $\mathbf{A}$ and ϕ', $\mathbf{A}'$ are to describe the same electromagnetic
field.

1.6 Establish the identity $\nabla^2(1/r) = -4\pi\delta^3(\mathbf{x})$ through the following
steps.

(a) Show that $\nabla^2(1/r)$ is zero for $r \neq 0$ by direct evaluation in spherical
     polar coordinates (with $\nabla^2 = (1/r^2)(\partial/\partial r)r^2(\partial/\partial r)$ in the absence
     of any angular dependence).
(b) Evaluate $\nabla^2[1/(r^2 + a^2)^{1/2}]$ where a is a constant.
(c) Show that the integral of $\nabla^2[1/(r^2 + a^2)^{1/2}]$ over all space (us-
     ing $\int d^3\mathbf{x} = 4\pi \int dr\, r^2$ in spherical polar coordinates for an angle-
     independent integrand) reduces to -4π.
(d) Compare the result of part (c) for $a = 0$ with the integral of
     $\nabla^2(1/r) = -4\pi\delta^3(\mathbf{x})$ over all space.

2

Cartesian Tensors

Preamble

We digress from the development of electromagnetic theory to introduce tensor algebra. Tensor algebra, or some equivalent mathematical tool, is needed to describe the response of an anisotropic medium and to describe stresses and other intrinsically tensorial quantities that appear in electromagnetic theory. Once mastered, tensor algebra is a useful alternative to vector algebra, and in many specific examples is simpler to use than vector algebra. The version of tensor algebra introduced here is for cartesian tensors in three dimensions.

2.1 Vector Components and Rotations

Consider a vector $\mathbf{V}$. The vector is described by three numbers, which are the components of the vector along a set of orthonormal basis vectors. Let the basis vectors be $\mathbf{e}_1$, $\mathbf{e}_2$, $\mathbf{e}_3$. The components of $\mathbf{V}$ are

$$V_i = \mathbf{V} \cdot \mathbf{e}_i, \quad \text{with} \quad i = 1, 2, 3. \tag{2.1}$$

The vector itself is written in the form

$$\mathbf{V} = \sum_{i=1}^{3} V_i \mathbf{e}_i. \tag{2.2}$$

In vector algebra one denotes a specific vector by a boldface symbol $\mathbf{V}$, and if one wishes to exhibit its components in a particular coordinate system one writes $\mathbf{V} = (V_1, V_2, V_3)$. In tensor algebra one describes a vector in terms of its components. Thus in the language of tensor algebra one refers to "the vector V_i". It is implicit that a particular set of basis vectors is chosen, and that the index i runs over the three components.

Here the components are labeled $1, 2, 3$, but they could equally as well be labeled x, y, z, and then i would run over x, y, z rather than $1, 2, 3$. The index here is denoted by i, but one could choose any other letter, such as j or s, and rewrite all indices in terms of the new letter without changing the meaning. In principle the labeling of indices is arbitrary, but it is convenient to restrict it to lower case italic letters as subscripts, and to use other letters or superscripts to denote labels that are not vector indices. Thus V_{Ai} and B_j^T are interpreted as the ith component of the vector $\mathbf{V}_A$ and the jth component of the vector $\mathbf{B}^T$ respectively.

The vectorial character of a quantity is defined by the way its components transform under a rotation. It is helpful to distinguish two viewpoints for rotations: these are the active and passive viewpoints. In an *active* rotation the coordinate system is held fixed and the vector itself is rotated, and in a *passive* rotation the vector is held fixed and the coordinate axes are rotated. In an active rotation the vector itself changes. Thus one might describe the initial vector as $\mathbf{V}$ and the vector after rotation as $\mathbf{V}'$. Then if V_i denotes the vector before the rotation, it is appropriate to describe the vector after the active rotation by V_i'.

Here only passive rotations are considered. After the passive rotation is performed, let the new set of basis vectors be $\mathbf{e}_{1'}, \mathbf{e}_{2'}, \mathbf{e}_{3'}$; note that the primes are added to the indices rather than to the basis vectors themselves. Then the components of the vector after the rotation are denoted by $V_{i'}$ with i' running over $1', 2', 3'$. The vector with respect to the new set of basis vectors is given by

$$\mathbf{V} = \sum_{i'=1'}^{3'} V_{i'} \mathbf{e}_{i'}. \tag{2.3}$$

2.2 Rotation Matrix

A formal way of defining a (passive) rotation is in terms of a rotation matrix. Let the rotation change the initial set of basis vectors $\mathbf{e}_1, \mathbf{e}_2, \mathbf{e}_3$ into $\mathbf{e}_{1'}, \mathbf{e}_{2'}, \mathbf{e}_{3'}$. The rotation matrix has components

$$R_{ii'} := \mathbf{e}_i \cdot \mathbf{e}_{i'}, \tag{2.4}$$

where $:=$ denotes a definition. The relation between the two sets of basis vectors is written

$$\mathbf{e}_i = \sum_{i'=1'}^{3'} (\mathbf{e}_i \cdot \mathbf{e}_{i'}) \, \mathbf{e}_{i'} = \sum_{i'=1}^{3'} R_{ii'} \mathbf{e}_{i'}, \tag{2.5}$$

$$\mathbf{e}_{i'} = \sum_{i=1}^{3} (\mathbf{e}_{i'} \cdot \mathbf{e}_i)\, \mathbf{e}_i = \sum_{i=1}^{3} R_{i'i} \mathbf{e}_i. \tag{2.6}$$

These relations are written in matrix notation:

$$\begin{pmatrix} \mathbf{e}_1 \\ \mathbf{e}_2 \\ \mathbf{e}_3 \end{pmatrix} = \begin{pmatrix} R_{11'} & R_{12'} & R_{13'} \\ R_{21'} & R_{22'} & R_{23'} \\ R_{31'} & R_{32'} & R_{33'} \end{pmatrix} \begin{pmatrix} \mathbf{e}_{1'} \\ \mathbf{e}_{2'} \\ \mathbf{e}_{3'} \end{pmatrix} = [R] \begin{pmatrix} \mathbf{e}_{1'} \\ \mathbf{e}_{2'} \\ \mathbf{e}_{3'} \end{pmatrix}, \tag{2.7}$$

$$\begin{pmatrix} \mathbf{e}_{1'} \\ \mathbf{e}_{2'} \\ \mathbf{e}_{3'} \end{pmatrix} = \begin{pmatrix} R_{1'1} & R_{1'2} & R_{1'3} \\ R_{2'1} & R_{2'2} & R_{2'3} \\ R_{3'1} & R_{3'2} & R_{3'3} \end{pmatrix} \begin{pmatrix} \mathbf{e}_1 \\ \mathbf{e}_2 \\ \mathbf{e}_3 \end{pmatrix} = [R]^{\mathrm{T}} \begin{pmatrix} \mathbf{e}_1 \\ \mathbf{e}_2 \\ \mathbf{e}_3 \end{pmatrix}, \tag{2.8}$$

where $[R]$ denotes the transformation matrix with components $R_{ii'}$ and $[R]^{\mathrm{T}}$ denotes its transpose with components $R_{i'i}$. The matrix $[R]$ is *orthogonal*, that is, its inverse is equal to its transpose because the basis vectors $\mathbf{e}_i$ and $\mathbf{e}_{i'}$ are chosen to be orthonormal.

In the formalism of cartesian tensors, a formal definition of a *vector* is as a set of three numbers which transform under a rotation in the same way as the basis vectors transform. Thus if V_i describes the components of a vector, then under a rotation the components transform to

$$V_{i'} = \sum_{i=1}^{3} R_{i'i} V_i. \tag{2.9}$$

A quantity that is unchanged (invariant) under a rotation is a *scalar*. A scalar has no index. A second rank *tensor* is a quantity that transforms like the outer product of two vectors. Thus a second rank tensor T_{ij} has two indices and under a rotation it transforms to

$$T_{i'j'} = \sum_{i=1}^{3} \sum_{j=1}^{3} R_{i'i} R_{j'j} T_{ij}. \tag{2.10}$$

In general a tensor of rank n (for example, $T_{i_1 i_2 \ldots i_n}$) has n indices $(i_1 i_2 \ldots i_n)$, and it transforms like the outer product of n vectors. A scalar is a tensor of rank 0, and a vector is a tensor of rank 1.

There are two special tensors whose numerical values are unchanged by a rotation. One of these is the unit tensor or *Kronecker delta*:

$$\delta_{ij} = \begin{cases} 1 & \text{for } i = j, \\ 0 & \text{for } i \neq j. \end{cases} \tag{2.11}$$

The Kronecker delta is defined in terms of the basis vectors by writing

$$\delta_{ij} := \mathbf{e}_i \cdot \mathbf{e}_j. \tag{2.12}$$

The orthogonality of the rotation matrix is expressed by

$$\sum_{i'=1'}^{3'} R_{ii'} R_{i'j} = \delta_{ij}, \quad \sum_{i=1}^{3} R_{i'i} R_{ij'} = \delta_{i'j'}. \tag{2.13}$$

The Kronecker delta is a symmetric tensor, that is, it satisfies $\delta_{ij} = \delta_{ji}$.

The other tensor that has the same value independent of rotations is the permutation symbol

$$\epsilon_{ijk} = \begin{cases} 1 & \text{for } ijk \text{ an even permutation of 123,} \\ -1 & \text{for } ijk \text{ an odd permutation of 123,} \\ 0 & \text{otherwise.} \end{cases} \qquad (2.14)$$

Alternatively the permutation symbol is defined in terms of the scalar triple product of the basis vectors (which are defined to form a right hand set)

$$\epsilon_{ijk} := \mathbf{e}_i \cdot \mathbf{e}_j \times \mathbf{e}_k. \qquad (2.15)$$

The permutation symbol is *antisymmetric* in all its indices; that is, it changes sign when any two neighboring indices are interchanged. Thus one has

$$\epsilon_{ijk} = -\epsilon_{jik} = -\epsilon_{ikj} = \epsilon_{kij}. \qquad (2.16)$$

2.3 Tensor Equations

In the foregoing equations, notice that the indices appear either singly or in identical pairs, and that whenever a pair of identical indices appears the sum over these indices is performed. The single indices are called *free* indices, and the pairs of identical indices are called *dummy* indices. The number of free indices is equal to the rank of the tensor.

A tensor equation also has a rank. Each element in the tensor equation must have the same number and kind of free indices. The number of free indices defines the rank of the tensor equation.

An important simplifying assumption adopted in tensor algebra is that the sum over dummy indices is implied. This *summation convention*, attributed to Einstein, is that one is to omit the $\sum$ in all tensor equations, and to note that a sum is included implicitly whenever a pair of identical indices is present. In a tensor equation the dummy indices are ignored when counting free indices to determine the rank.

Use of the summation convention allows one to form a scalar from a second rank tensor. From the tensor T_{ij} one may construct the scalar T_{ii}, which is just the trace of the tensor ($T_{ii} = T_{11}+T_{22}+T_{33}$). The proof that this quantity is a scalar follows from the orthogonality relations (2.13):

$$T_{i'i'} = R_{i'i}R_{i'j}T_{ij} = R_{ii'}R_{i'j}T_{ij} = \delta_{ij}T_{ij} = T_{ii}, \qquad (2.17)$$

where the definition (2.4) implies $R_{i'i} = R_{ii'}$. Note also the use of the

Kronecker delta in relabeling an index:

$$V_i = \delta_{ij} V_j. \tag{2.18}$$

The construction of the scalar T_{ii} from the second rank tensor T_{ij} is an example of a contraction. More generally, a *contraction* refers to the process of setting two free indices equal to each other, and thus converting them into a pair of dummy indices. This reduces the number of free indices by two, and so reduces the rank of the tensor or tensor equation by two. Another example of a contraction is the formation of a vectorial quantity from the outer product of a second rank tensor and a vector. Such relations appear whenever one describes a response in an anisotropic medium. For example,

$$J_i = \sigma_{ij} E_j \tag{2.19}$$

relates the induced current **J** to the electric field **E**, and where σ_{ij} is the conductivity tensor.

2.4 Formal Rules for Cartesian Tensor Equations

The foregoing introduction to cartesian tensors may be summarized and extended in the following set of rules for writing down tensor equations.

(1) Each element in a tensor equation is *either* a kernel symbol *or* a product of kernel symbols.

(2) The indices are italic subscripts which denote cartesian components, usually understood to label components $1, 2, 3$ or x, y, z. Indices may have affices, such as primes or numerical subscripts, and two indices are the same only if they have the same affix.

(3) Indices appear only either once or twice in any element in a tensor equation. Indices that appear once are *free* indices, and indices that appear twice are *dummy* indices.

(4) The *summation convention* is that the sum (over all cartesian components) over dummy indices is implied.

(5) A tensor equation relates elements each of which has the same number and kind of free indices.

(6) The *rank* of a tensor or of a tensor equation is the number of its free indices. A tensor of rank zero is a scalar and a tensor of rank one is a vector.

(7) The allowed manipulations of a tensor equation include: (i) relabeling free indices, (ii) relabeling dummy indices, and (iii) perform-

ing contractions (that is, converting two free indices into a pair of dummy indices).

There are several other definitions that are useful. A second rank tensor T_{ij} may be separated into its symmetric part $T_{(ij)}$ and its antisymmetric part $T_{[ij]}$:

$$T_{(ij)} := \tfrac{1}{2}(T_{ij} + T_{ji}), \quad T_{[ij]} := \tfrac{1}{2}(T_{ij} - T_{ji}). \tag{2.20}$$

A tensor that satisfies $T_{ij} = T_{ji}$ is said to be *symmetric* and a tensor that satisfies $T_{ij} = -T_{ji}$ is said to be *antisymmetric*. For a tensor of rank higher than two one may symmetrize or antisymmetrize over any two indices. The tensor is said to be completely symmetric only if the symmetry applies to all pairs of indices, and a tensor is said to be completely antisymmetric only if the antisymmetry applies to all pairs of neighboring indices.

For a tensor which is complex, it is usually more appropriate to separate into the hermitian part T_{ij}^{H} and the antihermitian T_{ij}^{A} part:

$$T_{ij}^{\mathrm{H}} := \tfrac{1}{2}(T_{ij} + T_{ji}^{*}), \quad T_{ij}^{\mathrm{A}} := \tfrac{1}{2}(T_{ij} - T_{ji}^{*}), \tag{2.21}$$

where the asterisk denotes complex conjugation.

2.5 The Vector Calculus

The vector calculus is rewritten in tensor notation by noting the basic relations for the scalar and vector products of two vectors $\mathbf{A}$ and $\mathbf{B}$:

$$\mathbf{A} \cdot \mathbf{B} = A_i B_i, \quad (\mathbf{A} \times \mathbf{B})_i = \epsilon_{ijk} A_j B_k. \tag{2.22a, b}$$

To evaluate a vector triple product one needs to rewrite a product of two permutation symbols in terms of Kronecker deltas. The basic identity is

$$\epsilon_{abc}\epsilon_{ijk} = \delta_{ai}\delta_{bj}\delta_{ck} + \delta_{ak}\delta_{bi}\delta_{cj} + \delta_{aj}\delta_{bk}\delta_{ci}$$
$$- \delta_{bi}\delta_{aj}\delta_{ck} - \delta_{bk}\delta_{ai}\delta_{cj} - \delta_{bj}\delta_{ak}\delta_{ci}. \tag{2.23}$$

An important identity follows from (2.23) by making a contraction:

$$\epsilon_{iab}\epsilon_{ijk} = \delta_{aj}\delta_{bk} - \delta_{ak}\delta_{bj}, \tag{2.24}$$

where the identities

$$\delta_{ab}\delta_{bc} = \delta_{ac}, \quad \delta_{ii} = 3 \tag{2.25}$$

are used. Performing further contractions on (2.24) leads to the identities

$$\epsilon_{ijc}\epsilon_{ijk} = 2\delta_{ck}, \quad \epsilon_{ijk}\epsilon_{ijk} = 6. \tag{2.26}$$

The evaluation of the vector triple product involves starting with the identity (2.22b), which is used twice, and then using (2.24):

$$\{\mathbf{A} \times (\mathbf{B} \times \mathbf{C})\}_i = \epsilon_{ijk} A_j (\mathbf{B} \times \mathbf{C})_k = \epsilon_{ijk} A_j \epsilon_{kab} B_a C_b$$
$$= (\delta_{ia}\delta_{jb} - \delta_{ib}\delta_{ja}) A_j B_a C_b = A_j C_j B_i - A_j B_j C_i$$
$$= \{(\mathbf{A} \cdot \mathbf{C})\,\mathbf{B} - (\mathbf{A} \cdot \mathbf{B})\,\mathbf{C}\}_i, \qquad (2.27)$$

where the final line involves using (2.22a).

In a similar way the identities of the differential vector calculus are rederived using the tensor formalism. The basic identities are for grad, div and curl. These involve the differential operator $\nabla = \partial/\partial\mathbf{x}$, where $\mathbf{x}$ is the position vector. One has

$$(\mathrm{grad}\,\phi)_i = \frac{\partial\phi}{\partial x_i}, \quad \mathrm{div}\,\mathbf{A} = \frac{\partial A_i}{\partial x_i}, \quad (\mathrm{curl}\,\mathbf{A})_i = \epsilon_{ijk}\frac{\partial A_k}{\partial x_j}. \quad (2.28a,b,c)$$

There are two other operators that appear in the differential vector calculus:

$$\mathbf{A} \cdot \mathrm{grad} = A_i \frac{\partial}{\partial x_i}, \quad \nabla^2 = \frac{\partial^2}{\partial x_i \partial x_i}. \qquad (2.29a,b)$$

The identities of the differential vector calculus are:

$$\mathrm{grad}\,(\phi_1\phi_2) = \phi_2\,\mathrm{grad}\,\phi_1 + \phi_1\,\mathrm{grad}\,\phi_2, \qquad (2.30)$$

$$\mathrm{div}\,(\phi\mathbf{A}) = \phi\,\mathrm{div}\,\mathbf{A} + \mathbf{A} \cdot \mathrm{grad}\,\phi, \qquad (2.31)$$

$$\mathrm{curl}\,(\phi\mathbf{A}) = \phi\,\mathrm{curl}\,\mathbf{A} + \mathrm{grad}\,\phi \times \mathbf{A}, \qquad (2.32)$$

$$\mathrm{curl}\,\mathrm{grad}\,\phi = 0, \qquad (2.33)$$

$$\mathrm{div}\,\mathrm{curl}\,\mathbf{A} = 0, \qquad (2.34)$$

$$\mathrm{curl}\,\mathrm{curl}\,\mathbf{A} = \mathrm{grad}\,\mathrm{div}\,\mathbf{A} - \nabla^2\mathbf{A}, \qquad (2.35)$$

$$\mathrm{grad}\,(\mathbf{A} \cdot \mathbf{B}) = (\mathbf{A} \cdot \mathrm{grad})\mathbf{B} + (\mathbf{B} \cdot \mathrm{grad})\mathbf{A}$$
$$+ \mathbf{A} \times \mathrm{curl}\,\mathbf{B} + \mathbf{B} \times \mathrm{curl}\,\mathbf{A}, \qquad (2.36)$$

$$\mathrm{div}\,(\mathbf{A} \times \mathbf{B}) = \mathbf{B} \cdot \mathrm{curl}\,\mathbf{A} - \mathbf{A} \cdot \mathrm{curl}\,\mathbf{B}, \qquad (2.37)$$

$$\mathrm{curl}\,(\mathbf{A} \times \mathbf{B}) = \mathbf{A}\,\mathrm{div}\,\mathbf{B} + (\mathbf{B} \cdot \mathrm{grad})\,\mathbf{A}$$
$$- \mathbf{B}\,\mathrm{div}\,\mathbf{A} - (\mathbf{A} \cdot \mathrm{grad})\,\mathbf{B}. \qquad (2.38)$$

Each of these identities may be derived using tensor notation. The following three examples illustrate the derivations. The proofs of (2.30)–(2.32) involve only the identifications (2.28) and the differentiation of a product, for example,

$$\left[\mathrm{curl}\,(\phi\mathbf{A})\right]_i = \epsilon_{ijk}\frac{\partial}{\partial x_j}(\phi A_k) = \epsilon_{ijk}\frac{\partial\phi}{\partial x_j}A_k + \phi\epsilon_{ijk}\frac{\partial A_k}{\partial x_j}$$
$$= (\mathrm{grad}\,\phi \times \mathbf{A})_i + \phi(\mathrm{curl}\,\mathbf{A})_i.$$

The proofs of (2.33) and (2.34) involve a more subtle argument. One has

$$\text{div curl }\mathbf{A} = \frac{\partial}{\partial x_i}\epsilon_{ijk}\frac{\partial A_k}{\partial x_j} = \epsilon_{ijk}\frac{\partial^2 A_k}{\partial x_i \partial x_j} = 0,$$

where in the final step the fact that ϵ_{ijk} is antisymmetric in jk and the double derivative is symmetric in jk is used. More generally, the contraction of a symmetric tensor $S_{jk} = S_{kj}$ with an antisymmetric tensor $A_{jk} = -A_{kj}$ gives zero. One has

$$S_{jk}A_{jk} = S_{kj}A_{jk} = S_{jk}A_{kj} = -S_{jk}A_{jk},$$

where in the first and last steps the symmetry and antisymmetry properties are used, respectively, and in the middle step the dummy indices j and k are interchanged. In the proof of (2.35) the identity (2.24) is used:

$$(\text{curl curl }\mathbf{A})_i = \epsilon_{ijk}\epsilon_{krs}\frac{\partial}{\partial x_j}\left(\frac{\partial A_s}{\partial x_r}\right) = (\delta_{ir}\delta_{js} - \delta_{is}\delta_{jr})\frac{\partial^2 A_s}{\partial x_j \partial x_r}$$

$$= \frac{\partial}{\partial x_i}\left(\frac{\partial A_s}{\partial x_s}\right) - \frac{\partial^2 A_i}{\partial x_s \partial x_s} = (\text{grad div }\mathbf{A} - \nabla^2\mathbf{A})_i\,.$$

The remaining proofs are somewhat more lengthy and include the differential operator (2.29a). For example, for (2.38) one has

$$\left[\text{curl}\,(\mathbf{A}\times\mathbf{B})\right]_i = \epsilon_{ijk}\epsilon_{krs}\frac{\partial(A_r B_s)}{\partial x_j}$$

$$= (\delta_{ir}\delta_{js} - \delta_{is}\delta_{jr})\left(A_r\frac{\partial B_s}{\partial x_j} + B_s\frac{\partial A_r}{\partial x_j}\right)$$

$$= A_i\frac{\partial B_j}{\partial x_j} + \left(B_j\frac{\partial}{\partial x_j}\right)A_i - B_i\frac{\partial A_j}{\partial x_j} - \left(A_j\frac{\partial}{\partial x_j}\right)B_i$$

$$= \left[\mathbf{A}\,\text{div}\,\mathbf{B} + (\mathbf{B}\cdot\text{grad}\,)\mathbf{A}\right.$$
$$\left. - \mathbf{B}\,\text{div}\,\mathbf{A} - (\mathbf{A}\cdot\text{grad}\,)\mathbf{B}\right]_i.$$

Exercise Set 2

2.1 Consider a rotation through an angle θ is two dimensions, as illustrated in the figure. Let $\mathbf{e}_1$ and $\mathbf{e}_2$ denote unit vectors along the x and y axes, respectively, and let $\mathbf{e}_1'$ and $\mathbf{e}_2'$ denote unit vectors along the x' and y' axes, respectively.

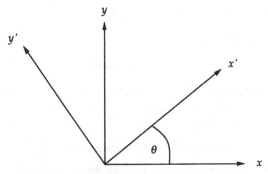

(i) Find the rotation matrix $[R]$ that links the two bases:

$$\begin{pmatrix} \mathbf{e}_1 \\ \mathbf{e}_2 \end{pmatrix} = [R] \begin{pmatrix} \mathbf{e}_1' \\ \mathbf{e}_2' \end{pmatrix}. \tag{E2.1}$$

(ii) Show that $[R]$ is orthogonal:

$$[R][R]^\mathrm{T} = I_2, \tag{E2.2}$$

where $[R]^\mathrm{T}$ is the transpose of the matrix $[R]$, and I_2 is the unit matrix in two dimensions.

2.2 Which of the following equations are acceptable as tensor equations, and which are not? Give reasons for each answer.

$$A_i = B_i + C_{ij}D_j + \epsilon_{irs}C_r D_{sj}, \tag{i}$$

$$A_i B_{jk} = C\epsilon_{ijk} + (D_i)^2 \delta_{jk}, \tag{ii}$$

$$A_i - B_j = C_{ij}, \tag{iii}$$

$$A_{ij'}D_{j'kr'} = C_{ikr'}. \tag{iv}$$

2.3 Use tensor notation to establish the following vector identities

$$\mathrm{grad}\,\mathbf{A}\cdot\mathbf{B} = (\mathbf{A}\cdot\mathrm{grad})\,\mathbf{B} + (\mathbf{B}\cdot\mathrm{grad})\,\mathbf{A}$$
$$+ \mathbf{A}\times\mathrm{curl}\,\mathbf{B} + \mathbf{B}\times\mathrm{curl}\,\mathbf{A}, \tag{i}$$

$$\mathrm{div}\,(\mathbf{A}\times\mathbf{B}) = \mathbf{B}\cdot\mathrm{curl}\,\mathbf{A} - \mathbf{A}\cdot\mathrm{curl}\,\mathbf{B}, \tag{ii}$$

$$\mathrm{curl\,grad}\,\phi = 0. \tag{iii}$$

2.4 An arbitrary second rank tensor T_{ij} is separated into its symmetric part $T_{ij}^{\mathrm{s}} = T_{ji}^{\mathrm{s}}$ and its antisymmetric part $T_{ij}^{\mathrm{a}} = -T_{ji}^{\mathrm{a}}$.

(a)   Exhibit this decomposition by identifying the components of T_{ij}^{s} and T_{ij}^{a} in terms of T_{ij}.

(b)   Let the vector **g** be defined by

$$g_i := \epsilon_{ijk} T_{jk}. \qquad (E2.3)$$

Show that only the antisymmetric part contributes to **g**:

$$g_i = \epsilon_{ijk} T_{jk}^{\text{a}}. \qquad (E2.4)$$

(c)   Establish the identity

$$T_{ij}^{\text{a}} = \tfrac{1}{2} \epsilon_{ijk} g_k. \qquad (E2.5)$$

2.5 Let **n** be an arbitrary unit vector, and let angular brackets denote an average over all possible orientations of **n**. To be specific, if **n** is written in terms of spherical polar coordinates

$$\mathbf{n} = (\sin\theta\cos\phi, \sin\theta\sin\phi, \cos\theta),$$

then the average of any function of **n**, $f(\mathbf{n})$ say, is of the form

$$\langle f(\mathbf{n}) \rangle = \frac{1}{4\pi} \int_0^{2\pi} d\phi \int_{-1}^{1} d\cos\theta\, f(\mathbf{n}). \qquad (E2.6)$$

One has $\langle \mathbf{n} \rangle = 0$.

(a)   Show that one has

$$\langle n_i n_j \rangle = \tfrac{1}{3}\delta_{ij}. \qquad (E2.7)$$

Remark: The result $(E2.7)$ follows directly from the following argument. The tensor $\langle n_i n_j \rangle$ cannot depend on any vector (because it is an average over all directions) and the only tensorial quantity with two indices that involves no direction is δ_{ij}. Hence $\langle n_i n_j \rangle$ must be proportional to δ_{ij}.

(b)   Writing

$$\langle n_i n_j \rangle = A\delta_{ij},$$

show that taking the trace gives $A = \tfrac{1}{3}$.

Hints: $n_i n_i = 1$, $\delta_{ii} = 3$.

(c)   Generalize this argument to evaluate $\langle n_i n_j n_k n_l \rangle$.

2.6 Let $\{\phi_i\}$, $\{\psi_i\}$ denote two orthonormal sets of complex functions defined on an interval $[0, d]$. The scalar product of two functions f and g is then defined by

$$f \cdot g := \frac{1}{d} \int_0^d dx\, f(x) g^*(x). \qquad (E2.8)$$

Use this definition to define the matrix $[N]$ with components N_{ij} satisfying

$$\phi_i = N_{ij}\psi_j = (\phi_i \cdot \psi_j)\psi_j. \qquad (E2.9)$$

(a) Expand $\phi_i \cdot \phi_j = \delta_{ij}$ and hence show that the product of $[N]$ and its hermitian adjoint $[N]^\dagger$ is the identity matrix I:

$$[N][N]^\dagger = I. \qquad (E2.10)$$

(b) Expand $\psi_i \cdot \psi_j = \delta_{ij}$ and hence establish the identity

$$[N]^\dagger[N] = I. \qquad (E2.11)$$

Remark: The matrix relations $(E2.10)$ and $(E2.11)$ are used to check the adequacy or completeness of one set of functions being used to expand a second set of functions.

3

The Stress Tensor and
Multipole Moments

Preamble

Two applications of the tensor formalism to electromagnetic fields *in vacuo* concern (1) the electromagnetic stresses and (2) multipole moments. These two applications are discussed here as illustrations of the use of the tensor formalism.

3.1 The Maxwell Stress Tensor

Maxwell's equations (1.1)–(1.4) are rewritten in tensor notation using the tensor representations of the differential vector operators, cf. §2.5. Thus in tensor notation Maxwell's equations are

$$\epsilon_{ijl}\frac{\partial E_l}{\partial x_j} = -\frac{\partial B_i}{\partial t}, \tag{3.1}$$

$$\epsilon_{ijl}\frac{\partial B_l}{\partial x_j} = \mu_0 J_i + \frac{1}{c^2}\frac{\partial E_i}{\partial t}, \tag{3.2}$$

$$\frac{\partial E_r}{\partial x_r} = \frac{\rho}{\varepsilon_0}, \tag{3.3}$$

$$\frac{\partial B_r}{\partial x_r} = 0. \tag{3.4}$$

Similarly, all the equations in Chapter 1 involving electromagnetic fields may be rewritten in tensor notation. Simply rewriting them in tensor notation has no effect on their interpretation; tensor notation is useful only insofar as it facilitates the derivation of new results.

An example where tensor notation is useful is in the derivation of the continuity equation for electromagnetic momentum. The continuity equation for electromagnetic energy is given by (1.7), and it is rederived by contracting (3.1) with $-B_i$, adding it to the contraction of (3.2) with $c^2 E_i$ and manipulating the resulting identity. The continuity equation for momentum is derived in a similar manner. Specifically, in (3.2) replace the dummy index j by r and then the free index i by j and contract with $\epsilon_{ijk} B_k/\mu_0$, then add the resulting expression to $\varepsilon_0 E_i$ times (3.3). After some manipulation, the result reduces to the form

$$\frac{\partial}{\partial x_j}(\mathbf{T}_{\text{EM}})_{ij} = \rho E_i + (\mathbf{J} \times \mathbf{B})_i + \frac{\partial}{\partial t}(\mathbf{P}_{\text{EM}})_i, \qquad (3.5)$$

where

$$\mathbf{P}_{\text{EM}} = \mathbf{E} \times \mathbf{B}/\mu_0 c^2, \qquad (3.6)$$

is the electromagnetic momentum density, which is equal to the energy flux (the Poynting vector) divided by c^2. The second rank tensor that appears in (3.5) is the Maxwell stress tensor

$$(\mathbf{T}_{\text{EM}})_{ij} = - \left(\tfrac{1}{2}\varepsilon_0 |\mathbf{E}|^2 + \tfrac{1}{2}|\mathbf{B}|^2/\mu_0\right) \delta_{ij} + \varepsilon_0 E_i E_j + B_i B_j/\mu_0. \qquad (3.7)$$

The terms on the right hand side of (3.5) give the time rate of increase of momentum per unit volume due to the forces exerted by charges, currents and electromagnetic fields, respectively.

The stress tensor (3.7) may be separated into electric and magnetic contributions. The terms proportional to δ_{ij} are unchanged by rotations and are thus identified as being isotropic forces per unit volume, that is, as pressures. There is a pressure $\tfrac{1}{2}\varepsilon_0 |\mathbf{E}|^2$ due to the electric field and a pressure $\tfrac{1}{2}|\mathbf{B}|^2/\mu_0$ due to the magnetic field. The other terms are interpreted as tensions along the electric and magnetic field lines. The tensions have the opposite signs to the corresponding pressure terms, and are twice their magnitude. Thus the stress due to either an electric field or a magnetic field is an intrinsically anisotropic quantity that is only properly represented by a tensor.

Some of the implications of (3.5) are illustrated by considering an electrically conducting fluid permeated by electric and magnetic fields. The interaction between the electromagnetic field and the fluid allows stresses to be transferred between them. In the absence of non-electromagnetic forces, the sum of the momenta of the field and the fluid must be conserved, and hence (3.5) implies there is a force per unit volume of the fluid equal to $-\partial \mathbf{P}_{\text{EM}}/\partial t$. Thus if the fluid is charged there is an electric body force $\rho \mathbf{E}$ on it, and if there is a current flowing there is a magnetic body force $\mathbf{J} \times \mathbf{B}$ on it. There is a fluid pressure in general and the gradient in this pressure contributes along with the electromagnetic stress to the total stress on the system. In a steady state the net stress, including the body forces, must be zero. For example, in the case $\mathbf{E} = 0$ and $\mathbf{J} = 0$ the sum of the magnetic and fluid pressures must be uniform; this property is the basis for the magnetic confinement of plasmas where the plasma pressure is high near the center of the confinement chamber and the plasma pressure decreases outwards being balanced by an increasing magnetic pressure. The existence of a magnetic pressure as well as a gas pressure allows a magnetic sound wave ("magnetoacoustic" wave) as well as a gas sound wave to propagate in the fluid. The magnetic tension implies that the magnetic field lines can act like stretched strings, and the elastic-type waves that are allowed by this property are called Alfvén waves. As a final example of the implications of (3.5) consider the effect of light (or other waves) when there is a gradient in the intensity, or rather in the wave energy density. Then (3.5) implies that the gradient of E^2 produces a body force on the medium through which the waves are propagating. This is called the *ponderomotive* force. It can cause self-focusing of light and other non-linear wave effects; these non-linear effects tend to cause a beam of light with uniform cross-section to break up into filaments of light which self-focus to high intensity.

Further continuity equations are derived from Maxwell's equations by considering higher rank tensor quantities, but such equations are of little physical interest. The only additional continuity equation that is usually considered is that for the angular momentum density in the electromagnetic field, cf. Exercise 3.2.

3.2 Multipole Moments as Cartesian Tensors

A second example of intrinsically tensorial quantities that arise in electromagnetic theory is that of multipole moments. A multipole expansion of a charge and current distribution, such as in an atom, a molecule or a macroscopic system, involves replacing the distributed charge and current by a set of multipoles at a specifically chosen origin. The distant field is then regarded as composed of the multipole fields. The fields of multipoles of increasing order fall off increasingly rapidly with distance. At sufficiently great distance the field that falls off most slowly with distance dominates, and this is the field due to the lowest non-vanishing multipole. The multipole expansion in this case is regarded as an expansion in the ratio of a distance parameter characteristic of the dimensions of the source, for example, the radius of the atom or molecule, and the distance r of the point of observation. A multipole expansion is also used in treating the emission and absorption of radiation. Then the relevant parameter is the ratio of the characteristic size of the system to the wavelength of the radiation; the multipole expansion is rapidly convergent only if this parameter is small.

Consider a charge and current distribution that is non-zero inside a volume V and is zero on the surface of V and outside V. A set of electric moments is defined by writing

$$Q = \int d^3x \, \rho(t, \mathbf{x}), \quad d_i(t) = \int d^3x \, x_i \rho(t, \mathbf{x}),$$

$$q_{ij}(t) = \int d^3x \, x_i x_j \rho(t, \mathbf{x}), \tag{3.8}$$

and so on. The moment with $x_{i_1} x_{i_2} \ldots x_{i_l}$ in the integrand, that is, the moment with l tensor indices, is referred to as the 2^l-multipole moment. The zeroth moment ($l = 0$) is the charge within the volume V, the first ($l = 1$) moment is the electric dipole moment, the second ($l = 2$) moment is the electric quadrupole moment, the third ($l = 3$) moment is the electric octupole moment, and so on. A subtle point is that all the traces of the moments $l \geq 2$ do not contribute to the electromagnetic field *in vacuo*. As a consequence it is conventional to redefine the moments $l \geq 2$ to remove all the traces. The conventional definition of the electric quadrupole moment in cartesian tensor form is

$$d_{ij}(t) = 3q_{ij}(t) - q_{ss}(t)\delta_{ij}, \tag{3.9}$$

where the factor of 3 is included by convention. By construction the trace of d_{ij} is zero, that is, $d_{ss} = 0$. As d_{ij} is a symmetric second rank tensor it would have six independent components in general, and the removal of the trace leaves five independent components. The 2^l-multipole moment (with traces retained) is symmetric and so has $\frac{1}{2}(l+1)(l+2)$ independent components; there are $\frac{1}{2}l(l-1)$ traces and when these are removed, the conventional traceless 2^l-multipole moment has $2l + 1$ remaining independent components.

It should be emphasized that a multipole moment is a mathematical construction and not a physical entity as such. In particular, the values of the multipole moments depend on the choice of origin. The situation here is directly analogous to the calculation of the moment of inertia for a system in mechanics; the moment of inertia is a moment of the mass distribution and it is well known that it depends on the choice of origin. Consider the case illustrated in Figure 3.1 where multipole expansions are made relative to two origins, O and O', separated by a displacement **a**. A given point at a displacement **x** from O is at a displacement

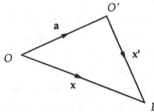

Fig. 3.1 The relation between the displacement vectors $\mathbf{x}$ and $\mathbf{x}'$ of a point P from origins O and O' separated by a displacement $\mathbf{a}$.

$\mathbf{x}' = \mathbf{x} - \mathbf{a}$ from O'. On defining a set of multipole moments relative to the origin O', denoted by primes, the counterparts of the definitions (3.8) imply

$$Q' = Q, \quad \mathbf{d}' = \mathbf{d} - Q\mathbf{a}, \quad q'_{ij} = q_{ij} - a_i d_j - d_i a_j + Q a_i a_j, \quad (3.10)$$

and so on. This dependence of the multipole expansion on the choice of origin tends to be obscured in practice, such as in the evaluation of the multipole moments of an atom or molecule, for basically two reasons. First, there is often a natural center about which it is favorable to make the expansion and it is then not appropriate to consider the expansion about other centers. Second, as is apparent from (3.10), the lowest non-vanishing moment is independent of the choice of origin, and hence the choice of origin is unimportant in determining this lowest moment. Thus, for example, for an uncharged system ($Q = 0$) the value of the electric dipole moment is independent of the choice of origin.

The magnetic multipole expansion involves terms of the form

$$\mu_a(t) = \int d^3\mathbf{x}\, J_a(t, \mathbf{x}), \quad \mu_{ia}(t) = \int d^3\mathbf{x}\, x_i J_a(t, \mathbf{x}),$$

$$\mu_{ija}(t) = \int d^3\mathbf{x}\, x_i x_j J_a(t, \mathbf{x}), \qquad (3.11)$$

and so on. The term $\mu_a(t)$ is related to the time-derivative of the electric dipole moment, as shown explicitly below; similarly, $\mu_{ia}(t)$ contains a term related to the time-derivative of the electric quadrupole moment, and so on.

The magnetic multipole moments are defined from the parts of the moments (3.11) that are antisymmetric under interchanges of the index a with each of the other indices. As with the electric moments, the magnetic moments are also defined conventionally to be traceless in the remaining indices. Specifically, the magnetic dipole moment $\mathbf{m}(t)$ is the axial vector formed from the antisymmetric part of $\mu_{ia}(t)$. (An *axial vector* differs from an ordinary or polar vector in its properties under reflections: under $\mathbf{x} \to -\mathbf{x}$ a scalar is unchanged, a pseudoscalar changes sign, a vector changes sign like $\mathbf{x}$ itself, an axial vector is unchanged, and so on.) The usual construction of such an axial vector is $m_i(t) = \frac{1}{2}\epsilon_{ijk}\mu_{jk}(t)$, and this gives

$$\mathbf{m}(t) = \frac{1}{2}\int d^3x \, \mathbf{x} \times \mathbf{J}(t,\mathbf{x}). \tag{3.12}$$

The first term in (3.11) is related to the time-derivative of the electric dipole moment as follows. Consider the identity obtained by multiplying the equation of charge continuity (1.6) by x_a and integrating over space:

$$\int d^3x \, x_a \frac{\partial\rho(t,\mathbf{x})}{\partial t} = -\int d^3x \, x_a \, \mathrm{div}\, \mathbf{J}(t,\mathbf{x}). \tag{3.13}$$

The left hand side is the time-derivative of the ath component of the electric dipole moment. The right hand side is evaluated by writing $\mathrm{div}\,\mathbf{J} = \partial J_s/\partial x_s$, and partially integrating over x_s. The integrated term gives zero, because the current is assumed to vanish on the surface of V, and the remaining term reduces to the first of (3.11); this follows from $\partial x_a/\partial x_s = \delta_{sa}$ and $\delta_{sa}J_s = J_a$. Hence one finds

$$\mu_a(t) = \partial d_a(t)/\partial t. \tag{3.14}$$

The second term in (3.11) is simplified in a similar manner. Writing

$$x_i J_a = \tfrac{1}{2}(x_i J_a + x_a J_i) + \tfrac{1}{2}(x_i J_a - x_a J_i),$$

the symmetric part is evaluated by repeating the foregoing procedure with x_a in (3.13) replaced by $x_i x_a$, and the antisymmetric part is written in terms of the magnetic dipole moment using (3.12). Thus one finds

$$\mu_{ia}(t) = \tfrac{1}{2}\partial q_{ia}(t)/\partial t + \epsilon_{ias}m_s(t). \tag{3.15}$$

In similar fashion the higher order moments (3.11) are rewritten in terms of time-derivatives of electric multipole moments plus magnetic moments.

3.3 Multipole Fields

The fields due to multipoles are inferred from (1.23) and (1.24) by expanding in $|\mathbf{x}'|/|\mathbf{x}|$ in the integrands. Only the static case is considered in the following. The expansion of the electric potential in the Coulomb gauge, cf. (1.23), gives

$$\phi(\mathbf{x}) = \frac{1}{4\pi\varepsilon_0}\int d^3x' \, \frac{\rho(\mathbf{x}')}{|\mathbf{x}-\mathbf{x}'|}$$

$$= \frac{1}{4\pi\varepsilon_0}\int d^3x' \, \rho(\mathbf{x}')\sum_{l=0}^{\infty}\frac{1}{l!}\left(-\mathbf{x}'\cdot\frac{\partial}{\partial\mathbf{x}}\right)^l\frac{1}{r}, \tag{3.16}$$

with $r = |\mathbf{x}|$. The definitions (3.8) then give

$$\phi(\mathbf{x}) = \frac{Q}{4\pi\varepsilon_0 r} - \mathbf{d} \cdot \frac{\partial}{\partial \mathbf{x}} \frac{1}{4\pi\varepsilon_0 r} + \tfrac{1}{2} q_{ij} \frac{\partial^2}{\partial x_i x_j} \frac{1}{4\pi\varepsilon_0 r} + \cdots$$

$$= \frac{Q}{4\pi\varepsilon_0 r} + \frac{\mathbf{d} \cdot \mathbf{x}}{4\pi\varepsilon_0 r^3} + \frac{q_{ij}(3x_i x_j - r^2 \delta_{ij})}{8\pi\varepsilon_0 r^5} + \cdots . \qquad (3.17)$$

Note that the trace of q_{ij} does not contribute and that the numerator in the final term in (3.17) is replaced by $d_{ij}x_i x_j$, with d_{ij} given by (3.9).

The expansion of the electric field in multipoles follows from (3.17) and $\mathbf{E} = -\mathrm{grad}\,\phi$. The field produced by the charge Q is just the Coulomb field

$$\mathbf{E} = \frac{Q\mathbf{x}}{4\pi\varepsilon_0 r^3}, \qquad (3.18)$$

which has an inverse square law dependence on r. The dipole field is

$$E_i = \frac{(3x_i x_j - \delta_{ij}r^2)d_j}{4\pi\varepsilon_0 r^5}, \qquad (3.19)$$

which has an inverse cube law dependence on r. In general the field due to a 2^l-multipole has a $1/r^{l+2}$ dependence on r. The electric field due to an electric quadrupole, in the traceless form (3.9), reduces to

$$E_i = \frac{(x_i \delta_{rs}r^2 + x_r \delta_{is}r^2 + x_s \delta_{ir}r^2 - 5x_i x_r x_s)d_{rs}}{8\pi\varepsilon_0 r^7}. \qquad (3.20)$$

In an analogous way the expansion of the vector potential, cf. (1.24), gives

$$\mathbf{A}(\mathbf{x}) = \frac{\mu_0}{4\pi} \int d^3x' \frac{\mathbf{J}(\mathbf{x}')}{|\mathbf{x} - \mathbf{x}'|}$$

$$= \frac{1}{4\pi\varepsilon_0} \int d^3x' \, \mathbf{J}(\mathbf{x}') \sum_{l=0}^{\infty} \frac{1}{l!} \left(-\mathbf{x}' \cdot \frac{\partial}{\partial \mathbf{x}} \right)^l \frac{1}{r}. \qquad (3.21)$$

The leading term in the static case is the magnetic dipole term, which arises from the $l = 1$ term in (3.21). On using (3.11) and (3.15) this leading term reduces to

$$\mathbf{A}(\mathbf{x}) = \frac{\mu_0}{4\pi} \int d^3x' \frac{\mathbf{J}(\mathbf{x}')}{|\mathbf{x} - \mathbf{x}'|} = \frac{\mu_0}{4\pi} \left(-\mathbf{m} \times \frac{\partial}{\partial \mathbf{x}} \frac{1}{r} + \cdots \right). \qquad (3.22)$$

The magnetic field then follows from $\mathbf{B} = \mathrm{curl}\,\mathbf{A}$. The dipole field is

$$B_i = \frac{\mu_0}{4\pi} \frac{(3x_i x_j - \delta_{ij}r^2)m_j}{r^5}. \qquad (3.23)$$

One application of (3.23) is to the magnetic fields of objects of astrophysical interest. The strength of the magnetic field of a planet or star is expressed in terms of the value of B on the surface of the object, and usually the value at the magnetic poles is specified. Let R be the radius of the planet or star. The polar value corresponds to $\mathbf{x}$ parallel to $\mathbf{m}$, in which case (3.23) gives

$$B_p = \frac{\mu_0}{4\pi} \frac{2|\mathbf{m}|}{R^3}. \tag{3.24}$$

The value of the field strength at the equator is half the value at the pole.

3.4 The Potential Energy Associated with a Multipole

The motion of a charged particle (charge q) in an electromagnetic field is determined by the equation of motion

$$d\mathbf{p}/dt = q(\mathbf{E} + \mathbf{v} \times \mathbf{B}), \tag{3.25}$$

where $\mathbf{p}$ and $\mathbf{v}$ are the momentum and velocity, respectively, of the particle. Consider a particle with no charge but other electromagnetic properties, for example, a neutral atom with an electric dipole moment or a neutron (which has a magnetic dipole moment). How does one determine the electromagnetic force on such a system? One way is to identify the the relevant potential energy; the force is given by minus the gradient of the potential.

The energy of an electric multipole in an electrostatic field is derived by making the multipole expansion in the interaction energy for the charge distribution interacting with an electric field. This energy is given by the integral of the potential energy per unit volume $\rho(\mathbf{x})\phi(\mathbf{x})$ over all space. On expanding about a point $\mathbf{x}_0$, one has

$$\phi(\mathbf{x}) = \phi(\mathbf{x}_0) + (\mathbf{x} - \mathbf{x}_0) \cdot \left[\frac{\partial \phi(\mathbf{x})}{\partial \mathbf{x}}\right]_{\mathbf{x}=\mathbf{x}_0}$$
$$+ \tfrac{1}{2}(\mathbf{x} - \mathbf{x}_0)_i (\mathbf{x} - \mathbf{x}_0)_j \left[\frac{\partial^2 \phi(\mathbf{x})}{\partial x_i \partial x_j}\right]_{\mathbf{x}=\mathbf{x}_0} + \cdots. \tag{3.26}$$

Then using $\mathbf{E}(\mathbf{x}) = -\partial\phi(\mathbf{x})/\partial\mathbf{x}$ and the definitions (3.8), the interaction energy reduces to

$$\int d^3\mathbf{x}\, \rho(\mathbf{x})\phi(\mathbf{x}) = Q\phi(\mathbf{x}_0) - \mathbf{d} \cdot \mathbf{E}(\mathbf{x}_0) - \frac{1}{6}d_{ij}\frac{\partial E_i(\mathbf{x}_0)}{\partial x_{0j}} + \cdots. \tag{3.27}$$

The leading term in the interaction energy for a current with a magnetostatic field is

$$-\int d^3\mathbf{x}\, \mathbf{J}(\mathbf{x}) \cdot \mathbf{A}(\mathbf{x}) = -\mathbf{m} \cdot \mathbf{B}(\mathbf{x}_0) + \cdots. \tag{3.28}$$

It follows from (3.27) that the force on an electric dipole is given by $\mathrm{grad}\,\mathbf{d} \cdot \mathbf{E}(\mathbf{x})$. Thus the force on an electric dipole depends on the gradient of the electric field. Similarly the force on a magnetic dipole depends on the gradient of the magnetic field. The force on an electric quadrupole depends on the second (spatial) derivative of the electric field.

One specific application is to the force exerted by one multipole on another. Suppose that the field in (3.27) or (3.28) is attributed to a second multipole, which is located at the origin with the notation used in (3.27) or (3.28). Consider the force between a multipole with label 1 at x_1 and a multipole with label 2 at x_2. It is convenient to identify the unlabeled moment appearing in (3.27) or (3.28) as that with subscript 1 and to attribute the field to the moment labeled with a subscript 2; the displacement x_0 is then reinterpreted as the displacement $x_1 - x_2$. The force exerted by one multipole on the another is then obtained by taking the gradient with respect to the appropriate displacement vector. For example, the force on an electric dipole d_1 at x_1 by another electric dipole d_2 at x_2 is given by $\mathrm{grad}_1\, d_1 \cdot E(x_1 - x_2)$ where $E(x)$ is identified as the electric field (3.19), with d replaced by d_2, and where grad_1 denotes the derivative with respect to x_1. A magnetic dipole–dipole interaction is treated in an analogous way.

3.5 Expansion in Spherical Harmonics

The use of cartesian tensors to describe multipole moments is convenient for the dipole moments but it becomes increasingly cumbersome for $l \geq 2$. There is an alternative description of multipole moments in terms of spherical tensor components, and this alternative is more convenient for the higher moments. For completeness this alternative description is summarized here.

The spherical tensor components for the electric multipole moments follow by evaluating the integral (1.23) using spherical polar coordinates. One introduces the polar coordinates r, θ and ϕ and r', θ' and ϕ' of x and x', respectively, relative to an arbitrary axis. Then one has $|x - x'| = (r^2 + r'^2 - 2rr' \cos \Theta)^{1/2}$, where Θ is the angle between x and x', with

$$\cos \Theta = \cos \theta \cos \theta' + \sin \theta \sin \theta' \cos(\phi - \phi'). \qquad (3.29)$$

There is a generating function for Legendre polynomials $P_l(x)$ that is used to evaluate this integral in terms of electric moments. The generating function is

$$\frac{1}{(1 - 2xt + t^2)^{1/2}} = \sum_{l=0}^{\infty} t^l P_l(x), \qquad (3.30)$$

and for the case $r' < r$ of relevance here, this gives

$$\frac{1}{|\mathbf{x} - \mathbf{x}'|} = \sum_{l=0}^{\infty} \frac{r'^l P_l(\cos \Theta)}{r^{l+1}}. \tag{3.31}$$

The expansion (3.31) is reexpressed in terms of spherical harmonics by using the addition theorem

$$P_l(\cos \Theta) = \frac{4\pi}{2l+1} \sum_{m=-l}^{l} Y_l^{*m}(\theta, \phi) Y_l^m(\theta', \phi'). \tag{3.32}$$

The spherical harmonics are defined by

$$Y_l^m(\theta, \phi) = \left(\frac{2l+1}{4\pi}\right)^{1/2} \left[\frac{(l-m)!}{(l+m)!}\right]^{1/2} P_l^m(\cos \theta) e^{im\phi}, \tag{3.33}$$

where $P_l^m(\xi)$ is an associated Legendre function defined by

$$P_l^m(\xi) = \begin{cases} (1-\xi^2)^{m/2} d^m P_l(\xi)/d\xi^m & \text{for } 0 \leq m \leq l, \\ (-1)^m [(l-m)!/(l+m)!] P_l^m(\xi) & \text{for } -l \leq m < 0 \end{cases} \tag{3.34}$$

The orthogonality condition for the spherical harmonics is

$$\int_0^{2\pi} d\phi \int_0^{\pi} d\theta \sin \theta \, Y_l^{*m}(\theta, \phi) Y_{l'}^{m'}(\theta, \phi) = \delta_{ll'} \delta^{mm'}. \tag{3.35}$$

On inserting (3.31) with (3.32) into (3.16), the integral reduces to the form

$$\phi(\mathbf{x}) = \sum_{l=0}^{\infty} \sum_{m=-l}^{l} \frac{d_l^m Y_l^m(\theta, \phi)}{4\pi\varepsilon_0 r^{l+1}}, \tag{3.36}$$

where the spherical tensor components of the multipole moments are identified as

$$d_l^m = \int_0^{2\pi} d\phi' \int_0^{\pi} d\theta' \sin \theta' \int_0^{\infty} dr' r'^{l+2} Y_l^{*m}(\theta', \phi') \rho(\mathbf{x}'). \tag{3.37}$$

Note that the 2^l-moment d_l^m has only the $2l+1$ components corresponding to $m = -l, -(l-1), \ldots, l+1, l$. These correspond to the traceless cartesian tensor forms of the multipole moments.

The relation between the cartesian and spherical tensor components for the dipole and quadrupole moments is deduced from (3.35), (3.36) and (3.37), cf. Exercise 3.8. There is no counterpart of the traces of the cartesian tensor moments q_{ij} etc. in spherical polar coordinates.

The first few spherical harmonics are

$$Y_0^0(\theta, \phi) = \left(\frac{1}{4\pi}\right)^{1/2},$$

$$Y_1^{\pm 1}(\theta, \phi) = \mp \left(\frac{3}{8\pi}\right)^{1/2} e^{\pm i\phi} \sin\theta = \mp \left(\frac{3}{8\pi}\right)^{1/2} \frac{(x \pm iy)}{r},$$

$$Y_1^0(\theta, \phi) = \left(\frac{3}{4\pi}\right)^{1/2} \cos\theta = \left(\frac{3}{4\pi}\right)^{1/2} \frac{z}{r},$$

$$Y_2^{\pm 2}(\theta, \phi) = \left(\frac{15}{32\pi}\right)^{1/2} e^{\pm 2i\phi} \sin^2\theta = \mp \left(\frac{15}{32\pi}\right)^{1/2} \frac{(x \pm iy)^2}{r^2},$$

$$Y_2^{\pm 1}(\theta, \phi) = \mp \left(\frac{15}{8\pi}\right)^{1/2} e^{\pm i\phi} \cos\theta \sin\theta = \mp \left(\frac{15}{8\pi}\right)^{1/2} \frac{(x \pm iy)z}{r^2},$$

$$Y_2^0(\theta, \phi) = \left(\frac{5}{16\pi}\right)^{1/2} (3\cos^2\theta - 1) = \left(\frac{5}{16\pi}\right)^{1/2} \frac{(2z^2 - x^2 - y^2)}{r^2}.$$

$$(3.38)$$

Exercise Set 3

3.1 Compare the magnetic pressure due to the terrestrial magnetic field at the surface of the Earth at one of the poles (take $B = 6 \times 10^{-5}$ T) with (a) atmospheric pressure (1 atm $= 1.01 \times 10^5$ Pa), and (b) the gas pressure near the top of the ionosphere where the particle number density is about 10^9 m^{-3} and the temperature is about 10^4 K.

3.2 The angular momentum density in the electromagnetic field is defined in terms of the momentum density (3.6) by

$$\mathbf{L}_{\text{EM}} = \mathbf{x} \times \mathbf{P}_{\text{EM}} = \mathbf{x} \times (\mathbf{E} \times \mathbf{B})/\mu_0 c^2. \qquad (E3.1)$$

Show that if the continuity equation for angular momentum is written in the form

$$\frac{\partial}{\partial t}(\mathbf{L}_{\text{EM}})_i + \frac{\partial}{\partial x_j}(\mathbf{M}_{\text{EM}})_{ij} = (\mathbf{S}_{\text{EM}})_i$$

then (3.5) implies

$$(\mathbf{M}_{\text{EM}})_{ij} = \epsilon_{irs}(\mathbf{T}_{\text{EM}})_{jr}x_s,$$

$$(\mathbf{S}_{\text{EM}})_i = -\rho\epsilon_{ijk}x_j E_k - J_i\,\mathbf{x}\cdot\mathbf{B} + B_i\,\mathbf{x}\cdot\mathbf{J}.$$

3.3 Consider an electric dipole **d** directed along the z axis.

(a) Write down the electric field components in cartesian coordinates from (3.19).

(b) Evaluate the components of the electric field in spherical polar coordinates by writing the potential in the form

$$\phi(\mathbf{x}) = \frac{\mathbf{d}\cdot\mathbf{x}}{4\pi\varepsilon_0 r^3} = \frac{d\cos\theta}{4\pi\varepsilon_0 r^2};$$

$$E_r = \frac{\partial\phi(\mathbf{x})}{\partial r}, \quad E_\theta = \frac{1}{r}\frac{\partial\phi(\mathbf{x})}{\partial\theta}, \quad E_\phi = \frac{1}{r\sin\theta}\frac{\partial\phi(\mathbf{x})}{\partial\phi}.$$

(c) Show that the results obtained in parts (a) and (b) are mutually compatible.

3.4 The field lines for any magnetic field are found by solving the parametric equations

$$\frac{dx}{B_x} = \frac{dy}{B_y} = \frac{dz}{B_z} \quad \text{or} \quad \frac{dr}{B_r} = \frac{rd\theta}{B_\theta} = \frac{r\sin\theta d\phi}{B_\phi}.$$

(a) Using the result of part (b) of the previous exercise, show that the magnetic field lines satisfy the following equations in spherical polar coordinates:

$$r = r_0\sin^2\theta, \quad \phi = \phi_0,$$

where r_0 and ϕ_0 are constants.

(b)   Give a physical interpretation of the constants r_0 and ϕ_0.

3.5 The generation of magnetic fields in stars and planets is not well understood, but it is strongly believed that there is a correlation between the angular momentum of the object and its magnetic moment. Use the following parameters (mass, period of rotation, radius and polar magnetic field B_p) to test this hypothesis by comparing the ratio of the magnetic moment to the angular momentum for the Earth, Jupiter and the Sun:

object	mass	period	radius	B_p
Earth	6.0×10^{24} kg	1 day	6.4×10^6 m	6.3×10^{-5} T
Jupiter	2.1×10^{27} kg	9 h 50 min	7.1×10^7 m	1.4×10^{-3} T
Sun	2.0×10^{30} kg	25.3 days	7.0×10^8 m	1.0×10^{-4} T

3.6 The force $\mathbf{F}$ and the torque $\mathbf{T}$ due to an electromagnetic field acting on a charge and current distribution are given by, respectively,

$$\mathbf{F} = \int d^3x \left[\rho(\mathbf{x})\mathbf{E}(\mathbf{x}) + \mathbf{J}(\mathbf{x}) \times \mathbf{B}(\mathbf{x})\right],$$
$$\mathbf{T} = \int d^3x \, \mathbf{x} \times \left[\rho(\mathbf{x})\mathbf{E}(\mathbf{x}) + \mathbf{J}(\mathbf{x}) \times \mathbf{B}(\mathbf{x})\right]. \qquad (E3.2)$$

(a)   By expanding in multipoles about the origin, show that the force implied by (3.2) reproduces that obtained by taking minus the gradient of the potential energy (3.27) plus (3.28).

(b)   Show that the torque on an electric dipole and that on a magnetic dipole are given by, respectively,

$$\mathbf{T} = \mathbf{d} \times \mathbf{E}, \quad \mathbf{T} = \mathbf{m} \times \mathbf{B}. \qquad (E3.3)$$

(c)   A particle has its magnetic moment $\mathbf{m}$ related to its angular momentum $\mathbf{L}$ by

$$\mathbf{m} = \Gamma \mathbf{L}.$$

Show that in the presence of a magnetic field, $\mathbf{L}$ precesses about $\mathbf{B}$ with angular frequency

$$\omega = \Gamma B.$$

3.7 Show that the relations between the cartesian tensor representations of the first three electric multipole moments and the corresponding spherical tensor components (3.37) are

$$d_0^0 = \left(\frac{1}{4\pi}\right)^{1/2} Q, \qquad d_2^{\pm2} = \left(\frac{5}{96\pi}\right)^{1/2} (d_{xx} \mp 2\mathrm{i}d_{xy} - d_{yy}),$$

$$d_1^{\pm1} = -\left(\frac{3}{8\pi}\right)^{1/2} (d_x \mp \mathrm{i}d_y), \quad d_2^{\pm1} = -\left(\frac{5}{24\pi}\right)^{1/2} (d_{xz} \mp \mathrm{i}d_{yz}),$$

$$d_1^0 = \left(\frac{3}{4\pi}\right)^{1/2} d_z, \qquad d_2^0 = \left(\frac{5}{16\pi}\right)^{1/2} d_{zz}.$$

3.8 Two electric dipoles are orthogonal to the line joining them. Assume that the dipole moments have magnitude d and that the distance between them is a. Find the force exerted by one dipole on the other when (a) the dipoles are parallel, (b) the dipoles are antiparallel, and (c) the dipoles are mutually orthogonal.

4

Fourier Transforms

Preamble

For many purposes it is both mathematically convenient and physically relevant to describe a physical quantity in terms of its Fourier transform, rather than in terms of its space-time dependence. Fourier transformations are introduced here and their general properties are described.

4.1 Fourier Integral Theorem

The mathematical basis for Fourier transformations is the Fourier integral theorem.

The Fourier integral theorem: For any function $f(t)$ such that

(1) $f(t)$ is sectionally continuous over $-\infty < t < \infty$,

(2) $f(t)$ is defined as

$$\lim_{\Delta \to 0} \tfrac{1}{2}\left[f(t_0 + \Delta) + f(t_0 - \Delta)\right]$$

at any point of discontinuity,

(3) $f(t)$ is amplitude integrable, that is,

$$\int_{-\infty}^{\infty} dt\, |f(t)| < \infty, \tag{4.1}$$

the following identity holds:

$$f(t) = \frac{1}{2\pi} \int_{-\infty}^{\infty} dy \int_{-\infty}^{\infty} dz\, f(z) e^{\pm iy(z-t)}. \tag{4.2}$$

This theorem is used to define Fourier transforms. For a function $f(t)$

of time t one defines the Fourier transform by

$$\tilde{f}(\omega) := \int_{-\infty}^{\infty} dt\, e^{i\omega t} f(t). \qquad (4.3)$$

Then (4.2) implies that the inverse transform is given by

$$f(t) = \int_{-\infty}^{\infty} \frac{d\omega}{2\pi}\, e^{-i\omega t} \tilde{f}(\omega). \qquad (4.4)$$

The choice of sign in the exponent in (4.3) is a matter of convention, but once this is chosen (4.2) implies that the opposite sign must appear in the exponent in (4.4). Similarly, it is a matter of convention as to where the factor 2π in (4.2) is to be included, and the convention adopted in (4.3) and (4.4) is that the 2π be associated with the inverse transform. The parameter ω is the angular frequency (units: s^{-1}). An alternative convention is to write $\omega = 2\pi\nu$, where ν is the cyclic frequency (units: Hz); the factor 2π then appears in both exponents.

For spatial Fourier transforms it is conventional to choose the opposite sign in the exponent. Let $A(\mathbf{x})$ be a function of position vector $\mathbf{x}$. Its Fourier transform is defined by

$$\tilde{A}(\mathbf{k}) := \int d^3x\, e^{-i\mathbf{k}\cdot\mathbf{x}}\, A(\mathbf{x}), \qquad (4.5)$$

where $\mathbf{k}$ is the *wave vector* and where the integral is over all space. The inverse transform is

$$A(\mathbf{x}) = \int \frac{d^3k}{(2\pi)^3}\, e^{i\mathbf{k}\cdot\mathbf{x}}\, \tilde{A}(\mathbf{k}). \qquad (4.6)$$

In cartesian coordinates d^3x denotes $dx\,dy\,dz$ and similarly d^3k denotes $dk_x\,dk_y\,dk_z$, and all integrals are over the range $-\infty$ to ∞.

For a function $F(t, \mathbf{x})$ of both time and position the Fourier transform is defined by combining (4.3) and (4.5):

$$\tilde{F}(\omega, \mathbf{k}) := \int dt\,d^3x\, e^{i(\omega t - \mathbf{k}\cdot\mathbf{x})}\, F(t, \mathbf{x}). \qquad (4.7)$$

The inverse transform is

$$F(t, \mathbf{x}) = \int \frac{d\omega\,d^3k}{(2\pi)^4}\, e^{-i(\omega t - \mathbf{k}\cdot\mathbf{x})}\, \tilde{F}(\omega, \mathbf{k}). \qquad (4.8)$$

It is convenient to introduce Fourier transforms to solve certain differential equations. The reason for this is that differential operators are replaced by algebraic operators, so that differential equations are replaced by algebraic equations. Let an arrow ($\rightarrow$) denote the operation of taking a Fourier transform in time and space:

$$F(t, \mathbf{x}) \rightarrow \tilde{F}(\omega, \mathbf{k}).$$

Then we have

$$\partial F(t,\mathbf{x})/\partial t \rightarrow -i\omega \tilde{F}(\omega,\mathbf{k}), \quad \text{grad}\, F(t,\mathbf{x}) \rightarrow i\mathbf{k}\tilde{F}(\omega,\mathbf{k}),$$
$$\text{div}\,\mathbf{V}(t,\mathbf{x}) \rightarrow i\mathbf{k}\cdot\tilde{\mathbf{V}}(\omega,\mathbf{k}), \quad \text{curl}\,\mathbf{V}(t,\mathbf{x}) \rightarrow i\mathbf{k}\times\tilde{\mathbf{V}}(\omega,\mathbf{k}). \tag{4.9}$$

Where no confusion should result, the tilde on the Fourier transformed quantities is omitted below. The tilde is included to denote that $\tilde{f}$ is a different mathematical function from f. In mathematics one uses different kernel functions (here $\tilde{f}$ and f) to denote different mathematical functions, whereas in physics one uses different kernel symbols to denote different physical quantities. Thus below f is used to refer to a physical quantity either in time and space or in terms of frequency and wave vector, with the two being distinguished by the arguments of the function. The tilde is included only if confusion might arise.

4.2 Properties of Fourier Transforms

Three general properties of Fourier transforms are the reality condition, the power theorem and the convolution theorem.

The Reality Condition: If $F(t,\mathbf{x})$ is real then its Fourier transform satisfies

$$F^*(\omega,\mathbf{k}) = F(-\omega,-\mathbf{k}). \tag{4.10}$$

The Power Theorem: If $F(t,\mathbf{x})$ and $G(t,\mathbf{x})$ are real functions, then one has

$$\int dt d^3\mathbf{x}\, F(t,\mathbf{x})G(t,\mathbf{x}) = \int \frac{d\omega d^3\mathbf{k}}{(2\pi)^4} F^*(\omega,\mathbf{k})G(\omega,\mathbf{k}). \tag{4.11}$$

The Convolution Theorem: The Fourier transform of a product of functions is equal to the convolution of their Fourier transforms, and the Fourier transform of the convolution of functions is equal to the product of their Fourier transforms.

The convolution of two functions $F(t,\mathbf{x})$ and $G(t,\mathbf{x})$ is

$$H(t,\mathbf{x}) = F(t,\mathbf{x})*G(t,\mathbf{x}) := \int dt' d^3\mathbf{x}'\, F(t-t',\mathbf{x}-\mathbf{x}')G(t',\mathbf{x}'). \tag{4.12}$$

The theorem implies

$$H(\omega,\mathbf{k}) = F(\omega,\mathbf{k})G(\omega,\mathbf{k}). \tag{4.13}$$

The theorem also implies that the Fourier transform of

$$J(t,\mathbf{x}) = F(t,\mathbf{x})G(t,\mathbf{x}) \tag{4.14}$$

is

$$J(\omega, \mathbf{k}) = F(\omega, \mathbf{k}) * G(\omega, \mathbf{k}) := \int \frac{d\omega' d^3 k'}{(2\pi)^4} F(\omega - \omega', \mathbf{k} - \mathbf{k}') G(\omega', \mathbf{k}').$$

(4.15)

Note that the convolution is defined without any factor $1/2\pi$ in terms of the time and space variables, and with one factor of $1/2\pi$ for each variable in Fourier space.

The proof of the reality condition follows by comparing (4.7) and its complex conjugate, with $F^*(t, \mathbf{x}) = F(t, \mathbf{x})$ by hypothesis:

$$\tilde{F}^*(\omega, \mathbf{k}) = \int dt d^3 x \, e^{-i(\omega t - \mathbf{k} \cdot \mathbf{x})} \, F(t, \mathbf{x})$$

$$= \tilde{F}(-\omega, -\mathbf{k}).$$

The proofs of the power theorem and the convolution theorem involve the use of Dirac δ-functions. The Dirac δ-function is proportional to the Fourier transforms of unity:

$$\pi \delta(\omega) = \int_{-\infty}^{\infty} dt \, e^{i\omega t}, \quad (2\pi)^3 \delta^3(\mathbf{k}) = \int d^3 x \, e^{-i\mathbf{k} \cdot \mathbf{x}}.$$

(4.16)

In cartesian coordinates $\delta^3(\mathbf{k})$ means $\delta(k_x)\delta(k_y)\delta(k_z)$.

The proof of the power theorem (4.11) involves writing the functions on the left in terms of their Fourier transforms, and carrying out the integrals using (4.16). For simplicity in writing, let us give the proof for functions of t only. One has

$$\int dt \, f(t) g(t) = \int dt \int \frac{d\omega}{2\pi} e^{-i\omega t} f(\omega) \int \frac{d\omega'}{2\pi} e^{-i\omega' t} g(\omega')$$

$$= \int \frac{d\omega}{2\pi} \int \frac{d\omega'}{2\pi} f(\omega) g(\omega') \, 2\pi \delta(\omega + \omega')$$

$$= \int \frac{d\omega}{2\pi} f(\omega) g(-\omega).$$

The final step involves using the reality condition to write $g(-\omega) = g^*(\omega)$. Alternatively, on replacing the variable of integration ω by $-\omega$, the integrand becomes $f^*(\omega)g(\omega)$, which is the form quoted in (4.11).

The proof of the convolution theorem is similar. Once again considering functions of time only, the Fourier transform of the convolution $h(t) := \int dt' \, f(t - t') g(t')$ is

$$h(\omega) = \int dt \, e^{i\omega t} \int dt' \int \frac{d\omega'}{2\pi} e^{-i\omega'(t-t')} f(\omega') \int \frac{d\omega''}{2\pi} e^{-i\omega'' t'} g(\omega'')$$

$$= \int \frac{d\omega'}{2\pi} \int \frac{d\omega''}{2\pi} f(\omega') g(\omega'') \, 2\pi \delta(\omega - \omega') \, 2\pi \delta(\omega' - \omega'')$$

$$= f(\omega) g(\omega).$$

The proof generalizes straightforwardly to multi-dimensional Fourier transforms.

4.3 The Dirac δ-function

The δ-function $\delta(x)$ was defined originally by Dirac as that function (a) which is equal to zero for $x \neq 0$, and (b) whose integral is unity. That is,

$$\delta(x) = \begin{cases} 0 & \text{for } x \neq 0, \\ \infty & \text{for } x = 0, \end{cases} \qquad \int dx\, \delta(x) = 1, \qquad (4.17)$$

where the integral is over a range from $x < 0$ to $x > 0$. The δ-function is an even function:

$$\delta(-x) = \delta(x). \qquad (4.18)$$

Suppose that the δ-function has as its argument a function $f(x)$. Then according to Dirac's definition, $\delta[f(x)]$ is zero except at the zeros of $f(x)$. Suppose that $f(x)$ has n simple zeros at $x = x_1, x_2, \ldots, x_n$. Then one has

$$\int_{-\infty}^{\infty} dx\, \delta[f(x)] = \sum_{i=1}^{n} \frac{1}{|f'(x_i)|}, \qquad (4.19)$$

with $f'(x) := \partial f(x)/\partial x$.

The proof of (4.19) is as follows. By hypothesis each zero is a simple zero, and so one has $f(x) = 0$ and $f'(x) \neq 0$ at each zero, and furthermore there is a finite separation between each zero. It follows that the integral in (4.19) separates into n segments each of which contains only one zero. Consider the segment of the integral that contains the zero at $x = x_i$. The δ-function is zero except at $x = x_i$ according to (4.17), and hence the range of integration may be shrunk to an arbitrarily small range about x_i: $x_i - \Delta < x < x_i + \Delta$. Within this range one makes a Taylor expansion about $x = x_i$,

$$f(x) = (x - x_i)f'(x_i) + \cdots.$$

Suppose that $f'(x_i)$ is positive. Then if one writes $y := (x - x_i)f'(x_i)$ the integral reduces to

$$\int_{x_i - \Delta}^{x_i + \Delta} dx\, \delta[(x - x_i)f'(x_i)] = \frac{1}{f'(x_i)} \int dy\, \delta(y) = \frac{1}{f'(x_i)},$$

which establishes (4.19) for cases with $f'(x_i) > 0$. For $f'(x_i) < 0$ one writes $y := (x - x_i)|f'(x_i)|$ and uses the property (4.18) to establish the result.

In practice one is usually concerned with integrals that have a non-trivial integrand, that is, with a function $g(x)$, say, also appearing in the integrand in (4.19). Provided that $g(x)$ is continuous at each zero of $f(x)$, (4.19) is generalized to

$$\int_{-\infty}^{\infty} dx \, g(x) \, \delta[f(x)] = \sum_{i=1}^{n} \frac{g(x_i)}{|f'(x_i)|}. \qquad (4.20)$$

This equation, with $f(\omega) = \omega$ and $g(\omega) = e^{-i\omega t}$, is used to show that the inverse transform of $2\pi\delta(\omega)$ is equal to unity, so that the two definitions (4.16) and (4.17) of $\delta(\omega)$ are consistent.

4.4 Truncations

The Dirac δ-function as defined by (4.16) is the Fourier transform of unity. However, the condition (4.1) for the Fourier integral theorem to apply is clearly violated for $f(t) = 1$. This implies that the Fourier transform of unity does not exist as an ordinary function. The δ-function may be defined as a *generalized function*, that is, as a sequence of ordinary functions. Operations involving the δ-function are given meaning by evaluating the limit of the operation applied to each member of the sequence. Loosely, the generalized function may be thought of as the limit of a sequence of ordinary functions. A variety of different sequences may be chosen, and an important property of a generalized function is that it is independent of the specific sequence chosen to define it.

A sequence of functions of ω that defines $\delta(\omega)$ is identified as the Fourier transform of a sequence of functions of t whose limit is unity. One example is the sequence obtained by *truncating*: the truncated form of a function $f(t)$ is

$$f_T(t) := \begin{cases} f(t) & \text{for } -T/2 < t < T/2, \\ 0 & \text{for } |t| > T/2. \end{cases} \qquad (4.21)$$

With this truncation the integral in (4.1) is necessarily finite and so the Fourier transforms exist for arbitrarily large but finite T. Another sequence that is used to define a truncated function involves truncating with an exponential. Consider

$$f_\eta(t) := f(t)e^{-\eta|t|} \qquad (4.22)$$

in the limit $\eta \to 0$. For a wide class of functions $f(t)$ that are not themselves amplitude integrable in the sense (4.1), the function $f_\eta(t)$ is amplitude integrable.

The Fourier transform of the truncated form of the unit function may be used to construct a sequence of functions that defines the δ-function. In the case of (4.21) one considers increasingly large values of T. The δ-function is then defined as the limit $T \to \infty$ of the sequence:

$$\delta(\omega) = \lim_{T \to \infty} \frac{\sin(\omega T/2)}{\pi \omega}. \tag{4.23}$$

With the choice (4.22) for the truncation, the appropriate sequence is

$$\delta(\omega) = \lim_{\eta \to 0} \frac{\eta}{\pi(\omega^2 + \eta^2)}. \tag{4.24}$$

There are two other discontinuous functions whose Fourier transforms define important generalized functions. One of these is the step function

$$H(t) := \begin{cases} 1 & \text{for } t > 0, \\ 0 & \text{for } t < 0. \end{cases} \tag{4.25}$$

The form (4.22) of truncation gives a Fourier transform

$$\tilde{H}_\eta(\omega) = \int_0^\infty dt\, e^{i\omega t - \eta t} = \frac{i}{\omega + i\eta}. \tag{4.26}$$

The generalized function is defined by

$$\frac{1}{\omega + i0} := \lim_{\eta \to 0} \frac{1}{\omega + i\eta}. \tag{4.27}$$

The other function is the sign function

$$\operatorname{sgn}(t) := \frac{t}{|t|} = \begin{cases} 1 & \text{for } t > 0, \\ -1 & \text{for } t < 0. \end{cases} \tag{4.28}$$

The form (4.22) of truncation gives a Fourier transform

$$\begin{aligned} \widetilde{\operatorname{sgn}}_\eta(\omega) &= \int_0^\infty dt\, e^{i\omega t - \eta t} - \int_{-\infty}^0 dt\, e^{i\omega t + \eta t} \\ &= \frac{i}{\omega + i\eta} + \frac{i}{\omega - i\eta} = \frac{2i\omega}{\omega^2 + \eta^2}. \end{aligned} \tag{4.29}$$

The generalized function in this case is called the *Cauchy principal value function*:

$$\wp\frac{1}{\omega} := \lim_{\eta \to 0} \frac{\omega}{\omega^2 + \eta^2} = \begin{cases} 1/\omega & \text{for } \omega \neq 0, \\ 0 & \text{for } \omega = 0. \end{cases} \tag{4.30}$$

The sequence (4.30) may be used to define $\wp(1/\omega)$, but there are alternative sequences that are equivalent to it. In particular, consider the sequence in which a function is equal to $1/\omega$ for $|\omega| > \eta$ and is equal to zero for $|\omega| < \eta$, with η allowed to approach zero. This alternative sequence inside an integral corresponds to the conventional definition of the Cauchy principal value of the integral. The equivalence of the two sequences is addressed in Exercise 4.8. More generally, a variety of sequences may be used to define each of the three generalized functions considered here. For formal purposes it is often convenient to choose gaussian functions, as described in Exercise 4.2.

4.5 Relations Involving Generalized Functions

There is a relation between the unit, step and sign functions:

$$H(t) = \tfrac{1}{2}[1 + \mathrm{sgn}(t)]. \tag{4.31}$$

The Fourier transform of (4.31) leads to a relation between the generalized functions. Specifically, it implies the *Plemelj formula*

$$\frac{1}{\omega + i0} = \wp\frac{1}{\omega} - i\pi\delta(\omega). \tag{4.32}$$

The Plemelj formula (4.32) is used in evaluating certain singular integrals.

There are further identities involving the step, unit and sign functions. These include

$$[\mathrm{sgn}(t)]^2 = 1, \quad [H(t)]^2 = H(t), \quad \mathrm{sgn}(t)H(t) = H(t). \tag{4.33}$$

The Fourier transforms of these are evaluated using the convolution theorem. The first of (4.33) leads to the identity

$$-\frac{1}{\pi^2}\int_{-\infty}^{\infty} d\omega' \, \wp\left(\frac{1}{\omega - \omega'}\right) \wp\left(\frac{1}{\omega'}\right) = \delta(\omega). \tag{4.34}$$

This identity is used in the inverse of Hilbert transforms, which is another class of integral transform (along with Fourier, Laplace and Mellin transforms). The Hilbert transform $f_{\mathrm{H}}(\omega)$ of $f(\omega)$ is defined by

$$f_{\mathrm{H}}(\omega) := -\frac{1}{\pi}\int_{-\infty}^{\infty} d\omega' \, \wp\left(\frac{1}{\omega - \omega'}\right) f(\omega'). \tag{4.35}$$

Then (4.34) implies that the inverse Hilbert transform is given by

$$f(\omega) = \int_{-\infty}^{\infty} d\omega' \, f(\omega')\delta(\omega - \omega') = \frac{1}{\pi}\int_{-\infty}^{\infty} d\omega' \, \wp\left(\frac{1}{\omega - \omega'}\right) f_{\mathrm{H}}(\omega'), \tag{4.36}$$

where (4.34) and (4.35) are used.

One often wishes to consider the Fourier transform of an idealized physical quantity, such as a plane wave, which does not vanish at infinity. The Fourier transform then does not exist as an ordinary function. Implicitly one assumes that a truncation is performed, and one uses the Fourier transforms ignoring the truncation. In practice this leads to a problem only when the square of a δ-function appears. The square of the δ-function needs to be interpreted in terms of the truncated function.

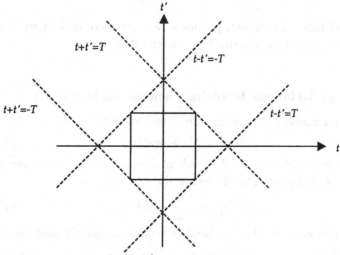

Fig. 4.1 In evaluating $[2\pi\delta(\omega)]^2$ using (4.37) the integral over the inner square $-T/2 < t < T/2$, $-T/2 < t' < T/2$ is evaluated using that over the outer (dashed) square $-T < t - t' < T$, $-T < t + t' < T$.

Using the form (4.21) of the truncation, one has

$$[2\pi\delta(\omega)]^2 = \int_{-T/2}^{T/2} dt\, e^{i\omega t} \int_{-T/2}^{T/2} dt'\, e^{i\omega t'}$$

$$= \tfrac{1}{2} \int_{-T}^{T} d(t - t') \int_{-T}^{T} d(t + t')\, e^{i\omega(t+t')} \qquad (4.37)$$

$$= T\, 2\pi\delta(\omega),$$

where the double integral is evaluated as outlined in Figure 4.1. The limit $T \to \infty$ is implicit in (4.37).

An alternative treatment of the square of the δ-function starts from the form (4.23). Let us write

$$D(\omega) := \frac{\sin^2(\omega T/2)}{\pi^2 \omega^2}. \qquad (4.38)$$

By inspection, one has $D(0) = T^2/4\pi^2$ and the ratio $D(\omega)/D(0)$ tends to zero for all $\omega \neq 0$ in the limit $T \to \infty$. Also one has

$$\int_{-\infty}^{\infty} d\omega\, D(\omega) = \frac{T}{2\pi} \int_{-\infty}^{\infty} dx\, \mathrm{sinc}^2 x = \frac{T}{2\pi}, \qquad (4.39)$$

with

$$\mathrm{sinc}\, x := \frac{\sin(\pi x)}{\pi x}. \qquad (4.40)$$

Now, as in (4.17), the $\delta(\omega)$ is defined as the limit of a sequence of functions each of which is zero for $\omega \neq 0$ and each of whose integral is equal to unity. It then follows that $D(\omega)$ is just $T/2\pi$ times $\delta(\omega)$, in accord with (4.37).

The square of the spatial δ-function is treated in an analogous way. The spatial functions are assumed to be truncated to an arbitrarily large but finite volume V. Then one has

$$[(2\pi)^3 \delta^3(\mathbf{k})]^2 = V\,(2\pi)^3 \delta^3(\mathbf{k}). \qquad (4.41)$$

There is an alternative derivation of (4.41) based on truncating to a box. The idea is to use a box normalization and allow the size of the box to tend to infinity. Let the box have sides of length L_x, L_y, L_z, so that its volume is $V = L_x L_y L_z$. Allowed solutions have nodes at the sides of the box, and so correspond to $k_x = 2\pi n_x / L_x$, $k_y = 2\pi n_y / L_y$, $k_z = 2\pi n_z / L_z$ with n_x, n_y, n_z arbitrary integers. The equality of two wavenumbers may be expressed in terms of the Kronecker delta (2.11). For example, $\delta_{n_x n_x'}$ implies equality of two wavenumbers in the x direction. In the limit $L_x \to \infty$, $\delta_{n_x n_x'}$ becomes $(2\pi/L_x)\delta(k_x - k_x')$ with $k_x' = 2\pi n_x'/L_x$. The Kronecker delta has only the values 0 and 1, each of which is equal to its square, and hence the Kronecker delta satisfies the identity

$$[\delta_{n_x n_x'}]^2 = \delta_{n_x n_x'}.$$

In the limit $L_x \to \infty$ this identity implies

$$[(2\pi/L_x)\delta(k_x - k_x')]^2 = (2\pi/L_x)\delta(k_x - k_x').$$

Taking the product of this latter identity with the corresponding identities in the y and z components, the result reduces to (4.41) with $\mathbf{k}$ replaced by $\mathbf{k} - \mathbf{k}'$.

Exercise Set 4

4.1 Write down the Fourier transforms of the following equations:

(a)   For $\mathbf{A}$, $\mathbf{J}$ denoting $\mathbf{A}(t, \mathbf{x})$, $\mathbf{J}(t, \mathbf{x})$:

$$\text{div curl}\mathbf{A} = 0,$$

$$\text{curl curl}\mathbf{A} = \text{grad div}\mathbf{A} - \nabla^2\mathbf{A},$$

$$\left(\frac{1}{c^2}\frac{\partial^2}{\partial t^2} - \nabla^2\right)\mathbf{A} = \mu_0\mathbf{J}.$$

(b)   With all vector functions of t, $\mathbf{x}$:

$$\text{grad}\,(\mathbf{A}\cdot\mathbf{B}) = (\mathbf{A}\cdot\text{grad})\mathbf{B} + (\mathbf{B}\cdot\text{grad})\mathbf{A}$$
$$+ \mathbf{A}\times\text{curl}\mathbf{B} + \mathbf{B}\times\text{curl}\mathbf{A},$$

$$\text{div}\,(\mathbf{A}\times\mathbf{B}) = \mathbf{B}\cdot\text{curl}\mathbf{A} - \mathbf{A}\cdot\text{curl}\mathbf{B},$$

$$\text{curl}\,(\mathbf{A}\times\mathbf{B}) = \mathbf{A}\,\text{div}\mathbf{B} + (\mathbf{B}\cdot\text{grad})\mathbf{A} - \mathbf{B}\,\text{div}\mathbf{A} - (\mathbf{A}\cdot\text{grad})\mathbf{B}.$$

Remark: You need to use the convolution theorem.

4.2 A sequence of functions tending to the unit function in the limit $T \to \infty$ is

$$f_T(t) = e^{-t^2/T^2}. \qquad\qquad (E4.1)$$

Show that this sequence gives rise to the representation

$$\delta(\omega) = \lim_{T\to\infty}\frac{T}{2\pi^{1/2}}e^{-\omega^2 T^2/4} \qquad\qquad (E4.2)$$

4.3 The following identity is given

$$\delta(\omega) = \lim_{t\to\infty}\frac{\sin(\omega t/2)}{\pi\omega}. \qquad\qquad (E4.3)$$

(a)   Show that $(E4.3)$ implies

$$\lim_{t\to\infty}\sin(\omega t) = 0 \quad\text{for all } \omega;$$

$$\lim_{t\to\infty}\cos(\omega t) = \begin{cases} 0 & \text{for } \omega \neq 0, \\ 1 & \text{for } \omega = 0. \end{cases}$$

(b)   Using $\cos(x - \pi/2) = \sin x$, show that

$$\lim_{t\to\infty}\frac{\cos(\omega t/2)}{\pi\omega} = -\delta(\omega).$$

(c)   Use the trigonometric addition formulas and the results of part (a) to show that, for all ω_1, ω_2,

$$\lim_{t\to\infty}\sin(\omega_1 t)\cos(\omega_2 t) = 0 = \lim_{t\to\infty}\cos(\omega_1 t)\sin(\omega_2 t).$$

Further, for $\omega_1 \neq \omega_2$, show that

$$\lim_{t\to\infty}\cos(\omega_1 t)\cos(\omega_2 t) = 0 = \lim_{t\to\infty}\sin(\omega_1 t)\sin(\omega_2 t).$$

4.4 By integrating by parts, evaluate the integral

$$\int_{-\infty}^{\infty} dx\, f(x)\, \frac{d}{dx}\mathrm{sgn}(x),$$

for a differentiable function $f(x)$ which, with its derivative, is assumed amplitude integrable.

(a)   Hence prove

$$\frac{d}{dx}\mathrm{sgn}(x) = 2\delta(x). \tag{E4.4}$$

(b)   Consider

$$F(\omega) = \int_{-\infty}^{\infty} dt\, \frac{e^{i\omega t}}{2\pi t}.$$

Show that (i) $F(\omega)$ is an odd function of ω, (ii) $dF(\omega)/d\omega = i\delta(\omega)$, and (iii) hence $F(\omega) = \frac{1}{2}i\,\mathrm{sgn}(\omega)$, and (iv)

$$\int_{0}^{\infty} dx\, \frac{\sin x}{x} = \frac{\pi}{2}. \tag{E4.5}$$

4.5 The Fourier transform of the step function $H(t)$ is $i/(\omega + i0)$.

(a)   Show that the Fourier transform of $H(-t)$ is $-i/(\omega - i0)$.

(b)   Let $\tilde{F}(\omega)$ be the Fourier transform of $F(t)$, $\tilde{F}^{+}(\omega)$ be the Fourier transform of $F^{+}(t) := H(t)F(t)$ and $\tilde{F}^{-}(\omega)$ be the Fourier transform of $F^{-}(t) := H(-t)F(t)$. Find expressions for $\tilde{F}^{+}(\omega)$ and $\tilde{F}^{-}(\omega)$ given $\tilde{F}(\omega)$.

(c)   Establish the identity

$$\tilde{F}^{+}(\omega) - \tilde{F}^{-}(\omega) = \frac{i}{\pi}\,\wp \int_{-\infty}^{\infty} d\omega'\, \frac{\tilde{F}(\omega')}{\omega - \omega'}. \tag{E4.6}$$

Remark: If $F^{-}(t)$ vanishes then $F(t) = F^{+}(t)$ is a *causal function*, and (E4.6), with $\tilde{F}(\omega) = \tilde{F}^{+}(\omega)$, gives the *dispersion integral* satisfied by a causal function.

4.6 Given that $f(x)$ is a periodic function with period 2π, it is expanded in the form

$$f(x) = \sum_{n=-\infty}^{\infty} c_n\, e^{inx}, \quad c_n = \frac{1}{\pi}\int_{-\pi}^{\pi} dx\, f(x) e^{-inx}. \tag{E4.7}$$

(a)   Apply these relations to establish the identity

$$\pi \sum_{m=-\infty}^{\infty} \delta(x - 2m\pi) = \sum_{n=-\infty}^{\infty} e^{inx}. \tag{E4.8}$$

(b)   Given a periodic function $f(t)$ with period d, show that

$$\tilde{f}(\omega) = \sum_{m=-\infty}^{\infty} c_m \, \delta(\omega - 2m\pi/d), \quad c_m = \frac{2\pi}{d} \int_0^d dt \, f(t) e^{i2m\pi t/d}.$$

$$(E4.9)$$

4.7 Suppose that the function $f(x)$ has only one zero in the range a to b, and that this zero is of order $m = 2$ and is at $x = x_0$, so that $f(x)$ is proportional to $(x - x_0)^2$ sufficiently near $x = x_0$. Suppose that $g(x)$ is continuous in (a, b).

(a)   Show that

$$\int_a^b dx \, g(x) \, \delta\big(f(x)\big)$$

diverges unless $g(x_0) = 0$.

(b)   Show that for $g(x_0) = 0$ the value of the integral is zero.

(c)   Show that the corresponding result is zero for order $m = 2n$ and give the result for order $m = 2n + 1$, where the degree of the zero of $g(x)$ in all cases is $m - 1$.

4.8 The conventional definition of the Cauchy principal value of an integral over $-\infty < \omega < \infty$ with the integrand having a pole at $\omega = 0$ involves replacing the integral by the sum of the integrals over $-\infty < \omega < -\eta$ and $\eta < \omega < \infty$ and taking the limit $\eta \to 0$. This is equivalent to inserting the generalized function

$$\wp\frac{1}{\omega} = \lim_{\eta \to 0} \begin{cases} 1/\omega & \text{for } |\omega| > \eta, \\ 0 & \text{for } |\omega| < \eta \end{cases} \qquad (E4.10)$$

inside the initial integral. The following exercise involves the details of showing that the Cauchy principal value of the integral is defined by inserting either $(E4.10)$ or (4.30) into the integrand, thereby establishing that $(E4.10)$ and (4.30) define the same generalized function. That is, if one defines

$$I_1 = \lim_{\eta \to 0} \left(\int_{-\infty}^{-\eta} + \int_{\eta}^{\infty} \right) d\omega \, \frac{F(\omega)}{\omega}, \quad I_2 = \lim_{\eta \to 0} \int_{-\infty}^{\infty} d\omega \, \frac{\omega F(\omega)}{\omega^2 + \eta^2},$$

$$(E4.11)$$

for any analytic function $F(\omega)$ then the identity $I_1 = I_2$ obtains.

(a)   Show that the difference between the two integrals reduces to

$$I_2 - I_1 = \lim_{\eta \to 0} \Bigg[-\eta^2 \int_{\eta}^{\infty} d\omega \, \frac{F(\omega) - F(-\omega)}{\omega(\omega^2 + \eta^2)}$$

$$+ \int_{-\eta}^{\eta} d\omega \, \frac{\omega F(\omega)}{\omega^2 + \eta^2} \Bigg]. \qquad (E4.12)$$

(b) Argue that in the limit $\eta \to 0$ the only possible non-zero contribution could be from the singular term at $\omega = 0$ in the first integral on the right hand side of $(E4.12)$.

(c) After inserting the expansion $F(\pm\omega) = F(0) \pm \omega F'(0) + \cdots$ for sufficiently small ω, show that one has

$$\lim_{\eta \to 0} \eta \int_{\eta}^{\infty} d\omega \, \frac{F(\omega) - F(-\omega)}{\omega(\omega^2 + \eta^2)} = \frac{\pi}{2} F'(0).$$

(d) Hence argue that one has $I_1 = I_2$ for any analytic function $F(\omega)$.

5

Greens Functions

Preamble

Using the mathematical tools developed in the previous chapter we now return to the development of electromagnetic theory. After discussing the Fourier transforms of Maxwell's equations, the solutions that are written down in Chapter 1 of Poisson's equation for the potentials of static fields in the Coulomb gauge and of d'Alembert's equation for the potentials of fluctuating fields in the Lorentz gauge are derived in terms of the relevant Greens functions. One new feature appears in solving Maxwell's equations in the temporal gauge: the Greens function involved is a tensor.

5.1 Fourier Transformed Fields

The Fourier transformed form of the set (1.1)–(1.4) of Maxwell's equations is

$$\mathbf{k} \times \mathbf{E}(\omega, \mathbf{k}) = \omega \mathbf{B}(\omega, \mathbf{k}), \qquad (5.1)$$

$$i\mathbf{k} \times \mathbf{B}(\omega, \mathbf{k}) = \mu_0 \mathbf{J}(\omega, \mathbf{k}) - i\omega \mathbf{E}(\omega, \mathbf{k})/c^2, \qquad (5.2)$$

$$\mathbf{k} \cdot \mathbf{E}(\omega, \mathbf{k}) = -i\rho(\omega, \mathbf{k})/\varepsilon_0, \qquad (5.3)$$

$$\mathbf{k} \cdot \mathbf{B}(\omega, \mathbf{k}) = 0. \qquad (5.4)$$

A somewhat surprising feature of the Fourier transformed form of Maxwell's equations is that the four equations are not independent: (5.4) is implied by (5.1). It is obvious on mathematical grounds that one cannot reduce four independent equations to three independent equations without loss of generality, and because the loss of generality is not immediately obvious, the result seems paradoxical. The resolution of the paradox is to note that (5.4) follows from (5.1) only for $\omega \neq 0$. For static fields one has $\omega = 0$ and then (5.1) and (5.4) are independent equations. Thus the loss of generality is that the Fourier transformed set of equations (5.1)–(5.4) cannot be used to describe static fields. Consequently it is relevant to distinguish between *static fields* and *fluctuating fields*. Static fields may be described in terms of potentials in the Coulomb gauge, in which case they satisfy Poisson's equation (1.18) and (1.19). Fluctuating fields are described in a variety of different ways. One description is in terms of the potentials in the Lorentz gauge, in which case the potentials satisfy (1.20) and (1.21), and another description is in terms of the temporal gauge, in which case the potentials satisfy (1.22). Solutions of these equations are derived below by introducing Greens functions and using Fourier transforms to evaluate them.

Before discussing these solutions it is of interest to reduce the Fourier transforms of Maxwell's equations to a single equation, using arguments similar to those given in Chapter 1. As already noted, after Fourier transforming and for $\omega \neq 0$, there are only three independent equations for a fluctuating field. Suppose that one reinterprets (5.1) as *defining* the magnetic field $\mathbf{B}(\omega, \mathbf{k})$ as a subsidiary field; that is, suppose that $\mathbf{B}(\omega, \mathbf{k})$ is regarded as equivalent to the transverse part of $\mathbf{E}(\omega, \mathbf{k})$:

$$\mathbf{B}(\omega, \mathbf{k}) = \mathbf{k} \times \mathbf{E}(\omega, \mathbf{k})/\omega. \tag{5.5}$$

(Note that *transverse* and *longitudinal* refer to components orthogonal and parallel, respectively, to $\mathbf{k}$.) This involves no loss of generality because the field $\mathbf{B}(\omega, \mathbf{k})$, now regarded as a subsidiary field, is completely determined by $\mathbf{E}(\omega, \mathbf{k})$ through (5.1). On eliminating $\mathbf{B}(\omega, \mathbf{k})$, only two of the set (5.1)–(5.4) are independent. One replaces one of the two remaining equations by the Fourier transform of the equation of charge continuity (1.6), that is, by

$$\omega \rho(\omega, \mathbf{k}) = \mathbf{k} \cdot \mathbf{J}(\omega, \mathbf{k}), \tag{5.6}$$

which is now regarded as defining the subsidiary quantity $\rho(\omega, \mathbf{k})$ in terms of $\mathbf{J}(\omega, \mathbf{k})$. In this way the set of Fourier transformed equations (5.1)–(5.4) is reduced to a single equation, referred to as the wave equation below, which relates the electric field to the current density, plus two subsidiary equations that give the magnetic field in terms of the electric field and the charge density in terms of the current density.

The wave equation has the form

$$\frac{\omega^2}{c^2}\mathbf{E}(\omega, \mathbf{k}) + \mathbf{k} \times [\mathbf{k} \times \mathbf{E}(\omega, \mathbf{k})] = -i\omega\mu_0\mathbf{J}(\omega, \mathbf{k}). \tag{5.7}$$

This equation is rederived by Fourier transforming (1.22), which applies in the temporal gauge. After Fourier transforming, (1.22) becomes

$$\frac{\omega^2}{c^2}\mathbf{A}(\omega, \mathbf{k}) + \mathbf{k} \times [\mathbf{k} \times \mathbf{A}(\omega, \mathbf{k})] = -\mu_0 \mathbf{J}(\omega, \mathbf{k}), \qquad (5.8)$$

and in the temporal gauge $\mathbf{A}(\omega, \mathbf{k})$ is related to $\mathbf{E}(\omega, \mathbf{k})$ by the Fourier transform of (1.8) with $\phi = 0$:

$$\mathbf{E}(\omega, \mathbf{k}) = i\omega\mathbf{A}(\omega, \mathbf{k}). \qquad (5.9)$$

The wave equation in the form (5.7) or (5.8) is used extensively below to discuss the electromagnetic field in a medium. The inclusion of a medium involves separating the current on the right hand side into an induced part and an extraneous part, and transferring the induced part to the left hand side of the equation. However, for the present we are concerned with the electromagnetic field *in vacuo*, and in this case the term on the right hand side of (5.7) and (5.8) is regarded as a source term.

5.2 Solution of Inhomogeneous Differential Equations

In §1.5 solutions of Poisson's equation and of d'Alembert's equation are simply written down. The solutions are derived by using Fourier transforms to evaluate the relevant Greens functions.

A Greens function is defined for a particular differential equation and a particular set of boundary conditions. Consider an nth order ordinary differential equation:

$$\hat{L}(z)F(z) = S(z), \qquad (5.10)$$

with

$$\hat{L}(z) = A_n\frac{\mathrm{d}^n}{\mathrm{d}z^n} + A_{n-1}\frac{\mathrm{d}^{n-1}}{\mathrm{d}z^{n-1}} + \cdots + A_0, \qquad (5.11)$$

and where $S(z)$ is the source term. A Greens function $G(z, z')$ is defined by replacing the source term by $\delta(z - z')$, that is, by replacing (5.10) by

$$\hat{L}(z)G(z, z') = \delta(z - z'). \qquad (5.12)$$

Provided that one can solve (5.12) for $G(z, z')$, the solution of (5.10) follows from

$$f(z) = \int \mathrm{d}z'\, G(z, z')S(z'). \qquad (5.13)$$

The proof follows by acting with the operator $\hat{L}(z)$ on (5.10) and using (5.12):

$$\hat{L}(z)f(z) = \int \mathrm{d}z'\, \hat{L}(z)G(z, z')S(z') = \int \mathrm{d}z'\, \delta(z - z')S(z') = S(z).$$

One solves (5.12) for a Greens function $G(z - z')$ by Fourier transforming. This procedure gives only a particular solution of (5.12) and ignores the boundary conditions. The boundary conditions may require that one add to the solution of (5.12) a solution of the homogeneous equation, or the boundary conditions may require that one integrate around poles in the Fourier transform $\tilde{G}(k)$ of $G(z - z')$ in a particular way.

The Fourier transform of (5.12) is

$$L(ik)\tilde{G}(k) = 1. \tag{5.14}$$

Provided that the differential operator $\hat{L}(z)$ has constant coefficients, it is replaced by the algebraic function $L(ik)$. The solution of (5.14) is trivial:

$$\tilde{G}(k) = \frac{1}{L(ik)}. \tag{5.15}$$

Then on inverting the Fourier transform one obtains

$$G(z - z') = \int \frac{dk}{2\pi} \frac{e^{ik(z-z')}}{L(ik)}. \tag{5.16}$$

In the case of the operator (5.11), $L(ik)$ is an nth order polynomial, and the integrand in (5.16) has n poles in the complex-k plane. The integral is evaluated by elementary methods in some cases, but one often needs to appeal to contour integration, cf. §8.3.

5.3 The Greens Function for Poisson's Equation

Let us apply this approach to the solution of Poisson's equation

$$\nabla^2 \phi(\mathbf{x}) = -\rho(\mathbf{x})/\varepsilon_0. \tag{5.17}$$

The Greens function corresponding to a unit charge density at the point $\mathbf{x}'$ is defined by

$$\nabla^2 G(\mathbf{x}, \mathbf{x}') = -\delta(\mathbf{x} - \mathbf{x}')/\varepsilon_0. \tag{5.18}$$

We look for a solution $G(\mathbf{x} - \mathbf{x}')$ by Fourier transforming. The Fourier transform $\tilde{G}(\mathbf{k})$ of $G(\mathbf{x} - \mathbf{x}')$ satisfies

$$-|\mathbf{k}|^2 \tilde{G}(\mathbf{k}) = -1/\varepsilon_0. \tag{5.19}$$

The Greens function is then given by

$$G(\mathbf{x}) = \frac{1}{\varepsilon_0} \int \frac{d^3\mathbf{k}}{(2\pi)^3} \frac{e^{i\mathbf{k}\cdot\mathbf{x}}}{|\mathbf{k}|^2}. \tag{5.20}$$

An implicit boundary condition is imposed here in that $G(\mathbf{x})$ is assumed to remain finite at infinity.

The integral in (5.20) is evaluated by introducing spherical polar coordinates k, θ, ϕ, with $k = |\mathbf{k}|$ and with θ the angle between $\mathbf{k}$ and $\mathbf{x}$, so that one has $\mathbf{k} \cdot \mathbf{x} = kr \cos \theta$ with $r = |\mathbf{x}|$. This corresponds to writing

$$\int d^3 k = \int_0^{2\pi} d\phi \int_{-1}^1 d\cos\theta \int_0^\infty dk\, k^2, \qquad (5.21)$$

Thus one finds

$$G(\mathbf{x}) = \frac{1}{(2\pi)^3 \varepsilon_0} \int_0^{2\pi} d\phi \int_{-1}^1 d\cos\theta \int_0^\infty dk\, e^{ikr \cos \theta}. \qquad (5.22)$$

The integral over ϕ is trivial and the integral over $\cos \theta$ is elementary, giving

$$G(\mathbf{x}) = \frac{4\pi}{(2\pi)^3 \varepsilon_0} \int_0^\infty dk\, \frac{\sin(kr)}{kr} = \frac{1}{4\pi\varepsilon_0 r}, \qquad (5.23)$$

where the final integral is evaluated using

$$\int_0^\infty dx\, \frac{\sin x}{x} = \frac{\pi}{2}. \qquad (5.24)$$

The solution of (5.18) is then as given by (1.21), viz.,

$$\phi(\mathbf{x}) = \frac{1}{4\pi\varepsilon_0} \int d^3 x'\, \frac{\rho(\mathbf{x}')}{|\mathbf{x} - \mathbf{x}'|}. \qquad (5.25)$$

The boundary conditions have not had to be specified in detail in deriving (5.25). One needs to add to $G(\mathbf{x} - \mathbf{x}')$, as given by (5.20) a function $F(\mathbf{x}, \mathbf{x}')$ which is a solution of $\nabla'^2 F(\mathbf{x}, \mathbf{x}') = 0$, and the boundary conditions determine the actual form of this function. The two most commonly used are the *Dirichlet boundary conditions* in which the potential is specified on a closed surface, and the *Neumann boundary conditions* in which the electric field is specified on a closed surface.

5.4 The Greens Function for d'Alembert's Equation

The Greens function for the wave equation for electromagnetic waves *in vacuo* is defined as a solution of

$$\left(\frac{1}{c^2} \frac{\partial^2}{\partial t^2} - \nabla^2 \right) G(t - t', \mathbf{x} - \mathbf{x}') = \mu_0\, \delta(t - t')\, \delta^3(\mathbf{x} - \mathbf{x}'). \qquad (5.26)$$

After Fourier transforming in time and space, (5.26) reduces to

$$\left(\frac{\omega^2}{c^2} - |\mathbf{k}|^2 \right) \tilde{G}(\omega, \mathbf{k}) = -\mu_0. \qquad (5.27)$$

The solution of (5.27) is

$$\tilde{G}(\omega, \mathbf{k}) = -\frac{\mu_0}{\omega^2/c^2 - |\mathbf{k}|^2}, \qquad (5.28)$$

and on inverting the Fourier transforms this gives

$$G(t, \mathbf{x}) = -\mu_0 \int \frac{d\omega d^3\mathbf{k}}{(2\pi)^4} \frac{e^{-i(\omega t - \mathbf{k}\cdot\mathbf{x})}}{\omega^2/c^2 - |\mathbf{k}|^2}. \tag{5.29}$$

Let us perform the ω-integral in (5.29) first. On writing

$$\frac{1}{\omega^2/c^2 - |\mathbf{k}|^2} = \frac{c}{2k} \left(\frac{1}{\omega - kc} - \frac{1}{\omega + kc} \right) \tag{5.30}$$

it is evident that the integrand has poles at $\omega = \pm kc$. Suppose that we add infinitesimal imaginary parts $+i0$ to ω, which corresponds to imposing the causal condition. Then the Fourier representation of the step function, cf. (4.26),

$$H(t) = \int_{-\infty}^{\infty} \frac{d\omega}{2\pi} \frac{ie^{-i\omega t}}{\omega + i0}$$

implies

$$\int_{-\infty}^{\infty} \frac{d\omega}{2\pi} \frac{e^{-i\omega t}}{\omega \pm kc + i0} = -i\, H(t)\, e^{\pm ikct}, \tag{5.31}$$

where the variable of integration is changed from ω to $\omega \pm kc$ in evaluating the integral. Let us denote by $G_r(t, \mathbf{x})$ the particular form of the Greens function implied by inserting (5.30) and (5.31) into (5.29); the subscript r denotes the *retarded* Greens function. The resulting expression reduces to

$$G_r(t, \mathbf{x}) = \frac{ic\mu_0}{2} H(t) \int \frac{d^3\mathbf{k}}{(2\pi)^3} \frac{e^{i\mathbf{k}\cdot\mathbf{x}}}{k} \left(e^{-ikct} - e^{ikct} \right)$$

$$= \frac{c\mu_0}{4\pi^2 r} H(t) \int_0^{\infty} dk \, \sin(kr)\sin(kct), \tag{5.32}$$

where the angular integrals are evaluated as in (5.23). One has

$$\sin(kr)\sin(kct) = \tfrac{1}{2}\{\cos[k(r - ct)] - \cos[k(r + ct)]\},$$

$$\int_0^{\infty} dk \, \cos[k(r \pm ct)] = \frac{1}{2}\int_{-\infty}^{\infty} dk \, e^{ik(r \pm ct)} = \pi\delta(r \pm ct),$$

and hence (5.32) reduces to

$$G_r(t, \mathbf{x}) = \frac{c\mu_0}{4\pi r} H(t) \left[\pi\delta(r - ct) - \pi\delta(r + ct)\right] = \frac{\mu_0 \delta(t - r/c)}{4\pi r}. \tag{5.33}$$

The function obtained by making the replacement $\omega \to \omega - i0$, rather than the replacements $\omega \to \omega + i0$ made above, leads to the *advanced* Greens function

$$G_a(t, \mathbf{x}) = \frac{\mu_0 \delta(t + r/c)}{4\pi r}. \tag{5.34}$$

The Greens function $G(t - t', \mathbf{x} - \mathbf{x}')$ describes the electromagnetic field at $\mathbf{x}$ at t associated with a source at $\mathbf{x}'$ at t'. For the retarded Greens function, the δ-function in (5.33) implies $t' = t_r$ where

$$t_r := t - \frac{|\mathbf{x} - \mathbf{x}'|}{c} \tag{5.35}$$

is the *retarded time*. Thus the field at $\mathbf{x}$ at t is due to the source at $\mathbf{x}'$ at a time t' earlier than t by the light propagation time from $\mathbf{x}$ to $\mathbf{x}'$. For the advanced Greens function, the time t is earlier than t' by the light propagation time. One might conclude that the advanced Greens function is of no interest because it clearly violates the causal condition. However, it does play a role in some theories, but it is unclear from a physical viewpoint why this should be so.

The Greens Function for the Temporal Gauge

In the temporal gauge the electromagnetic fields satisfy the wave equation (5.8). The Greens function in this case differs from the two cases considered above in that it is a tensor $\tilde{G}_{ij}(\omega, \mathbf{k})$. In tensor form (5.8) becomes

$$\left[\left(\frac{\omega^2}{c^2} - |\mathbf{k}|^2 \right) \delta_{ij} + k_i k_j \right] A_j(\omega, \mathbf{k}) = -\mu_0 J_i(\omega, \mathbf{k}). \tag{5.36}$$

The corresponding Greens function is defined to satisfy

$$\left[\left(\frac{\omega^2}{c^2} - |\mathbf{k}|^2 \right) \delta_{ij} + k_i k_j \right] \tilde{G}_{jl}(\omega, \mathbf{k}) = -\mu_0 \delta_{il}. \tag{5.37}$$

One solves (5.36) by writing it in matrix form and inverting the matrix. It then follows from (5.37) that $\tilde{G}_{ij}$ is $-\mu_0$ times the inverse of the matrix form of $(\omega^2/c^2 - |\mathbf{k}|^2)\delta_{ij} + k_i k_j$. In this way one finds

$$\tilde{G}_{ij}(\omega, \mathbf{k}) = -\frac{\mu_0}{\omega^2/c^2 - |\mathbf{k}|^2} \left(\delta_{ij} - \frac{c^2}{\omega^2} k_i k_j \right). \tag{5.38}$$

It is possible to invert the Fourier transform and construct the Greens function $G_{ij}(t, \mathbf{x})$ corresponding to (5.38) but this Greens function is an operator rather than a function and it is of little interest in practice.

To solve (5.8) it is convenient to separate into longitudinal and transverse parts. The solution is then of the form

$$\begin{aligned} \mathbf{k} \cdot \mathbf{A}(\omega, \mathbf{k}) &= -\frac{\mu_0 c^2}{\omega^2} \mathbf{k} \cdot \mathbf{J}(\omega, \mathbf{k}), \\ \mathbf{k} \times \mathbf{A}(\omega, \mathbf{k}) &= -\frac{\mu_0}{\omega^2/c^2 - |\mathbf{k}|^2} \mathbf{k} \times \mathbf{J}(\omega, \mathbf{k}). \end{aligned} \tag{5.39}$$

The singular terms corresponding to $\omega^2/c^2 - |\mathbf{k}|^2 = 0$ occur only in the transverse part of the field. This is interpreted in terms of the emission of electromagnetic radiation. Electromagnetic waves are transverse, that is, they have $\mathbf{k} \cdot \mathbf{A} = 0$, and they satisfy a dispersion relation $\omega^2 =$

$|\mathbf{k}|^2 c^2$, so that they correspond to the singular terms in (5.39). The longitudinal part is not involved in the emission of radiation and involves no singularity at $\omega^2/c^2 - |\mathbf{k}|^2 = 0$. The longitudinal part of the Fourier transformed field is related to the inductive part of the field in coordinate space; for example, a Coulomb field is purely longitudinal in Fourier space. However, this does not imply that the transverse part of the field is entirely associated with radiation. On the contrary, the radiation field is associated only with the singular terms in the transverse part of the field.

Exercise Set 5

5.1 Find the Greens function for Poisson's equation in two dimensions.

(a)   Using a Fourier transform derive the form

$$G(\mathbf{x}) = \frac{1}{4\pi^2 \varepsilon_0} \int_{-\infty}^{\infty} dk_x \int_{-\infty}^{\infty} dk_y \, \frac{e^{i(k_x x + k_y y)}}{(k_x^2 + k_y^2)}. \qquad (E5.1)$$

(b)   Write $k_x x + k_y y = kr \cos\theta$ and integrate from 0 to 2π over θ using

$$\int_{-\pi}^{\pi} d\theta \, e^{ikr \cos\theta} = 2\pi J_0(kr), \qquad (E5.2)$$

where $J_0(z)$ is a Bessel function. You should obtain $G(\mathbf{x}) = G(r)$ with

$$G(r) = \frac{1}{2\pi\varepsilon_0} \int_0^{\infty} dk \, \frac{J_0(kr)}{k}. \qquad (E5.3)$$

(c)   The integral $(E5.3)$ diverges logarithmically. A convergent result is obtained by requiring that $G(r)$ vanish at $r = a$:

$$G(r) = \frac{1}{2\pi\varepsilon_0} \int_0^{\infty} dk \, \frac{J_0(kr) - J_0(ka)}{k}.$$

Using the result of Exercise 5.2, prove

$$G(r) = \frac{1}{2\pi\varepsilon_0} \ln\left(\frac{a}{r}\right). \qquad (E5.4)$$

5.2 Consider Frullani's integral:

$$I(a, b) = \int_0^{\infty} dx \, \frac{[f(ax) - f(bx)]}{x}, \qquad (E5.5)$$

where a, b are positive constants, and $f(x)$ is continuous at $x = 0$. Write

$$I(a, b) = \lim_{\varepsilon \to 0} \left[\int_{\varepsilon}^{\infty} dx \, \frac{f(ax)}{x} - \int_{\varepsilon}^{\infty} dx \, \frac{f(bx)}{x} \right],$$

and prove

$$I(a, b) = \lim_{\varepsilon \to 0} \int_{\varepsilon a}^{\varepsilon b} dx \, \frac{f(x)}{x} = f(0) \ln\left(\frac{b}{a}\right). \qquad (E5.6)$$

5.3 Consider the retarded Greens function for d'Alembert's equation in two dimensions.

$$\left(\frac{1}{c^2} \frac{\partial^2}{\partial t^2} - \nabla^2 \right) G(t, \mathbf{x}) = \mu_0 \, \delta(t) \, \delta^2(\mathbf{x}).$$

(a)   Use the Fourier transform method and the causality argument to derive

$$G_r(t, \mathbf{x}) = \frac{c\mu_0 H(t)}{2\pi} \int_0^{\infty} dk \, J_0(kr) \sin(kct).$$

(b) Use the standard integral $(a > 0, b > 0)$

$$\int_0^\infty dx\, e^{ibx} J_0(ax) = \frac{H(a-b)}{(a^2-b^2)^{1/2}} + i\frac{H(b-a)}{(b^2-a^2)^{1/2}} \qquad (E5.7)$$

to show

$$G_R(t,\mathbf{x}) = \frac{c\mu_0 H(t)H(ct-r)}{2\pi(c^2t^2-r^2)^{1/2}}. \qquad (E5.8)$$

Remark: G_r may be considered to result from a line source in three dimensions; the parts out of the plane $(z \ne 0)$ emit electromagnetic waves with a smaller component of velocity in the plane $z = 0$ than those parts in the plane $z = 0$. The waves from near $z = 0$ produce the singularity at $r = ct$; the waves from large z produce the "tail" in the form $(E5.8)$ which is absent in the three-dimensional case.

5.4 The following problem concerns the solution of

$$\frac{d^2 f(y)}{dy^2} + k_0^2 f(y) = f(0)\delta(y). \qquad (E5.9)$$

(a) Using the identity $(d/dy)[\mathrm{sgn}(y)] = 2\delta(y)$ derived in Exercise 4.4, show that $G(y) = e^{ik_0|y|}$ satisfies

$$\left[\frac{d^2}{dy^2} + k_0^2\right] G(y) = 2ik_0\delta(y), \qquad (E5.10)$$

where k_0 is a constant.

(b) In solving $(E5.9)$, split $f(y)$ into an even part $f_e(y) = f_e(-y)$ and an odd part $f_o(y) = -f_o(-y)$. Show that a general solution of the equation is made up from

$$f_e(y) = A\left[e^{ik_0|y|} + \left(\frac{2ik_0 - 1}{2ik_0 + 1}\right)e^{-ik_0|y|}\right],$$

$$f_o(y) = B\sin(k_0 y),$$

where A and B are arbitrary constants.

PART TWO

Electromagnetic Responses of Media

A medium with electromagnetic properties modifies an electromagnetic field imposed on it. The response of some media may be described satisfactorily in terms of induced dipole moments, and this is particularly the case for the response to static fields. The response to a fluctuating field may also be described in terms of the induced current density. This alternative description is used widely in plasma physics and is emphasized in the approach adopted here.

In Part Two the nature of electromagnetic responses is discussed in Chapter 6 and general properties of response tensors are summarized in Chapter 7. An understanding of the material in these two chapters is important in the discussion of waves in media in Part Three. The remaining three Chapters in Part Two are more of the nature of reference material. Although much of the material in these Chapters is referred to and used in the remainder of the book, a detailed understanding of this material is not essential before proceeding to Parts Three, Four and Five.

6

The Response of a Medium

Preamble

There are two different descriptions of the electromagnetic response of a medium. One, which is the older description, involves introducing the polarization and magnetization of the medium and was originally applied to the response to static fields. This method may be generalized to apply to fluctuating fields but it becomes cumbersome and ill-defined for sufficiently general media. The other description is based on the use of Fourier transforms and so applies only to fluctuating fields. The Fourier transform description is used widely in plasma physics, and although it is less familiar than the other description when applied to dielectrics and magnetizable media, it is simpler and no less general than the older description.

6.1 Static Responses

The response of a medium to a static uniform electromagnetic field is described in terms of induced dipole moments. On a microscopic level a static uniform electric field polarizes the atoms or molecules. The *polarization* **P** is defined as the induced electric dipole moment per unit volume. A medium which becomes polarized in this way is called a *dielectric*.

The response of a magnetizable medium to a static uniform magnetic field is attributed to induced magnetic dipole moments and is described in terms of the *magnetization* **M**, which is defined as the induced magnetic dipole moment per unit volume. Magnetizable media are classified as paramagnetic or diamagnetic depending on whether the magnetization is parallel or antiparallel, respectively, to the applied magnetic field. ("Ferromagnetism" is regarded as an extreme form of paramagnetism in which the magnetization remains when the inducing magnetic field is turned off.) From a classical point of view one expects the induced current to obey Lenz' law and be such that the associated magnetic field opposes the original field; this would imply that all media should be diamagnetic, contrary to the existence of paramagnetism and ferromagnetism. Paramagnetism is associated with the alignment of the intrinsic magnetic dipoles (due to the spin of the electron) in the medium by the imposed magnetic field.

The description in terms of induced dipole moments is strictly valid only for the case of static uniform fields. The assumption that the fields are static implies a complete separation of electric and magnetic effects. Suppose that we relax the assumption that the imposed field is uniform, but still assume that it is static (the field then has a spatial derivative but not a temporal derivative). The responses can then include higher order multipole moments and magnetoelectric responses (a polarization induced by a magnetic field or a magnetization induced by an electric field). These more general responses, discussed in §6.2, are excluded for uniform fields simply because, for example, a quadrupole moment is induced by the gradient in the field, and an octupole moment is induced by the second derivative of the field. For non-uniform static fields the response in terms of $\mathbf{P}$ and $\mathbf{M}$ remains approximately valid for any medium in which the multipole expansion converges sufficiently rapidly, as is the case for many media.

In a similar way, the static response remains approximately valid when the assumption that the fields are static is relaxed to allow sufficiently slowly varying fields. However, this raises the question as to when a field is to be regarded as static or slowly varying, and when it is to be regarded as fluctuating. In practice, the distinction between slowly varying and fluctuating fields involves a separation of time scales, and no general condition can be specified. Usually there are natural frequencies associated with a medium, and a field is regarded as slowly varying if it varies on a time scale that is long compared with the inverse of the lowest relevant natural frequency of the medium. An extreme example is that of a vacuum in quantum electrodynamics where the response is due to virtual electron–positron pairs; the only natural frequency is then that obtained by dividing the rest energy mc^2 of the electron by Planck's constant $\hbar$, and this corresponds to a γ-ray energy of 0.511 MeV, so that even fields at optical frequencies (energies $\lesssim 1\,\text{eV}$) and higher are regarded as slowly varying in this case. Another extreme example is an electrolyte (a solution containing free ions) in which case there is no static electric response (in the sense implied by the foregoing discussion) because all the ions are freely accelerated by a static electric field; this acceleration is limited by collisions so that all ions reach a terminal drift velocity and the response in this case is a constant current and not a polarization.

For many media there is at least an approximate proportionality between the response and the imposed field. When such a proportionality is assumed to apply, the medium is said to be *linear*. The linear response of a medium is described by a response function defined as the constant of proportionality between the response and the imposed field.

It is conventional to introduce two additional fields when describing a static response. These are the *electric induction* (or *electric displacement*) $\mathbf{D}$ and the *magnetic field strength* $\mathbf{H}$. They are defined by writing

$$\mathbf{D} := \varepsilon_0 \mathbf{E} + \mathbf{P}, \quad \mathbf{H} := \mathbf{B}/\mu_0 - \mathbf{M}. \tag{6.1}$$

The traditional description of the response of an isotropic medium to static fields is through the relations

$$\mathbf{D} = \varepsilon \mathbf{E}, \quad \mathbf{H} = \mathbf{B}/\mu, \tag{6.2}$$

where the response functions are the *dielectric permittivity* ε and the *magnetic permeability* μ . Alternatively, the response is described by

$$\mathbf{P} = \varepsilon_0 \chi^e \mathbf{E}, \quad \mathbf{M} = \chi^m \mathbf{B}/\mu_0, \tag{6.3}$$

where χ^e is the *electric susceptibility* and χ^m is related to the *magnetic susceptibility*. (The magnetic susceptibility is defined as $\mu/\mu_0 - 1$, due to the disturbance being described by the field $\mathbf{H}$ rather than $\mathbf{B}$ in (6.3); this is an unfortunate legacy of an outdated convention.) The *dielectric constant* K, which is dimensionless, is defined by writing

$$\varepsilon = K\varepsilon_0, \quad K = 1 + \chi^e. \tag{6.4}$$

A corresponding dimensionless constant is sometimes defined for the magnetic response.

The underlying procedure for describing the response in terms of $\mathbf{P}$ and $\mathbf{M}$ involves a multipole expansion truncated at the dipole terms. Quite generally, the response may be described in terms of the induced charge density ρ_{ind} and the induced current density $\mathbf{J}_{\text{ind}}$. These are related to the polarization and magnetization by

$$\rho_{\text{ind}} = -\operatorname{div}\mathbf{P}, \quad \mathbf{J}_{\text{ind}} = \partial\mathbf{P}/\partial t + \operatorname{curl}\mathbf{M}. \tag{6.5}$$

In a theory where $\mathbf{P}$ and $\mathbf{M}$ are defined, (6.5) is a prescription for constructing ρ_{ind} and $\mathbf{J}_{\text{ind}}$. In other theories where ρ_{ind} and $\mathbf{J}_{\text{ind}}$ are calculated in an independent way, (6.5) defines $\mathbf{P}$ and $\mathbf{M}$, but the definition is incomplete and ambiguous.

To see how (6.5) arises, note first that the electric dipole moment $\mathbf{d}(t)$ and the magnetic dipole moment $\mathbf{m}(t)$ of any charge and current distribution are defined by (3.8) and (3.12), *viz.*,

$$\mathbf{d}(t) = \int d^3\mathbf{x}\,\mathbf{x}\,\rho(t,\mathbf{x}), \quad \mathbf{m}(t) = \tfrac{1}{2}\int d^3\mathbf{x}\,\mathbf{x}\times\mathbf{J}(t,\mathbf{x}), \qquad (6.6)$$

respectively. These are leading terms in a multipole expansion which is related to the spatial Fourier transform by expanding the exponentials in

$$\rho(t,\mathbf{k}) = \int d^3\mathbf{x}\,e^{-i\mathbf{k}\cdot\mathbf{x}}\,\rho(t,\mathbf{x}), \qquad (6.7)$$

$$\mathbf{J}(t,\mathbf{k}) = \int d^3\mathbf{x}\,e^{-i\mathbf{k}\cdot\mathbf{x}}\,\mathbf{J}(t,\mathbf{x}). \qquad (6.8)$$

That is, one makes the expansion

$$e^{-i\mathbf{k}\cdot\mathbf{x}} = 1 - i\mathbf{k}\cdot\mathbf{x} + \tfrac{1}{2}(i\mathbf{k}\cdot\mathbf{x})^2 + \cdots. \qquad (6.9)$$

On inserting the expansion (6.9) into (6.7), the unit term in (6.9) gives the net electric charge, which is assumed to be zero here, the linear term is related to the electric dipole moment, the quadratic term to the electric quadrupole moment, and so on. On inserting the expansion (6.9) into (6.8), the unit term is related to the time-derivative of the electric dipole moment, cf. (3.14), the symmetric and antisymmetric parts of the next term are related to the electric quadrupole moment and magnetic dipole moment, respectively, cf. (3.15), and so on. Then on writing $\mathbf{d}(t) = \int d^3\mathbf{x}\,\mathbf{P}(t,\mathbf{x})$ and $\mathbf{m}(t) = \int d^3\mathbf{x}\,\mathbf{M}(t,\mathbf{x})$, one finds that if only the electric and magnetic dipole terms are retained, then (6.7) and (6.8) with (6.9) are approximated by

$$\rho(t,\mathbf{k}) = \int d^3\mathbf{x}\,[-i\mathbf{k}\cdot\mathbf{P}(t,\mathbf{x})], \qquad (6.10)$$

$$\mathbf{J}(t,\mathbf{k}) = \int d^3\mathbf{x}\,[\partial\mathbf{P}/\partial t + i\mathbf{k}\times\mathbf{M}]. \qquad (6.11)$$

In evaluating the electric dipole term in (6.11) one needs to use the charge continuity equation and to partially integrate (ignoring surface terms). Finally on inverting the spatial Fourier transform, (6.10) and (6.11) lead to (6.5).

The foregoing derivation involves approximations that are not valid in general. It is not obvious that higher order terms in the expansion (6.9) can be neglected, and indeed in some media it is important to retain the electric quadrupole term, as discussed in §6.2. Moreover, there is no guarantee that the multipole expansion converges rapidly, or that it converges at all. There is also a more subtle difficulty: the multipole expansion is incomplete because certain terms in the expansion (6.9) are omitted; specifically the traces of all multipole moments are defined to be zero, cf. §3.2. The traces are not zero in general, and even if the multipole expansion does converge, it needs to be supplemented by the trace terms for the result to be completely general.

One case where the multipole expansion does not converge is in an electrolyte or in a thermal plasma. Then the Debye shielding effect causes the response of the medium to become arbitrarily large at arbitrarily small distances, that is, at arbitrarily large k. The multipole expansion is based on (6.9), and applies only for $|\mathbf{k} \cdot \mathbf{x}| \ll 1$, which corresponds to sufficiently small k. This expansion is invalid in the large-k limit required to treat Debye shielding.

6.2 Properties of Media

The electromagnetic properties of media are reflected in mathematical properties of the response functions. The simplest media, in the sense of §6.1, are described by their dielectric permittivity ε or dielectric constant K and by their magnetic permeability μ. Actual media are often more complicated and may exhibit any of the following phenomena: non-linearity, anisotropy, dispersion, optical activity, spatial dispersion, gyrotropy, and magnetoelectric responses. Let us consider each of these briefly.

Non-linearity: Non-linear effects, such as hysteresis, may be included in the response functions for sufficiently simple media. However, in general one needs to expand in the electromagnetic field and to treat the linear and non-linear terms separately. An expansion of the response in powers of the applied field defines a hierarchy of non-linear response functions, cf. §6.4. In the following summary it is assumed that only the linear response is retained.

Anisotropy: A medium is anisotropic if the response is not in the same direction as the disturbance. Then relations such as (6.3) need to be replaced by tensor relations:

$$P_i = \varepsilon_0 \chi_{ij}^{e} E_j, \quad M_i = \chi_{ij}^{m} B_j / \mu_0. \tag{6.12}$$

Thus the susceptibilities and other response functions become tensors.

Dispersion: Dispersion is the property that the response is a function of frequency. Dispersion is attributed to the existence of one or more natural frequencies in the response of the medium. Formally, dispersion is included by allowing all the quantities in (6.12) to be functions of ω. All media are necessarily dispersive, but for many media, such as glass, water and air, there is a broad range of natural frequencies, causing the dispersion to be weak, at least over the optical spectral range. Weak dispersion requires rather special conditions and in this sense it is surprising that so many familiar media are weakly dispersive. This is attributed to a selection effect: weak dispersion is related to transparency (dispersion and absorption are related through the Kramers–Kronig relations) and transparent materials are of obvious special interest in optics. Thus, dispersion is weak in transparent materials such as glass essentially because one selects materials that are transparent over a substantial frequency range, and this property is physically related to weak dispersion.

Dispersion is associated with the response at time t being due to disturbances at times $t' \neq t$. The *causal condition* is that cause must precede effect, and this requires that the response at t be due to disturbances only at $t' \leq t$. In the limit of arbitrarily slowly varying fields, the response becomes nearly instantaneous with the disturbance, and dispersion is then unimportant.

Natural Optical Activity: In some media the electric quadrupole response is significant, and this is the cause of a phenomenon called *natural optical activity*. Optically active media have an intrinsic handedness that is determined by the sense of rotation of the plane of polarization of light passing through the medium. The media are classified as *dextrorotatory* and *levorotatory*. The most familiar example of such a medium is a solution of dextrose. The inclusion of an electric quadrupole term implies a term in the response function that is linear in **k**.

Spatial Dispersion: Spatial dispersion is the property that the response is a function of **k**. Two specific types of spatial dispersion mentioned above are Debye shielding and natural optical activity; these involve specific functional forms of the dependence of the response functions on **k**. More generally, the response function is regarded as an arbitrary function of **k** (as well as of ω if the medium is also dispersive). Spatial dispersion is associated with the presence of natural lengths in the medium, such as the dimensions of atoms, interatomic spacings or of inhomogeneities. Although all media are technically spatially dispersive, the effect is weak for wavelengths that are long compared with the natural lengths, that is, for $kL \ll 1$, where $2\pi/k$ is the wavelength and L is a natural length in the medium.

Spatial dispersion is associated with non-local responses. That is, the response at a point $\mathbf{x}$ is due to disturbances at other points $\mathbf{x'} \neq \mathbf{x}$. The neglect of spatial dispersion implies a local response, such that the spatial variations of the response and the disturbance are identical.

An important point to note is that when discussing the response of a medium there is no implied relation between ω and $\mathbf{k}$. For electromagnetic waves and for other waves there is a relation between ω and $\mathbf{k}$ (the "dispersion relation" for the waves) but this does not imply any connection between dispersion and spatial dispersion. In principle, one studies the temporal response of a medium, and hence the dispersion, by subjecting it to a spatially uniform field that oscillates in time, and one studies the spatial response, and hence its spatial dispersion, by subjecting it to a static field that is spatially periodic. Dispersion and spatial dispersion are independent effects.

Gyrotropy: The presence of a magnetic field can cause a medium to exhibit the Faraday effect, that is, a rotation of the plane of polarization of radiation passing through the medium. Such media are said to be gyrotropic. Gyrotropy is an important feature of an ionized gas in a magnetic field, and is of particular relevance to the ionosphere, to astrophysical plasmas and to magnetically confined laboratory plasmas.

Magnetoelectric responses: In (6.3) the possibility is neglected that a magnetic disturbance may cause the medium to become polarized or an electric disturbance cause the medium to become magnetized. Such responses can occur in principle and media that exhibit them are said to be magnetoelectric. To include magnetoelectric responses two further terms need to be included in (6.12), a term relating the polarization to the magnetic field and a term relating the magnetization to the electric field.

A subtle point is that magnetoelectric responses correspond to a particular form of spatial dispersion. Once one has Fourier transformed in space (as well as in time) the magnetic field is completely specified by the transverse part of the electric field, and hence it is not possible to separate electric and magnetic disturbances in a unique way.

Other responses: There are further types of response that are not of direct relevance in the following discussion, but which should be mentioned. First there are modifications to high-frequency responses due to various kinds of stress imposed on a system. One example is the effect of a static electric field on the high-frequency response. For an otherwise isotropic medium the effect is quadratic in the field, and leads to a response tensor of the form $K_{ij}(\omega) = K^{(0)}(\omega)\delta_{ij} + \beta_{ij}^{(\mathrm{E})}(\omega)E_i E_j$, implying that the medium becomes uniaxial with principal axis along $\mathbf{E}$. This is called the *Kerr effect*. The corresponding magnetic effect, in which the static $\mathbf{E}$ is replaced by a static $\mathbf{B}$ is the *Cotton–Mouton effect*. There are analogous modifications to $K_{ij}(\omega)$ due to elastic deformation of the medium, to sheared velocity flows in a fluid medium (*streaming birefringence* or the *Maxwell effect*), and so on. One such effect that is important in the following discussion is a modification to $K_{ij}(\omega)$ that is linear in the static $\mathbf{B}$; this leads to the *Faraday effect*, which is attributed to the natural wave modes of the medium being elliptically (or circularly) polarized.

There are also other types of response in which only one of the disturbance and the response is electromagnetic in character. Perhaps the most familiar are the *thermoelectric effect*, in which an electric current flows when a system is subjected to a temperature gradient, and the *piezoelectric effect*, in which a crystal subjected to a mechanical stress becomes polarized. The existence of the inverse effects is implied by the Onsager relations. Thus a thermoelectric medium develops a heat flow in an electric field, and a piezoelectric medium develops a strain in the presence of an electric field.

6.3 Two Alternative Descriptions of the Linear Response

There are two different ways of introducing a linear response function for an arbitrary medium. One is based on generalizing the response functions in (6.3), as outlined above. The other involves starting from the Fourier transform of the induced current and introducing a single response tensor that is a function of ω and $\mathbf{k}$.

The most general form of a linear response in terms of the polarization and the magnetization is for a medium which is anisotropic, dispersive, spatially dispersive and magnetoelectric. This general form involves the relations

$$P_i(\omega, \mathbf{k}) = \chi_{ij}^{\mathrm{e}}(\omega, \mathbf{k})\varepsilon_0 E_j(\omega, \mathbf{k}) + \chi_{ij}^{\mathrm{em}}(\omega, \mathbf{k})B_j(\omega, \mathbf{k})/\mu_0 c,$$
$$M_i(\omega, \mathbf{k}) = \chi_{ij}^{\mathrm{me}}(\omega, \mathbf{k})\varepsilon_0 c E_j(\omega, \mathbf{k}) + \chi_{ij}^{\mathrm{m}}(\omega, \mathbf{k})B_j(\omega, \mathbf{k})/\mu_0. \tag{6.13}$$

Thus when anisotropy, dispersion and spatial dispersion, and magnetoelectric responses are included one appears to require four second-rank tensors that are functions of ω and $\mathbf{k}$ to describe a general response using this theory. The Onsager relations, cf. §7.4, imply a relation between the magnetoelectric response tensors χ_{ij}^{em} and χ_{ij}^{me}, so that only three different 3-tensors need be introduced.

Although (6.13) is a general form of the response of a medium to fluctuating fields, it is ill defined. As mentioned above, magnetoelectric responses may be regarded as a specific form of spatial dispersion, and the inclusion of magnetoelectric response tensors that are functions of $\mathbf{k}$ as well as of ω introduces redundancy. Strictly, a magnetoelectric medium is one in which the response is of the form (6.13) with all the response tensors independent of $\mathbf{k}$.

One difficulty with (6.13) is that the electric field $\mathbf{E}(\omega, \mathbf{k})$ already includes the magnetic field $\mathbf{B}(\omega, \mathbf{k})$ through (5.5), and hence it is not clear how the separation into electric and magnetic fields is to be made on the right hand side of (6.13). Thus, once one has Fourier transformed in both space and time, the separation into electric and magnetic disturbances is ill defined. Also, the separation of the induced current into electric and magnetic parts in the Fourier transform of (6.5) is ill defined, that is,

$$\mathbf{J}_{\mathrm{ind}}(\omega, \mathbf{k}) = -i\omega\mathbf{P}(\omega, \mathbf{k}) + i\mathbf{k} \times \mathbf{M}(\omega, \mathbf{k}) \qquad (6.14)$$

does not specify how the separation into $\mathbf{P}(\omega, \mathbf{k})$ and $\mathbf{M}(\omega, \mathbf{k})$ is to be made. In particular one could define $\mathbf{M}(\omega, \mathbf{k})$ to be zero without loss of generality; then all the response would be included in $\mathbf{P}(\omega, \mathbf{k})$. As the full disturbance is already included in $\mathbf{E}(\omega, \mathbf{k})$, (6.13) is replaced by a single equation which relates $\mathbf{P}(\omega, \mathbf{k})$, now defined by (6.14) with $\mathbf{M}(\omega, \mathbf{k})$ identically zero, to $\mathbf{E}(\omega, \mathbf{k})$. This form is equivalent to the description of the response in the alternative method discussed below.

The use of the polarization and magnetization to describe the response of a medium is useful only for media which are not spatially dispersive or which have particular forms of spatial dispersion, such as in natural optical activity.

The basis for this alternative description is that, after Fourier transforming, both the disturbance and the response are each described completely in terms of a single electric field. (Note that this excludes static fields because they are excluded on Fourier transforming.) A general response is described completely in terms of the relation between the Fourier transform of the induced current density $\mathbf{J}_{\mathrm{ind}}(\omega, \mathbf{k})$ and the electromagnetic disturbance as described by $\mathbf{E}(\omega, \mathbf{k})$. The choice of $\mathbf{E}(\omega, \mathbf{k})$ leads to a relation

$$(J_{\mathrm{ind}})_i(\omega, \mathbf{k}) = \sigma_{ij}(\omega, \mathbf{k})E_j(\omega, \mathbf{k}), \qquad (6.15)$$

where $\sigma_{ij}(\omega, \mathbf{k})$ is the *conductivity tensor*. The point that is to be emphasized is that (6.15) contains the same information as (6.13). The seemingly greater generality of (6.13) is an illusion because neither the separation of the disturbance into $\mathbf{E}(\omega, \mathbf{k})$ and $\mathbf{B}(\omega, \mathbf{k})$ nor the separation of the disturbance into $\mathbf{P}(\omega, \mathbf{k})$ and $\mathbf{M}(\omega, \mathbf{k})$ is properly defined when one Fourier transforms in space as well as in time.

The conductivity tensor in the general sense used in (6.15) includes both non-dissipative and dissipative processes. The more familiar use of "electrical conductivity" or its inverse the "resistivity" to describe ohmic dissipation of electric currents refers only to the dissipative part of the conductivity in the general sense used in (6.15). In fact the "electrical conductivity" corresponds to the low-frequency limit of the dissipative part of the electrical conductivity, cf. Exercise 6.3.

The choice of the vector potential $\mathbf{A}(\omega, \mathbf{k})$ in the temporal gauge in place of the electric field $\mathbf{E}(\omega, \mathbf{k})$ in (6.15) leads to a response of the form

$$(J_{\text{ind}})_i (\omega, \mathbf{k}) = \alpha_{ij}(\omega, \mathbf{k}) A_j(\omega, \mathbf{k}). \tag{6.16}$$

The tensor α_{ij} has no widely accepted name; *polarization response tensor* would reflect an analogy with the "vacuum polarization tensor" in quantum electrodynamics, however, in optics it is conventional to use "polarization tensor" with a different meaning in reference to the polarization of transverse waves. The relation (5.9), *viz.*, $\mathbf{E}(\omega, \mathbf{k}) = i\omega \mathbf{A}(\omega, \mathbf{k})$, implies that $\alpha_{ij}(\omega, \mathbf{k})$ is equal to $i\omega$ times $\sigma_{ij}(\omega, \mathbf{k})$.

In place of the conductivity tensor σ_{ij} or the polarization tensor α_{ij} it is often more convenient to use the *equivalent dielectric tensor* K_{ij} to describe the response. The equivalent dielectric tensor is defined to be dimensionless and to reduce to the unit tensor *in vacuo*. One defines K_{ij} by writing

$$K_{ij}(\omega, \mathbf{k}) = \delta_{ij} + \frac{i}{\omega \varepsilon_0} \sigma_{ij}(\omega, \mathbf{k})$$

$$= \delta_{ij} + \frac{1}{\omega^2 \varepsilon_0} \alpha_{ij}(\omega, \mathbf{k})$$

$$= \delta_{ij} + \chi_{ij}(\omega, \mathbf{k}). \tag{6.17}$$

The *equivalent permittivity tensor* $\varepsilon_{ij} = \varepsilon_0 K_{ij}$ differs from the dielectric tensor only in its dimensions. The forms (6.17) correspond to describing the response in terms of an equivalent polarization, which is defined by setting $\mathbf{M} = 0$ in (6.14), and then defining an equivalent electric induction through (6.1). The final form in (6.17) corresponds to introducing the equivalent susceptibility tensor $\chi_{ij}(\omega, \mathbf{k})$.

It should be emphasized that the equivalent dielectric tensor is not the same as the dielectric tensor introduced in (6.4). To emphasize the distinction between the description of the response in terms of Fourier transforms in space as well as time and the older method discussed above, it is instructive to derive the equivalent dielectric tensor corresponding to the simple models of media described by (6.2), (6.3), (6.12) and so on. For example, the equivalent dielectric tensor that corresponds to (6.2) is

$$K_{ij}(\omega, \mathbf{k}) = (\varepsilon/\varepsilon_0)\, \delta_{ij} + (k^2 c^2/\omega^2)\, (1 - \mu_0/\mu)\, (\delta_{ij} - \kappa_i \kappa_j), \qquad (6.18)$$

with

$$\boldsymbol{\kappa} := \mathbf{k}/k, \quad k := |\mathbf{k}|. \qquad (6.19)$$

A more general relation corresponds to the response described by (6.13) with all the susceptibilities functions of ω but not of $\mathbf{k}$; the corresponding equivalent dielectric tensor is

$$K_{ij}(\omega, \mathbf{k}) = \delta_{ij} + \frac{1}{\varepsilon_0} \left\{ \chi_{ij}^{\mathrm{e}}(\omega) - \frac{k_r k_b c^2}{\omega^2}\, \epsilon_{irs} \chi_{sa}^{\mathrm{m}}(\omega) \epsilon_{abj} \right.$$
$$\left. + \frac{k_s c}{\omega} \left[\chi_{ir}^{\mathrm{em}}(\omega) \epsilon_{rsj} - \epsilon_{isr} \chi_{rj}^{\mathrm{me}}(\omega) \right] \right\}. \quad (6.20)$$

It follows from (6.18) or (6.20) that the equivalent dielectric tensor is equal to the dielectric tensor, defined in the older sense, only for a dielectric which is not spatially dispersive and which has no magnetic properties. The equivalent dielectric tensor includes the magnetic as well as the electric response in general.

When the response is written in terms of Fourier transforms in space as well as in time, Maxwell's equations are reduced to a single equation called the *wave equation*. The wave equation is written down by making the separation into induced (ind) and extraneous (ext) parts of the current, $\mathbf{J} = \mathbf{J}_{\mathrm{ind}} + \mathbf{J}_{\mathrm{ext}}$ and including the induced part, as written in the form (6.16), on the left hand side of (5.8). The result is written in the form

$$\Lambda_{ij}(\omega, \mathbf{k}) A_j(\omega, \mathbf{k}) = -\frac{\mu_0 c^2}{\omega^2}\, (J_{\mathrm{ext}})_i(\omega, \mathbf{k}), \qquad (6.21)$$

with

$$\Lambda_{ij}(\omega, \mathbf{k}) := \frac{k^2 c^2}{\omega^2}\, (\kappa_i \kappa_j - \delta_{ij}) + K_{ij}(\omega, \mathbf{k}). \qquad (6.22)$$

6.4 Non-linear Response Tensors

Non-linear effects that involve waves are of wide interest in a variety of contexts. Examples include self-focusing of light, degenerate four-wave mixing and frequency doubling in optics and laser physics; cross modulation of radio waves in the ionosphere; wave decay, modulational instabilities and wave collapse in plasma physics; and photon splitting and photon–photon scattering in quantum electrodynamics. These various effects are attributed to non-linear responses of the media in which they occur. Formally, non-linear response tensors are defined by expanding the response in powers of the disturbance. The form of this expansion is written down here and some qualitative remarks on its significance are made.

There are two natural starting points for a non-linear expansion. One is relevant to non-linear optics where spatial dispersion is unimportant. On expanding the polarization in powers of the electric field one obtains

$$P_i(\omega)/\varepsilon_0 = \chi_{ij}(\omega)\, E_j(\omega) + \int d\omega^{(2)}\, \chi_{ijl}^{(2)}(\omega, \omega_1, \omega_2)\, E_j(\omega_1)\, E_l(\omega_2)$$

$$+ \int d\omega^{(3)}\, \chi_{ijlm}^{(3)}(\omega, \omega_1, \omega_2, \omega_3)\, E_j(\omega_1)\, E_l(\omega_2)\, E_m(\omega_3) + \cdots,$$

$$(6.23)$$

with

$$d\omega^{(n)} = d\omega_1 \ldots d\omega_n\, \delta(\omega - \omega_1 - \ldots - \omega_n), \qquad (6.24)$$

and so on. The third-rank tensor $\chi_{ijl}^{(2)}(\omega, \omega_1, \omega_2)$ is the quadratic non-linear susceptibility and the fourth rank tensor $\chi_{ijlm}^{(3)}(\omega, \omega_1, \omega_2, \omega_3)$ is the cubic non-linear susceptibility. The δ-functions in (6.24) imply that $\chi_{ijl}^{(2)}$ is not an independent function of all three arguments, and that it is a function of only two frequency differences, for example, $\omega - \omega_1$ and $\omega - \omega_2$. Similarly, $\chi_{ijlm}^{(3)}$ is a function of three frequency differences.

The quadratic response allows the beating of two waves with frequencies ω_1 and ω_2 to form a third wave with frequency $\omega = \omega_1 + \omega_2$. The particular case of frequency doubling corresponds to $\omega_1 = \omega_2$. There is a kinematic constraint on such processes that can be very restrictive in practice: all waves must satisfy a beat condition $\mathbf{k} = \mathbf{k}_1 + \mathbf{k}_2$ as well as $\omega = \omega_1 + \omega_2$ and must also satisfy appropriate dispersion relations. Furthermore, the quadratic non-linear susceptibility is zero for many media. A subtle but powerful argument implies that $\chi_{ijl}^{(2)}$ is zero for an isotropic medium. The argument concerns the tensor indices. An arbitrary tensor may be written as a sum of terms consisting of scalar functions times tensorial factors constructed from the available vectors and tensors. For an isotropic medium there is no natural vector in the problem, except $\mathbf{k}$ and this is not available if the medium is not spatially dispersive (so that the response tensor does not depend on $\mathbf{k}$). Thus the tensor indices can appear only in the form of constant tensors that are always available; these are the unit tensor δ_{ij} or the permutation symbol ϵ_{ijl}. There is no way that one can construct a tensor with three indices from δ_{ij} without involving a vector, and as there is no vector in the problem there can be no term constructed from δ_{ij}. It is apparent from (6.23) that $\chi_{ijl}^{(2)}(\omega, \omega_1, \omega_2)$ is unchanged by permuting the labels j, 1 and l, 2. This is inconsistent with $\chi_{ijl}^{(2)}$ being antisymmetric in j and l, and hence excludes any term that is dependent on ϵ_{ijl}. Thus there is no possible way of constructing an acceptable third-rank tensor for the quadratic nonlinear response of an isotropic medium that is not spatially dispersive. In an anisotropic medium there is at least one characteristic direction associated with the medium. Examples include the principal axis of a uniaxial crystal and the normal to the plane of asymmetry in a medium which is not reflection symmetric. In such media $\chi_{ijl}^{(2)}$ is non-zero in general.

The term in (6.23) involving the cubic response tensor allows the existence of such processes as frequency trebling, degenerate four-wave mixing, wave–wave scattering and, depending on its sign, self-focusing of light. The cubic response tensor is non-zero for all media; it is even non-zero for the vacuum itself.

In a spatially dispersive medium, and in an ionized gas in particular, the so-called weak-turbulence expansion is of the induced current in powers of the amplitude of the Fourier transformed electromagnetic field. Choosing to describe the electromagnetic field in terms of the vector potential in the temporal gauge, this expansion has the form

$$(J_{\text{ind}})_i(k) = \alpha_{ij}(k) A_j(k) + \int d\lambda^{(2)} \alpha_{ijl}^{(2)}(k, k_1, k_2) A_j(k_1) A_l(k_2)$$

$$+ \int d\lambda^{(3)} \alpha_{ijlm}^{(3)}(k, k_1, k_2, k_3) A_j(k_1) A_l(k_2) A_m(k_3) + \cdots,$$

$$(6.25)$$

where the nth order convolution integral is defined by

$$d\lambda^{(n)} = \frac{d^4k_1}{(2\pi)^4} \frac{d^4k_2}{(2\pi)^4} \cdots \frac{d^4k_n}{(2\pi)^4} (2\pi)^4 \delta^4(k - k_1 - k_2 - \cdots - k_n), \quad (6.26)$$

with

$$d^4k = d\omega d^3\mathbf{k}, \quad \delta^4(k) = \delta(\omega)\delta^3(\mathbf{k}). \quad (6.27)$$

The expansion (6.25) includes the linear response tensor $\alpha_{ij}(k)$, the quadratic response tensor $\alpha_{ijl}^{(2)}(k, k_1, k_2)$ and the cubic response tensor $\alpha_{ijlm}^{(3)}(k, k_1, k_2, k_3)$. Unlike the case of non-linear optics, the quadratic response tensor is non-zero for all plasmas, including isotropic plasmas. The important distinction between the non-linear response of a plasma and the non-linear response of media of interest in non-linear optics is that plasmas are spatially dispersive.

Exercise Set 6

6.1 A medium is described by the response functions $\varepsilon(\omega)$ and $\mu^{-1}(\omega)$ in

$$\mathbf{D} = \varepsilon(\omega)\mathbf{E}, \quad \mathbf{H} = \mu^{-1}(\omega)\mathbf{B}.$$

Construct the equivalent dielectric tensor $K_{ij}(\omega, \mathbf{k})$ in terms of $\varepsilon(\omega)$ and $\mu^{-1}(\omega)$. You should obtain the result (6.18).

6.2 Consider a medium in which the response is purely electrostatic and is described by the induced charge density $\rho(\omega, \mathbf{k})$. Suppose that a response function $X(\omega, \mathbf{k})$ is defined by writing

$$\rho(\omega, \mathbf{k}) = X(\omega, \mathbf{k})\phi(\omega, \mathbf{k}), \qquad (E6.1)$$

where $\phi(\omega, \mathbf{k})$ is the scalar potential in the Coulomb gauge. Find an expression that relates $X(\omega, \mathbf{k})$ to the components of the equivalent dielectric tensor $K_{ij}(\omega, \mathbf{k})$.

6.3 Let an electron gas be described as a fluid with a fluid velocity $\mathbf{v}$ and an equation of motion

$$m_e d\mathbf{v}/dt = -e\mathbf{E} - \nu_e m_e \mathbf{v}, \qquad (E6.2)$$

where ν_e is a collision frequency.

(a) Construct the equivalent dielectric tensor $K_{ij}(\omega)$ for a cold unmagnetized plasma including the effects of collisions as follows. Fourier transform $(E6.2)$, to find $\mathbf{v}(\omega)$, insert the result into

$$\mathbf{J}_{\text{ind}}(\omega) = -en_e\mathbf{v}(\omega),$$

and hence identify $\sigma_{ij}(\omega)$ and ω_p in

$$K_{ij}(\omega) = \delta_{ij}\left(1 - \omega_p^2/\omega(\omega + i\nu_e)\right). \qquad (E6.3)$$

(b) Show that in the limit $\omega \to 0$, $\sigma_{ij}(\omega)$ remains real and finite. Hence show that the electrical resistivity η, which is the inverse of the conductivity, is given by

$$\eta = \nu_e/\varepsilon_0\omega_p^2. \qquad (E6.4)$$

6.4 Consider an isotropic magnetoelectric medium whose response is described in the form (6.13) with all the susceptibilities of the form $\chi_{ij}(\omega, \mathbf{k}) = \chi(\omega)\delta_{ij}$.

(a) Use (2.24) in (6.22) to derive an explicit expression for the corresponding equivalent dielectric tensor.

(b) Show that if the medium has no magnetic properties ($\chi^m = 0$) then its response is of the form

$$K_{ij}(\omega, \mathbf{k}) = K(\omega)\delta_{ij} + \eta(\omega)\epsilon_{ijr}k_r \qquad (E6.5)$$

that is characteristic of an optically active medium, and identify
the rotatory part of the response $\eta(\omega)$.

(c)   Does this imply that naturally optical active media are necessarily
magnetoelectric and vice versa?

6.5 Derive the form (6.20) of the dielectric tensor using the response
in the form (6.13) for magnetoelectric media; you may also need to use
(5.5), (6.14) and (6.17).

6.6 Use the identity (2.23), *viz.*,

$$\epsilon_{abc}\epsilon_{ijk} = \delta_{ai}\delta_{bj}\delta_{ck} + \delta_{ak}\delta_{bi}\delta_{cj} + \delta_{aj}\delta_{bk}\delta_{ci}$$
$$- \delta_{bi}\delta_{aj}\delta_{ck} - \delta_{bk}\delta_{ai}\delta_{cj} - \delta_{bj}\delta_{ak}\delta_{ci},$$

to derive an alternative form for (6.20) that does not involve a product
of two permutation symbols.

7

General Properties of Response Tensors

Preamble

The response tensors for a medium have mathematical properties that reflect basic physical requirements and constraints. These mathematical properties and some of their implications are described here, specifically for the equivalent dielectric tensor $K_{ij}(\omega, \mathbf{k})$, henceforth called simply the "dielectric tensor" where no confusion should result.

7.1 Positive and Negative Frequencies

It is conventional to describe a wave in terms of a positive frequency, but when one Fourier transforms both positive and negative frequencies appear. This leads one to question what role, if any, negative frequencies play. To address this question it is helpful to start by considering the nature of the response in terms of dependences on time and space. The point is that the response tensor is the Fourier transform of a real quantity and this has relevant mathematical implications.

Suppose that one inverts the Fourier transformed form

$$D_i(\omega, \mathbf{k}) = \varepsilon_0 K_{ij}(\omega, \mathbf{k}) E_j(\omega, \mathbf{k}) \qquad (7.1)$$

to find a description of the response in terms of of t and $\mathbf{x}$. The convolution theorem (4.12) with (4.13) implies that (7.1) corresponds to

$$D_i(t, \mathbf{x}) = \varepsilon_0 \int \mathrm{d}t' \int \mathrm{d}^3\mathbf{x}' \, \hat{K}_{ij}(t - t', \mathbf{x} - \mathbf{x}') E_j(t', \mathbf{x}'). \qquad (7.2)$$

Thus the response at t, $\mathbf{x}$ is due to the disturbance at other points t', $\mathbf{x}'$. Dispersion is associated with a response that is not localized in time ($t' \neq t$), and spatial dispersion is associated with a response that is not localized in space ($\mathbf{x} \neq \mathbf{x}'$). The response is expressed as an integral over $dt'd^3\mathbf{x}'$ involving a kernel function $\hat{K}_{ij}(t, \mathbf{x})$, which is the inverse Fourier transform of $K_{ij}(\omega, \mathbf{k})$; the hat denotes that $\hat{K}_{ij}(t, \mathbf{x})$ is an operator (an integro-differential operator) in general, rather than an ordinary function operating on the disturbance at t', $\mathbf{x}'$. This kernel operator contains the information on how the disturbance at t', $\mathbf{x}'$ contributes to the response at t, $\mathbf{x}$.

The relevant point for present purposes is that all functions in the convolution integral (7.2), including this kernel operator, are real. It follows that $K_{ij}(\omega, \mathbf{k})$ is the Fourier transform of a real quantity, and hence that it must satisfy the reality condition (4.10):

$$K_{ij}^*(\omega, \mathbf{k}) = K_{ij}(-\omega, -\mathbf{k}). \tag{7.3}$$

An implication of (7.3) is that negative frequencies contain the same information as positive frequencies. This allows one to choose to describe actual disturbances only in terms of positive frequencies.

7.2 Separation into Dissipative and Non-dissipative Parts

The linear response of a medium includes both non-dissipative and dissipative parts. One example is for light in a transparent medium such as glass or water; the non-dissipative part of the response of the medium leads to the refractive index being different from zero, and the dissipative part of the response leads to absorption of the light in the medium. Another example is at arbitrarily low frequencies in an electrolyte or a plasma where the non-dissipative part of the response leads to Debye shielding and the dissipative part corresponds to an electrical resistivity causing dissipation of any current in the medium. Let us consider how the separation into non-dissipative and dissipative parts is to be made given an arbitrary dielectric tensor $K_{ij}(\omega, \mathbf{k})$.

The tensor $K_{ij}(\omega, \mathbf{k})$ is not real in general, and it may be separated it into hermitian and antihermitian parts, cf. (2.21),

$$K_{ij}(\omega, \mathbf{k}) = K_{ij}^{\mathrm{H}}(\omega, \mathbf{k}) + K_{ij}^{\mathrm{A}}(\omega, \mathbf{k}),$$
$$K_{ij}^{\mathrm{H}}(\omega, \mathbf{k}) = \tfrac{1}{2}\left[K_{ij}(\omega, \mathbf{k}) + K_{ji}^*(\omega, \mathbf{k})\right], \tag{7.4}$$
$$K_{ij}^{\mathrm{A}}(\omega, \mathbf{k}) = \tfrac{1}{2}\left[K_{ij}(\omega, \mathbf{k}) - K_{ji}^*(\omega, \mathbf{k})\right].$$

The two parts satisfy

$$K_{ji}^{\mathrm{H}*}(\omega, \mathbf{k}) = K_{ij}^{\mathrm{H}}(\omega, \mathbf{k}), \qquad K_{ji}^{\mathrm{A}*}(\omega, \mathbf{k}) = -K_{ij}^{\mathrm{A}}(\omega, \mathbf{k}). \tag{7.5}$$

We now argue that the separation (7.4) corresponds to a separation into dissipative and non-dissipative parts. This is established by applying the separation to the response tensor $\alpha_{ij}(\omega, \mathbf{k})$, which is related to $K_{ij}(\omega, \mathbf{k})$ by (6.17). According to (1.7) the source term for the electromagnetic energy is $-\mathbf{J} \cdot \mathbf{E}$. Let us evaluate the total energy transferred to the electromagnetic field by interpreting the current $\mathbf{J}$ as the induced current. Then, using the power theorem (4.11), one has

$$- \int \mathrm{d}t \mathrm{d}^3 \mathbf{x} \, \mathbf{J}_{\mathrm{ind}}(t, \mathbf{x}) \cdot \mathbf{E}(t, \mathbf{x}) = - \int \frac{\mathrm{d}\omega \mathrm{d}^3 \mathbf{k}}{(2\pi)^4} \, \mathbf{J}_{\mathrm{ind}}(\omega, \mathbf{k}) \cdot \mathbf{E}^*(\omega, \mathbf{k})$$

$$= - \int \frac{\mathrm{d}\omega \mathrm{d}^3 \mathbf{k}}{(2\pi)^4} \, \tfrac{1}{2} \left[\mathbf{J}_{\mathrm{ind}}(\omega, \mathbf{k}) \cdot \mathbf{E}^*(\omega, \mathbf{k}) + \mathbf{J}_{\mathrm{ind}}^*(\omega, \mathbf{k}) \cdot \mathbf{E}(\omega, \mathbf{k}) \right], \quad (7.6)$$

where, in the second form, the variables of integration are replaced by their negative values, the reality condition is used, and then the integral is replaced by half the sum of the two alternative forms. The response is now written in the form (6.16) and the electric field is written in terms of the vector potential in the temporal gauge using (5.9). Then the energy transferred becomes

$$- \int \mathrm{d}t \mathrm{d}^3 \mathbf{x} \, \mathbf{J}_{\mathrm{ind}}(t, \mathbf{x}) \cdot \mathbf{E}(t, \mathbf{x}) = \int \frac{\mathrm{d}\omega \mathrm{d}^3 \mathbf{k}}{(2\pi)^4} \, i\omega \, \alpha_{ij}^{\mathrm{A}}(\omega, \mathbf{k}) A_i^*(\omega, \mathbf{k}) A_j(\omega, \mathbf{k}),$$

$$(7.7)$$

where the definition of the antihermitian part (7.4) is used. It follows that the energy transferred involves only the antihermitian part of the response tensor α_{ij}. Hence dissipation or any other energy transferring process involves α_{ij}^{A}. Conversely, α_{ij}^{H} makes no contribution to dissipative processes, and it describes the non-dissipative aspects of the response.

It follows from the relation (6.17) between K_{ij} and α_{ij} that it is also the antihermitian part of K_{ij} that describes the dissipative part of the response. However, because of a factor i in the relation (6.17) between K_{ij} and σ_{ij}, it is the hermitian part of σ_{ij} that describes the dissipative part of the response.

The *non-dissipative* part of the response, described by $K_{ij}^{\mathrm{H}}(\omega, \mathbf{k})$, corresponds to the time-reversible part of the response. It describes *reactive* or *inductive* processes. The *dissipative* part of the response, described by $K_{ij}^{\mathrm{A}}(\omega, \mathbf{k})$ corresponds to the time-irreversible part of the response. It describes *resistive* or *absorptive* processes.

7.3 The Kramers–Kronig Relations

The response tensor is a causal function. That is, the response tensor relates a response ($\mathbf{J}_{\text{ind}}$) at a time t due to an electromagnetic disturbance ($\mathbf{E}$ or $\mathbf{A}$) at earlier times $t' < t$. As pointed out in Chapter 4, a causal function $f(t)$ satisfies the identity $f(t) = H(t)f(t)$. The convolution theorem (4.13) implies that the Fourier transform of $f(t)$ then satisfies $\tilde{f}(\omega) = \tilde{H}(\omega) * \tilde{f}(\omega)$, with $\tilde{H}(\omega) = i/(\omega + i0)$ according to (4.26) and (4.27). Applying this identity to the dielectric tensor minus the vacuum contribution δ_{ij} gives

$$K_{ij}(\omega, \mathbf{k}) - \delta_{ij} = \int_{-\infty}^{\infty} \frac{d\omega'}{2\pi} \frac{i}{\omega - \omega' + i0} \left[K_{ij}(\omega', \mathbf{k}) - \delta_{ij} \right]. \qquad (7.8)$$

The integral in (7.8) is reduced using the Plemelj formula (4.32), which takes the following form here:

$$\frac{1}{\omega - \omega' + i0} = \wp \frac{1}{\omega - \omega'} - i\pi\delta(\omega - \omega'). \qquad (7.9)$$

The semiresidue (from the term involving the δ-function) gives a contribution equal to $\frac{1}{2}[K_{ij}(\omega, \mathbf{k}) - \delta_{ij}]$, which is taken over to the left hand side, so that (7.8) reduces to

$$K_{ij}(\omega, \mathbf{k}) - \delta_{ij} = \frac{i}{\pi} \wp \int_{-\infty}^{\infty} d\omega' \frac{K_{ij}(\omega', \mathbf{k}) - \delta_{ij}}{\omega - \omega'}. \qquad (7.10)$$

A separation of (7.10) into hermitian and antihermitian parts gives

$$K_{ij}^{\text{H}}(\omega, \mathbf{k}) - \delta_{ij} = \frac{i}{\pi} \wp \int_{-\infty}^{\infty} d\omega' \frac{K_{ij}^{\text{A}}(\omega', \mathbf{k})}{\omega - \omega'},$$

$$K_{ij}^{\text{A}}(\omega, \mathbf{k}) = \frac{i}{\pi} \wp \int_{-\infty}^{\infty} d\omega' \frac{K_{ij}^{\text{H}}(\omega', \mathbf{k}) - \delta_{ij}}{\omega - \omega'}, \qquad (7.11)$$

which are the *Kramers–Kronig relations*.

The Kramers–Kronig relations imply that the dissipative and non-dissipative parts are not independent of each other. Referring to the definition (4.35) of a Hilbert transform, one sees that (7.11) implies that the dissipative and non-dissipative parts are, apart from a factor of i, Hilbert transforms of each other. Thus if one calculates the non-dissipative part using some theory or other, then one may construct the dissipative part from it by taking the Hilbert transform. Similarly, if one has some experimental data on the reactive part of a response as a function of ω, then one may infer a constraint on the dissipative part from the Hilbert transform.

A subtle point is that integral relations such as (7.8) should apply in principle to any form of the response and so could be applied to either $\sigma_{ij}(\omega, \mathbf{k})$ or $\alpha_{ij}(\omega, \mathbf{k})$, which are related to $K_{ij}(\omega, \mathbf{k})$ by (6.17). However, some care is required with the conditions for the Hilbert transform to exist, and in practice, the relation (7.8) must be applied only to the forms

·of the response tensor that satisfy the appropriate conditions. One of these conditions is that the response function approach zero for $\omega \to \infty$ faster than $1/\omega$. This condition is not satisfied by $K_{ij}(\omega, \mathbf{k})$ itself, and for most media the tensor obtained by subtracting δ_{ij} from it varies as $1/\omega^2$ for $\omega \to \infty$. It is for this reason that the term δ_{ij} is subtracted from K_{ij} before applying the integral relation (7.8).

To illustrate the mathematical implications of (7.11) consider the case of a non-dissipative medium, that is, a medium with $K_{ij}^{\mathrm{A}}(\omega, \mathbf{k}) = 0$. It then follows that $K_{ij}^{\mathrm{H}}(\omega, \mathbf{k})$ can depend on ω only through functions of ω whose Hilbert transform is zero. Such functions include constants and functions of the form $1/(\omega - \omega_n)$, where ω_n is independent of ω. Thus a non-dissipative response is severely constrained in mathematical form in that it consists of a sum of constants and functions of the form $1/(\omega - \omega_n)$.

7.4 The Onsager Relations

There is another set of relations that must be satisfied by every response tensor. These arise from the time-reversal invariance properties of the equations of mechanics used in the derivation of the response tensor. Any theory, whether it be classical or quantum mechanical, used to derive the response tensor involves an equation of motion. Let us consider Newton's equation; the same conclusions follow from Schrödinger's equation. If one reverses the sense of time in

$$d\mathbf{p}/dt = q(\mathbf{E} + \mathbf{v} \times \mathbf{B}),$$

then the equation returns to its original form if one also reverses the sense of $\mathbf{p}$ and of $\mathbf{B}$. That is, the equation is invariant under the transformation

$$t \to -t, \quad \mathbf{p} \to -\mathbf{p}, \quad \mathbf{B} \to -\mathbf{B}.$$

When one calculates a response tensor, an integral over $\mathbf{p}$ is performed, or equivalently over $-\mathbf{p}$. It follows that the invariance property of the equation of motion implies relations on the time-reversal properties of the response tensor, and that such relations involve the reversal of the sense of any static magnetic field.

The non-dissipative and dissipative parts of a response are even and odd functions, respectively, under reversal of the sense of time, which corresponds to $\omega \to -\omega$ after Fourier transforming. Hence one has

$$K_{ij}^{\mathrm{H}}(-\omega,\mathbf{k})\Big|_{-\mathbf{B}} = K_{ij}^{\mathrm{H}}(\omega,\mathbf{k})\Big|_{\mathbf{B}}, \quad K_{ij}^{\mathrm{A}}(-\omega,\mathbf{k})\Big|_{-\mathbf{B}} = -K_{ij}^{\mathrm{A}}(\omega,\mathbf{k})\Big|_{\mathbf{B}},$$

$$(7.12)$$

where the dependence on the sign of $\mathbf{B}$ is shown explicitly. The two relations (7.12) are combined into one by using the reality condition (7.3):

$$K_{ij}(\omega,-\mathbf{k})\Big|_{-\mathbf{B}} = K_{ji}(\omega,\mathbf{k})\Big|_{\mathbf{B}}, \qquad (7.13)$$

which is a standard form of the *Onsager relations*.

To illustrate the implications of the Onsager relations, let us consider some specific cases.

Isotropic non-spatially-dispersive medium: For a medium which is not spatially dispersive and is unmagnetized (that is, it has no static magnetic field or is uninfluenced by any such field), (7.13) implies

$$K_{ij}(\omega) = K_{ji}(\omega). \qquad (7.14)$$

Thus the response tensor is symmetric. This case is discussed in more detail in §7.5.

Magnetized medium: For a magnetized medium one is free to choose axes that are determined by the directions of $\mathbf{B}$ and $\mathbf{k}$. A convenient choice is such that $\mathbf{B}$ is along the 3 axis, and $\mathbf{k}$ is in the 1–3 plane. Then the three axes $\mathbf{e}_1$, $\mathbf{e}_2$, $\mathbf{e}_3$ are along the directions $\mathbf{B} \times (\mathbf{k} \times \mathbf{B})$, $\mathbf{k} \times \mathbf{B}$, $\mathbf{B}$. On reversing the signs of $\mathbf{B}$ and $\mathbf{k}$, as in (7.13), the axes change to $-\mathbf{e}_1$, $\mathbf{e}_2$, $-\mathbf{e}_3$, that is, the 1 axis and the 3 axis reverse direction and the 2 axis is unchanged. It then follows that with this choice of axes, (7.13) implies

$$K_{12}(\omega,\mathbf{k}) = -K_{21}(\omega,\mathbf{k}), \quad K_{13}(\omega,\mathbf{k}) = K_{31}(\omega,\mathbf{k}),$$
$$K_{23}(\omega,\mathbf{k}) = -K_{32}(\omega,\mathbf{k}). \qquad (7.15)$$

Optically active medium: For an isotropic optically active medium, one may treat the dependence on $\mathbf{k}$ by adding a term linear in $\mathbf{k}$ to the dielectric tensor for an isotropic dielectric. The tensorial quantities that are available to construct such a term are k_i, δ_{ij} and ϵ_{ijk}, and the only independent combinations consistent with the Onsager relations are δ_{ij} and $\epsilon_{ijr}k_r$. Hence the response tensor must be of the form

$$K_{ij}(\omega,\mathbf{k}) = K(\omega)\delta_{ij} + \eta(\omega)\epsilon_{ijr}k_r, \qquad (7.16)$$

where $K(\omega)$ and $\eta(\omega)$ are two functions that depend on the particular medium. The form (7.16) applies to an isotropic optically active medium in which only the first term in an expansion in **k** is retained. The more general case where the dependence on **k** is arbitrary is discussed in §7.5.

Magnetoelectric medium: For magnetoelectric media the response in the form (6.13) is written in terms of the equivalent dielectric tensor by using the relations (5.1), (6.14) and (6.17). This gives

$$K_{ij}(\omega, \mathbf{k}) = \delta_{ij} + \left\{ \chi_{ij}^{e}(\omega, \mathbf{k}) - \frac{k_r k_b c^2}{\omega^2} \epsilon_{irs} \chi_{sa}^{m}(\omega, \mathbf{k}) \epsilon_{abj} \right.$$

$$\left. + \frac{k_s c}{\omega} \left[\chi_{ir}^{em}(\omega, \mathbf{k}) \epsilon_{rsj} - \epsilon_{isr} \chi_{rj}^{me}(\omega, \mathbf{k}) \right] \right\}. \quad (7.17)$$

On applying the Onsager relation (7.13) to (7.17), one finds that the magnetoelectric responses are related:

$$\chi_{ij}^{em}(\omega, -\mathbf{k}) \Big|_{-\mathbf{B}} = -\chi_{ji}^{me}(\omega, \mathbf{k}) \Big|_{\mathbf{B}}. \quad (7.18)$$

For a medium which is isotropic and in which the susceptibility tensors do not depend on **k** or **B**, one replaces each $\chi_{ji}^{em}(\omega, \mathbf{k})$ by $\chi^{em}(\omega)\delta_{ji}$, and similarly for the other tensors in (7.17). An interesting implication is that the magnetoelectric terms then reduce to the form of the correction term in (7.16) that is associated with optical activity. Thus optical activity and magnetoelectric responses are associated phenomena, at least in the case of isotropic media. The magnetic response, from the χ^m term in (7.17), is proportional to k^2. Thus, if an expansion in k can be truncated at k^2, then the terms independent of k are attributed to the electric response, the terms linear in k to the magnetoelectric response and the terms quadratic in k to the magnetic response. However, if terms of higher order in k are significant, such a separation is ill defined.

Isotropic spatially-dispersive medium: For an isotropic dielectric that is spatially dispersive, three independent functions are needed to describe the response. To see this, first note that any response tensor $K_{ij}(\omega, \mathbf{k})$ may be written as a sum of terms each of which is a function of ω and k times a second-rank tensor that (a) is constructed from the available vectors and tensors, and (b) satisfies the Onsager relations. For an isotropic spatially dispersive medium, the only available vector is **k**, and the only available tensors are the constant tensors δ_{ij} and ϵ_{ijk}. From these one can construct only three independent second rank tensors that satisfy the Onsager relations, which in this case require that the tensors be unchanged under the transformation $\mathbf{k} \to -\mathbf{k}$, $i \leftrightarrow j$. One choice consists of $k_i k_j$, δ_{ij} and $\epsilon_{ijr} k_r$. A more convenient choice involves the unit vector **κ** along **k**:

$$\boldsymbol{\kappa} = \mathbf{k}/k, \quad k = |\mathbf{k}|. \quad (7.19)$$

A suitable choice consists of $\kappa_i \kappa_j$, $(\delta_{ij} - \kappa_i \kappa_j)$ and $\epsilon_{ijr}\kappa_r$. Then the dielectric tensor for an arbitrary isotropic medium has the form

$$K_{ij}(\omega, \mathbf{k}) = K^L(\omega, k)\, \kappa_i \kappa_j + K^T(\omega, k)\,(\delta_{ij} - \kappa_i \kappa_j) + iK^R(\omega, k)\epsilon_{ijr}\kappa_r,$$
(7.20)

where $K^L(\omega, k)$, $K^T(\omega, k)$ and $K^R(\omega, k)$ are the *longitudinal* part, the *transverse* part and the *rotatory* part, respectively. For plasmas the rotatory part is identically zero; it is non-zero in optically active media.

7.5 The Principal Axes of Anisotropic Crystals

In optics spatial dispersion and the effect of any ambient magnetic field on the response of a medium can usually be ignored. The Onsager relations then require that the dielectric tensor be symmetric, cf. (7.14). One can separate media whose dielectric tensor satisfies (7.14) into three classes: isotropic, uniaxial and biaxial dielectrics.

Dielectrics that are isotropic have no preferred orientation of the atoms. In gases, liquids and amorphous solids it is obvious that no preferred orientation is to be expected. The response tensor for such a medium is of the form

$$K_{ij}(\omega) = K(\omega)\, \delta_{ij},$$
(7.21)

where $K(\omega)$ is the dielectric constant. Thus in the isotropic case the response of the medium is described by a single function $K(\omega)$. This might be compared with the situation in the presence of spatial dispersion, when (7.20) shows that three independent functions are required. When spatial dispersion is absent one has $K^L = K^T$ and $K^R = 0$ in (7.20).

For an anisotropic dielectric the response is described by two or three independent functions depending on whether the medium is uniaxial or biaxial. To see how these functions arise it is helpful to use a result from the theory of matrices. One writes the tensor $K_{ij}(\omega)$ as a square matrix, which is symmetric according to (7.14). Any symmetric matrix may be diagonalized. This may be interpreted as follows. A symmetric tensor $K_{ij}(\omega)$ involves six independent functions in general. However, the orientation of the coordinate axes is arbitrary and three of these functions are needed to specify the orientation of the coordinate axes relative to preferred directions in the anisotropic medium. The diagonalization procedure corresponds to rotating the coordinate axes along the preferred directions in the medium. The relevant directions are called the *principal axes* of the medium. There may be zero, one or two principal axes, corresponding to isotropic, uniaxial and biaxial media, respectively. (The names uniaxial and biaxial refer to the number of optic axes, where an optic axis is a direction along which the two natural modes have equal refractive indices.)

Dielectrics in which two of the diagonal components are equal and different from the third are *uniaxial*. One may think of a uniaxial crystal as having a cylindrical symmetry; if the coordinate axes are chosen so that e_3 is along the *principal axis*, which corresponds to the axis of the cylinder, then the cylindrical symmetry about this axis implies that the response is diagonal in the two-dimensional subspace orthogonal to the principal axis. Then $K_{11}(\omega)$ and $K_{22}(\omega)$ are equal and are different from $K_{33}(\omega)$. If all three diagonal components are unequal the medium is *biaxial*.

It is appropriate to write down the dielectric tensor for anisotropic crystals for coordinate axes oriented along the principal axes. For a uniaxial crystal one writes

$$K_{ij}(\omega) = \begin{pmatrix} K_\perp(\omega) & 0 & 0 \\ 0 & K_\perp(\omega) & 0 \\ 0 & 0 & K_\parallel(\omega) \end{pmatrix}. \tag{7.22}$$

For a biaxial crystal the general form is

$$K_{ij}(\omega) = \begin{pmatrix} K_1(\omega) & 0 & 0 \\ 0 & K_2(\omega) & 0 \\ 0 & 0 & K_3(\omega) \end{pmatrix}. \tag{7.23}$$

One is still free to choose e_3 so that $K_3(\omega)$ has some preferred property, for example, such that $K_3(\omega)$ is the smallest or the largest of the three components.

For an isotropic medium that is spatially dispersive, the response tensor (7.20) may be written in matrix form by choosing $\mathbf{k}$ along the 3 axis. One then has

$$K_{ij}(\omega,k) = \begin{pmatrix} K^T(\omega,k) & iK^R(\omega,k) & 0 \\ -iK^R(\omega,k) & K^T(\omega,k) & 0 \\ 0 & 0 & K^L(\omega,k) \end{pmatrix}. \tag{7.24}$$

Exercise Set 7

7.1 The Kramers–Kronig relations in the form (7.11) involve an integral over both positive and negative frequencies. The response function for negative frequencies is rewritten in terms of positive frequencies using (7.3). Show that for an unmagnetized medium the Kramers–Kronig relations may be written in the form

$$K_{ij}^{H}(\omega, \mathbf{k}) - \delta_{ij} = \frac{2i}{\pi} \wp \int_0^\infty d\omega'\, \omega' \frac{K_{ij}^{A}(\omega', \mathbf{k})}{\omega^2 - \omega'^2}, \qquad (E7.1a)$$

$$K_{ij}^{A}(\omega, \mathbf{k}) = \frac{2i\omega}{\pi} \wp \int_0^\infty d\omega' \frac{\left[K_{ij}^{H}(\omega', \mathbf{k}) - \delta_{ij}\right]}{\omega^2 - \omega'^2}. \qquad (E7.1b)$$

7.2 The Kramers–Kronig relations in the form $(E7.1a, b)$ are used to derive *sum rules*. At sufficiently high frequencies most media respond to an electromagnetic disturbance like a free electron gas, so that the hermitian part of the response tensor is assumed of the form

$$\lim_{\omega \to \infty} \left[K_{ij}^{H}(\omega, \mathbf{k}) - \delta_{ij}\right] = -\frac{\omega_p^2}{\omega^2} \delta_{ij}. \qquad (E7.2)$$

Also at sufficiently low frequencies, the dissipative processes in many media are described in terms of a static electrical conductivity $\sigma_{ij}(0, \mathbf{k})$:

$$\lim_{\omega \to 0} K_{ij}^{A}(\omega, \mathbf{k}) = \frac{i\sigma_{ij}(0, \mathbf{k})}{\varepsilon_0 \omega}. \qquad (E7.3)$$

(a)   Show that $(E7.2)$ in $(E7.1b)$ implies the sum rule

$$\int_0^\infty d\omega\, \omega K_{ij}^{A}(\omega, \mathbf{k}) = \frac{i\pi}{2} \omega_p^2 \delta_{ij}. \qquad (E7.4)$$

(b)   Apply the Kramers–Kronig relations to $\omega^2 \left[K_{ij}^{H}(\omega, \mathbf{k}) - \delta_{ij}\right]$ rather than to $K_{ij}^{H}(\omega, \mathbf{k}) - \delta_{ij}$, derive counterparts of $(E7.1)$, and, using $(E7.3)$, derive the sum rule

$$\int_0^\infty d\omega \left[K_{ij}^{H}(\omega, \mathbf{k}) - \delta_{ij}\right] = -\frac{\pi}{2\varepsilon_0} \sigma_{ij}(0, \mathbf{k}). \qquad (E7.5)$$

Remarks: (i) In solid state physics $(E7.4)$ is sometimes used to determine the number of electrons contributing to a specific absorption band by integrating under the absorption peak measured experimentally. (ii) Formally the constant term in $(E7.2)$ should be subtracted in deriving $(E7.5)$; the resulting additional term in $(E7.5)$ is equal to zero, this is shown by evaluating its contribution in $(E7.1b)$.

7.3 Show that if a system is rotating with angular velocity $\mathbf{\Omega}$, then the Onsager relations in the form (7.13) are replaced by

$$K_{ij}(\omega, -\mathbf{k})\Big|_{-\mathbf{B}, -\mathbf{\Omega}} = K_{ji}(\omega, \mathbf{k})\Big|_{\mathbf{B}, \mathbf{\Omega}}, \qquad (E7.6)$$

Hint: Consider the time-reversal properties of a particle in a rotating coordinate system.

7.6 Consider (7.17) in the case of an isotropic medium in which the χs do not depend on $\mathbf{k}$ or $\mathbf{B}$, and for which each $\chi_{ij}(\omega, \mathbf{k})$ is replaced by a $\chi(\omega)\delta_{ij}$, so that there are four scalar response functions $\chi^e(\omega)$, $\chi^{em}(\omega)$, $\chi^{me}(\omega)$, $\chi^m(\omega)$. Show that (7.17) reduces to

$$K_{ij}(\omega, \mathbf{k}) = \delta_{ij} \left[1 + \chi^e(\omega) + \frac{k^2 c^2}{\omega^2} \chi^m(\omega) \right]$$
$$- \frac{2c}{\omega} \epsilon_{ijs} k_s \chi^{em}(\omega) - \frac{c^2}{\omega^2} k_i k_j \chi^m(\omega). \qquad (E7.7)$$

7.7 Consider an anisotropic dielectric with dielectric tensor $K_{ij}(\omega) = K_{ji}(\omega)$ in an arbitrary coordinate system. The diagonal components of the dielectric tensor in the form (7.23) are given by the solutions for $\lambda = K_1$, K_2, K_3 of the characteristic equation

$$\det \left[K_{ij} - \lambda \delta_{ij} \right] = 0. \qquad (E7.8)$$

Argue that the solutions satisfy

$$K_1 K_2 K_3 = \det \left[K_{ij} \right], \quad K_1 + K_2 + K_3 = K_{ss},$$
$$K_1 K_2 + K_1 K_3 + K_2 K_3 =$$
$$K_{11} K_{22} + K_{11} K_{33} + K_{22} K_{33} - K_{12}^2 - K_{13}^2 - K_{23}^2.$$

Hint: Consider the characteristic equation in the form $(K_1 - \lambda)(K_2 - \lambda)(K_3 - \lambda) = 0$.

7.8 A magnetized plasma exhibits the Faraday effect; specifically, at high frequencies the response tensor is approximated by the form

$$K_{ij}(\omega) = K^{(0)}(\omega)\delta_{ij} + \eta_{ijk}(\omega)B_k, \qquad (E7.9)$$

where $\mathbf{B}$ is the static field.

(a) What do the Onsager relations imply on the form of $\eta_{ijk}(\omega)$?
(b) Is the corresponding electric effect allowed, that is, if B_k in $(E7.9)$ is replaced by the components E_k of a static effect field, is the resulting form acceptable?
(c) If your answer to part (b) is affirmative, what restrictions do the Onsager relations imply on the corresponding tensor $\eta_{ijk}(\omega)$?

8

Analytic Properties of
Response Functions

Preamble

In order to appreciate the significance of the mathematical properties of response functions it is helpful to appeal to the theory of contour integration. The causal condition implies that the response function has specific analytic properties as a function of complex ω. The relevant properties appear naturally in terms of a Laplace transform. We digress here to introduce both Laplace transforms and contour integration and to discuss the analytic properties of the response functions.

8.1 Complex Frequencies and Analytic Continuation

In the definition (4.3) of the Fourier transform $\tilde{f}(\omega)$ of a function $f(t)$ it is implicit that the frequency ω is real. However, for some purposes it is appropriate to consider $\tilde{f}(\omega)$ as a function of complex ω. Specifically, consider a disturbance that is oscillating at a frequency ω_0 and is decaying in time with a decay rate $\gamma/2$, so that the temporal variation is as $e^{-\gamma t/2}\cos(\omega_0 t)$. (It is convenient to include the factor $\frac{1}{2}$ in the decay rate $\gamma/2$ of the amplitude so that the energy, which varies as the square of the amplitude, decays at the rate γ.) On Fourier transforming such a disturbance, assuming it to be created at $t = 0$, one finds

$$\int_0^\infty dt\, e^{i\omega t}\, e^{-\gamma t/2}\cos(\omega_0 t) = \frac{i}{2}\left[\frac{1}{\omega - \omega_0 + i\gamma/2} + \frac{1}{\omega + \omega_0 + i\gamma/2}\right].$$

$$(8.1)$$

It follows that such a disturbance may be regarded as having Fourier components at the complex frequencies $\omega = \pm\omega_0 - i\gamma/2$. In the language of the theory of functions, the function on the right hand side of (8.1) consists of two isolated poles in the lower half ω plane.

More generally, an arbitrary response function $\tilde{f}(\omega)$ is regarded as a function of complex ω. The foregoing example suggests that if the medium is stable, in the sense that it does not allow any growing disturbances, then all its poles should be in the lower half ω plane or on the real-ω axis. It turns out that even for unstable media, when one or more poles or other singularities is in the upper half ω plane, the causal condition requires that there exists a region of the upper half ω plane that is free of all singularities. This region is denoted by $\mathrm{Im}\,\omega > \Gamma$, where Γ is a finite constant. This is the basic analytic property of causal functions. It is discussed here first from an elementary point of view assuming $\Gamma = 0$, then using Laplace transforms in §8.2, and further in terms of contour integration in §8.4.

By definition a causal function $f(t)$ vanishes at $t < 0$. Hence its Fourier transform, cf. (4.3), is written

$$\tilde{f}(\omega) = \int_0^\infty \mathrm{d}t\, \mathrm{e}^{i\omega t}\, f(t). \tag{8.2}$$

Consider $\tilde{f}(\omega)$ as a function of complex $\omega = \mathrm{Re}\,\omega + i\mathrm{Im}\,\omega$. For $\mathrm{Im}\,\omega > 0$ the integral has a finite value because the factor $\mathrm{e}^{-\mathrm{Im}\,\omega t}$ tends to zero for $t \to \infty$. Moreover, on differentiating with respect to ω this exponential factor ensures that all the derivatives of $\tilde{f}(\omega)$ are also finite. By definition, a function is said to be an *analytic* function in a region of the complex plane, if the function and all its derivatives are finite in that region. It follows that a causal function $\tilde{f}(\omega)$ is analytic in the upper half ω plane.

This property obviously implies that there are no poles in the upper half ω plane, because the function would be singular at a pole. It also implies that the function has no branch points in the upper half ω plane so that a causal function is single valued in the upper half ω plane. The proof of this property follows by considering various kinds of branch point. For example, a logarithmic function $\ln(\omega - \Omega)$ and a power-law function $(\omega - \Omega)^\alpha$, with $\alpha > 0$ and not an integer, are multi-valued functions with a branch point at $\omega = \Omega$. The derivative of the logarithmic function and a sufficiently high order derivative of the power-law function are singular $\omega = \Omega$. However, a causal function and all its derivatives are finite at all points in the upper half ω plane, and hence any branch point $\omega = \Omega$ cannot be in the upper half ω plane for a causal function.

8.2 Laplace Transforms

Historically the correct imposition of the causal condition in the kinetic theory of plasmas by Landau in 1946 occurred nearly a decade after the kinetic theory was first formulated. Landau's approach is based on the use of the Laplace transform which, unlike the Fourier transform, takes the causal condition into account automatically. Moreover, the procedure for inverting the the Laplace transform involves an explicit prescription as to how to integrate around singularities in the complex plane. This property led Landau to identify the so-called Landau damping process for plasma waves.

The Laplace transform of a function $f(t)$ of time t is defined by

$$F(s) := \int_0^\infty dt\, e^{-st}\, f(t). \tag{8.3}$$

Comparison of (8.3) with the Fourier transform (8.2) shows that the Laplace transform is related to the temporal Fourier transform $\tilde{f}(\omega)$ by $F(s) = \tilde{f}(is)$, or $\tilde{f}(\omega) = F(-i\omega)$. In terms of complex variables s and ω are related by a rotation in the complex ω plane. There is an important difference in the range of integration between the Fourier transform and the Laplace transform. The integral for a Fourier transform is over all t, as in (4.3), and the integral is restricted to the range $0 < t < \infty$ in (8.2) only because $f(t)$ is assumed to vanish for $t < 0$. On the other hand, for the Laplace transform (8.1) the integral is over $0 < t < \infty$ by definition, and hence the Laplace transform does not depend on the value of $f(t)$ for $t < 0$. Thus the Laplace and Fourier transforms are equivalent (apart from the rotation in the complex plane) only for functions that vanish for $t < 0$.

The specific condition for the inverse of the transform (8.3) is

$$f(t) = \frac{1}{2\pi i} \int_{\Gamma - i\infty}^{\Gamma + i\infty} ds\, e^{st}\, F(s), \tag{8.4}$$

where Γ is to be chosen so that the contour of integration is to the right of all singularities of $F(s)$ in the complex s plane. The contour is indicated in Figure 8.1. "To the right of all singularities in the complex s plane" corresponds to a contour above all singularities in the complex ω plane.

The condition that one is to integrate above all singularities in the complex ω plane when inverting the Fourier transform of a causal function is a statement of the *Landau condition*. When the singularities are on the real-ω axis, the Landau condition reduces to using the Plemelj formula.

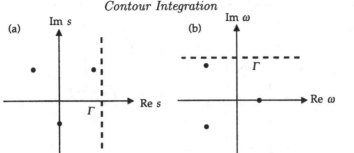

Fig. 8.1 (a) A contour parallel to the imaginary-s axis transforms into
(b) a contour parallel to the real-ω axis.

8.3 Contour Integration

Contour integration provides an alternative way of treating singular
functions. The basic result used is the following.

The Theorem of Residues: If a function $f(z)$ is analytic throughout a
contour C and its interior except at a number of poles inside the contour,
then one has

$$\int_C \mathrm{d}z\, f(z) = 2\pi\mathrm{i} \sum_i R_i, \tag{8.5}$$

where $\sum_i R_i$ denotes the sum of the residues of the function $f(z)$ at
those poles situated at $z = z_i$ inside the contour C.

It is implicit that the integral is to be taken in the positive sense,
which is counterclockwise around the contour C; if the integral is taken
clockwise then the opposite sign appears on the right hand side of (8.5).

The value of the integral (8.5) is unchanged by changes in the contour
C provided that any deformations in the contour do not cause any pole
to cross the contour. In particular one deforms the contour C so that
it reduces to the contour C' that consists of a sum of circles around
the poles, plus canceling connecting lines between them, as illustrated
in Figure 8.2. This type of deformation is used in the proof of the
theorem of residues. The integral around a circle centered on a pole is
evaluated in an elementary way. Specifically, a function $f(z)$ that has
a simple pole at $z = z_i$ is approximated sufficiently near the pole by
$f(z) \approx R_i/(z - z_i)$,

$$R_i = \lim_{z \to z_i} (z - z_i)f(z), \tag{8.6}$$

where R_i is the residue at $z = z_i$, and the radius of the circle around
which one integrates is made arbitrarily small so that this approximation
becomes arbitrarily accurate. On introducing a new variable of integration $z' = z - z_i$ the integral around the circle C_i centered on $z = z_i$

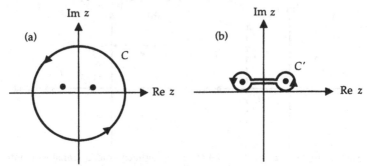

Fig. 8.2 (a) The integral around C enclosing poles (large dots)
is evaluated by (b) deforming to the contour C'.

is evaluated by writing $z' = |z'|e^{i\theta'}$, $dz' = id\theta'|z'|e^{i\theta'}$ and integrating
around the circle $0 < \theta' < 2\pi$ in the positive sense. Thus one obtains

$$\int_{C_i} \frac{dz'}{z'} = 2\pi i. \tag{8.7}$$

On summing over the contribution from all poles, this leads to the result
(8.5).

Contour integration may be used to represent the Fourier transform of
the step function and to give an alternative interpretation of the Plemelj
formula (4.32). Consider the integral representation of the step function
implied by (4.26):

$$H(t) = \lim_{\eta \to 0} \int_{-\infty}^{\infty} \frac{d\omega}{2\pi} \frac{i}{\omega + i\eta} e^{-i\omega t}. \tag{8.8}$$

There is a simple pole at $\omega = -i\eta$, which approaches the real-ω axis from
below as η approaches zero. The contour of integration is along the real-
ω axis above the pole. This contour is closed by including a semicircle
at infinity, as indicated in Figure 8.3. The semicircle is chosen either in
the upper half plane or the lower half plane, with the choice such that
the exponential factor gives zero. For the semicircle in the upper half
plane one has $\text{Im}\,\omega \to +\infty$ and then $e^{-i\omega t} \propto e^{\text{Im}\,\omega t}$ gives zero for $t < 0$.
Thus the contour is to be closed in the upper half ω plane for $t < 0$.
For the semicircle in the lower half plane one has $\text{Im}\,\omega \to -\infty$ and then
$e^{-i\omega t} \propto e^{-|\text{Im}\,\omega|t}$ gives zero for $t > 0$. Thus one closes the contour with
a semicircle in the lower half ω plane for $t > 0$.

When the contour is closed in the upper half plane it does not enclose
the pole, and then (8.5) implies that the integral gives zero, which applies
for $t < 0$. On the other hand, when the contour is enclosed in the lower

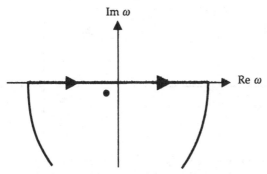

Fig. 8.3 The contour in (8.8) passes above the pole at $\omega = -i0$,
and closes in the lower half plane for $t > 0$.

half plane the pole is inside the contour. The residue at the pole is

$$R = \lim_{\eta \to 0} \frac{i}{2\pi} e^{-i(-i\eta)t} = \frac{i}{2\pi}. \tag{8.9}$$

The contour is traced in the negative sense (clockwise) and hence the
integral is equal to $-2\pi i$ times the residue, which with (8.9) gives unity.
Thus the integral on the right hand side of (8.8) is equal to zero for $t < 0$
and to unity for $t > 0$, as required for its identification with the step
function $H(t)$.

The Plemelj formula (4.32) has an interpretation in terms of contour
integration, as illustrated in Figure 8.4. Suppose that (4.32) is in the
integrand of an integral over the range $-\infty < \omega < \infty$. The left hand
side corresponds to an integral along the real axis above the pole, which
is infinitesimally below the real axis at $\omega = -i0$. The right hand side
corresponds to the pole being on the real axis. The integral then consists
of three parts: from $-\infty$ to $-\eta$ along the real axis, along a semicircle in
the upper half plane from $-\eta$ to η, and then from η to ∞ along the real
axis. The two segments along the real axis give the Cauchy principal
value of the integral in the limit $\eta \to 0$:

$$\wp \int_{-\infty}^{\infty} d\omega \, \frac{f(\omega)}{\omega - \omega_0} := \lim_{\eta \to 0} \left(\int_{-\infty}^{\omega_0 - \eta} + \int_{\omega_0 + \eta}^{\infty} \right) d\omega \, \frac{f(\omega)}{\omega - \omega_0}. \tag{8.10}$$

The definition (8.10) of the Cauchy principal value involves a sequence
$\eta \to 0$ of integrals. Alternatively, the integral is extended to the full
range $-\infty < \omega < \infty$ and the generalized function $\wp[1/(\omega - \omega_0)]$, defined by (4.30), is inserted in the integrand. The generalized function
is equal to unity except in the range $|\omega - \omega_0| \ll \eta$ where it is small,
and so including $\wp[1/(\omega - \omega_0)]$, as defined by (4.30), in the integrand is
equivalent to the definition (8.10) of the principal value of the integral.

The semicircular section of the contour above the pole contributes

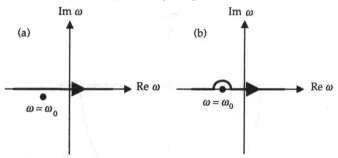

Fig. 8.4 The equivalence of evaluating an integral using the contours
(a) and (b) implies the Plemelj formula.

half the value of a complete circle around the pole. The integral is taken
in the negative sense, and hence the semicircular section of the contour
gives $-\mathrm{i}\pi$ times the residue, or equivalently $-2\mathrm{i}\pi$ times half the residue
at the pole. From the latter viewpoint, the final term in (4.32) is referred
to as the *semiresidue* term.

8.4 Poles in the Upper Half ω Plane

The implications of the causal condition on the analytic properties of a
response function follow from the inversion of the Laplace transform and
the assumption that the response function falls off faster than $1/\omega$ for
$\omega \to \infty$ so that the integral around a semicircle at infinity gives zero. It
then follows that the integral around a contour above all the singularities
and closed by a semicircle at infinity in the upper half plane gives zero.
The singularities that are poles not too far from the real-ω axis are
interpreted in terms of the dispersion relations for the weakly damped
natural wave modes of the medium. If one writes $\omega = \omega_M(\mathbf{k}) - \mathrm{i}\frac{1}{2}\gamma_M(\mathbf{k})$
for a pole then $\omega = \omega_M(\mathbf{k})$ corresponds to the dispersion relation and
$\gamma_M(\mathbf{k})$ corresponds to the absorption coefficient for the wave mode M.
If the waves are damped then $\gamma_M(\mathbf{k})$ is positive and the corresponding
pole is in the lower half plane. The contour in the upper half plane is
drawn along the real axis if all waves are damped.

An unstable system is one in which there is at least one growing wave
mode. A growing wave corresponds to a negative absorption coefficient
so that the corresponding pole is in the upper half ω plane. This case
is to be treated by analytic continuation from the case where all poles
are in the lower half ω plane. The analytic properties are defined in the
case where all singularities are in the lower half plane and one is to treat
a pole in the upper half plane as though it were initially in the lower

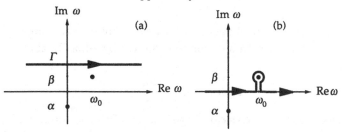

Fig. 8.5 The contour in (a) above a pole at $\operatorname{Im}\omega = -\beta > 0$
may be deformed into the contour (b).

half plane and enters the upper half plane by a slow variation in some
parameter or other. Initially the contour is above all singularities and
as a pole wanders into the upper half plane in this way the contour is
to be deformed so that it is always above the pole. Thus when there
is a pole in the upper half plane the contour is to be drawn above the
pole and this may be done either by choosing a line parallel to the real
ω-axis above the pole or by deforming the contour around the pole, as
illustrated in Figure 8.5.

To illustrate some of the implications of these points consider the
following idealized example. The response of a particular medium is
assumed to involve two independent processes, one is a damping independent of frequency, described by a response function of the form
$A/(\omega + i\alpha)$ and the other involves a damped resonance at a frequency
ω_0, and is described by $B/(\omega - \omega_0 + i\beta)$, where A, B, ω_0, α and β are
constants. This type of response is characteristic of a classical damped
oscillator, as discussed in §9.3 below. More general types of response
may be built up by summing or integrating over responses of the forms
$A/(\omega + i\alpha)$ and $B/(\omega - \omega_0 + i\beta)$. The total response in this simple example is assumed to be the convolution of the responses to these two
independent processes:

$$K(\omega) = \int_{-\infty}^{\infty} \frac{\mathrm{d}\omega'}{2\pi} \frac{A}{\omega - \omega' + i\alpha} \frac{B}{\omega' - \omega_0 + i\beta}. \tag{8.11}$$

For $\alpha > 0$, $\beta > 0$ the integral in (8.11) is evaluated by separating the
integrand into a sum of simple poles. The integral becomes

$$K(\omega) = \frac{AB}{\omega - \omega_0 + i(\alpha + \beta)} \int_{-\infty}^{\infty} \frac{\mathrm{d}\omega'}{2\pi} \left(\frac{1}{\omega - \omega' + i\alpha} + \frac{1}{\omega' - \omega_0 + i\beta} \right), \tag{8.12}$$

which gives $\ln[(\omega' - \omega_0 + i\beta)/(\omega' - \omega - i\alpha)]$ evaluated between $\omega' = -\infty$
and $\omega' = \infty$. In the limit $\omega' \to -\infty$ the numerator and denominator in
the logarithm both give contributions of $i\pi$, and in the limit $\omega' \to \infty$

they both give infinitesimal contributions. Thus over the range $\omega' = -\infty$ to $\omega' = \infty$ the logarithmic function changes by $-i2\pi$. In this way (8.11) gives

$$K(\omega) = \frac{-iAB}{\omega - \omega_0 + i(\alpha + \beta)}. \tag{8.13}$$

There is a simple physical interpretation of the denominator in (8.13). The imaginary part describes dissipation and it follows that in the presence of two separate dissipation mechanisms, the net rate of dissipation is the sum of the rates due to the two independent responses. Thus the rate at which an electromagnetic disturbance damps in a medium is determined by the sum of the damping rates due to all the dissipation processes in the medium.

Now consider the case where one or both of α and β are negative. Even though one or both poles are now in the upper half plane, the result (8.13) still applies. This is implied directly by the requirement that the case of a pole in the upper half plane is to be obtained by analytic continuation of the case where the pole is in the lower half plane. Thus, if (8.13) is the appropriate result for $\alpha > 0$, $\beta > 0$ then it must continue if either α or β becomes negative. The formal derivation of this result by carrying out the integral is outlined in Exercise 8.5.

Thus a system that is subject to two or more dissipation processes has a net dissipation rate that is the sum of the independent dissipation rates. If one process has negative dissipation and so tends to make the system unstable, the net dissipation is negative only if this negative contribution exceeds in magnitude the sum of all the positive contributions from other dissipation processes. Physically this is as expected. For example, a lossy laser amplifies light only if the growth rate due to the lasing process exceeds the dissipation rate associated with the lossiness.

The mathematical requirement of analytic continuation into the upper half ω plane may be understood physically by considering a system which is initially stable and which becomes unstable due to a slow variation of some parameter. In a maser or laser this parameter is the overpopulation of the metastable state that leads to maser or laser action, and in a plasma it is a streaming velocity or a pitch angle anisotropy. As this parameter is varied, the location of the pole corresponding to the wave mode that becomes unstable starts from the lower half ω plane, reaches the real ω axis at the point of marginal stability and moves into the upper half ω plane as the system becomes unstable. One encounters conceptual problems if an unstable system is not defined in this way. Specifically, one can follow the evolution of an unstable system only from some initial time when it must be "turned on". In a sense the motion of a pole from the lower half ω plane to the upper half ω plane reflects the effect of this "turning on" of the unstable system due to the variation of a relevant parameter. In contrast, a stable system is well defined in terms of its properties at $t \rightarrow -\infty$, because the absence of any growing modes implies that all the transients die away, and it is this property that implies that such a system has no poles in the upper half ω plane.

A detailed discussion of the analytic properties of response functions is outside the scope of this book. Let us end with one remark that relates the foregoing discussion to this mathematically subtle aspect of existing theory. The concept of the motion of poles in the complex ω plane, as some parameter is varied, is particularly important in the theory of plasma instabilities. The response tensor for a plasma involves transcendental functions and these can have an infinite number of poles in the complex plane. Most of these poles are of no physical interest. Interesting effects are associated with the movement of two poles towards each other. One simple case is when two real poles, representing two undamped wave modes, move together and become a complex conjugate pair of poles. One of this complex conjugate pair corresponds to a growing mode and the other corresponds to a damped mode. More generally, instabilities appear when two poles coalesce and then move apart again having changed their mathematical character, for example, by moving from one sheet of the complex plane to another.

Exercise Set 8

8.1 The arguments given in §8.1 imply that a function that vanishes for $t < 0$ has no singularities in the upper half ω plane, and a similar argument implies that a function that vanishes for $t > 0$ has no singularities in the lower half ω plane. What analytic property is satisfied by a function that is non-zero only for a finite time, for example, a function that satisfies $f(t) = 0$ for $|t| > T/2$, where T is a constant?

8.2 Given that $F(s)$ is the Laplace transform of a function $f(t)$ prove the following.

(a) For $a > 0$, the Laplace transform of $f(at)$ is $F(s/a)/a$.

(b) If $f(t) = 0$ for all $t < 0$, and $b > 0$, the Laplace transform of $f(t-b)$ is $e^{-bs}F(s)$.

(c) If $f_1(t) = 0$ and $f_2(t) = 0$ for all $t < 0$, then the Laplace transform of their convolution $f_1 * f_2$ is $F_1(s)F_2(s)$.

8.3 The full-range and half-range Laplace transforms of a function $f(t)$ are defined by

$$F_{\mathrm{F}}(s) := \int_{-\infty}^{\infty} dt\, e^{-st} f(t), \quad F(s) := \int_{0}^{\infty} dt\, e^{-st} f(t). \qquad (E8.1)$$

Let $F_1(s)$ be the half-range Laplace transform of $f(-t)$.

(a) Given that $\tilde{f}(\omega)$ is the Fourier transform of $f(t)$, establish the identities

$$\tilde{f}(\omega) = F_{\mathrm{F}}(-i\omega) = F(-i\omega) + F_1(i\omega). \qquad (E8.2)$$

(b) Find the Fourier transforms of $tf(t)$, $f(t)/t$, $df(t)/dt$. (You may assume $|tf(t)| \to 0$ for $t \to \pm\infty$.)

8.4 Use contour integration to evaluate the integral (5.24), *viz.*,

$$\int_{0}^{\infty} dx\, \frac{\sin x}{x} = \frac{\pi}{2}.$$

Specifically consider the imaginary part of $\int_{C} d\omega\, (e^{i\omega}/\omega)$ around a contour C that consists of an infinite semicircle in the upper half plane, and a line along the real axis except for an infinitesimal semicircular deviation into the upper half plane above the pole at $\omega = 0$.

8.5 The result (8.13) for the response of a compound system follows from (8.11) in the case where both dissipation rates α and β are positive. Analytic continuation requires that (8.13) apply when either α or β is negative. The following exercise concerns a formal derivation of this fact.

(a) Consider the case where one or both of α, β are negative. Show that the result (8.13) follows from (8.11) provided that one applies the convolution by integrating along a line parallel to the ω' axis at $\mathrm{Im}\,\omega = \Gamma$, above both poles as illustrated in Figure 8.5(a).

(b) By analytic continuation the contour in Figure 8.5(a) is deformed to that illustrated in Figure 8.5(b), in which case the integral is along the real axis, with the principal value to be taken at $\omega = \omega_0$, plus the integral around the pole at $\omega = \omega_0 - i\beta$ as illustrated. Evaluate this integral and show that it reproduces the result (8.13).

9

Response Tensors for Idealized Media

Preamble

The response tensor for a dielectric depends on the polarizability of the individual atoms and molecules. In a "dense" isotropic medium, where "dense" is not a well-defined concept, the relation between the dielectric constant and the polarizability is given by the Lorenz–Lorentz relation. The polarizability needs to be calculated quantum mechanically, but many of the features of the response of a dielectric may be inferred from a classical model of forced oscillators.

9.1 The Polarizability of Atoms and Molecules

A simple model for the response of dielectric materials is based on assuming that the response results from induced electric dipole moments in the medium. One distinguishes between three classes of polarization on a microscopic level in different types of media. One class consists of media in which the polarization is attributed to deformation of atoms or molecules so that the mean centers of the positive and negative charges become slightly separated, implying that the atoms or molecules develop induced dipole moments. A second class of media consists of those in which the individual particles have intrinsic dipole moments. These moments are randomly oriented in the absence of an external field and become partially aligned in the presence of an external field. The responses for these two classes of media exist for static fields as well as for oscillating fields. The third class of media consists of charges that are free to move and such media become polarized in the sense that there is a net displacement between the positive and negative charges. The response of such media is meaningful only for oscillating fields. For the present we consider only media of the first class; that is, we assume that on a microscopic level the polarization of the medium is interpreted in terms of a collection of induced electric dipoles.

The *polarizability* of the atom or molecule is defined as the relation between the induced electric dipole moment, $\mathbf{d}(\omega)$, and the electric field:

$$d_i(\omega) = a_{ij}(\omega)E_j(\omega). \tag{9.1}$$

Any spatial variation of $\mathbf{E}$ is irrelevant in the dipole model. The polarization $\mathbf{P}$ is defined as the induced dipole moment per unit volume, and so one has

$$P_i(\omega) = n a_{ij}(\omega)E_j(\omega), \tag{9.2}$$

where n is the number density of the atoms or molecules. If there is a collection of atoms or molecules of various species, with an arbitrary species labeled α, then the polarization is

$$P_i(\omega) = \sum_\alpha n_\alpha a_{ij}^{(\alpha)}(\omega)E_j(\omega), \tag{9.3}$$

where the sum is over all species in the medium, and where n_α is the number density of species α.

9.2 The Lorenz–Lorentz Equation

There is an important point that needs to be discussed with some care relating to the response in a medium where it is attributed to induced dipoles. For such media the macroscopic response is described adequately by (6.17) only if the density of the medium is sufficiently low. An alternative relation provides a better description for some "dense" dielectrics.

The additional effect included where the density of induced dipoles is relatively high is the contribution of the polarization to the electric field on the right hand side of (9.2). Suppose that the divergence of the polarization is non-zero on a microscopic level. According to (6.5) there is then an induced charge density equal to $-\operatorname{div}\mathbf{P}$, and this charge density gives rise to a contribution to the electric field which needs to be included in $\mathbf{E}$ on the right hand side of (9.2). Thus the electric field in (9.2) is regarded as composed of two parts: the imposed electric field $\mathbf{E}^{(0)}$ and an electric field $\mathbf{E}^{(d)}$ due to the effects of the induced dipoles:

$$\mathbf{E} = \mathbf{E}^{(0)} + \mathbf{E}^{(d)}. \tag{9.4}$$

The field $\mathbf{E}^{(d)}(\mathbf{x})$ is evaluated using $-\operatorname{grad}\phi(\mathbf{x})$ with $\phi(\mathbf{x})$ given by the solution (1.23) of Poisson's equation and with the charge density in Poisson's equation identified as $-\operatorname{div}\mathbf{P}(\mathbf{x})$ in accord with (6.5). One has

$$\mathbf{E}^{(d)}(\mathbf{x}) = \operatorname{grad}\left[\frac{1}{4\pi\varepsilon_0}\int d^3x' \frac{\operatorname{div}'\mathbf{P}(\mathbf{x}')}{|\mathbf{x}-\mathbf{x}'|}\right], \qquad (9.5)$$

where the prime in $\operatorname{div}'\mathbf{P}$ implies that the divergence is to be taken with respect to the primed coordinate. The field (9.5) is separated into a part due to the non-singular contribution to the integral from dipoles at $\mathbf{x}' \neq \mathbf{x}$ and a part due to the singular contribution to the integral at $\mathbf{x}' = \mathbf{x}$. We now argue that the former part vanishes and that the latter part gives

$$\mathbf{E}^{(d)}(\mathbf{x}) = \mathbf{P}(\mathbf{x})/3\varepsilon_0. \qquad (9.6)$$

An outline of the proof that (9.5) implies (9.6) is as follows. First note that the gradient operator in (9.5) acts only on $1/|\mathbf{x}-\mathbf{x}'|$, and that on taking it inside the integral the gradient with respect to $\mathbf{x}$ is replaced by minus the gradient with respect to $\mathbf{x}'$. Next introduce tensor notation, writing $\operatorname{div}'\mathbf{P} = \partial P_s/\partial x'_s$, and then partially integrate with respect to x'_s to find

$$E_i^{(d)}(\mathbf{x}) = -\frac{1}{4\pi\varepsilon_0}\int d^3x'\, P_s(\mathbf{x}')\frac{\partial^2}{\partial x'_s \partial x'_i}\frac{1}{|\mathbf{x}-\mathbf{x}'|}. \qquad (9.7)$$

The differentiation in (9.7) gives, cf. Exercise 9.1,

$$\frac{\partial^2}{\partial x'_s \partial x'_i}\frac{1}{|\mathbf{x}-\mathbf{x}'|} = -\frac{4\pi}{3}\delta_{is}\,\delta(\mathbf{x}-\mathbf{x}')$$

$$-\left[\frac{\delta_{is}}{|\mathbf{x}-\mathbf{x}'|^3} - \frac{3(x_i - x'_i)(x_s - x'_s)}{|\mathbf{x}-\mathbf{x}'|^5}\right]. \qquad (9.8)$$

The singular part in (9.8) gives (9.6) and the non-singular part vanishes, cf. Exercise 9.2.

On combining (9.3), (9.4) and (9.6), one obtains

$$P_i(\omega) = \sum_\alpha n_\alpha a_{ij}^{(\alpha)}(\omega)\left[E_j^{(0)}(\omega) + P_j(\omega)/3\varepsilon_0\right]. \qquad (9.9)$$

For an isotropic medium, with $a_{ij}^{(\alpha)}(\omega) = a^{(\alpha)}(\omega)\delta_{ij}$, let us define a "susceptibility" $\tilde{\chi}(\omega)$ that does not include the field $\mathbf{E}^{(d)}(\omega)$. Specifically, we use (9.9) to write

$$\tilde{\chi}(\omega) = \sum_\alpha \frac{n_\alpha a^{(\alpha)}(\omega)}{\varepsilon_0}, \quad K(\omega) = \tilde{\chi}(\omega)\{1 + [K(\omega) - 1]/3\}. \qquad (9.10)$$

In terms of the dielectric constant $K(\omega) = 1 + \chi(\omega)$, (9.10) implies

$$\frac{K(\omega) - 1}{K(\omega) + 2} = \frac{\tilde{\chi}(\omega)}{3} = \frac{1}{3\varepsilon_0} \sum_\alpha n_\alpha a^{(\alpha)}(\omega), \qquad (9.11)$$

which is sometimes called the *Clausius–Mossotti relation*. When applied to the response at optical frequencies the relation (9.11) is called the *Lorenz–Lorentz equation*, in which case it is conventional to replace $K(\omega)$ by $n^2(\omega)$ in (9.11). As we are interested in the response to oscillating fields, we refer to (9.11) as the Lorenz–Lorentz equation.

The Lorenz–Lorentz equation is derived here for a system whose response on a microscopic level is attributed to induced dipole moments. For a system composed of particles each with an intrinsic dipole moment the response is attributed to partial alignment of the otherwise randomly oriented dipoles; the foregoing arguments may be modified to include this case, and the Lorenz–Lorentz equation is again applicable. However, even though the response of a plasma (that is, a medium that consists of free charges) may be expressed in terms of an induced polarization (§10.1), there is no $\mathbf{E}^{(d)}$ and the Lorenz–Lorentz equation does not apply to a plasma. The additional electric field $\mathbf{E}^{(d)}$ is present only for a medium composed of neutral particles which develop an induced dipole moment. This is because it is the divergence of $\mathbf{P}$ that leads to a microscopic charge density in the immediate vicinity of each induced dipole. For a medium composed of free charges, the perturbation due to $\mathbf{E}$ causes the charges to move but it does not introduce a new type of charge density on a microscopic scale. One way of thinking of the distinction between these two cases is that in a medium composed of neutral particles that become polarized the response on a microscopic level is thought of as an induced *displacement* between positive and negative charges producing an induced polarization, and in a medium composed of free charges the response on a microscopic level is thought of as an induced *motion* of the free charges in opposite directions producing an induced current.

An alternative derivation of the Lorenz–Lorentz equation is based on the picture illustrated in Figure 9.1. The dielectric medium is assumed to be polarized due to being between two charged capacitor plates. The relation (6.5), *viz.*, $\rho_{\text{ind}} = -\text{div}\,\mathbf{P}$ implies that at any sharp boundary the medium has a surface charge (that is, a charge per unit area) $\sigma_P = P_n$, where P_n is the normal component of $\mathbf{P}$. The relation $\sigma_P = P_n$ gives the surface charge on the actual surface of the medium, and it is thought of as existing on any artificial surface within the medium,

Response Tensors for Idealized Media

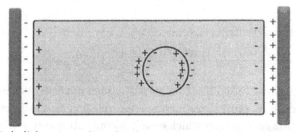

Fig. 9.1 A dielectric medium lies between two charged capacitor plates.
The polarization of the medium produces a surface charge
on the actual surface or any artificial internal surface.

with the surface charge having equal and opposite signs on either side
of the surface. Consider a sphere around a particular point within the
medium. Intuitively this surface is thought of as cutting through the
induced charge distributions of the atoms or molecules, giving a surface
charge on the inside of the surface and an equal but opposite surface
charge on the outside of the surface. The surface charge on the outside
of the surface produces no electric field within the surface. The surface
charge on the inside of the surface produces an electric field inside the
sphere, and this electric field is $\mathbf{E}^{(d)}$ as given by (9.6), cf. Exercise 9.3.

The Lorenz–Lorentz equation relates the macroscopic fields to the
microscopic fields, and it includes a response to the response field itself. As a consequence some thought is required in applying the causal
condition to (9.11). On the one hand, there is no formal problem because $K(\omega)$, as defined by (9.11) has poles at the resonances determined
by the microphysics, as in a dilute dielectric, plus additional poles at
$K(\omega) + 2 = 0$ which are not of physical importance. There is, however,
a question of the time scale on which the field $\mathbf{E}^{(d)}$ can be set up on
the macroscopic scale. In the remainder of this book we bypass this
point and regard the Lorenz–Lorentz equation as an alternative algorithm to (6.17) for relating the macroscopic fields to the microscopic
fields. Finally note that in the derivation of the Lorenz–Lorentz equation it is assumed implicitly that only dipole fields are needed to model
the response of the medium, and for some media this is an inadequate
approximation. Except where stated otherwise, the relation between
the macroscopic and microscopic fields is assumed to be described by
the relations (6.17) rather than by (9.11).

9.3 A System of Forced Classical Oscillators

For some purposes the response of dielectrics may be described by a classical model based on a damped oscillator. Consider a classical oscillator that corresponds to a mass m with charge q at a displacement $\mathbf{X}(t)$ from its mean position. Let the frequency of the oscillator be ω_0, and let it be damped with a decay constant Γ. The oscillator is assumed to be forced by an electric field $\mathbf{E}(t)$. The equation of motion is

$$\ddot{\mathbf{X}}(t) + \Gamma\dot{\mathbf{X}}(t) + \omega_0^2\mathbf{X}(t) = \frac{q\mathbf{E}(t)}{m}, \tag{9.12}$$

where a dot denotes a time derivative and a double dot denotes a second time derivative. After Fourier transforming in time, the solution of (9.12) is

$$\mathbf{X}(\omega) = \frac{q\mathbf{E}(\omega)}{m(\omega_0^2 - \omega^2 - i\omega\Gamma)}. \tag{9.13}$$

The Fourier transform of the electric dipole moment is $\mathbf{d}(\omega) = q\mathbf{X}(\omega)$.

From (9.13) one identifies the polarizability in this damped classical oscillator model as

$$a_{ij}(\omega) = \frac{q^2}{m(\omega_0^2 - \omega^2 - i\omega\Gamma)}\,\delta_{ij}. \tag{9.14}$$

It is straightforward to generalize to a set of oscillators of various species. Let an arbitrary species be labeled α and have charge q_α, mass m_α, natural frequency $\omega_{\alpha 0}$ and number density n_α. The expression for the dielectric tensor in this case is given by

$$K_{ij}(\omega) = K(\omega)\delta_{ij}, \quad K(\omega) = 1 + \sum_\alpha \frac{\omega_{p\alpha}^2}{\omega_{\alpha 0}^2 - \omega^2 - i\omega\Gamma_\alpha}, \tag{9.15}$$

where

$$\omega_{p\alpha}^2 = \frac{q_\alpha^2 n_\alpha}{\varepsilon_0 m_\alpha} \tag{9.16}$$

is the square of the *plasma frequency* for species α, and where the damping rate Γ_α is different for each oscillator α. Note that because the system is composed of free charges the Lorenz–Lorentz equation (9.11) is not applicable.

In the high-frequency limit the dielectric tensor approaches the limit

$$K(\omega) \approx 1 - \frac{\omega_p^2}{\omega^2}, \quad \omega_p^2 = \sum_\alpha \omega_{p\alpha}^2. \tag{9.17}$$

This corresponds to all the oscillators acting like free charged particles. It follows that at sufficiently high frequencies the response of a dielectric becomes like that of a free electron gas (§10.1).

At low frequencies, ignoring the dissipation, an expansion of (9.15) in powers of the ratio ω^2 to the square of one of the natural frequencies

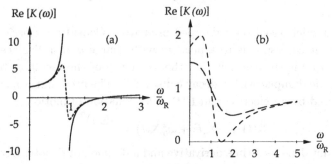

Fig. 9.2 (a) The real part of the dielectric constant (9.19): for $\Gamma = 0$ (solid curve), for $\Gamma/2\omega_R = 0.1$ (dashed curve). (b) As in (a) for $\Gamma/2\omega_R = 0.5$ (medium dashes), and $\Gamma/2\omega_R = 1$ (long dashes).

$\omega_{\alpha 0}$ gives

$$K(\omega) \approx 1 + \sum_{\alpha} \frac{\omega_{p\alpha}^2}{\omega_{\alpha 0}^2} \left(1 + \frac{\omega^2}{\omega_{\alpha 0}^2} + \cdots \right). \qquad (9.18)$$

This implies that the dielectric constant at low frequencies is greater than unity and that there is a weak dispersive term proportional to ω^2. When reexpressed as an expansion of the refractive index n with $n^2 = K(\omega)$ in inverse powers of the wavelength $\lambda = 2\pi c/n\omega$, an expansion of the form (9.18) is called a *Cauchy formula*.

The form (9.15) for the response of a system of forced classical oscillators may be used to describe the response of some dielectrics, despite the fact that a proper treatment needs to be quantum mechanical (§9.3). To apply (9.15) to a system of atoms or molecules, the natural frequencies $\omega_{\alpha 0}$ are reinterpreted in terms of the transition frequencies of the atoms or molecules, the sum over α is reinterpreted as the sum over all transition frequencies, and the parameters $\omega_{p\alpha}^2$ are assumed to incorporate the "oscillator strengths" of the various transitions. With these reinterpretations (9.15) is a realistic model for an isotropic dielectric, but it cannot be used to treat an anisotropic medium.

The properties of the response of a dielectric at high and at low frequencies are of the forms (9.17) and (9.18), respectively. Near a resonance, where ω is equal to one of the natural frequencies $\omega_{\alpha 0}$, (9.15) implies that the response becomes a strong function of frequency. Consider an example where there is only one natural frequency ω_R, or there are several natural frequencies but they are sufficiently well separated that near any one of them the others are negligible. Then for ω suffi-

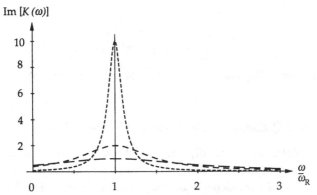

Fig. 9.3 As in Figure 9.2 but for the imaginary part of $K(\omega)$.
The four curves are for the four cases in Figure 9.2.

ciently close to ω_R, the response is approximated by

$$K(\omega) = 1 + \frac{\Omega}{\omega_R - \omega - i\Gamma/2}, \tag{9.19}$$

where Ω and Γ are constants.

The real part of $K(\omega)$, as given by (9.11), is

$$\mathrm{Re}\,[K(\omega)] = 1 + \frac{\Omega(\omega_R - \omega)}{(\omega_R - \omega)^2 + \Gamma^2/4}. \tag{9.20}$$

For $\Gamma = 0$, $\mathrm{Re}\,[K(\omega)]$ has a *resonance* at $\omega = \omega_R$ where it passes through infinity. As illustrated in Figure 9.2, for $\omega < \omega_R$ the singularity at $+\infty$ is approached as the difference $\omega - \omega_R$ decreases, and as $\omega = \omega_R$ is passed the value of $\mathrm{Re}\,[K(\omega)]$ jumps from $+\infty$ to $-\infty$. For $\omega > \omega_R$ the value of $\mathrm{Re}\,[K(\omega)]$ continues to increase back towards zero through negative values. In this case $\mathrm{Re}\,[K(\omega)]$ is a monotonically increasing function of ω. A region with $d\mathrm{Re}\,[K(\omega)]/d\omega > 0$ is referred to as a region of *positive* or *normal dispersion*. The inclusion of a small $\Gamma \neq 0$ washes out the singularity and introduces a region of *negative* or *anomalous dispersion* with $d\mathrm{Re}\,[K(\omega)]/d\omega < 0$. As the value of Γ is increased the resonance becomes increasingly washed out, as illustrated in Figure 9.2.

The imaginary part of (9.11) gives

$$\mathrm{Im}\,[K(\omega)] = \frac{\Omega\Gamma/2}{(\omega_R - \omega)^2 + \Gamma^2/4}. \tag{9.21}$$

Some examples are illustrated in Figure 9.3. The imaginary part describes dissipation, and when plotted as a function of ω as in Figure 9.3, it defines a *Lorentzian line profile* for the dissipation process. By taking the limit $\Gamma \to 0$ in (9.19) and (9.21) one finds that the Lorentzian line profile reduces to a δ-function line profile in the limit $\Gamma \to 0$ according to

$$\lim_{\Gamma \to 0} \frac{\Gamma/2}{(\omega_R - \omega)^2 + \Gamma^2/4} = \pi\delta(\omega_R - \omega). \tag{9.22}$$

The sequence in (9.22) corresponds to (4.24) with η replaced by $\Gamma/2$.

9.4 Quantum Calculation of Polarizability

A realistic treatment of the dipolar model requires that the problem be treated quantum mechanically. The quantum mechanical calculation involves the use of time-dependent perturbation theory. The following is a brief outline of a standard quantum mechanical treatment of the response in the dipole approximation.

The perturbed Hamiltonian in the dipole approximation is

$$\hat{H}^{(1)}(t) = -\hat{\mathbf{d}} \cdot \mathbf{E}(t), \tag{9.23}$$

where all dependences on $\mathbf{x}$ are suppressed. Let q label an arbitrary eigenstate of the unperturbed Hamiltonian with energy E_q, and let q' label another eigenstate. In the interaction picture (denoted by subscript I), the states evolve only due to the influence of the perturbed Hamiltonian, and the density operator $\hat{\rho}_I(t)$, whose matrix elements are the density matrix, also evolves only due to the effect of the perturbed Hamiltonian:

$$i\hbar \frac{d\hat{\rho}_I(t)}{dt} = -\big[\hat{\rho}_I(t), \hat{H}_I^{(1)}(t)\big]. \tag{9.24}$$

We solve (9.24) by taking matrix elements between states q and q' and Fourier transforming. In the interaction picture, a matrix element of an operator $\hat{A}$ between eigenstates of the unperturbed Hamiltonian evolves according to

$$A_{qq'}(t) = A_{qq'}e^{i\omega_{qq'}t}, \quad \omega_{qq'} = (E_q - E_{q'})/\hbar. \tag{9.25}$$

The matrix element of (9.24) is, now omitting subscripts I,

$$i\hbar \frac{d}{dt}\big[\rho_{qq'}(t)e^{i\omega_{qq'}t}\big] = -\sum_{q''}\big[\rho_{qq''}(t)H_{q''q'}^{(1)}(t) - H_{qq''}^{(1)}(t)\rho_{q''q'}(t)\big]e^{i\omega_{qq'}t}. \tag{9.26}$$

The density matrix is expanded in the powers of the perturbation. The zeroth order terms are diagonal with $\rho_{qq}^{(0)}$ giving the probability of the state q being occupied:

$$\sum_q \rho_{qq}^{(0)} = 1. \tag{9.27}$$

The Fourier transform of the first order term is obtained directly from the Fourier transform of (9.26):

$$\rho_{qq'}^{(1)}(\omega) = \frac{\rho_{qq}^{(0)} - \rho_{q'q'}^{(0)}}{\hbar(\omega - \omega_{qq'})} \mathbf{d}_{qq'} \cdot \mathbf{E}(\omega),$$ (9.28)

where $\mathbf{d}_{qq'}$ is the matrix element of the electric dipole operator. The expectation value of the induced electric dipole moment is given by

$$\langle \mathbf{d}^{(1)} \rangle(\omega) = \sum_{q,q'} \mathbf{d}_{qq'} \rho_{q'q}^{(1)}(\omega).$$ (9.29)

The polarizability is identified directly from (9.28) and (9.29):

$$a_{ij}(\omega) = \sum_{q,q'} \frac{\rho_{qq}^{(0)} - \rho_{q'q'}^{(0)}}{\hbar} \frac{(d_i)_{qq'}(d_j)_{q'q}}{\omega_{q'q} - \omega}$$

$$= \sum_{q,q'} \frac{\rho_{qq}^{(0)}}{\hbar} \left[\frac{(d_i)_{qq'}(d_j)_{q'q}}{\omega_{q'q} - \omega} + \frac{(d_j)_{qq'}(d_i)_{q'q}}{\omega_{q'q} + \omega} \right].$$ (9.30)

The identification of the polarizability completes the formal quantum mechanical calculation. The polarizability is used to describe the response, as discussed in §9.1.

One step in writing down the response of a medium is to relate the diagonal component $\rho_{qq}^{(0)}$ of the density matrix to the occupation number for the state q. The *occupation number* n_q is defined as the product $N\rho_{qq}^{(0)}$, where N is the total number of atoms or molecules of the relevant species. With $N = nV$, where V is the volume of the system, this gives $n_q = nV\rho_{qq}^{(0)}$. Then (9.2) reduces to

$$P_i(\omega) = \sum_{q,q'} \frac{n_q - n_{q'}}{V\hbar} \frac{(d_i)_{qq'}(d_j)_{q'q}}{\omega_{q'q} - \omega} E_j(\omega),$$ (9.31)

where the contribution of only one species is retained. One could redefine n_q to incorporate the power of V by introducing another symbol to denote n_q/V, which would be interpreted as the occupation number density of the state q, but for a later purpose we prefer to use the form (9.31).

In practice, one is usually concerned only with electronic states, in which case the matrix elements $\mathbf{d}_{qq'}$ of the electric dipole moment are given by $-e\mathbf{x}_{qq'}$, where $\mathbf{x}$ is the position vector of an electron. The diagonal components of the polarizability (9.28) are then written

$$a_{xx} = \sum_{q,q'} \frac{e^2}{m} \frac{\rho_{qq}^{(0)} f_{qq'}^x}{\omega_{qq'}^2 - \omega^2}, \quad a_{yy} = \sum_{q,q'} \frac{e^2}{m} \frac{\rho_{qq}^{(0)} f_{qq'}^y}{\omega_{qq'}^2 - \omega^2},$$

$$a_{zz} = \sum_{q,q'} \frac{e^2}{m} \frac{\rho_{qq}^{(0)} f_{qq'}^z}{\omega_{qq'}^2 - \omega^2},$$

where

$$f^x_{qq'} = \frac{2m}{\hbar} \, \omega_{qq'} |x_{qq'}|^2, \quad f^y_{qq'} = \frac{2m}{\hbar} \, \omega_{qq'} |y_{qq'}|^2, \quad f^z_{qq'} = \frac{2m}{\hbar} \, \omega_{qq'} |z_{qq'}|^2$$
(9.32)

are the *oscillator strengths*. A formal property of the oscillator strengths is that they satisfy a sum rule: the *Thomas–Reiche–Kuhn sum rule* for a system with Z electrons requires

$$\sum_q f^x_{qq'} = Z,$$
(9.33)

for all values of q'.

The dissipative part of the response is neglected in the foregoing treatment. The antihermitian part of the response tensor is attributed to the finite lifetime of excited states, and this is included in the following approximate way. In the neighborhood of a particular transition, $\omega \approx \omega_{qq'}$ say, the finite lifetime of the two states involved leads to a Lorentzian line profile. Let there be a net decay rate $\Gamma_{qq'}$ which is included in replacing $\omega_{qq'} - \omega$ by $\omega_{qq'} - \omega - i\Gamma_{qq'}/2$ in the first form on the right hand side of (9.30). This replacement leads to a frequency dependence that has the same functional form as for the damping ($\Gamma \neq 0$) in the classical oscillator (9.19) sufficiently close to the natural frequency, and it is only close to the natural frequency that the replacement is a valid approximation.

Exercise Set 9

9.1 Establish the identity (9.8), *viz.*,

$$\frac{\partial^2}{\partial x'_s \partial x'_i} \frac{1}{|\mathbf{x} - \mathbf{x}'|} = -\frac{4\pi}{3} \delta_{is} \delta(\mathbf{x} - \mathbf{x}')$$
$$- \left[\frac{\delta_{is}}{|\mathbf{x} - \mathbf{x}'|^3} - \frac{3(x_i - x'_i)(x_s - x'_s)}{|\mathbf{x} - \mathbf{x}'|^5} \right],$$

in the following way.

(a) For $\mathbf{x}' \neq \mathbf{x}$ establish the identity by direct differentiation.
(b) Argue that the singular term may be treated without loss of generality by assuming $\mathbf{x} = 0$.
(c) Generalize the arguments given in Exercise 1.6 to prove

$$\frac{\partial^2}{\partial x_i \partial x_j} \frac{1}{r} = -\frac{4\pi}{3} \delta_{ij} \delta(\mathbf{x}).$$

9.2 Show that if $\mathbf{P}(\mathbf{x}')$ in (9.7) is independent of $\mathbf{x}'$ on a macroscopic scale then the integral over the non-singular term in (9.8) does not contribute to the integral in (9.7).

9.3 Derive the field (9.6),*viz.*, $\mathbf{E}^{(d)}(\mathbf{x}) = \mathbf{P}(\mathbf{x})/3\varepsilon_0$, by calculating the electric field inside the sphere illustrated in Figure 9.1 as follows.

(a) Set up spherical polar coordinates centered on the center of the sphere and with polar axis ($\theta = 0$) directed along $\mathbf{P}$. Show that the surface charge on the inside of the sphere is $\sigma_P = P \cos \theta$.
(b) Write this surface charge as the charge density $\rho = P \cos \theta \, \delta(r - r_0)$, where r_0 is the radius of the sphere, and use the solution (1.23) of Poisson's equation to find $\mathbf{E}^{(d)}(\mathbf{x}) = -\text{grad} \, \phi(\mathbf{x})$ inside the sphere.
(c) Show that the integral in (1.23) for $\mathbf{x} = 0$, which gives $\mathbf{E}^{(d)}$ at the center of the sphere, implies that $\mathbf{E}^{(d)}$ is directed along the polar axis and is equal in magnitude to

$$E^{(d)} = \frac{1}{4\pi\varepsilon_0} 2\pi \int_{-1}^{1} d\cos\theta \, P\cos^2\theta = \frac{P}{3\varepsilon_0}. \qquad (E9.1)$$

(d) Argue that this value of $\mathbf{E}^{(d)}$ is independent of $\mathbf{x}$.

9.4 A reference work gives the following approximate expression (Cauchy formula) for the refractive index n of dry air at $15\,^\circ\text{C}$ as a function of the wavelength λ in angstroms ($1\,\text{Å} = 10^{-10}\,\text{m}$):

$$(n - 1) \times 10^7 = 2726 + \frac{12.3}{\lambda^2 \times 10^{-8}} + \frac{0.38}{\lambda^4 \times 10^{-16}}.$$

(a) Assuming that the refractive index is the square root of the dielectric constant, find the corresponding approximate expression for $K(\omega)$ as a function of $\omega = 2\pi c/n\lambda$ up to the term proportional to ω^4.

(b) Assume that the leading terms arise from an expansion of the form (9.15) with only one type of oscillator, described by ω_p and ω_0 with no label α. Estimate the values of the parameters ω_p and ω_0.

(c) Assuming that the original formula applies at optical frequencies, is the expansion justified in view of your calculated value of ω_0?

9.5 Consider a system of classical damped oscillators with a dielectric constant of the form (9.19), *viz.*,

$$K(\omega) = 1 + \Omega/(\omega_0 - \omega - i\Gamma),$$

where Ω, ω_0, Γ are constants.

(a) Under what condition does the real part of $K(\omega)$ become negative?

(b) Find the range over which the dispersion is negative, that is, where the real part of $K(\omega)$ is a decreasing function of ω.

9.6 The dielectric constant (9.19) is unacceptable for describing the response at arbitrary frequencies because it does not satisfy the reality condition $K(-\omega) = K^*(\omega)$.

(a) Show that the response function (9.15) for damped classical oscillators does satisfy the reality condition.

(b) Assuming that there is only one oscillator, show that the form (9.19) is obtained by expanding in partial fractions and neglecting the term centered on a negative resonant frequency.

(c) Identify the relation between ω_R in (9.19) and ω_0 in (9.15).

(d) Replot the curves in Figure 9.2 using (9.15) in place of (9.19).

10

Response Tensors for Plasmas

Preamble

The term "plasma" usually means an ionized gas. Many other media show plasma-like responses, especially at high frequencies where even bound electrons respond to an electromagnetic disturbance as though they were free electrons. Metals can exhibit a plasma-like response due to the electrons in the conduction band acting rather like free electrons. The characteristic feature of a plasma is the presence of free or effectively free electrons, whose number density n_e defines a *plasma frequency* $\omega_p = (e^2 n_e / \varepsilon_0 m_e)^{1/2}$, where m_e is the mass of the electron. For example, in laboratory plasma machines the plasma frequency is typically of order 1 GHz, in the ionosphere its maximum value is 3–10 MHz, and in a metal it is at X-ray frequencies.

10.1 The Magnetoionic Theory

One of the earliest theories of the response of a plasma was developed in the early 1930s to describe the propagation of radio waves in the iono-sphere. In this theory the thermal motion of the electrons is neglected, that is, the electrons are assumed *cold* and the effect of the Earth's magnetic field is taken into account. The theory is sometimes referred to as the *Appleton–Hartree theory*, following its development separately by Hartree and Appleton. The more formal name is the *magnetoionic theory*, which includes a now outdated meaning of "ion" that is inconsistent with modern usage where "ion" excludes electrons. The only ionized particles assumed present in the magnetoionic theory are electrons. It is usually assumed that there is a uniform background charge density due to positive ions so that the plasma is charge-neutral, and that these ions play no role in the response. In one sense this is an unnecessary assumption; the theory applies to a non-neutral plasma in which there are no positive ions. In practice the neglect of the motion of the ions is valid only at sufficiently high frequencies, and at lower frequencies the magnetoionic theory needs to be generalized to cold plasma theory (§10.2) in which the motion of the ions is included.

In the magnetoionic theory the electrons are treated as a continuous fluid with number density n_e and fluid velocity $\mathbf{v}$. The equation of fluid motion is

$$m_e \mathrm{d}\mathbf{v}(t)/\mathrm{d}t = -e\big[\mathbf{E}(t) + \mathbf{v}(t) \times \mathbf{B_0}\big] - \nu_e m_e \mathbf{v}, \qquad (10.1)$$

where the spatial dependence of $\mathbf{E}(t)$ is unimportant, and where the subscript 0 is added temporarily to $\mathbf{B_0}$ to emphasize that it is the static field. The final term is a frictional drag assumed to exerted on the electrons by ions, with ν_e the electron–ion collision frequency.

The electromagnetic response of the electron gas may be described in terms of the Fourier transform of the polarization $\mathbf{P}(\omega)$. This is found as follows. The fluid velocity in (10.1) is replaced by $\mathbf{v} = \mathrm{d}\mathbf{X}(t)/\mathrm{d}t$, where $\mathbf{X}(t)$ is the displacement vector of the electron. The total time derivative in (10.1) is replaced according to

$$\mathrm{d}/\mathrm{d}t = \partial/\partial t + \mathbf{v} \cdot \mathrm{grad}, \qquad (10.2)$$

and (10.1) is linearized by omitting the convective derivative term from (10.2). The resulting linearized form of (10.1) is Fourier transformed in time, giving

$$-m_e \omega(\omega + \mathrm{i}\nu_e)\mathbf{X}(\omega) - \mathrm{i}\omega e \mathbf{X}(\omega) \times \mathbf{B} = -e\mathbf{E}(\omega), \qquad (10.3)$$

where the subscript 0 on $\mathbf{B_0}$ is now omitted. The polarization $\mathbf{P}(\omega)$ is defined as the induced dipole moment per unit volume, and hence is identified as

$$\mathbf{P}(\omega) = -en_e \mathbf{X}(\omega), \qquad (10.4)$$

with $\mathbf{X}(\omega)$ determined by the solution of (10.3). It is convenient to introduce the unit vector $\mathbf{b}$ along $\mathbf{B}$. Then, after multiplying by en_e, (10.3) with (10.4) gives

$$\omega(\omega + \mathrm{i}\nu_e)\mathbf{P}(\omega) + \mathrm{i}\omega\Omega_e \mathbf{P}(\omega) \times \mathbf{b} = -\varepsilon_0 \omega_p^2 \mathbf{E}(\omega), \qquad (10.5)$$

where

$$\Omega_e = eB/m_e, \qquad \omega_p = (n_e e^2/\varepsilon_0 m_e)^{1/2} \qquad (10.6)$$

are the electron cyclotron frequency and the electron plasma frequency, respectively.

It is conventional to introduce the *magnetoionic parameters*

$$X = \omega_p^2/\omega^2, \quad Y = \Omega_e/\omega, \quad Z = \nu_e/\omega. \tag{10.7}$$

Then in a coordinate system in which **b** is along the z axis, (10.5) may be written in the matrix form

$$\begin{pmatrix} U & iY & 0 \\ -iY & U & 0 \\ 0 & 0 & U \end{pmatrix} \begin{pmatrix} P_x(\omega) \\ P_y(\omega) \\ P_z(\omega) \end{pmatrix} = -\varepsilon_0 X \begin{pmatrix} E_x(\omega) \\ E_y(\omega) \\ E_z(\omega) \end{pmatrix}, \tag{10.8}$$

with $U = 1+iZ$. On inverting (10.8) one obtains an equation of the form $P_i(\omega) = \varepsilon_0 \chi_{ij}(\omega) E_j(\omega)$, which allows one to identify the susceptibility tensor $\chi_{ij}(\omega)$ and hence to identify the dielectric tensor using (6.17), which implies $K_{ij}(\omega) = \delta_{ij} + \chi_{ij}(\omega)$. The resulting expression for the dielectric tensor is

$$K_{ij}(\omega) = \begin{pmatrix} S & -iD & 0 \\ iD & S & 0 \\ 0 & 0 & P \end{pmatrix}, \tag{10.9}$$

$$S = 1 - \frac{UX}{U^2 - Y^2}, \quad D = -\frac{XY}{U^2 - Y^2}, \quad P = 1 - \frac{X}{U}. \tag{10.10}$$

In the absence of collisions one has $Z = 0$ and $U = 1$, and then the dielectric tensor (10.9) is hermitian. Hence the only source of dissipation in the magnetoionic theory is due to the collisional or frictional term in (10.1).

10.2 Cold Plasmas

The generalization of magnetoionic theory to include the motion of the ions was made by Aström nearly two decades later. In the following treatment of cold plasma theory, in which collisions are neglected, the method used is formally equivalent to that used in §10.1 but it is written and expressed in a different way.

A *cold plasma* is formally defined as an ionized gas in which the effect of thermal motions of the particles is neglected. The plasma is composed of charged particles of various species, usually electrons and one or more species of ions. Let an arbitrary species be labeled α, with $\alpha =$ e for electrons, and $\alpha =$ i for ions. Species α is described by the charge q_α and mass m_α of the particles, and by its number density n_α. These are combined into two natural frequencies for each species: the *plasma frequency* for species α, $\omega_{p\alpha}$, and the *cyclotron frequency* for species α, Ω_α:

$$\omega_{p\alpha} := (q_\alpha^2 n_\alpha / \varepsilon_0 m_\alpha)^{1/2}, \quad \Omega_\alpha := |q_\alpha| B / m_\alpha. \tag{10.11}$$

The *plasma frequency* ω_p is defined by

$$\omega_p^2 := \sum_\alpha \omega_{p\alpha}^2, \tag{10.12}$$

where the sum is over all species α. The plasma frequency is usually close to the electron plasma frequency and the two are often taken to be synonymous. It is also convenient to introduce the sign of the charge for each species:

$$\eta_\alpha := q_\alpha / |q_\alpha|. \tag{10.13}$$

The natural frequencies of the ions are much lower than the natural frequencies of the electrons, with the cyclotron frequency being smaller by the ratio m_e / m_i and the plasma frequency by the square root of this ratio for singly charged ions. In most plasmas of interest the highest natural frequency is the plasma frequency, with the electron cyclotron frequency being comparable with or greater than the plasma frequency only in strongly magnetized plasmas. The ion plasma frequency (or frequencies when there are multiple species of ions) is below the electron cyclotron frequency except in plasmas with weak magnetic fields, in which case the magnetic field can be neglected for many purposes. The ion cyclotron frequencies are the lowest natural frequencies. At frequencies below all its natural frequencies, a plasma may be treated as a magnetized fluid. The relevant fluid theory, for a magnetized, electrically conducting fluid, is called the *magnetohydrodynamic theory* (or hydromagnetic theory), often abbreviated *MHD theory*.

The cold plasma model is also a fluid model, but unlike MHD theory, each species is regarded as an independent fluid. Let $\mathbf{v}_\alpha(t)$ be the fluid velocity for species α. The equation of fluid motion for species α is approximated by its linearized form

$$m_\alpha \mathrm{d}\mathbf{v}_\alpha(t)/\mathrm{d}t = q_\alpha \mathbf{E}(t) + q_\alpha \mathbf{v}_\alpha(t) \times \mathbf{B}, \tag{10.14}$$

where the spatial dependence of $\mathbf{E}(t)$ is unimportant, and where $\mathbf{B}$ is the magnetostatic field. After Fourier transforming in time, converting to tensor notation and moving the final term in (10.14) over to the left hand side, one obtains

$$im_\alpha(\omega \delta_{ij} - i\eta_\alpha \Omega_\alpha \epsilon_{ijk} b_k) v_{\alpha j}(\omega) = -q_\alpha E_i(\omega). \tag{10.15}$$

To solve (10.15) for $\mathbf{v}_\alpha(\omega)$, let us define a tensor $\tau_{ij}^{(\alpha)}(\omega)$ that is the inverse of the tensor $\delta_{ij} - i\eta_\alpha \Omega_\alpha \epsilon_{ijk} b_k / \omega$ on the left hand side of (10.15):

$$(\delta_{ir} - i\eta_\alpha \Omega_\alpha \epsilon_{irk} b_k / \omega)\tau_{rj}^{(\alpha)}(\omega) = \delta_{ij}. \qquad (10.16)$$

Explicit evaluation gives

$$\tau_{ij}^{(\alpha)}(\omega) = \frac{1}{\omega^2 - \Omega_\alpha^2}(\omega^2 \delta_{ij} - \Omega_\alpha^2 b_i b_j + i\eta_\alpha \omega \Omega_\alpha \epsilon_{ijk} b_k)$$

$$= \begin{pmatrix} \omega^2/(\omega^2 - \Omega_\alpha^2) & i\eta_\alpha \omega \Omega_\alpha /(\omega^2 - \Omega_\alpha^2) & 0 \\ -i\eta_\alpha \omega \Omega_\alpha /(\omega^2 - \Omega_\alpha^2) & \omega^2/(\omega^2 - \Omega_\alpha^2) & 0 \\ 0 & 0 & 1 \end{pmatrix}.$$

$$(10.17)$$

The solution is then

$$v_{\alpha i}(\omega) = (iq_\alpha/m_\alpha \omega)\,\tau_{ij}^{(\alpha)}(\omega)E_j(\omega). \qquad (10.18)$$

In this multi-fluid model, the induced current is identified as

$$\mathbf{J}_{\text{ind}}(\omega) = \sum_\alpha q_\alpha n_\alpha \mathbf{v}_\alpha(\omega). \qquad (10.19)$$

On inserting (10.18) into (10.19) one identifies the conductivity tensor and hence the dielectric tensor using (6.17).

The resulting expression for the dielectric tensor is of the form (10.9), specifically,

$$K_{ij}(\omega) = (\delta_{ij} - b_i b_j)\,S(\omega) + b_i b_j\,P(\omega) - i\epsilon_{ijk} b_k\,D(\omega)$$

$$= \begin{pmatrix} S(\omega) & -iD(\omega) & 0 \\ iD(\omega) & S(\omega) & 0 \\ 0 & 0 & P(\omega) \end{pmatrix}. \qquad (10.20)$$

The functions $S(\omega)$ and $D(\omega)$ are given by

$$S(\omega) := \tfrac{1}{2}[R_+(\omega) + R_-(\omega)], \quad D(\omega) := \tfrac{1}{2}[R_+(\omega) - R_-(\omega)],$$

$$R_\pm(\omega) := 1 - \sum_\alpha \frac{\omega_{p\alpha}^2}{\omega^2}\frac{\omega}{\omega \pm \eta_\alpha \Omega_\alpha}, \quad P(\omega) := 1 - \sum_\alpha \frac{\omega_{p\alpha}^2}{\omega^2}.$$

$$(10.21)$$

The dielectric tensor (10.21) reduces to that for the magnetoionic theory, cf. (10.10), when the contributions from the ions are neglected in (10.21) and the effect of collisions is neglected in (10.10).

10.3 Isotropic Thermal Plasmas

The response tensor $K_{ij}(\omega, \mathbf{k})$ for an isotropic medium that is spatially dispersive may be written as a sum of three terms, cf. (7.20), corresponding to the longitudinal, transverse and rotatory parts of the response. The rotatory part of the response is zero for a plasma. (It is non-zero

only in optically active media.) Hence the equivalent dielectric tensor
for an isotropic plasma is of the form

$$K_{ij}(\omega, \mathbf{k}) = K^{\mathrm{L}}(\omega, k)\,\kappa_i\kappa_j + K^{\mathrm{T}}(\omega, k)\,(\delta_{ij} - \kappa_i\kappa_j), \qquad (10.22)$$

where $K^{\mathrm{L}}(\omega, k)$ and $K^{\mathrm{T}}(\omega, k)$ are the *longitudinal* and *transverse* parts,
respectively. These functions $K^{\mathrm{L}}(\omega, k)$ and $K^{\mathrm{T}}(\omega, k)$ are evaluated here
for a *thermal* plasma, that is, one in which the velocity distribution of
particles for each species is Maxwellian.

The response of a thermal plasma was first discussed in detail by Tonks
and Langmuir in 1929. They were interested in wavelike oscillations in
a plasma formed by an electric arc discharge, and they showed that
their theory implied two types of longitudinal waves. These waves are
now called Langmuir waves and ion acoustic (or ion sound) waves, cf.
Chapter 13. The correct treatment of the dissipative part of the response
tensor was not given until much later, specifically by Landau in 1946 as
discussed in §8.2.

Let us assume that each species of particle has a Maxwellian distribution of velocities corresponding to a temperature T_α, with these
temperatures not necessarily being equal. It is convenient to set Boltzmann's constant equal to unity so that T_α has the units of energy. In
laboratory plasma physics it is common to use energy units to express
temperatures: a temperature of $1\,\mathrm{eV}$ corresponds to 1.16×10^4 K. The
temperatures of plasmas vary widely in different contexts. In laboratory
plasma devices, a low temperature is $\approx 1\,\mathrm{eV}$, which is roughly the temperature at which gases become ionized, and the highest temperatures
are $\approx 10\,\mathrm{keV}$, which is roughly the ignition temperature required for
controlled nuclear fusion. In astrophysics, it is traditional to enumerate
temperatures in kelvin. For example, the temperature of the plasma in
the solar photosphere is $\approx 10^4$ K and this rises to $\approx 10^6$ K on average in
the solar corona, with some regions reaching $> 10^7$ K. In a solar flare
the plasma temperature can increase to between 10^8 K and 10^9 K.

The derivation of explicit expressions for the response of a thermal
plasma involves the use of kinetic theory. A derivation is outlined in
§10.4. The result is

$$K^{\mathrm{L}}(\omega, k) = 1 + \sum_\alpha \frac{1}{k^2\lambda_{\mathrm{D}\alpha}^2}\left[1 - \phi(y_\alpha) + \mathrm{i}(\pi)^{1/2}y_\alpha e^{-y_\alpha^2}\right], \,(10.23)$$

$$K^{\mathrm{T}}(\omega, k) = 1 - \sum_\alpha \frac{\omega_{\mathrm{p}\alpha}^2}{\omega^2}\left[\phi(y_\alpha) - \mathrm{i}(\pi)^{1/2}y_\alpha e^{-y_\alpha^2}\right], \qquad (10.24)$$

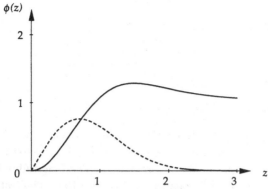

Fig. 10.1 The real part (solid curve) and the imaginary part
(dotted curve) of the plasma dispersion function (10.27).

where

$$\lambda_{\mathrm{D}\alpha} := V_\alpha/\omega_{\mathrm{p}\alpha}, \quad V_\alpha := (T_\alpha/m_\alpha)^{1/2}, \tag{10.25}$$

are the *Debye length* for species α and the *thermal speed* for species α,
respectively. The function $\phi(z)$, with argument

$$y_\alpha = \omega/2^{1/2} k V_\alpha \tag{10.26}$$

here, is a form of the *plasma dispersion function*.

The plasma dispersion function $\overline{\phi}(z)$ is defined by

$$\overline{\phi}(z) := -\frac{z}{(\pi)^{1/2}} \int_{-\infty}^{\infty} \frac{\mathrm{d}t\,\mathrm{e}^{-t^2}}{t - (z + \mathrm{i}0)}. \tag{10.27}$$

Its real part $\phi(z)$ is also given by

$$\phi(z) = 2z\mathrm{e}^{-z^2} \int_0^z \mathrm{d}t\,\mathrm{e}^{t^2}. \tag{10.28}$$

The power series expansion gives

$$\phi(z) = 2z^2 - 4z^4/3 + \cdots \quad \text{for } z^2 \ll 1, \tag{10.29}$$

and the asymptotic expansion gives

$$\phi(z) = 1 + \frac{1}{2z^2} + \frac{3}{4z^4} + \cdots \quad \text{for } z^2 \gg 1. \tag{10.30}$$

The real and imaginary parts of the plasma dispersion function are plotted in Figure 10.1.

In the limit of low frequencies, $\omega \ll k V_\alpha$ for all α, (10.23) with (10.24),
and (10.29) gives

$$K^{\mathrm{L}}(\omega, k) = 1 + \frac{1}{k^2\lambda_{\mathrm{D}}^2}, \quad K^{\mathrm{T}}(\omega, k) = \mathrm{i}\left(\frac{\pi}{2}\right)^{1/2} \sum_\alpha \frac{\omega_{\mathrm{p}}^2}{\omega k V_\alpha} + 1 - \frac{1}{k^2\lambda_{\mathrm{D}}^2}, \tag{10.31}$$

where the *Debye length* λ_D is defined by

$$\lambda_D^{-2} = \sum_\alpha \lambda_{D\alpha}^{-2}. \tag{10.32}$$

At high frequencies, $\omega \gg \omega_p$, (10.22) reduces to

$$K_{ij}(\omega, \mathbf{k}) = K(\omega)\delta_{ij}, \quad K(\omega) = \left(1 - \frac{\omega_p^2}{\omega^2}\right), \tag{10.33}$$

which corresponds to the response of a cold unmagnetized plasma.

The longitudinal and transverse responses in the low frequency limit (10.31) have simple physical interpretations. The longitudinal response describes the Debye shielding effect, cf. Exercise 10.4. The Debye shielding effect causes the potential due to a test charge q, assumed to be at the origin for simplicity, to be modified from the value $\phi(r) = q/4\pi\varepsilon_0 r$ that it would have *in vacuo* to $\phi(r) = (q/4\pi\varepsilon_0 r)e^{-r/\lambda_D}$. This effect is attributed to ambient charges of like sign to the test charge being repelled slightly from it and charges of opposite sign to the test charge being attracted slightly towards it so that over a distance of order λ_D a space charge density corresponding to a total charge $-q$ is present around the test charge and shields out its Coulomb field. This is the simplest example of the self-consistent field in the plasma. The situation is sometimes described in terms of the "bare" field of the test charge being "dressed" by the induced charge density in the plasma to produce the self-consistent field. The presence of the shielding charge can have important physical consequences. For example for Thomson scattering, which is scattering of radiation by electrons, when the wavelength satisfies $\lambda \gtrsim \lambda_D$ the scattering is modified in a major way with the Coulomb fields of electrons being shielded out so that scattering by electrons is ineffective, but with the shielding fields around ions acting somewhat like electrons, so that the scattering is actually attributed to ions.

The transverse part of the response (10.31) diverges at low frequencies. However, the conductivity remains finite, as may be seen by comparing (6.17) and the low-frequency limit of (10.31). The resulting expression for the conductivity is $\sigma = (\pi/2)^{1/2}\sum_\alpha q_\alpha^2 n_\alpha/km_\alpha V_\alpha$. This conductivity is real and, because no collisions are included, it describes a collisionless form of dissipation. The existence of collisionless dissipation caused some controversy in the early development of plasma theory, and a similar controversy arose when the theory of collisionless shock waves was first developed. The underlying question is whether or not a true dissipation can exist in the absence of an obvious randomizing process such as collisions. It is now accepted that collisionless dissipation processes occur widely in plasmas, and are due to a variety of microscopic processes. The dissipation described by the conductivity σ implied by (10.31) is due to Landau damping.

10.4 The Vlasov Equation

The derivation of the response tensor for a thermal plasma involves the use of plasma kinetic theory. As mentioned above, the first detailed derivation was by Tonks and Langmuir in 1929 and the imaginary part was first treated correctly by Landau in 1946. The formal ideas of kinetic theory were discussed by Vlasov in 1938.

The basic equations of kinetic theory are called the *Vlasov equations*. These consist of the following:

(1) A set of equations which is formally identical to Boltzmann's equation for each species of particles:

$$\left\{ \frac{\partial}{\partial t} + \mathbf{v} \cdot \frac{\partial}{\partial \mathbf{x}} + q_\alpha [\mathbf{E}(t, \mathbf{x}) + \mathbf{v} \times \mathbf{B}(t, \mathbf{x})] \cdot \frac{\partial}{\partial \mathbf{p}} \right\} f_\alpha(\mathbf{p}, t, \mathbf{x}) = 0, \quad (10.34)$$

where $f_\alpha(\mathbf{p}, t, \mathbf{x})$ is the single particle distribution function for species α.

(2) A pair of equations giving the charge and current density in terms of the single particle distribution functions:

$$\rho(t, \mathbf{x}) = \sum_\alpha q_\alpha \int d^3\mathbf{p} \, f_\alpha(\mathbf{p}, t, \mathbf{x}),$$

$$\mathbf{J}(t, \mathbf{x}) = \sum_\alpha q_\alpha \int d^3\mathbf{p} \, \mathbf{v} \, f_\alpha(\mathbf{p}, t, \mathbf{x}), \quad (10.35)$$

where the sum is over all species of particle.

(3) Maxwell's equations with the source terms identified with those in (10.35).

Note that $\mathbf{E}$ and $\mathbf{B}$ in the Vlasov equations are the *self-consistent* fields, which are functionals of the distribution functions. Despite its apparent simplicity, the set of equations of the form (10.34) is a set of non-linear coupled integro-differential equations for the distribution functions f_α.

The linear response is found by expanding the distribution functions in powers of the electromagnetic disturbance,

$$f_\alpha(\mathbf{p}, t, \mathbf{x}) = f_\alpha^0(\mathbf{p}) + f_\alpha^{(1)}(\mathbf{p}, t, \mathbf{x}) + \cdots, \quad (10.36)$$

where $f_\alpha^0(\mathbf{p})$ is the unperturbed distribution function. One then linearizes (10.34) and Fourier transforms it. This gives

$$i(\omega - \mathbf{k} \cdot \mathbf{v}) f_\alpha^{(1)}(\mathbf{p}, \omega, \mathbf{k}) = q_\alpha \int d^3 \mathbf{p} \frac{[(\omega - \mathbf{k} \cdot \mathbf{v})\delta_{sj} + k_s v_j]}{\omega}$$

$$\times E_j(\omega, \mathbf{k}) \frac{\partial}{\partial p_s} f_\alpha^0(\mathbf{p}). \tag{10.37}$$

The solution of (10.37) for $f_\alpha^{(1)}(\mathbf{p}, \omega, \mathbf{k})$ is inserted in the linearized and Fourier transformed form of (10.35) to find the induced current as a linear function of $\mathbf{E}$, allowing one to identify the conductivity tensor through (6.15) and the dielectric tensor through (6.17).

The resulting expression is

$$K_{ij}(\omega, \mathbf{k}) = \delta_{ij} + \sum_\alpha \frac{q_\alpha^2}{\varepsilon_0 \omega^2} \int d^3 \mathbf{p} \frac{v_i}{\omega - \mathbf{k} \cdot \mathbf{v}}$$

$$\times [(\omega - \mathbf{k} \cdot \mathbf{v})\delta_{sj} + k_s v_j] \frac{\partial}{\partial p_s} f_\alpha^0(\mathbf{p}). \tag{10.38}$$

For an isotropic medium $f_\alpha^0(\mathbf{p})$ depends only on the magnitude $p := |\mathbf{p}|$ and not on its direction. In this case $K_{ij}(\omega, \mathbf{k})$ must be of the form (10.22). The longitudinal and transverse parts are obtained from (10.38) using

$$K^{\mathrm{L}}(\omega, k) = \kappa_i \kappa_j K_{ij}(\omega, \mathbf{k}), \quad K^{\mathrm{T}}(\omega, k) = \tfrac{1}{2}(\delta_{ij} - \kappa_i \kappa_j) K_{ij}(\omega, \mathbf{k}). \tag{10.39}$$

The next steps in the derivations of (10.23), (10.24) involve making the non-relativistic approximation $\mathbf{p} = m\mathbf{v}$, and substituting the Maxwellian distribution

$$f_\alpha^0(\mathbf{p}) = \frac{n_\alpha e^{-v^2/2V_\alpha^2}}{(2\pi)^{3/2} m_\alpha^3 V_\alpha^3}. \tag{10.40}$$

The resulting explicit expressions are

$$K^{\mathrm{L}}(\omega, k) = 1 - \sum_\alpha \frac{\omega_{p\alpha}^2}{\omega^2 V_\alpha^2} \int d^3 \mathbf{v} \frac{\omega}{\omega - \mathbf{k} \cdot \mathbf{v}}$$

$$\times \left(\frac{\mathbf{k} \cdot \mathbf{v}}{k}\right)^2 \frac{e^{-v^2/2V_\alpha^2}}{(2\pi)^{3/2} V_\alpha^3}, \tag{10.41}$$

$$K^{\mathrm{T}}(\omega, k) = 1 - \sum_\alpha \frac{\omega_{p\alpha}^2}{2\omega^2 V_\alpha^2} \int d^3 \mathbf{v} \frac{\omega}{\omega - \mathbf{k} \cdot \mathbf{v}}$$

$$\times \left[v^2 - \left(\frac{\mathbf{k} \cdot \mathbf{v}}{k}\right)^2\right] \frac{e^{-v^2/2V_\alpha^2}}{(2\pi)^{3/2} V_\alpha^3}. \tag{10.42}$$

The final step is to carry out the integrals in (10.41) and (10.42). The integrals over the components of $\mathbf{v}$ orthogonal to $\mathbf{k}$ are elementary and the integral over the component along $\mathbf{k}$ is performed using the definition (10.27) of the plasma dispersion function.

Kinetic theory may be used to derive the non-linear response tensors defined in §6.4, cf. (6.25). One regards (10.37) as the first order term in a perturbation expansion $(f = f^{(0)} + f^{(1)} + f^{(2)} + \cdots)$ in powers of the amplitude of the electromagnetic field. One solves for the higher order terms in a step by step way as in any perturbation theory. This perturbation is also applied to the current $(\mathbf{J} = \mathbf{J}^{(1)} + \mathbf{J}^{(2)} + \cdots)$ in the Fourier transform of (10.35). The higher order currents then define the non-linear response tensors through (6.25).

Exercise Set 10

10.1 The electron plasma frequency ω_p, the electron cyclotron frequency Ω_e, and the electron Debye length λ_{De} have the following numerical values

$$\omega_p = 5.64 n_e^{1/2}\,\mathrm{s}^{-1}, \quad \Omega_e = 1.76 \times 10^{11} B\,\mathrm{T}, \quad \lambda_{De} = 69 T_e^{1/2} n_e^{-1/2}\,\mathrm{m},$$
$$(E10.1)$$

where n_e is the electron density per cubic meter, B is the magnetic field in tesla, and T_e is the electron temperature in kelvin. Estimate the numerical values of these quantities for the parameters listed in the table below. (Note that Boltzmann's constant corresponds to $1/1.16 \times 10^4\,\mathrm{eV\,K^{-1}}$.)

plasma	n_e	B	T_e
fusion plasma	$10^{20}\,\mathrm{m}^{-3}$	$10\,\mathrm{T}$	$10^4\,\mathrm{eV}$
cool laboratory plasma	$10^{20}\,\mathrm{m}^{-3}$	$3\,\mathrm{T}$	$1\,\mathrm{eV}$
ionosphere	$10^{13}\,\mathrm{m}^{-3}$	$30\,\mu\mathrm{T}$	$10^4\,\mathrm{K}$
solar corona	$10^{10}\,\mathrm{cm}^{-3}$	$30\,\mathrm{G}$	$3 \times 10^6\,\mathrm{K}$
interplanetary plasma	$10^7\,\mathrm{m}^{-3}$	$5\,\mathrm{nT}$	$10^5\,\mathrm{K}$
surface of an X-ray pulsar	$10^{24}\,\mathrm{cm}^{-3}$	$10^{12}\,\mathrm{G}$	$10^8\,\mathrm{K}$

10.2 The following exercise concerns the construction of the tensor $\tau_{rj}^{(\alpha)}(\omega)$, as given by (10.16).

(a) By noting that $\mathbf{b}$ is the only vector in the problem, justify the assumption that $\tau_{rj}^{(\alpha)}(\omega)$ must be of the form

$$\tau_{rj}^{(\alpha)}(\omega) = A(\omega)\delta_{rj} + B(\omega)b_r b_j + C(\omega)\epsilon_{rjk}b_k.$$

(b) Insert this form into

$$(\delta_{ir} - \mathrm{i}\eta_\alpha \Omega_\alpha \epsilon_{irk} b_k/\omega)\tau_{rj}^{(\alpha)}(\omega) = \delta_{ij},$$

and evaluate $A(\omega)$, $B(\omega)$, $C(\omega)$ by equating like terms.

(c) Check the result by writing (10.16) in matrix form and inverting the matrix corresponding to $(\delta_{ir} - \mathrm{i}\eta_\alpha \Omega_\alpha \epsilon_{irk} b_k/\omega)$.

10.3 Consider the response of a charge-neutral cold plasma at low frequencies. The plasma is assumed to be composed of electrons and various species of positive ions with charge $q_i = Z_i e$, mass $m_i = A_i m_{\mathrm{proton}}$ and number density n_i.

(a) Show that the charge neutrality condition $n_e = \sum_i Z_i n_i$ implies

$$\frac{\omega_p^2}{\Omega_e} = \sum_i \frac{\omega_{pi}^2}{\Omega_i}. \qquad (E10.2)$$

(b) With the Alfvén speed defined by $v_A = B/(\mu \rho_M)^{1/2}$, where ρ_M is the mass density, show that if the mass of an electron is neglected compared to that of an ion, then one has

$$\sum_i \frac{\omega_{pi}^2}{\Omega_i^2} = \frac{c^2}{v_A^2}. \qquad (E10.3)$$

(c) Show that in the limit of low frequencies ($\omega \to 0$) that in the form (10.20) with (10.21) for cold plasma dielectric tensor one has

$$S \to 1 + \frac{c^2}{v_A^2}, \quad D \to 0, \quad P \to -\infty. \qquad (E10.4)$$

10.4 The following exercise concerns Debye shielding in a plasma.

(a) Use the Greens function (5.23), *viz.*, $G(\mathbf{x}) = 1/4\pi\varepsilon_0 r$ to show that the potential in the Coulomb gauge due to a charge q at $\mathbf{x} = \mathbf{x}_0$, corresponding to a charge density $\rho(\mathbf{x}) = q\delta^3(\mathbf{x} - \mathbf{x}_0)$, is

$$\phi(\mathbf{x}) = \frac{q}{4\pi\varepsilon_0 |\mathbf{x} - \mathbf{x}_0|}. \qquad (E10.5)$$

(b) Generalize the Greens function in part (a) to include the effect of the medium as follows. Write div $\mathbf{D} = \rho_{\text{ext}}$ for the relation between the electric induction and the extraneous charge density. Fourier transform and use the relation $D_i(\omega, \mathbf{k}) = \varepsilon_0 K_{ij}(\omega, \mathbf{k}) E_j(\omega, \mathbf{k})$ and the explicit form (10.22) with (10.31). Hence show that the generalized Greens function is

$$\tilde{G}(\mathbf{k}) = \frac{1}{\varepsilon_0 (k^2 + \lambda_D^{-2})}. \qquad (E10.6)$$

(c) Show that (E10.6) corresponds to

$$G(\mathbf{x}) = \frac{e^{-r/\lambda_D}}{4\pi\varepsilon_0 r}, \qquad (E10.7)$$

(d) Hence show that (E10.2) is replaced by

$$\phi(\mathbf{x}) = \frac{q e^{-|\mathbf{x} - \mathbf{x}_0|/\lambda_D}}{4\pi\varepsilon_0 |\mathbf{x} - \mathbf{x}_0|}. \qquad (E10.8)$$

10.5 Consider a simple model for Debye shielding in which the plasma particles are described by Boltzmann distributions around a test charge. Let the plasma be charge-neutral, consisting of electrons and one species of singly charged ions with number densities $n_e(\mathbf{x})$ and $n_i(\mathbf{x})$ respectively.

(a)   Show that for a charge q at rest at the origin the potential in the Coulomb gauge is given by

$$\nabla^2\phi(\mathbf{x}) = -\frac{1}{\varepsilon_0}\left[q\delta^3(\mathbf{x}) - en_e(\mathbf{x}) + en_i(\mathbf{x})\right]. \qquad (E10.9)$$

(b)   Justify writing $n_e(\mathbf{x}) = n_0 e^{e\phi(\mathbf{x})/\Theta_e}$ for the electrons and $n_i(\mathbf{x}) = n_0 e^{-e\phi(\mathbf{x})/\Theta_i}$ for the ions.

(c)   Show that for $|e\phi(\mathbf{x})| \ll \Theta_e, \Theta_i$, the potential at $\mathbf{x} \neq 0$ satisfies

$$(\nabla^2 - \lambda_D^{-2})\phi(\mathbf{x}) = 0. \qquad (E10.10)$$

(d)   Show that $\phi(\mathbf{x}) = qe^{-r/\lambda_D}/4\pi\varepsilon_0 r$ is a solution of $(E10.10)$.

10.6 The following exercise concerns the real and imaginary parts of the plasma dispersion function $\phi(z)$ for real z.

(a)   Show that the real part of $\phi(z)$, defined by

$$\mathrm{Re}\left[\phi(z)\right] = -\frac{z}{(\pi)^{1/2}}\,\wp\!\int_{-\infty}^{\infty}\frac{dt\,e^{-t^2}}{t - z}$$

satisfies the differential equation

$$\frac{d\phi(z)}{dz} = \frac{\phi(z)}{z} + 2z[1 - \phi(z)]. \qquad (E10.11)$$

(b)   Show that the solution of $(E10.11)$ that gives the correct value for $\phi(0)$ is

$$\phi(z) = 2ze^{-z^2}\int_0^z dt\,e^{t^2}. \qquad (E10.12)$$

(c)   Show that the imaginary part of $\phi(z)$,

$$\mathrm{Im}\left[\phi(z)\right] = -(\pi)^{1/2}ze^{-z^2}, \qquad (E10.13)$$

also satisfies $(E10.11)$.

10.7 The following exercise concerns the evaluation of the integrals in (10.41) and (10.42) leading to the results (10.23) and (10.24).

(a)   Choose coordinate axes such that the z axis is along $\mathbf{k}$, and carry out the integrals over v_x and v_y in (10.41) and (10.42).

(b)   Use the definition (10.27) to show

$$\int_{-\infty}^{\infty}\frac{dv_z}{(2\pi)^{1/2}V}\frac{e^{-v_z^2/2V^2}}{\omega - kv_z} = \frac{\phi(\omega/2^{1/2}kV)}{\omega}. \qquad (E10.14)$$

(c)   Evaluate the other integral, which differs from $(E10.14)$ in having a factor v_z^2 in the numerator, using (E10.14) and $v_z^2 = [(\omega - kv_z) - \omega]^2/k^2$.

(d) Hence derive (10.23) and (10.24).

10.8 The relativistically correct forms of the fluid equations for a collisionless electron gas with fluid velocity $\mathbf{v}$ and number density n_e are

$$\left(\frac{\partial}{\partial t} + \mathbf{v} \cdot \frac{\partial}{\partial \mathbf{x}}\right) \mathbf{p} = -e\left(\mathbf{E} + \mathbf{v} \times \mathbf{B}\right), \qquad (E10.15)$$

$$\frac{\partial n_e}{\partial t} + \text{div}\,(\mathbf{v}n_e) = 0, \qquad (E10.16)$$

with $\mathbf{p} = m_e\gamma\mathbf{v}$, $\gamma = (1 - v^2/c^2)^{-1/2}$.

(a) Write (E10.15) in the form $D\mathbf{p}/Dt = \mathbf{F}$ and show that this implies $D\mathbf{v}/Dt = (1/m\gamma)\,(\mathbf{F} - \mathbf{v}\,\mathbf{v} \cdot \mathbf{F}/c^2)$, and hence

$$\left(\frac{\partial}{\partial t} + \mathbf{v} \cdot \frac{\partial}{\partial \mathbf{x}}\right) \mathbf{v} = -\frac{e}{m\gamma}\left(\mathbf{E} - \frac{\mathbf{v}\,\mathbf{v} \cdot \mathbf{E}}{c^2} + \mathbf{v} \times \mathbf{B}\right). \qquad (E10.17)$$

(b) Linearize equations (E10.16) and (E10.17) by writing

$$\mathbf{v} = \mathbf{v}_0 + \mathbf{v}^{(1)}, \quad n_e = n_{e0} + n_e^{(1)}, \qquad (E10.18)$$

assuming that the terms with subscript 0 are constant and the terms with superscript (1) are of first order in the amplitude of the electric field, with the magnetic field also of first order.

(c) Fourier transform the linearized equations and find $\mathbf{v}^{(1)}(\omega, \mathbf{k})$ and $n_e^{(1)}(\omega, \mathbf{k})$ in terms of $\mathbf{E}(\omega, \mathbf{k})$.

(d) Show that the first order current density is given by

$$\mathbf{J}^{(1)} = -e\left(n_{e0}\mathbf{v}^{(1)} + n_e^{(1)}\mathbf{v}_0\right). \qquad (E10.19)$$

(e) Insert the solutions from part (b) into (E10.19), hence identify the conductivity tensor $\sigma_{ij}(\omega, \mathbf{k})$, and show that the corresponding dielectric tensor is

$$K_{ij}(\omega, \mathbf{k}) = \delta_{ij} - \frac{\omega_{\mathrm{p}}^2}{\omega^2}\left[\delta_{ij} + \frac{k_i v_{0j} + v_{0i} k_j}{\omega - \mathbf{k} \cdot \mathbf{v}_0} + \frac{(k^2 - \omega^2/c^2)v_{0i}v_{0j}}{(\omega - \mathbf{k} \cdot \mathbf{v}_0)^2}\right],$$
$$(E10.20)$$

with $\omega_{\mathrm{p}}^2 = n_{e0}e^2/\varepsilon_0 m_e\gamma_0$, $\gamma_0 = (1 - v_0^2/c^2)^{-1/2}$.

Remark: If the non-relativistic approximation is made at the outset, by setting $\gamma = 1$ in $\mathbf{p} = m_e\gamma\mathbf{v}$, the resulting expression for $K_{ij}(\omega, \mathbf{k})$ differs from that obtained by setting $\gamma_0 = 1$ in (E10.20) in that the term $(k^2 - \omega^2/c^2)$ is replaced by k^2.

PART THREE

Wave Properties

The formal theory of waves is developed by solving the wave equation. The condition for a solution to exist leads to a dispersion equation, and each specific solution of this equation is called the dispersion relation for a particular wave mode. An arbitrary wave mode is referred to as "the mode M". The polarization vector for the mode M is defined as a unimodular vector along the direction of the electric vector found by solving the wave equation for waves in the mode M. Specific examples of wave modes are discussed for isotropic media, anisotropic crystals and cold magnetized plasmas. Transverse waves in a isotropic medium correspond to two degenerate wave modes, and the description of their polarization is discussed separately.

11

Wave Dispersion and Polarization

Preamble

The properties of waves are found by solving the relevant wave equation. Here we are concerned with plane wave solutions of the electromagnetic equations in the presence of a medium. The wave equation may then be written down in terms of Fourier transforms. The condition for a solution to exist leads to a relation between the frequency ω and wave vector $\mathbf{k}$ called the dispersion relation. Each different dispersion relation defines a different wave mode. Another property needed to describe a wave mode is its polarization vector, which is defined as a unimodular vector along the direction of the electric field in the wave.

11.1 The Wave Equation

By a "wave" is meant here a plane wave solution of Maxwell's equations, with or without the response of a medium depending on the context. An actual wave in a medium is described in terms of a superposition of plane waves through the use of Fourier transforms.

The form of the wave equation adopted as the starting point is (5.8), which follows from the Fourier transformed form of Maxwell's equations in the temporal gauge. Let us repeat the steps in the derivation of this equation. This involves writing the electric and magnetic fields in terms of the vector potential in the temporal gauge:

$$\mathbf{E}(\omega, \mathbf{k}) = i\omega \mathbf{A}(\omega, \mathbf{k}), \quad \mathbf{B}(\omega, \mathbf{k}) = i\mathbf{k} \times \mathbf{A}(\omega, \mathbf{k}). \qquad (11.1)$$

Then Maxwell's equations reduce to the form given by (5.8), *viz.*,

$$\frac{\omega^2}{c^2} \mathbf{A}(\omega, \mathbf{k}) + \mathbf{k} \times [\mathbf{k} \times \mathbf{A}(\omega, \mathbf{k})] = -\mu_0 \mathbf{J}(\omega, \mathbf{k}). \qquad (11.2)$$

The separation $\mathbf{J} = \mathbf{J}_{\mathrm{ind}} + \mathbf{J}_{\mathrm{ext}}$ of the current into induced and extraneous parts is made, and the induced part is rewritten in terms of the response tensor using (6.16), *viz.*,

$$J_{\mathrm{ind}\,i}(\omega, \mathbf{k}) = \alpha_{ij}(\omega, \mathbf{k}) A_j(\omega, \mathbf{k}). \tag{11.3}$$

Then (11.2) reduces to the *inhomogeneous wave equation*

$$\Lambda_{ij}(\omega, \mathbf{k}) A_j(\omega, \mathbf{k}) = -\frac{\mu_0 c^2}{\omega^2} J_{\mathrm{ext}\,i}(\omega, \mathbf{k}), \tag{11.4}$$

with

$$\Lambda_{ij}(\omega, \mathbf{k}) = \frac{|\mathbf{k}|^2 c^2}{\omega^2}(\kappa_i \kappa_j - \delta_{ij}) + K_{ij}(\omega, \mathbf{k}). \tag{11.5}$$

where $\boldsymbol{\kappa} = \mathbf{k}/|\mathbf{k}|$ is a unit vector along $\mathbf{k}$, and where the equivalent dielectric tensor is introduced using (6.17), *viz.*,

$$K_{ij}(\omega, \mathbf{k}) = \delta_{ij} + \frac{1}{\omega^2 \varepsilon_0} \alpha_{ij}(\omega, \mathbf{k}) \tag{11.6}$$

The homogeneous wave equation is obtained by omitting the source terms from (11.4). There is an explicit source term in the extraneous current on the right hand side, and this is omitted. There is also an implicit source term from the dissipative part of the response, which is associated with the antihermitian part of $K_{ij}(\omega, \mathbf{k})$ or of $\Lambda_{ij}(\omega, \mathbf{k})$. Dissipation is associated with a sink of field energy, and sources and sinks are regarded as equivalent. This sink term is also omitted. The *homogeneous wave equation* then becomes

$$\Lambda_{ij}^{\mathrm{H}}(\omega, \mathbf{k}) A_j(\omega, \mathbf{k}) = 0. \tag{11.7}$$

Where no confusion should result, the superscript H, which denotes the hermitian part, is omitted in the following discussion.

11.2 The Dispersion Equation and Dispersion Relations

The homogeneous wave equation (11.7) is regarded as three simultaneous equations for the three cartesian components of $\mathbf{A}(\omega, \mathbf{k})$. This set of equations is equivalent to a set of three real homogeneous equations. Formally one establishes this fact by arguing that, apart from an arbitrary phase factor multiplying $\mathbf{A}(\omega, \mathbf{k})$, the Onsager relations imply that it is possible to choose the relative phases of the components of $\mathbf{A}(\omega, \mathbf{k})$ so that (11.7) reduces to a set of three real homogeneous equations. However, it is not important to rewrite (11.7) explicitly so that all terms are real. It suffices to note that the condition for a solution of three real equations to exist is that the determinant of the coefficients vanish, and that this condition also applies to (11.7).

Thus one regards $\Lambda_{ij}(\omega, \mathbf{k})$ in (11.7) as the matrix of coefficients. The determinant of $\Lambda_{ij}(\omega, \mathbf{k})$ is written as

$$\Lambda(\omega, \mathbf{k}) := \det\left[\Lambda_{ij}(\omega, \mathbf{k})\right], \tag{11.8}$$

where "det" denotes the determinant. The determinant of an hermitian matrix is real, and hence $\Lambda(\omega, \mathbf{k})$ is real here. The reality condition (4.10) for Fourier transforms then implies

$$\Lambda(-\omega, -\mathbf{k}) = \Lambda^*(\omega, \mathbf{k}) = \Lambda(\omega, \mathbf{k}). \tag{11.9}$$

The condition for a solution of the homogeneous wave equation to exist thus reduces to

$$\Lambda(\omega, \mathbf{k}) = 0, \tag{11.10}$$

which is referred to as *the dispersion equation.*

The dispersion equation is an algebraic relation between the frequency ω and the components of the wave vector $\mathbf{k}$. A particular solution of (11.10) defines a particular *wave mode*, and the particular solution is called the *dispersion relation* for that particular wave mode. For the purpose of discussion, an arbitrary wave mode is called "the mode M" here. Thus one refers to the dispersion relation for the mode M.

There is an enormous variety of wave modes in different media, especially in plasmas and to a lesser extent in solids. Unfortunately there is no systematic way of naming the wave modes. Some names are descriptive of the wave (magnetoacoustic mode, extraordinary mode), some are of historical significance (Alfvén mode, Langmuir mode) and some are descriptive of the medium (magnetoionic modes, MHD modes). In some cases there are several different names for the one wave mode (whistler waves in plasma physics and helicon waves in solid state physics). Even "wave mode" itself is used with two distinct meanings: here "wave mode" refers to a type of wave motion, whereas in laser physics, for example, a "mode" refers to an eigenmode of oscillation of the laser. (An eigenmode is specified in terms an integer l_z that specifies one of the allowed wavenumbers $k_z = l_z 2\pi/L_z$ along the axis of the laser of length L_z.) This unfortunate historical legacy cannot be avoided and one simply has to memorize the names of various wave modes.

One solves the dispersion equation in various ways by making different choices of the independent and dependent variables. One choice of independent variable is ω, in which case one solves (11.10) for ω as a function of $\mathbf{k}$. The dispersion relation for the mode M is then written in the form

$$\omega = \omega_M(\mathbf{k}), \tag{11.11}$$

where one has $\Lambda(\omega_M(\mathbf{k}), \mathbf{k}) = 0$. Usually one solves for ω as a function of the magnitude $k := |\mathbf{k}|$ and of the angle or angles used to define the direction $\boldsymbol{\kappa}$ along $\mathbf{k}$. The direction $\boldsymbol{\kappa}$ itself is referred to as the *direction of wave propagation*, the *normal to the wave front* and the *wave normal direction*. Many examples of dispersion relations may be cited from different branches of physics. The following examples are for isotropic media: electromagnetic waves *in vacuo* have $\omega = kc$, sound waves have $\omega = kc_s$ where c_s is the speed of sound, internal gravity waves (of which water waves are a special case) have $\omega = (gk)^{1/2}$ where g is the acceleration due to gravity, and rotons in a superfluid have $\omega = \omega_0 + (k - k_0)^2/2a$ where ω_0, k_0 and a are constants. In the case of an isotropic medium the dispersion relations are independent of the direction $\boldsymbol{\kappa}$ of wave propagation. In an anisotropic medium such as a uniaxial crystal or a magnetized plasma there is one fixed direction in the medium and the dispersion relations depend on the angle between $\boldsymbol{\kappa}$ and this fixed direction in general. In a biaxial crystal there are two independent fixed directions and the dispersion relations depend on two angles in general.

Another choice of independent variable that is often made in electromagnetic theory is the *refractive index* $n = kc/\omega$. One usually solves for the square of the refractive index as a function of ω and of appropriate angles. In this case the dispersion relation for waves in the mode M is written

$$\frac{k^2 c^2}{\omega^2} = n_M^2(\omega, \boldsymbol{\kappa}). \tag{11.12}$$

The relation (11.9) implies that there are pairs of positive-frequency and negative-frequency solutions. One is free to choose the solutions in the forms (11.11) and (11.12) such that the following relations are satisfied:

$$\omega_M(-\mathbf{k}) = -\omega_M(\mathbf{k}), \quad n_M^2(-\omega, -\boldsymbol{\kappa}) = n_M^2(\omega, \boldsymbol{\kappa}). \tag{11.13}$$

In principle, one should distinguish between forward-propagating waves and backward-propagating waves by regarding them as separate modes, $M+$ and $M-$ say. However, this is only necessary for some formal purposes, and for most purposes this subtlety may be ignored.

11.3 Polarization Vectors

When the dispersion relation for the wave mode M is satisfied, there exists a solution of the homogeneous wave equation. One then solves the wave equation (11.7) for the vector potential $\mathbf{A}$ or for the electric field $\mathbf{E}$. However, the magnitude and phase of the solution are arbitrary, and so one can solve only for the relative magnitudes and phases of the three components of either vector. It is convenient to require that the solution be normalized to unity, so that one solves for a unimodular vector along either $\mathbf{A}$ or $\mathbf{E}$. This solution is called the *polarization vector* for the mode M. The polarization vector characterizes the directions and relative phases of the fields in the waves. Given the polarization vector, one has the relative magnitudes and phases of the components of the electric field, and the same information on the magnetic field in a wave in the mode M then follows from (11.1). Similarly, the same information on the induced current in a wave in the mode M follows by inserting the polarization vector in place of $\mathbf{A}$ in (11.3).

In the case where the components of $\mathbf{k}$ are chosen as the independent variables, the polarization vector for the mode M is $\mathbf{e}_M(\mathbf{k})$:

$$\mathbf{e}_M(\mathbf{k}) = \frac{\mathbf{A}(\omega_M(\mathbf{k}), \mathbf{k})}{|\mathbf{A}(\omega_M(\mathbf{k}), \mathbf{k})|}. \qquad (11.14)$$

The polarization vector is complex in general and its normalization condition is

$$\mathbf{e}_M(\mathbf{k})\mathbf{e}_M^*(\mathbf{k}) = 1. \qquad (11.15)$$

In general a polarization vector has both longitudinal and transverse parts. A strictly *longitudinal* polarization vector corresponds to $\mathbf{e} = \boldsymbol{\kappa}$. Waves whose polarization vector is longitudinal are called *longitudinal waves*. Examples of longitudinal waves that are important in plasma physics are Langmuir waves and ion sound waves. A *transverse* polarization vector corresponds to $\mathbf{e} \cdot \boldsymbol{\kappa} = 0$. Waves whose polarization vector is strictly transverse are called *transverse waves*. In general a transverse polarization vector is elliptical; circular and linear (or planar) polarization are special cases.

A formal procedure for constructing the polarization vector follows the standard procedure for solving a set of simultaneous algebraic equations. When the determinant of the coefficients is zero, solutions are found by constructing the matrix of cofactors. Let the cofactor of $\Lambda_{ij}(\omega, \mathbf{k})$ be $\lambda_{ji}(\omega, \mathbf{k})$. By definition the cofactors satisfy

$$\Lambda_{ij}(\omega, \mathbf{k})\lambda_{jl}(\omega, \mathbf{k}) = \delta_{il}\Lambda(\omega, \mathbf{k}). \qquad (11.16)$$

One may write (11.16) in the matrix form

$$\begin{pmatrix} \Lambda_{11} & \Lambda_{12} & \Lambda_{13} \\ \Lambda_{21} & \Lambda_{22} & \Lambda_{23} \\ \Lambda_{31} & \Lambda_{32} & \Lambda_{33} \end{pmatrix} \begin{pmatrix} \lambda_{11} & \lambda_{12} & \lambda_{13} \\ \lambda_{21} & \lambda_{22} & \lambda_{23} \\ \lambda_{31} & \lambda_{32} & \lambda_{33} \end{pmatrix} = \begin{pmatrix} \Lambda & 0 & 0 \\ 0 & \Lambda & 0 \\ 0 & 0 & \Lambda \end{pmatrix}, \qquad (11.17)$$

where the arguments $(\omega, \mathbf{k})$ are omitted for simplicity in writing. For $\Lambda = 0$ a vector with components equal to the elements of any of the columns of the matrix λ_{ij} is a solution of $\Lambda_{ij} A_j = 0$. This implies that $\lambda_{ij}(\omega_M(\mathbf{k}), \mathbf{k})$ is proportional to $e_{Mi}(\mathbf{k})$. The fact that $\lambda_{ij}(\omega_M(\mathbf{k}), \mathbf{k})$ is hermitian then implies that it is also proportional to $e^*_{Mj}(\mathbf{k})$:

$$\lambda_{ij}(\omega_M(\mathbf{k}), \mathbf{k}) \propto e_{Mi}(\mathbf{k}) e^*_{Mj}(\mathbf{k}).$$

The constant of proportionality is determined by taking the trace of this equation and noting the normalization condition (11.15). This gives

$$\lambda_{ij}(\omega_M(\mathbf{k}), \mathbf{k}) = \lambda_{ss}(\omega_M(\mathbf{k}), \mathbf{k}) e_{Mi}(\mathbf{k}) e^*_{Mj}(\mathbf{k}). \tag{11.18}$$

The foregoing discussion leads to the following prescription for constructing the polarization vector for waves in the mode M:

(i)   Construct the matrix of cofactors $\lambda_{ij}(\omega, \mathbf{k})$.

(ii) Insert the dispersion relation for waves in the mode M, for example, in the form $\omega = \omega_M(\mathbf{k})$.

(iii) Choose any column of $\lambda_{ij}(\omega_M(\mathbf{k}), \mathbf{k})$ and identify it as the components of a solution of the wave equation.

(iv) Normalize this solution using (11.14) to find the polarization vector.

This defines the polarization vector to within an arbitrary phase factor.

As noted above, positive-frequency and negative-frequency solutions appear in pairs, with their dispersion relations related by (11.13). The reality condition and the hermitian property imply

$$\lambda_{ij}(-\omega, -\mathbf{k}) = \lambda_{ji}(\omega, \mathbf{k}),$$

and then (11.18) leads to the relation

$$\mathbf{e}_M(-\mathbf{k}) = \mathbf{e}^*_M(\mathbf{k}) \tag{11.19}$$

between the polarization vectors for the solutions at positive and negative frequencies.

11.4 Damping of Waves

Damping of waves is due to dissipative processes in the medium. Dissipation is treated by including the antihermitian part $K^A_{ij}(\omega, \mathbf{k})$ of the dielectric tensor. Wave damping and energetics are discussed in detail in Chapter 15. The following treatment is somewhat heuristic.

Damping is treated by allowing the frequency to have an imaginary part $i\omega_I$. The amplitude of the wave then has a secular variation as $e^{\omega_I t}$. It is convenient to define the *absorption coefficient* $\gamma_M(\mathbf{k})$ for waves in the mode M such that the energy in the waves damps as $e^{-\gamma_M(\mathbf{k})t}$. The energy in the waves is proportional to the square of the wave amplitude, and hence one makes the identification $\gamma_M(\mathbf{k}) = -2\omega_{\mathrm{I}}$.

The imaginary parts of the frequency are found as follows. In the dispersion equation (11.10), make the replacements $\omega \rightarrow \omega + i\omega_{\mathrm{I}}$ and $\Lambda_{ij}^{\mathrm{H}} \rightarrow \Lambda_{ij}^{\mathrm{H}} + K_{ij}^{\mathrm{A}}$. Then expand the determinant (11.8) to first order in the dissipative terms. The zeroth order term reproduces (11.10) and then the first order terms relate ω_{I} to K_{ij}^{A}. The expansion gives

$$\det\left[\Lambda_{ij}^{\mathrm{H}}(\omega + i\omega_{\mathrm{I}}, \mathbf{k}) + K_{ij}^{\mathrm{A}}(\omega + i\omega_{\mathrm{I}}, \mathbf{k})\right] = \Lambda(\omega, \mathbf{k})$$
$$+ i\omega_{\mathrm{I}}\frac{\partial}{\partial\omega}\Lambda(\omega, \mathbf{k}) + \lambda_{ji}(\omega, \mathbf{k})K_{ij}^{\mathrm{A}}(\omega, \mathbf{k}) + \cdots. \quad (11.20)$$

Thus one finds

$$\omega_{\mathrm{I}} = \frac{i\lambda_{ji}(\omega, \mathbf{k})K_{ij}^{\mathrm{A}}(\omega, \mathbf{k})}{\partial\Lambda(\omega, \mathbf{k})/\partial\omega}. \quad (11.21)$$

Equation (11.21) is to be evaluated at a solution of $\Lambda(\omega, \mathbf{k}) = 0$, and we consider the solution $\omega = \omega_M(\mathbf{k})$ for waves in the mode M. Then on using (11.18) and making the identification $\gamma_M(\mathbf{k}) = -2\omega_{\mathrm{I}}$, (11.21) reduces to

$$\gamma_M(\mathbf{k}) = -2i\omega_M(\mathbf{k})R_M(\mathbf{k})e_{Mi}^*(\mathbf{k})e_{Mj}(\mathbf{k})K_{ij}^{\mathrm{A}}(\omega, \mathbf{k}), \quad (11.22)$$

with

$$R_M(\mathbf{k}) = \left[\frac{\lambda_{ss}(\omega, \mathbf{k})}{\omega\partial\Lambda(\omega, \mathbf{k})/\partial\omega}\right]\Bigg|_{\omega = \omega_M(\mathbf{k})}. \quad (11.23)$$

The meaning and significance of $R_M(\mathbf{k})$ is discussed in Chapter 15, where it is argued that it is to be interpreted as the ratio of the electric to total energy in the waves. The physical significance of the result (11.23) is also discussed in Chapter 15.

11.5 Explicit Forms for $\Lambda(\omega, \mathbf{k})$ and $\lambda_{ij}(\omega, \mathbf{k})$

For both formal and practical purposes it is convenient to have explicit expressions for $\Lambda(\omega, \mathbf{k})$ and $\lambda_{ij}(\omega, \mathbf{k})$. First we write down relations for these quantities in terms of $\Lambda_{ij}(\omega, \mathbf{k})$. Then we introduce the explicit form (11.5) in terms of the equivalent dielectric tensor $K_{ij}(\omega, \mathbf{k})$ and write down expressions for $\Lambda(\omega, \mathbf{k})$ and $\lambda_{ij}(\omega, \mathbf{k})$ in terms of $K_{ij}(\omega, \mathbf{k})$. For convenience in writing the arguments $(\omega, \mathbf{k})$ are omitted in this section.

The construction of the determinant Λ and of the matrix of cofactors λ_{ij} of Λ_{ij} is straightforward using matrix theory. In tensor notation formal relations between these quantities are as follows:

$$\Lambda = \tfrac{1}{6}\epsilon_{abc}\epsilon_{ijl}\Lambda_{ia}\Lambda_{jb}\Lambda_{lc}, \tag{11.24}$$

$$\lambda_{ij} = \tfrac{1}{2}\epsilon_{iab}\epsilon_{jrs}\Lambda_{ra}\Lambda_{sb}, \tag{11.25}$$

An alternative form of (11.24) is

$$\Lambda\epsilon_{abc} = \epsilon_{ijl}\Lambda_{ia}\Lambda_{jb}\Lambda_{lc}. \tag{11.26}$$

The equivalence of (11.24) and (11.26) follows using (2.26). Another formal relation is

$$\lambda_{ij}\lambda_{rs} = \lambda_{is}\lambda_{rj} + \Lambda\epsilon_{ira}\epsilon_{jsb}\Lambda_{ba}. \tag{11.27}$$

This relation for $\Lambda = 0$ is interpreted in matrix language as implying that λ_{ij} is of rank one. This follows from three facts. First, the *rank* of a matrix is the highest order of its minors (that is, of submatrices obtained by deleting an equal number of rows and columns, the order being the remaining number of rows or columns) which has a non-vanishing determinant. Second, $\lambda_{ij}\lambda_{rs} - \lambda_{is}\lambda_{rj}$ is the determinant of a minor of λ_{ij} of order two. Thus for $\Lambda = 0$, Λ_{ij} is of rank two, and λ_{ij} is of rank one. Third, a matrix of rank one may be written as the outer product of a column matrix with its (hermitian) transpose. This column matrix, after normalization, gives the polarization vector here, cf. (11.18).

With Λ_{ij} given in terms of the dielectric tensor by, cf. (11.5),

$$\Lambda_{ij} = n^2(\kappa_i\kappa_j - \delta_{ij}) + K_{ij}, \tag{11.28}$$

one has

$$\Lambda = n^4\kappa_i\kappa_j K_{ij} - n^2(\kappa_i\kappa_j K_{ij}K_{ss}$$
$$- \kappa_i\kappa_j K_{is}K_{sj}) + \det[K_{ij}], \tag{11.29}$$

$$\lambda_{ij} = n^4\kappa_i\kappa_j - n^2(\kappa_i\kappa_j K_{ss} + \delta_{ij}\kappa_r\kappa_s K_{rs} - \kappa_i\kappa_s K_{sj} - \kappa_j\kappa_s K_{is})$$
$$+ \tfrac{1}{2}\delta_{ij}[(K_{ss})^2 - K_{rs}K_{sr}] + K_{is}K_{sj} - K_{ss}K_{ij}. \tag{11.30}$$

A specific case that is important in practice is an anisotropic medium with only one natural direction, such as a uniaxial crystal or a magnetized medium. In this case one is free to choose the axes such that this characteristic direction in the medium is along the 3 axis and such that $\mathbf{k}$ is in the 1–3 plane at an angle θ to the 3 axis. This corresponds to

$$\boldsymbol{\kappa} = (\sin\theta, 0, \cos\theta). \tag{11.31}$$

For a magnetized medium the Onsager relations imply the symmetry properties $K_{21} = -K_{12}$, $K_{31} = K_{13}$, $K_{32} = -K_{23}$, cf. (7.15). In this case one has

$$\Lambda_{ij} = \begin{pmatrix} K_{11} - n^2 \cos^2 \theta & K_{12} & K_{13} + n^2 \sin \theta \cos \theta \\ -K_{12} & K_{22} - n^2 & K_{23} \\ K_{13} + n^2 \sin \theta \cos \theta & -K_{23} & K_{33} - n^2 \sin^2 \theta \end{pmatrix}.$$

(11.32)

This form also applies to uniaxial and biaxial crystals provided the coordinates axes are oriented along the principal axes, in which case the off-diagonal elements of K_{ij} are equal to zero. The determinant of (11.32) reduces to

$$\Lambda = An^4 - Bn^2 + C,$$

(11.33)

with

$$A = K_{11} \sin^2 \theta + K_{33} \cos^2 \theta + 2K_{13} \sin \theta \cos \theta,$$
$$B = AK_{22} + K_{11}K_{33} - K_{13}^2 + \left(K_{12} \sin \theta - K_{23} \cos \theta \right)^2,$$
$$C = K_{11}K_{22}K_{33} + K_{11}K_{23}^2 + K_{33}K_{12}^2 - K_{22}K_{13}^2 + 2K_{12}K_{23}K_{13}.$$

(11.34)

Note that (11.29) is a quadratic equation for n^2 only if the medium is not spatially dispersive; for a spatially dispersive medium, the coefficients A, B, and C are functions of k and hence of n through the dependence of K_{ij} on $\mathbf{k}$.

The matrix of cofactors is

$\lambda_{ij} =$

$$\begin{pmatrix} n^4 \sin^2 \theta - n^2(K_{22} \sin^2 \theta + K_{33}) + K_{22}K_{33} + K_{23}^2 \\ -n^2 \sin \theta(K_{12} \sin \theta - K_{23} \cos \theta) + K_{12}K_{33} + K_{23}K_{13} \\ n^4 \sin \theta \cos \theta - n^2(K_{22} \sin \theta \cos \theta - K_{13}) + K_{12}K_{23} - K_{13}K_{22} \end{pmatrix}$$

$$\begin{matrix} n^2 \sin \theta(K_{12} \sin \theta - K_{23} \cos \theta) - K_{12}K_{33} - K_{23}K_{13} \\ -n^2(K_{11} \sin^2 \theta + K_{33} \cos^2 \theta + 2K_{13} \sin \theta \cos \theta) + K_{11}K_{33} - K_{13}^2 \\ n^2 \cos \theta(K_{12} \sin \theta - K_{23} \cos \theta) + K_{11}K_{23} + K_{12}K_{13} \end{matrix}$$

$$\begin{pmatrix} n^4 \sin \theta \cos \theta - n^2(K_{22} \sin \theta \cos \theta - K_{13}) + K_{12}K_{23} - K_{13}K_{22} \\ -n^2 \cos \theta(K_{12} \sin \theta - K_{23} \cos \theta) - K_{11}K_{23} - K_{12}K_{13} \\ n^4 \cos^2 \theta - n^2(K_{11} + K_{22} \cos^2 \theta) + K_{11}K_{22} + K_{12}^2 \end{pmatrix}.$$

(11.35)

Exercise Set 11

11.1 In an isotropic plasma the dielectric tensor is of the form

$$K_{ij}(\omega, \mathbf{k}) = K^{L}(\omega, k)\kappa_i\kappa_j + K^{T}(\omega, k)(\delta_{ij} - \kappa_i\kappa_j).$$

The tensors $\Lambda_{ij}(\omega, \mathbf{k})$ and $\lambda_{ij}(\omega, \mathbf{k})$ are separated into longitudinal and transverse parts in similar manner.

(a)   Using the relation (11.28), show $\Lambda^{L} = K^{L}$, $\Lambda^{T} = K^{T} - n^2$.

(b)   Show $\Lambda = \Lambda^{L}(\Lambda^{T})^{2}$.

(c)   By separating λ_{ij} into longitudinal and transverse parts, and using $\Lambda_{ij}\lambda_{jl} = \Lambda\delta_{il}$, show $\lambda_{ij} = \Lambda^{T}[\Lambda^{T}\kappa_i\kappa_j + \Lambda^{L}(\delta_{ij} - \kappa_i\kappa_j)]$.

11.2 The determinant of the product of two square matrices is equal to the product of their determinants.

(a)   Use (11.24) and (11.26) with $\Lambda_{ij} = A_{ir}B_{rj}$ to prove this result for 3×3 matrices.

(b)   Hence or otherwise prove that the determinant of λ_{ij} is equal to Λ^2.

Remark: Note that for $\Lambda = 0$, λ_{ij} has a double zero, so that Λ_{ij} is of rank two (or less) and λ_{ij} is of rank one (or zero).

11.3 The following problem concerns a derivation of (11.27), *viz.*,

$$\lambda_{ij}\lambda_{rs} = \lambda_{is}\lambda_{rj} + \Lambda\epsilon_{ira}\epsilon_{jsb}\Lambda_{ba}.$$

(a)   Show that an alternative definition to (11.25) for the cofactor λ_{ij} of Λ_{ji} is

$$\Lambda_{ij}\Lambda_{rs} - \Lambda_{is}\Lambda_{rj} = \epsilon_{ira}\epsilon_{jsb}\lambda_{ba}. \qquad (E11.1)$$

(b)   Show that the cofactor L_{ij} of λ_{ji} is

$$L_{ij} = \Lambda_{ij}\Lambda. \qquad (E11.2)$$

Hint: You may need to use the fact that the determinant of λ_{ij} is equal to Λ^2.

(c)   Apply (E11.1) to λ_{ij} and use (E11.2) to derive (11.27).

11.4 Use (11.27) to prove that for $\Lambda = 0$ all three columns of λ_{ij} are proportional to each other. Specifically, prove the relations (for $\Lambda = 0$)

$$\lambda_{11} : \lambda_{21} : \lambda_{31} = \lambda_{12} : \lambda_{22} : \lambda_{32} = \lambda_{13} : \lambda_{23} : \lambda_{33}.$$

11.5 Show

$$\frac{\partial \Lambda}{\partial \Lambda_{ij}} = \lambda_{ji}, \quad \frac{\partial \lambda_{ij}}{\partial \Lambda_{rs}} = \epsilon_{isa}\epsilon_{jrb}\Lambda_{ba}.$$

11.6 The *Cayley–Hamilton theorem* is that a matrix satisfies its own characteristic equation. The characteristic equation for Λ_{ij} is

$$F_\Lambda(x) = \det[\Lambda_{ij} - x\delta_{ij}] = 0. \qquad (E11.3)$$

The Cayley–Hamilton theorem is then given by replacing x by Λ_{ij}, x^2 by $\Lambda_{ia}\Lambda_{aj}$, and so on, in $F_\Lambda(x) = 0$.

(a)　Show that the Cayley–Hamilton theorem implies that

$$\Lambda\delta_{ij} - \lambda_{ss}\Lambda_{ij} + \Lambda_{ss}\Lambda_{ia}\Lambda_{aj} - \Lambda_{ia}\Lambda_{ab}\Lambda_{bj} = 0. \qquad (E11.4)$$

(b)　Use $(E11.4)$ and $\Lambda_{ij}\lambda_{jl} = \Lambda\delta_{il}$ to derive the identities

$$\Lambda = \tfrac{1}{3}\Lambda_{sa}\Lambda_{ab}\Lambda_{bs} - \tfrac{1}{2}\Lambda_{ss}\Lambda_{ab}\Lambda_{ba} + \tfrac{1}{6}\left(\Lambda_{ss}\right)^3, \qquad (E11.5)$$

$$\lambda_{ij} = \tfrac{1}{2}\delta_{ij}\left[\left(\Lambda_{ss}\right)^2 - \Lambda_{sa}\Lambda_{as}\right] + \Lambda_{ia}\Lambda_{aj} - \Lambda_{ss}\Lambda_{ij}. \qquad (E11.6)$$

(c)　Insert the form (11.28) into $(E11.5)$ and $(E11.6)$ and hence derive (11.29) and (11.30).

11.7 Consider two different natural modes $M1$, $M2$ in a given medium in which dissipation is neglected. Let the dispersion relations be $\omega_{M1}(\mathbf{k})$, $\omega_{M2}(\mathbf{k})$ and the polarization vectors be $\mathbf{e}_{M1}(\mathbf{k})$, $\mathbf{e}_{M2}(\mathbf{k})$. The following problem emphasizes that the "orthogonality" of natural modes does not imply $\mathbf{e}_{M2}^*(\mathbf{k}) \cdot \mathbf{e}_{M1}(\mathbf{k}) = 0$.

(a)　Using an argument similar to that associated with the interpretation of (7.5), show that the presence of waves in one mode leads to no work being done in exciting waves in the other mode only if the following "orthogonality" relation is satisfied:

$$e_{M2i}^*(\mathbf{k})\left[K_{ij}^{\mathrm{H}}(\omega,\mathbf{k}) - \delta_{ij}\right]e_{M1j}(\mathbf{k}) = 0. \qquad (E11.7)$$

(b)　Use $(E11.7)$ and the wave equation to show

$$(n^2 - 1)\mathbf{e}_{M2}^*(\mathbf{k}) \cdot \mathbf{e}_{M1}(\mathbf{k}) = n^2 \boldsymbol{\kappa} \cdot \mathbf{e}_{M1}(\mathbf{k})\,\boldsymbol{\kappa} \cdot \mathbf{e}_{M2}(\mathbf{k}). \qquad (E11.8)$$

11.8 The dissipative part of the response tensor is usually neglected in deriving dispersion relations and polarization vectors. The following problem concerns solutions of the wave equation with the dissipative part included.

(a)　Show that the reality condition requires

$$\Lambda_{ij}(\omega,\mathbf{k}) = \Lambda_{ij}^*(-\omega^*, -\mathbf{k}^*).$$

(b)　Consider a solution of $\Lambda(\omega,\mathbf{k}) = 0$ for $\omega = \omega_M(\mathbf{k})$ found by searching the complex ω plane for real $\mathbf{k}$. Show that there must be a second solution corresponding to $-\mathbf{k}$ with $\omega = -\omega_M^*(\mathbf{k})$, so that

the real part of the frequency is opposite to that of the given solution, and the imaginary part of the frequency is equal to that of
the given solution.

(c)   Show that in the absence of dissipation the polarization vector may
be chosen to satisfy $\mathbf{e}_M(\mathbf{k}) = \mathbf{e}_M^*(-\mathbf{k})$.

Remark: In the absence of dissipation the Onsager relations imply that
the components of $\mathbf{e}_M(\mathbf{k})$ may be chosen either real or imaginary; in the
presence of dissipation the polarization vector is intrinsically complex,
and the modes are not orthogonal.

11.9 Justify the following statement: For an arbitrary wave the vec-
tors $\mathbf{D}(\omega, \mathbf{k})$ and $\mathbf{B}(\omega, \mathbf{k})$ are strictly transverse, but $\mathbf{E}(\omega)$ may have a
longitudinal component in general.
Hint: Identify $D_i(\omega, \mathbf{k})$ as $\varepsilon_0 K_{ij}(\omega, \mathbf{k})E_j(\omega, \mathbf{k})$ and appeal to the wave
equation.

12

Waves in Anisotropic Crystals

Preamble

The general theory of wave properties is applied to waves in isotropic and anisotropic dielectrics.

12.1 Waves in Isotropic Dielectrics

For an isotropic dielectric which is not spatially dispersive, the equivalent dielectric tensor reduces to a dielectric constant:

$$K_{ij}(\omega) = K(\omega)\delta_{ij}. \tag{12.1}$$

In this case the dispersion equation in the form (11.10) reduces to

$$K(\omega)\left[n^2 - K(\omega)\right]^2 = 0. \tag{12.2}$$

This leads to the familiar dispersion relation

$$n = [K(\omega)]^{1/2} \tag{12.3}$$

for transverse waves in an isotropic dielectric.

The polarization of transverse waves is considerably more complicated to treat than the polarization of waves in an anisotropic medium. This is due to the transverse wave mode being two degenerate wave modes. It is usually convenient to think of these as two degenerate linearly polarized modes, with polarization vectors in the x and y directions when $\mathbf{k}$ is along the z axis. A general polarization vector for a transverse wave may then be constructed from a linear combination (with complex coefficients) of these two orthogonal polarizations. Such a polarization vector describes an elliptical polarization. However, the most general state of the polarization of transverse waves cannot be described by a single polarization vector. For example, unpolarized radiation cannot be described by a polarization vector. The polarization of transverse waves is discussed in detail in Chapter 14.

If spatial dispersion is included, the general form of the equivalent dielectric tensor for an isotropic medium is given by (7.20), *viz.*,

$$K_{ij}(\omega, \mathbf{k}) = K^{\mathrm{L}}(\omega, k)\kappa_i\kappa_j + K^{\mathrm{T}}(\omega, k)(\delta_{ij} - \kappa_i\kappa_j) + \mathrm{i}K^{\mathrm{R}}(\omega, k)\epsilon_{ijl}\kappa_l,$$
(12.4)

where $K^{\mathrm{L}}(\omega, k)$, $K^{\mathrm{T}}(\omega, k)$ and $K^{\mathrm{R}}(\omega, k)$ are the longitudinal, transverse and rotatory parts. The rotatory part is non-zero in an optically active medium; waves in such media are discussed below following the discussion of waves in anisotropic crystals. In the absence of the rotatory part, the dispersion equation reduces to

$$K^{\mathrm{L}}(\omega, k)\left[n^2 - K^{\mathrm{T}}(\omega, k)\right]^2 = 0.$$
(12.5)

There are two types of solution of (12.5): solutions of the *longitudinal dispersion equation*

$$K^{\mathrm{L}}(\omega, k) = 0,$$
(12.6)

and solutions of the *transverse dispersion equation*

$$n^2 = K^{\mathrm{T}}(\omega, k).$$
(12.7)

The solutions (12.6) and (12.7) describe longitudinal waves and transverse waves, respectively. The *longitudinal* part of the polarization is the component along $\mathbf{k}$ and the *transverse* part is the part orthogonal to $\mathbf{k}$. It is only in an isotropic medium that all polarizations are either strictly longitudinal or strictly transverse.

The polarization vector for longitudinal waves is

$$\mathbf{e}_{\mathrm{L}}(\mathbf{k}) = \boldsymbol{\kappa}.$$
(12.8)

Examples of longitudinal waves are familiar in plasma physics. One example is Langmuir waves, which are essentially longitudinal electron plasma oscillations, and another example is ion sound waves, which are longitudinal waves like sound waves except that the restoring force is due to the electric field rather than to a perturbation in the gas pressure. The properties of these waves are discussed in §13.1.

An illustration of the use of the equivalent dielectric tensor and of the foregoing formalism is the following application. A particular model of an isotropic medium is one with a dielectric permittivity $\varepsilon(\omega)$ and a magnetic permeability $\mu(\omega)$, that is, with a response described by $\mathbf{D}(\omega) = \varepsilon(\omega)\mathbf{E}(\omega)$ and $\mathbf{H}(\omega) = \mathbf{B}(\omega)/\mu(\omega)$. Then from (6.18) one obtains

$$K^{\mathrm{L}}(\omega, k) = \frac{\varepsilon(\omega)}{\varepsilon_0}, \quad K^{\mathrm{T}}(\omega, k) = \frac{\varepsilon(\omega)}{\varepsilon_0} + \frac{k^2 c^2}{\omega^2}\left[1 - \frac{\mu_0}{\mu(\omega)}\right]. \quad (12.9)$$

In this case the solution (12.7) gives

$$n^2 = \frac{\varepsilon(\omega)\mu(\omega)}{\varepsilon_0 \mu_0} = \varepsilon(\omega)\mu(\omega)c^2. \quad (12.10)$$

The interesting point is to compare this derivation of the dispersion relation with the derivation within the framework of the older theory based on separate electric and magnetic responses. This point is pursued in Exercise 12.1.

12.2 Waves in Uniaxial Crystals

As discussed in §7.4, crystals are classified as isotropic, uniaxial and biaxial as far as their electromagnetic properties are concerned. Uniaxial and biaxial crystals are *birefringent*. This name refers to the property that a light ray incident on such a crystal splits (in general) into two rays within the crystal, and these rays follow different ray paths. There are then two emerging rays. It is found that the emerging rays are linearly polarized. The interpretation of birefringence is in terms of two different natural wave modes in the medium. The case of uniaxial crystals is discussed in this section and the more complicated case of biaxial crystals is discussed in §12.5.

The dielectric tensor for a uniaxial crystal with the principal axis oriented along the 3 axis is given by (7.22), *viz.*,

$$K_{ij}(\omega) = \begin{pmatrix} K_\perp(\omega) & 0 & 0 \\ 0 & K_\perp(\omega) & 0 \\ 0 & 0 & K_\parallel(\omega) \end{pmatrix}. \quad (12.11)$$

The other axes are chosen so that $\mathbf{k}$ is in the 1–3 plane at an angle θ to the principal axis; that is,

$$\boldsymbol{\kappa} = (\sin\theta, 0, \cos\theta). \quad (12.12)$$

This case provides a relatively simple but non-trivial example of the general theory of wave properties developed in Chapter 11. One has

$$\Lambda_{ij}(\omega, \mathbf{k}) = \begin{pmatrix} K_\perp(\omega) - n^2 \cos^2\theta & 0 & n^2 \sin\theta\cos\theta \\ 0 & K_\perp(\omega) - n^2 & 0 \\ n^2 \sin\theta\cos\theta & 0 & K_\parallel(\omega) - n^2 \sin^2\theta \end{pmatrix}. \quad (12.13)$$

The dispersion equation is given by setting the determinant of (12.13)

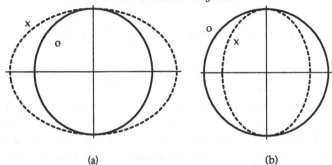

(a)   (b)

Fig. 12.1 Polar plots of the refractive index for uniaxial systems with
(a) $K_\parallel/K_\perp = 2$ and (b) $K_\parallel/K_\perp = 0.5$. Solid curves for
the o mode and dashed curves for the x mode.

equal to zero:

$$\left[K_\perp(\omega) - n^2\right]\left\{K_\perp(\omega)K_\parallel(\omega) - n^2\left[K_\perp(\omega)\sin^2\theta + K_\parallel(\omega)\cos^2\theta\right]\right\} = 0. \tag{12.14}$$

There are two solutions of (12.14). One is

$$n^2 = n_o^2(\omega) = K_\perp(\omega), \tag{12.15}$$

which is the dispersion equation for the *ordinary mode*. The other solution is

$$n^2 = n_x^2(\omega, \theta) = \frac{K_\perp(\omega)K_\parallel(\omega)}{K_\perp(\omega)\sin^2\theta + K_\parallel(\omega)\cos^2\theta}, \tag{12.16}$$

which is the dispersion equation of the *extraordinary mode*.

Uniaxial media are classified as *positive* for $K_\parallel > K_\perp$ (for example, quartz) and *negative* for $K_\parallel < K_\perp$ (for example, calcite). In the former case one has $n_x^2(\omega, \theta) \geq n_o^2(\omega, \theta)$ and in the latter case one has $n_x^2(\omega, \theta) \leq n_o^2(\omega, \theta)$, with the equalities applying only for $\sin\theta = 0$. For some purposes it is helpful to represent the refractive indices on a polar plot as illustrated in Figure 12.1. In three dimensions the refractive index surface for the ordinary mode is a sphere and that for the extraordinary mode is like an ellipsoid but is not actually ellipsoidal, with the shape being oblate for a positive and prolate for a negative uniaxial crystal. The curves plotted in Figure 12.1 correspond to a cut through any constant-ϕ plane of such a three-dimensional plot. For a uniaxial crystal the plots are axially symmetric and so all relevant information is contained in a polar plot of the form illustrated in Figure 12.1. (This is not the case for a biaxial crystal because the refractive indices are functions of ϕ.) The refractive index surface for each mode is useful for making a geometric construction of the group velocity (§12.3).

The polarization vectors for the two modes are constructed from any column of the matrix of cofactors λ_{ij} of Λ_{ji}. The matrix of cofactors is λ_{ij}

$$= \begin{pmatrix} (K_\perp - n^2)(K_\parallel - n^2 \sin^2 \theta) & 0 & -(K_\perp - n^2)n^2 \sin\theta\cos\theta \\ 0 & A & 0 \\ -(K_\perp - n^2)n^2 \sin\theta\cos\theta & 0 & (K_\perp - n^2)(K_\perp - n^2 \cos^2 \theta) \end{pmatrix}.$$
(12.17)

with $A = K_\perp K_\parallel - n^2(K_\perp \sin^2 \theta + K_\parallel \cos^2 \theta)$, and where arguments are omitted for simplicity in writing. On inserting the dispersion relation for the ordinary mode, the first and last columns are indeterminate, and the second column implies, after normalization, the polarization vector for the ordinary mode:

$$\mathbf{e}_o(\mathbf{k}) = (0, 1, 0).$$
(12.18)

Thus the ordinary mode is strictly transverse, and is linearly polarized along the direction orthogonal to both $\mathbf{k}$ and to the principal axis (the 3 axis here) of the crystal. On inserting the dispersion equation for the extraordinary mode into (12.17), the middle column becomes indeterminate, and the first and third columns are proportional to each other, as they must be. After normalization, either column gives the polarization vector for the extraordinary mode:

$$\mathbf{e}_x(\mathbf{k}) = \frac{(K_\parallel \cos\theta, 0, -K_\perp \sin\theta)}{(K_\parallel^2 \cos^2 \theta + K_\perp^2 \sin^2 \theta)^{1/2}}.$$
(12.19)

Note that the extraordinary mode is not strictly transverse; its longitudinal component of polarization is given by

$$\boldsymbol{\kappa} \cdot \mathbf{e}_x(\mathbf{k}) = \frac{(K_\parallel - K_\perp)\sin\theta\cos\theta}{(K_\parallel^2 \cos^2 \theta + K_\perp^2 \sin^2 \theta)^{1/2}},$$
(12.20)

which is zero for propagation along the principal axis ($\sin\theta = 0$) and normal to the principal axis ($\cos\theta = 0$) but not otherwise.

The directions of other vector fields associated with the waves follow from the polarization vector $\mathbf{e}$ for any wave. In particular, the magnetic field for the wave is along $\boldsymbol{\kappa} \times \mathbf{e}$ and the ith component of the electric induction $\mathbf{D}$ is proportional to $K_{ij}e_j$. Both these vectors are strictly transverse (cf. Exercise 11.9); the magnetic field for one mode is orthogonal to the electric induction for that mode, and the magnetic field or the electric induction for one mode is orthogonal to the corresponding field for the other mode. The polarization vectors for the ordinary and extraordinary modes of a uniaxial crystal are also orthogonal, as follows trivially from (12.18) and (12.19). However, this is not the case for the natural wave modes of an arbitrary medium. For example, in a biaxial crystal the polarization vectors for both modes have a longitudinal component in general, and these two components preclude the polarization vectors being orthogonal. The orthogonality of the polarization vectors for the two modes of a uniaxial crystal is a special case due to the ordinary mode being strictly transverse.

The foregoing remarks raise a question concerning the meaning of "orthogonal modes". The eigenmodes of any system are orthogonal in some meaningful sense, as is familiar in the context of orthogonal states in quantum mechanics for example. Clearly, the foregoing remarks imply that the orthogonality cannot mean the orthogonality of the polarization vectors. Physically, the important property of the natural modes of a linear medium is that the presence of waves in one mode does not cause the excitation of waves in any other mode. The source term for the energy is $\mathbf{J} \cdot \mathbf{E}$, cf. (1.7), and the orthogonality of the modes actually requires that the electric field in one mode be orthogonal to the electric current in the other mode, cf. Exercise 11.7. The implied orthogonality relation (E11.7) is satisfied, as follows by direct calculation using (12.11), (12.18) and (12.19).

12.3 The Group Velocity

The direction of energy propagation for waves in any wave mode is identified as the *ray direction* for that mode. When considering the propagation of wave energy in an anisotropic medium, it is often convenient to think in terms of the propagation of quasi-particles or wave quanta. (The wave quanta of transverse waves are called *photons*, and those of some other specific types of waves are also given specific names, but for the present we use the general term "wave quantum".) This idea is independent of quantum mechanics and the use of quantum mechanical language is no more than a matter of convenience. The important point is that there is a direct analogy between the propagation of rays in optics and of particles in mechanics. Historically, it is interesting that Hamilton developed Hamiltonian mechanics (in the middle of the nineteenth century) first in the context of optics and only later realized that the same equations could be applied to mechanics. Hamiltonian mechanics is the basis of modern quantum mechanics and statistical mechanics. Nowadays Hamiltonian mechanics is more familiar in the particle context than in its original context of optics. The analogy involves the following replacements: particle → wave quantum, momentum → wave vector $\mathbf{k}$, energy → wave frequency ω, and velocity → group velocity.

(The Hamiltonian function for waves in the mode M is then identified as $\omega_M(\mathbf{k}; \mathbf{x})$, the group velocity is identified as in (12.21) below and the other Hamiltonian equation $d\mathbf{k}/dt = -\partial\omega_M(\mathbf{k}; \mathbf{x})/\partial\mathbf{x}$ determines refraction of the ray along the ray path in an inhomogeneous medium.) Thus the group velocity is identified as the velocity of energy propagation. However, some care is required in interpreting this statement because it is based on the foregoing analogy and does not necessarily correspond to what is measured in any experiment; specifically, one has no reason to expect that the measurement of a signal velocity in a medium will correspond to the group velocity. The group velocity determines the direction of energy propagation, that is, the ray direction, in a steady state when the energy flow from source to detector does not depend on time.

It is convenient to discuss the evaluation of the group velocity for the specific case of waves in a uniaxial crystal, as this relatively simple example includes most of the important subtleties.

The group velocity for waves in the mode M is defined by

$$\mathbf{v}_{gM}(\mathbf{k}) = \frac{\partial\omega_M(\mathbf{k})}{\partial\mathbf{k}}. \tag{12.21}$$

In cartesian coordinates one has

$$\mathbf{v}_{gM}(\mathbf{k}) = \left(\frac{\partial\omega_M(\mathbf{k})}{\partial k_x}, \frac{\partial\omega_M(\mathbf{k})}{\partial k_y}, \frac{\partial\omega_M(\mathbf{k})}{\partial k_z}\right). \tag{12.22}$$

When the dispersion relation is given in term of spherical polar coordinates in $\mathbf{k}$-space, it is desirable to evaluate the group velocity directly in polar coordinates. This is achieved as follows. Let the polar coordinates be k, θ and ϕ, with

$$k_x = k\sin\theta\cos\phi, \quad k_y = k\sin\theta\sin\phi, \quad k_x = k\cos\theta. \tag{12.23}$$

Then the unit vectors in the $\mathbf{k}$, θ and ϕ directions are

$$\boldsymbol{\kappa} = (\sin\theta\cos\phi, \sin\theta\sin\phi, \cos\theta),$$
$$\boldsymbol{\theta} = (\cos\theta\cos\phi, \cos\theta\sin\phi, -\sin\theta), \tag{12.24}$$
$$\boldsymbol{\phi} = (-\sin\phi, \cos\phi, 0).$$

The resulting expression for the group velocity in polar coordinates is

$$\mathbf{v}_{gM}(\mathbf{k}) = \frac{\partial\omega_M(\mathbf{k})}{\partial k}\boldsymbol{\kappa} + \frac{1}{k}\frac{\partial\omega_M(\mathbf{k})}{\partial\theta}\boldsymbol{\theta} + \frac{1}{k\sin\theta}\frac{\partial\omega_M(\mathbf{k})}{\partial\phi}\boldsymbol{\phi}. \tag{12.25}$$

Note that (12.12) corresponds to the case where the wave is in the plane $\phi = 0$; in the following discussion $\phi = 0$ is assumed.

Further modification of (12.25) is required when the dispersion relation is given in the form $n^2 = n_M^2(\omega, \theta)$. It is necessary to change

variables, from the components (cartesian or polar) of $\mathbf{k}$ as the independent variables and the frequency as the dependent variable, to ω and the
angles θ and ϕ as the independent variables and n^2 as the dependent
variable. To achieve this let us write $k = \omega n/c$ and appeal to the rules
of partial differentiation to write

$$\frac{\partial \omega_M(\mathbf{k})}{\partial k} = \frac{c}{\partial[\omega n_M(\omega,\theta)]/\partial\omega}, \tag{12.26}$$

$$\frac{\partial \omega_M(\mathbf{k})}{\partial \theta} = -\frac{\partial[\omega n_M(\omega,\theta)]/\partial\theta}{\partial[\omega n_M(\omega,\theta)]/\partial\omega}. \tag{12.27}$$

In (12.27) the chain rule for partial derivatives is used. In the chain
rule, each partial derivative involves two of the variables ω, k, θ, with
the third being held constant. Suppose, for example, that k varies with
θ such that $\omega(k,\theta)$ remains constant; then

$$\delta\omega = \left(\frac{\partial\omega}{\partial k} + \frac{\partial\omega}{\partial\theta}\right)\delta\theta = 0$$

leads to

$$\frac{\partial\omega}{\partial\theta} = -\frac{\partial\omega}{\partial k}\bigg/\frac{\partial k}{\partial\theta}, \quad \text{or} \quad \frac{\partial\omega}{\partial\theta}\frac{\partial\theta}{\partial k}\frac{\partial k}{\partial\omega} = -1. \tag{12.28}$$

With (12.26) and (12.27), (12.25) becomes (for $\phi = 0$)

$$\mathbf{v}_{gM}(\mathbf{k}) = \frac{c}{\partial[\omega n_M(\omega,\theta)]/\partial\omega}\left[\boldsymbol{\kappa} - \frac{1}{n_M(\omega,\theta)}\frac{\partial n_M(\omega,\theta)}{\partial\theta}\boldsymbol{\theta}\right]. \tag{12.29}$$

The direction of energy propagation is along the group velocity and it
is often convenient to describe this in terms of polar angles. Here we are
considering only the uniaxial case explicitly, and then the group velocity
and the wave normal direction $\boldsymbol{\kappa}$ are in the same azimuthal plane, so
that the azimuthal angle ϕ is the same for both $\mathbf{v}_{gM}$ and $\boldsymbol{\kappa}$. Let us
define the *ray angle* θ_{gM}, which is implicitly a function of ω and θ, as
the angle between the group velocity and the principal axis, then one
has

$$\mathbf{v}_{gM}(\mathbf{k}) = |\mathbf{v}_{gM}(\mathbf{k})|\,(\sin\theta_{gM}, 0, \cos\theta_{gM}). \tag{12.30}$$

For the ordinary mode ($M = \text{o}$) one finds $\theta_{go} = \theta$, as for waves in an
isotropic medium. However, for the extraordinary mode ($M = \text{x}$) one
finds

$$\tan\theta_{gx} = \frac{K_\perp(\omega)}{K_\parallel(\omega)}\tan\theta. \tag{12.31}$$

Except for propagation along ($\theta = 0$) and perpendicular ($\theta = \pi/2$) to
the principal axis, the ray direction is different from the wave normal
direction.

There is a graphical construction of the group velocity based on the

Waves in Anisotropic Crystals

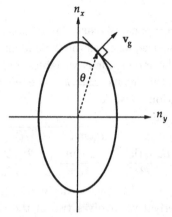

Fig. 12.2 Polar plot of the refractive index $(n_x, n_y) = (n\cos\theta, n\sin\theta)$.
The dashed vector defines the wave normal direction.
The group velocity $\mathbf{v_g}$ is normal to the tangent of the curve.

expression (12.29). Suppose that one plots the refractive index $n_M(\omega, \theta)$ as a function of θ for fixed ω, as in Figure 12.1 for the modes of a uniaxial crystal and as illustrated schematically in Figure 12.2 for a single wave mode. Consider a point on the refractive index curve. The radius vector from the origin to this point is along $\boldsymbol{\kappa}$ and is at the angle θ to the polar axis. The refractive index is equal to kc/ω and the radius vector to this point is regarded as representing the vector $\mathbf{k}$ in units of ω/c. Now according to a geometric interpretation of the definition (12.21), the group velocity is along the normal to the surface $\omega_M(\mathbf{k}) = $ constant, and on changing variables this corresponds to the surfaces of constant refractive index. Thus the group velocity is along the normal to the refractive index surface, as illustrated in Figure 12.2. An alternative derivation of this result for a non-dispersive system is outlined in Exercise 12.7.

12.4 Waves in Optically Active Media

The waves in an isotropic medium are either transverse or longitudinal, and the waves in an anisotropic crystal have a linearly polarized transverse part. There are two classes of media where the natural modes have a specific handedness, that is, where the electric vector rotates. These are optically active media and magnetoactive media. Optically active media are discussed here. An important example of a magnetoactive medium is a magnetized plasma, as discussed in §13.2.

The handedness of wave polarization is defined as a screw sense of the direction of rotation of the electric vector in the wave relative to the direction $\boldsymbol{\kappa}$ of wave propagation. Unfortunately this convention is not used universally, although older conventions are now rarely used. Other notable conventions that are sometimes still encountered are: (i) in older astronomical references the handedness is defined as a screw sense relative to the line of sight, which is opposite to the direction of wave propagation, and (ii) in older versions of the magnetoioinic theory the handedness is defined as a screw sense relative to the direction of the ambient magnetic field, which is the same as the modern convention when the wave is propagating at an acute angle to the magnetic field and opposite to it when this angle is oblique. A further source of possible confusion occurs with the convention adopted with the sign of the exponential when Fourier transforming. The modern convention is to assume that fields oscillate as $e^{-i\omega t}$, but some authors adopt the opposite sign, which corresponds effectively to complex conjugating. The polarization vector for a circular or elliptically polarized wave is intrinsically complex, cf. (12.37) below, and the interpretation of whether a given mathematical expression for a polarization vector represents left or right handedness reverses on changing the sign convention from $e^{-i\omega t}$ to $e^{i\omega t}$.

The dielectric tensor for an optically active medium has the form (7.16), *viz.*,

$$K_{ij}(\omega, \mathbf{k}) = K^{\mathrm{L}}(\omega, k)\,\kappa_i \kappa_j + K^{\mathrm{T}}(\omega, k)\,(\delta_{ij} - \kappa_i \kappa_j) + iK^{\mathrm{R}}(\omega, k)\epsilon_{ijr}\kappa_r,$$
(12.32)

On choosing the 3 axis to be along $\boldsymbol{\kappa}$ in this case, one finds the matrix representations

$$\Lambda_{ij} = \begin{pmatrix} K^{\mathrm{T}} - n^2 & iK^{\mathrm{R}} & 0 \\ -iK^{\mathrm{R}} & K^{\mathrm{T}} - n^2 & 0 \\ 0 & 0 & K^{\mathrm{L}} \end{pmatrix}.$$
(12.33)

and

$$\lambda_{ij} = \begin{pmatrix} K^{\mathrm{L}}(K^{\mathrm{T}} - n^2) & iK^{\mathrm{L}}K^{\mathrm{R}} & 0 \\ -iK^{\mathrm{L}}K^{\mathrm{R}} & K^{\mathrm{L}}(K^{\mathrm{T}} - n^2) & 0 \\ 0 & 0 & (K^{\mathrm{T}} - n^2)^2 - (K^{\mathrm{R}})^2 \end{pmatrix}.$$
(12.34)

where arguments are omitted for simplicity in writing. The determinant of (12.33) gives the dispersion equation

$$\Lambda = K^{\mathrm{L}}\big[(K^{\mathrm{T}} - n^2)^2 - (K^{\mathrm{R}})^2\big] = 0.$$
(12.35)

The possible solutions of (12.35) correspond to longitudinal waves satisfying $K^L = 0$, which usually has no solutions of interest, and to transverse waves satisfying $(K^T - n^2)^2 - (K^R)^2 = 0$. One can usually justify regarding K^T and K^R as functions only of ω, and then there are two transverse wave modes with dispersion relations

$$n^2 = n_\pm^2 = K^T(\omega) \pm |K^R(\omega)|.$$ (12.36)

On substituting $n^2 = n_\pm^2$ into (12.34), either of the first two columns, after normalization, give the polarization vectors

$$\mathbf{e}_+ = \mathbf{e}_R = 2^{-1/2}(1, i, 0), \quad \mathbf{e}_- = \mathbf{e}_L = 2^{-1/2}(1, -i, 0),$$ (12.37)

where $\mathbf{e}_R$ and $\mathbf{e}_L$ denote right and left hand circular polarizations, respectively. As already stated, the handedness of the polarization is defined as the screw sense of the electric vector relative to the wave normal direction $\boldsymbol{\kappa}$.

Radiation that is incident on an optically active medium splits into components with opposite circular polarizations in the medium, and due to the different refractive indices, these components develop a phase difference that increases linearly with distance along the ray path. (The ray paths for the two components are different but this difference is usually small and is ignored in the present discussion.) If the incident radiation is linearly polarized, the introduction of this phase difference corresponds to a rotation of the plane of linear polarization. Optically active media are classified as *dextrorotatory* and *levorotatory* depending on whether the sense of this rotation has a right or left hand screw sense relative to $\boldsymbol{\kappa}$.

There is a notable distinction between rotation of the plane of polarization of radiation incident on an optically active medium and of radiation incident on a magnetoactive medium (the *Faraday effect*). In an optically active medium the sense of rotation is a screw sense relative to $\boldsymbol{\kappa}$, and so, if the ray is reflected at some point, the phase difference and hence the angle of the plane of polarization continue to increase. On the other hand, in a magnetoactive medium the sense of rotation is a screw sense relative to the direction of the ambient magnetic field, and so, if the ray is reflected at some point, then the phase difference decreases and the angle of the plane of polarization decreases back towards its initial value.

12.5 Waves in Biaxial Crystals

The properties of waves in a biaxial crystal differ in form from those of a uniaxial crystal because of the lack of axial symmetry: the dispersion relations and polarization vectors depend explicitly on the azimuthal angle ϕ. This leads to a considerable increase in the algebraic complexity in generalizing from the uniaxial to the biaxial case. The discussion here is restricted to some general remarks, to writing down the dispersion equation in different ways and to a remark on the group velocity.

One notable distinction between the uniaxial and biaxial cases concerns the principal axes and the optic axes. An *optic axis* is defined as a direction along which the two natural modes have the same refractive index. The names *uniaxial* and *biaxial* actually refer to the number of optic axes. As is evident from Figure 12.1, the optic axis of a uniaxial crystal coincides with its principal axis. For a biaxial crystal the optic axes are not necessarily related to the principal axes and it is important to distinguish between them. For propagation along an optic axis an anisotropic crystal has two degenerate states of polarization. An interesting implication of this for a biaxial crystal is the existence of *conical refraction*, which was predicted by Hamilton and confirmed experimentally soon after it was predicted, cf. Exercise 12.6.

The dielectric tensor for a biaxial crystal is of the form (7.23) in the coordinate system in which the axes are oriented along the principal axes. The corresponding matrix of coefficients of the wave equation is

$$
\Lambda_{ij} = \begin{pmatrix} K_x - n^2(1 - \kappa_x^2) & n^2 \kappa_x \kappa_y & n^2 \kappa_x \kappa_z \\ n^2 \kappa_y \kappa_x & K_y - n^2(1 - \kappa_y^2) & n^2 \kappa_y \kappa_z \\ n^2 \kappa_z \kappa_x & n^2 \kappa_z \kappa_y & K_z - n^2(1 - \kappa_z^2) \end{pmatrix}
$$
(12.38)

where dependences on ω are omitted, and where the axes are labeled x, y, z rather than 1, 2, 3. It is straightforward to evaluate the determinant of (12.38). The dispersion equation given by setting the determinant of (12.38) equal to zero is

$$
\left(K_x \kappa_x^2 + K_y \kappa_y^2 + K_z \kappa_z^2\right) n^4 - \left[K_y K_z (1 - \kappa_x^2)\right.
$$
$$
\left. + K_x K_z (1 - \kappa_y^2) + K_x K_y (1 - \kappa_z^2)\right] n^2 + K_x K_y K_z = 0.
$$
(12.39)

It is sometimes convenient to write the dispersion equation (12.39) in terms of the characteristic speeds v_x, v_y and v_z introduced by writing

$$
K_x = \frac{c^2}{v_x^2}, \quad K_y = \frac{c^2}{v_y^2}, \quad K_z = \frac{c^2}{v_z^2},
$$
(12.40)

where dependences on ω are omitted. Also the refractive index is expressed in terms of the phase speed v_ϕ:

$$
n^2 = \frac{c^2}{v_\phi^2}.
$$
(12.41)

The dispersion equation reduces to

$$\frac{\kappa_x^2}{v_\phi^2 - v_x^2} + \frac{\kappa_y^2}{v_\phi^2 - v_y^2} + \frac{\kappa_z^2}{v_\phi^2 - v_z^2} = 0. \qquad (12.42)$$

The evaluation of the group velocity and the ray angle for waves in a biaxial crystal is cumbersome in general. In the particular case of a non-dispersive medium there is a simple way of writing down the ray properties by appealing to a "duality" between wave and ray properties, cf. Exercise 12.7. Let the group velocity for either mode be written in the form $\mathbf{v_g} = v_g \mathbf{g}$, so that $\mathbf{g}$ is a unit vector, then duality applied to (12.42) implies that the group velocity satisfies

$$\frac{v_x^2 g_x^2}{v_g^2 - v_x^2} + \frac{v_y^2 g_y^2}{v_g^2 - v^2} + \frac{v_z^2 g_z^2}{v_g^2 - v_z^2} = 0. \qquad (12.43)$$

However, (12.43) is satisfied only for rays in a non-dispersive medium.

Exercise Set 12

12.1 Rederive the dispersion relation (12.10), *viz.*, $n^2 = \varepsilon(\omega)\mu(\omega)c^2$, as follows.

(a) Show when the induced current is written in terms of the polarization and magnetization as in (6.5) and the fields $\mathbf{D}$ and $\mathbf{H}$ are introduced as in (6.1) that the Fourier transformed of Maxwell's equations without extraneous source terms reduces to the form

$$\mathbf{k} \times \mathbf{H}(\omega) = -\omega\mathbf{D}(\omega), \quad \mathbf{k} \cdot \mathbf{D}(\omega) = 0,$$

together with $\mathbf{k} \times \mathbf{E}(\omega) = \omega\mathbf{B}(\omega)$ and $\mathbf{k} \cdot \mathbf{B}(\omega) = 0$.

(b) Then assuming $\mathbf{D}(\omega) = \varepsilon(\omega)\mathbf{E}(\omega)$ and $\mathbf{H}(\omega) = \mathbf{B}(\omega)/\mu(\omega)$, show that these equations imply $n^2 = \varepsilon(\omega)\mu(\omega)c^2$.

12.2 In the presence of a static electromagnetic field the vacuum becomes birefringent, due to the *vacuum polarization*. The response of the vacuum to a static electromagnetic field is described by susceptibility tensors χ_{rs}^{e} and χ_{rs}^{m} with

$$\chi_{11}^{e} = \chi_{22}^{e} = -\frac{2\alpha}{45\pi}\left(\frac{B}{B_c}\right)^2, \quad \chi_{33}^{e} = \frac{\alpha}{9\pi}\left(\frac{B}{B_c}\right)^2,$$

$$\chi_{11}^{m} = \chi_{22}^{m} = -\frac{2\alpha}{45\pi}\left(\frac{B}{B_c}\right)^2, \quad \chi_{33}^{m} = -\frac{2\alpha}{15\pi}\left(\frac{B}{B_c}\right)^2, \tag{$E12.1$}$$

with $\alpha = e^2/4\pi\varepsilon_0\hbar c \approx 1/137$, $B_c = m_e^2 c^2/\hbar e \approx 4.4 \times 10^9$ T and where the magnetostatic field is along the 3 axis.

(a) Show that the refractive indices of the two wave modes in the magnetized vacuum are given by, for $B \ll B_c$,

$$n^2 = 1 + \frac{(11 \pm 3)\alpha}{90\pi}\left(\frac{B}{B_c}\right)^2 \sin^2\theta, \tag{$E12.2$}$$

(b) Show that the polarization vectors are given by

$$\mathbf{e}_+ = (-\cos\theta, 0, \sin\theta), \quad \mathbf{e}_- = (0, 1, 0). \tag{$E12.3$}$$

where θ is the angle between $\mathbf{k}$ and the magnetostatic field.

12.3 In a uniaxial crystal the group velocity of the ordinary mode is along the wave-normal direction $\boldsymbol{\kappa}$, but this is not the case for the extraordinary mode.

(a) Let the group velocity for the extraordinary mode be at an angle θ_{gx} to the principal axis. Show that this ray angle is give by

$$\tan\theta_{gx} = \tan\theta \frac{K_\perp(\omega)}{K_\parallel(\omega)}. \tag{$E12.4$}$$

(b)   The direction of the Poynting vector is along $\mathrm{Re}\,(\mathbf{E}^* \times \mathbf{B}) \propto \mathrm{Re}\,(\boldsymbol{\kappa} - \mathbf{e}^* \cdot \boldsymbol{\kappa}\,\mathbf{e})$. Show that the ray direction and the direction of the Poynting vector coincide.

Remark: The group velocity is parallel to the Poynting vector in non-spatially dispersive media.

12.4 Total internal reflection occurs when the angle ξ between the ray direction and the normal to the surface satisfies $n\sin\xi > 1$. Determine the conditions on the angle ξ_{gx} between the ray direction and the normal to the surface of a uniaxial crystal for total internal reflection to occur for the extraordinary mode: (a) for the principal axis parallel to the surface and the wave normal in the plane defined by the principal axis and the normal to the surface, (b) for the principal axis parallel to the surface and the wave normal orthogonal to the plane defined by the principal axis and the normal to the surface, and (c) for the principal axis orthogonal to the surface.

12.5 An *optic axis* of a crystal is defined as a wave-normal direction in which the two modes have equal refractive indices or phase speeds.

(a)   Show that in a uniaxial crystal the optic axis coincides with the principal axis.

(b)   Show that in a biaxial crystal that the two directions $\mathbf{c}$, $|\mathbf{c}|^2 = 1$, found by solving

$$\left[c_i c_j \left(K_{rr}K_{ij} - K_{ir}K_{rj}\right)\right]^2 = 4 c_r c_s K_{rs}\det\left[K_{ij}\right], \qquad (E12.5)$$

for the direction cosines c_x, c_y, c_z define the two optic axes.

12.6 The rays corresponding to wave normals directed along an optic axis $\mathbf{c}$ of a biaxial crystal exhibit a phenomenon called *conical refraction*. Along an optic axis the two wave modes are degenerate, and hence both the dispersion equation $\Lambda = 0$ and its derivative vanish simultaneously. The group velocity is defined formally (using the chain rule) as $\partial\omega/\partial\mathbf{k} = -(\partial\Lambda/\partial\mathbf{k})/(\partial\Lambda/\partial\omega)$, and as both numerator and denominator vanish in this degenerate case, the usual derivation of the group velocity leads to an indeterminate result.

(a)   In the degenerate case one has $\lambda_{ij} = 0$ and hence Λ_{ij} is of rank one and may be written as the outer product of a vector with itself. Let the unit vector along this direction be $\mathbf{h}$. Show

$$h_i = \frac{c_a K_{ai}}{(c_r c_s K_{rs}\Lambda_{tt})}^{1/2}. \qquad (E12.6)$$

(b) Show that the two degenerate modes allow any polarization vector **e** that satisfies

$$\mathbf{e} = \cos\psi\,\mathbf{e}^1 + \sin\psi\,\mathbf{e}^2,$$

where ψ is an arbitrary angle, and with

$$\mathbf{e}^1 := \frac{\mathbf{h} \times \mathbf{c}}{|\mathbf{h} \times \mathbf{c}|}, \quad \mathbf{e}^2 := = \mathbf{h} \times \mathbf{e}^1.$$

(c) Use the fact that in a non-spatially dispersive medium the group velocity (the ray direction) is along the Poynting vector to show that the most general form for the ray direction for waves directed along the optic axis **c** is

$$\mathbf{v_g} \propto \mathbf{c}\cos^2\psi + \mathbf{h}\times\mathbf{c}\sin\psi\cos\psi + \mathbf{h}\cdot\mathbf{c}\,\mathbf{h}\sin^2\psi.$$

Remark: The directions given by allowing ψ to vary in this formula define the cone of rays in conical refraction corresponding to the single wave normal direction.

12.7 In a non-dispersive crystal there is a duality relation that allows one to write down ray properties by inspection from wave properties. The group velocity in a non-dispersive medium is given by the ratio of the Poynting vector to the total wave energy, $W(\mathbf{k}) = \mathbf{D}^* \cdot \mathbf{E}$, $D_i = \varepsilon_0 K_{ij} E_i$, or by $|\mathbf{B}|^2/\mu_0$.

(a) Show that Maxwell's equations with $\mathbf{B} = \mu_0\mathbf{H}$ imply

$$\omega\mathbf{D} = -\mathbf{k} \times \mathbf{H}, \quad \omega\mathbf{B} = \mathbf{k} \times \mathbf{E}. \qquad (E12.7)$$

(b) Use $(E12.10)$ to show

$$\frac{\mathbf{k}}{\omega} = \frac{\mathbf{D}^* \times \mathbf{B}}{\mathbf{D}^* \cdot \mathbf{E}}.$$

(c) Given the form of the wave energy stated above, show that for a non-dispersive medium the group velocity is given by

$$\mathbf{v_g} = \frac{\mathbf{E}^* \times \mathbf{H}}{\mathbf{E}^* \cdot \mathbf{D}}.$$

(d) Let **g** be a unit vector along the ray direction. Show, for a non-dispersive medium, that the group speed and the phase speed $v_\phi = \omega/k$ are related by

$$\boldsymbol{\kappa} \cdot \mathbf{g} = \frac{v_\phi}{v_g}, \qquad (E12.8)$$

(e)   Hence show that there are two complementary sets of equations

$$\omega \mathbf{D} = -\mathbf{k} \times \mathbf{H}, \qquad \mathbf{E} = -\mathbf{v_g} \times \mathbf{B},$$
$$\omega \mathbf{B} = \mathbf{k} \times \mathbf{E}, \qquad \mathbf{H} = \mathbf{v_g} \times \mathbf{D},$$
$$D_i = \varepsilon_0 K_{ij} E_j, \qquad E_i = \varepsilon_0^{-1} K_{ij}^{-1} D_j, \qquad (E12.9)$$
$$\frac{\mathbf{k}}{\omega} = \frac{\mathbf{D}^* \times \mathbf{B}}{\mathbf{D}^* \cdot \mathbf{E}} \qquad \mathbf{v_g} = \frac{\mathbf{E}^* \times \mathbf{H}}{\mathbf{E}^* \cdot \mathbf{D}}$$

(f)   Hence establish the *duality relation* that under the following inter-
      changes the sets of equations $(E12.9)$ are interchanged:

$$\mathbf{D} \leftrightarrow \mathbf{E}, \quad \mathbf{B} \leftrightarrow \mathbf{H}, \quad \mathbf{k}/\omega \leftrightarrow \mathbf{v_g}, \quad \varepsilon_0 K_{ij} \leftrightarrow \varepsilon_0^{-1} K_{ij}^{-1}, \quad c \to c^{-1}.$$
$$(E12.10)$$

(g)   Use the duality relations to infer the equation (12.43) for the rays
      from the dispersion equation in the form (12.42).

13

Waves in Plasmas

Preamble

Plasmas can support a great variety of wave motions. For many purposes it suffices to have knowledge of three classes of waves, two of which are discussed here. These are waves in isotropic thermal plasmas, and waves in cold magnetized plasmas. The third class of waves are the MHD waves (MHD is short for magnetohydrodynamics), which are derived within the framework of a fluid model for the plasma. The MHD waves are not discussed in detail here.

13.1 Waves in Isotropic Thermal Plasmas

An *isotropic* plasma is defined to be a plasma (a) with no ambient magnetic field (it is *unmagnetized*), and (b) in which all species of particles have a Maxwellian distribution of velocities (or its relativistic generalization if relativistic effects are included). In any isotropic medium the waves are either longitudinal or transverse (§12.1). The longitudinal waves satisfy the longitudinal dispersion equations (12.6), *viz.*, $K^L(\omega, k) = 0$, and the transverse waves satisfy the transverse dispersion equations (12.7), *viz.*, $n^2 = K^T(\omega, k)$. The longitudinal and transverse parts of the dielectric tensor for an isotropic thermal plasma are given by (10.23) and (10.24), respectively.

Langmuir Waves

There are two solutions of the longitudinal dispersion equation that are important in practice. These are for Langmuir waves, which involve only the motion of the electrons, and ion sound waves (also called ion acoustic waves) that are associated with motion of the ions. As mentioned in §10.3, both these wave modes were identified by Tonks and Langmuir in 1929 in what is now recognized as the first major article in the development of modern plasma theory.

Let us assume that the phase speed of the waves of interest is much greater than the thermal speed V_e of electrons. This corresponds to the *long wavelength limit* $k\lambda_D \ll 1$. The dispersion equation $K^L(\omega, k) = 0$ involves the plasma dispersion function $\phi(y_\alpha)$, with $y_\alpha = \omega/kV_\alpha$, for each species α of particles. The expansion (10.30) of the plasma dispersion function is inserted into the expression (10.23) for $K^L(\omega, k)$. Assuming that V_e is much greater than the thermal speed of the ions (which assumption is invalid only if the ions have a temperature that is very much greater than the electron temperature), the contribution of the ions is negligible in comparison with that of the electrons. The solution of $K^L(\omega, k) = 0$ then gives the dispersion relation for Langmuir waves. An explicit expression for the dispersion relation is found using a perturbation approach in which the leading term is $\omega^2 = \omega_p^2$. On retaining the next term in the expansion (10.30), with this term evaluated at $\omega^2 = \omega_p^2$, one finds that in the dispersion relation $\omega = \omega_L(k)$ for Langmuir waves one has the following approximation:

$$\omega_L^2(k) = \omega_p^2 + 3k^2 V_e^2. \tag{13.1}$$

The range of validity of (13.1) is restricted by the use of the expansion (10.30) of the plasma dispersion function $\phi(y_\alpha)$. From the plot of the plasma dispersion function in Figure 10.1, it is apparent that the function is greater than unity only when its argument is greater than about unity, and it follows from (10.23) that the real part of $K^L(\omega, k)$ is strictly positive for $\phi(y_\alpha) \leq 1$. Hence there can be no Langmuir waves for $y_e \lesssim 1$, which corresponds to $\omega \lesssim 2^{1/2} kV_e$. It turns out that Landau damping, cf. (13.6) below, is very strong for phase speeds $v_\phi = \omega/k$ close to V_e, and as a consequence Langmuir waves exist as weakly damped propagating waves satisfying (13.1) only for v_ϕ greater than several times V_e.

A simple model for electron plasma oscillations is based on a fluid description of the electrons. In the fluid model the electrons are described by their number density, which is separated into an average and a fluctuating part associated with the waves, and their flow velocity $\mathbf{v}$ which has only a fluctuating part. The current density in the fluid model is

$$\mathbf{J} = -n_e e \mathbf{v}, \tag{13.2}$$

and the equation of fluid motion is

$$m_e \frac{d\mathbf{v}}{dt} = -e\left(\mathbf{E} + \mathbf{v} \times \mathbf{B}\right), \qquad (13.3)$$

where $\mathbf{E}$ and $\mathbf{B}$ are the fluctuating electric and magnetic fields in the waves. Langmuir waves are longitudinal and have $\mathbf{B} = 0$ so that the final term in (13.3) is absent for Langmuir waves. The other equations in the model are (1.3) and (1.6), *viz.*,

$$\operatorname{div}\mathbf{E} = \frac{\rho}{\varepsilon_0}, \quad \frac{\partial \rho}{\partial t} + \operatorname{div}\mathbf{J} = 0. \qquad (13.4a, b)$$

To find the wave equation for electron plasma oscillations, one takes the derivative of (13.4a) with respect to t, uses (13.2) and then (13.3) to find that both ρ and $\operatorname{div}\mathbf{E}$ vary harmonically:

$$\frac{\partial^2 \rho}{\partial t^2} + \frac{e^2 n_e}{\varepsilon_0 m_e}\rho = 0, \quad \frac{\partial^2 \operatorname{div}\mathbf{E}}{\partial t^2} + \frac{e^2 n_e}{\varepsilon_0 m_e}\operatorname{div}\mathbf{E} = 0. \qquad (13.5a, b)$$

These equations imply that the the electron density and the electric field oscillate at the plasma frequency $\omega_p = (e^2 n_e / \varepsilon_0 m_e)^{1/2}$.

It is possible to extend this model to include thermal effects, but there is no simple convincing way of accounting for the factor of 3 in the final term in (13.1), cf. Exercise 13.1.

The usual interpretation of electron plasma oscillations is based on the foregoing fluid model: electron plasma oscillations are attributed to a charge separation between the electrons and a background distribution of effectively immobile ions. If in localized regions the average position of the electrons becomes slightly displaced relative to the ions, so that there are peaks and troughs in the electron number density, then there is an electric field due to the charge separation that tends to accelerate the electrons back to their mean position relative to the ions. An oscillation is set up with the localized enhancements (and depletions) in the electron density oscillating at ω_p. These are the electron plasma oscillations, and the inclusion of thermal effects causes these oscillations to become Langmuir waves, which propagate as well as oscillate.

An interesting point concerning the foregoing interpretation of electron plasma oscillations is that the presence of the ions is unnecessary. Although it is useful to think in terms of an electric field due to charge separation between the electrons and ions, in fact the charge separation is due to the electron density deviating from its mean value, so that the electric potential is a fluctuating function of position. Langmuir waves are independent of the ions and exist even in a pure electron gas with no ions.

The absorption coefficient for Langmuir waves follows from (11.22) with the imaginary part of $K^L(\omega, k)$ given by (10.23). One finds

$$\gamma_L(k) = 2R_L(k)\omega_L(k)\,\mathrm{Im}\,\left[K^L(\omega_L(k), k)\right]\Big|_{\mathrm{Re}\,\left[K^L(\omega, k)\right]=0}$$

$$\approx \left(\frac{\pi}{2}\right)^{1/2} \frac{\omega_p^4}{k^3 V_e^3}\, e^{-\{\omega_L(k)\}^2/2k^2 V_e^2}, \tag{13.6}$$

where (11.23) is used with

$$R_L(k) = \left\{\frac{\omega\partial\mathrm{Re}\,\left[K^L(\omega, k)\right]}{\partial\omega}\right\}^{-1}\Bigg|_{\mathrm{Re}\,\left[K^L(\omega, k)\right]=0} \approx \frac{1}{2}. \tag{13.7}$$

The absorption coefficient (13.6) describes *Landau damping* of Langmuir waves. One physical interpretation of Landau damping is that it is due to absorption of the waves by the thermal electrons through the inverse of Cerenkov emission. Landau damping exists in the absence of collisions, and is the simplest example of a collisionless damping mechanism.

Note that for $\omega/k \geq V_e$, Landau damping is strictly zero and the non-zero damping implied by (13.6) is non-physical. In fact the damping implied by (13.6) for $\omega/k \geq V_e$ is due to the Maxwellian distribution (10.40) implying a non-zero probability of finding particles with $v \geq c$. These non-physical electrons lead to the non-physical Landau damping implied by (13.6) for $\omega/k \geq c$.

Langmuir waves are also damped due to the effect of electron-ion collisions. Such damping is interpreted in terms of a frictional loss, with the collisions causing a frictional drag on the electrons. An alternative physical interpretation is that the damping is due to absorption by the thermal electrons through the inverse of bremsstrahlung. On evaluating the imaginary part of the dielectric constant using $(E6.3)$ for $\nu_e \ll \omega$, one finds

$$\mathrm{Im}\,\left[K^L(\omega)\right] = \mathrm{Im}\,\left[K^T(\omega)\right] \approx \nu_e \omega_p^2/\omega^3. \tag{13.8}$$

The absorption coefficient for collisional damping of Langmuir waves implied by (11.22) and (13.8) is

$$[\gamma_L(k)]_{\mathrm{coll}} \approx \nu_e. \tag{13.9}$$

In practice, Landau damping usually dominates for Langmuir waves with phase speeds only slightly greater than the thermal speed of electrons but falls off very rapidly with increasing phase speed $v_\phi = \omega/k$ due to the exponential factor in (13.6), and collisional damping dominates at sufficiently high phase speeds.

Ion Sound Waves

For completeness let us mention the other longitudinal wave mode of an isotropic thermal plasma, which is the ion sound mode. Assuming that there is only one species of ions, this mode is treated by assuming $y_e \ll 1$ and $y_i \gg 1$ in the the expression (10.23) for $K^L(\omega, k)$. Then using the expansions (10.29) and (10.30) for the electrons and ions, respectively, one finds

$$\text{Re}\left[K^L(\omega, k)\right] \approx 1 + \frac{1}{k^2 \lambda_{De}^2} - \frac{\omega_{pi}^2}{\omega^2}. \tag{13.10}$$

The resulting dispersion relation for ion sound waves is

$$\omega = \omega_s(k) \approx \frac{k v_s}{[1 + k^2 \lambda_{De}^2]^{1/2}}, \tag{13.11}$$

where $v_s := \omega_{pi} \lambda_{De}$ is the *ion sound speed*. In the limit $k \lambda_{De} \ll 1$ the dispersion relation (13.11) reduces to $\omega \approx k v_s$, which is characteristic of a sound wave. The ratio of electric to total energy is

$$R_s(k) \approx \frac{1}{2} \left[\frac{\omega_s(k)}{\omega_{pi}}\right]^2, \tag{13.12}$$

and the absorption coefficient is

$$\gamma_s(k) \approx \left(\frac{\pi}{2}\right)^{1/2} \omega_s(k) \left\{ \frac{v_s}{V_e} + \left[\frac{\omega_s(k)}{k V_i}\right]^3 e^{-[\omega_s(k)]^2/2k^2 V_i^2} \right\}. \tag{13.13}$$

The two terms inside the curly brackets in (13.13) are due to Landau damping by thermal electrons and by thermal ions, respectively. The damping by thermal ions is strong for $v_s \approx V_i$, and ion sound waves exist as weakly damped waves only for $v_s \gg V_i$, which requires $Z_i T_e \gg T_i$, where $Z_i e$ is the ionic charge.

Transverse Waves

It is well known that transverse waves propagating in a cold electronic plasma have a refractive index $n = (1 - \omega_p^2/\omega^2)^{1/2}$. The derivation of this result for a thermal plasma requires some care. The dispersion equation for transverse waves is $n^2 = K^T(\omega, k)$, and the expression (10.24) for $K^T(\omega, k)$ involves the plasma dispersion function $\phi(y_\alpha)$, with $y_\alpha = \omega/k V_\alpha$, for each species α of particles. The point that requires care is to show that the only solution of the dispersion equation is in the limit $y_\alpha \gg 1$ for all α, cf. Exercise 13.2.

The transverse part of the dielectric tensor for sufficiently high phase speeds reduces to

$$K^T(\omega, k) = 1 - \omega_p^2/\omega^2. \tag{13.14}$$

The derivation of (13.14) follows by making the approximation (10.30) in (10.24), where the expansion is effectively in powers of the square of the ratio of the thermal speed to the speed of light. Thus, provided relativistic effects are ignored, the dispersion relation for transverse waves is written in the forms

$$n^2 = n_T^2(\omega) = 1 - \omega_p^2/\omega^2, \quad \omega^2 = \omega_T^2(k) = \omega_p^2 + k^2 c^2. \qquad (13.15)$$

A simple model that reproduces (13.15) is similar to the fluid model used above to treat Langmuir waves. The model involves the fluid equations (13.2) and (13.3) with the final term in (13.3) neglected on the grounds that it leads only to a relativistic correction. The other equations in the model are two of Maxwell's equations:

$$\text{curl}\,\mathbf{E} = -\frac{\partial \mathbf{B}}{\partial t}, \quad \text{curl}\,\mathbf{B} = \mu_0 \mathbf{J} + \frac{1}{c^2}\frac{\partial \mathbf{E}}{\partial t}, \qquad (13.16a,b)$$

and the requirement that the waves be transverse:

$$\text{div}\,\mathbf{E} = 0. \qquad (13.17)$$

One takes the time-derivative of (13.2), uses (13.3) to eliminate $d\mathbf{v}/dt$, substitutes the result into the time-derivative of (13.16b), and eliminates $\mathbf{B}$ using (13.16a). Finally one uses (2.35) to replace $\text{curl}\,\text{curl}\,\mathbf{E}$ by $-\nabla^2\mathbf{E}$ after noting (13.17). The resulting equation is

$$\nabla^2\mathbf{E} - \frac{1}{c^2}\frac{\partial^2\mathbf{E}}{\partial t^2} = \mu_0\frac{\partial\mathbf{J}}{\partial t} = \frac{1}{\varepsilon_0 c^2}\frac{e^2 n_e}{m_e}\mathbf{E}. \qquad (13.18)$$

On assuming $\mathbf{E} \propto e^{-i(\omega t - \mathbf{k}\cdot\mathbf{x})}$ and using the definition of the plasma frequency ($\omega_p^2 = e^2 n_e/\varepsilon_0 m_e$), (13.18) reproduces the dispersion relation (13.15).

The transverse waves (13.15) have $R_T(k) = \frac{1}{2}$, and because their phase speed is greater than the speed of light, there is no Landau damping. The absorption coefficient for collisional damping follows from (11.22) and (13.8); expressed as a function of frequency it is

$$[\gamma_T(k)]_{\text{coll}} \approx \nu_e\frac{\omega_p^2}{\omega^2}. \qquad (13.19)$$

13.2 Dispersion Relations for the Magnetoionic Waves

The *magnetoionic waves* are the natural wave modes in the magnetoionic theory. Historically, the magnetoionic theory was developed to describe the propagation of radio waves in the ionosphere, and a knowledge of the magnetoionic wave modes is essential in understanding a wide variety of phenomena in ionospheric physics, space physics and radio astronomy.

For many purposes the effect of collisions on the magnetoionic waves is unimportant, and they are neglected here. Then there are only two plasma parameters in the magnetoionic theory: the electron plasma frequency $\omega_p = (e^2 n_e / \varepsilon_0 m_e)^{1/2}$ and the electron cyclotron frequency $\Omega_e = eB/m_e$. These are combined with the wave frequency ω into the two magnetoionic parameters $X = \omega_p^2 / \omega^2$ and $Y = \Omega_e / \omega$ introduced in (10.7). The dielectric tensor is given by (10.9) with (10.10), which in the absence of collisions reduces to

$$K_{ij} = \begin{pmatrix} S & -iD & 0 \\ iD & S & 0 \\ 0 & 0 & P \end{pmatrix}, \qquad (13.20)$$

$$S = \frac{1 - X - Y^2}{1 - Y^2}, \quad D = \frac{-XY}{1 - Y^2}, \quad P = 1 - X. \qquad (13.21)$$

In deriving the wave properties it is convenient to choose a specific coordinate system: the coordinate axes are chosen such that the magnetic field is along the 3 axis and $\mathbf{k}$ is in the 1–3 plane at an angle θ to the 3-axis. The matrix of coefficients in the wave equation is

$$\Lambda_{ij} = \begin{pmatrix} S - n^2 \cos^2 \theta & -iD & n^2 \sin \theta \cos \theta \\ iD & S - n^2 & 0 \\ n^2 \sin \theta \cos \theta & 0 & P - n^2 \sin^2 \theta \end{pmatrix}, \qquad (13.22)$$

The determinant of this matrix gives the dispersion equation which is written in the form

$$\Lambda = An^4 - Bn^2 + C = 0, \qquad (13.23)$$

with

$$A = S \sin^2 \theta + P \cos^2 \theta = [1 - X - Y^2 + XY^2 \cos^2 \theta]/[1 - Y^2],$$

$$\begin{aligned} B &= (S^2 - D^2) \sin^2 \theta + PS(1 + \cos^2 \theta) \\ &= [2(1 - X)^2 - 2Y^2 + XY^2(1 + \cos^2 \theta)]/[1 - Y^2], \end{aligned} \qquad (13.24)$$

$$C = P(S^2 - D^2) = (1 - X)[(1 - X)^2 - Y^2]/[1 - Y^2].$$

The coefficients A, B and C in (13.24) do not depend on $\mathbf{k}$, and hence (13.23) is a quadratic equation for n^2. The solutions may be written in the form

$$n^2 = n_\pm^2 = \frac{B \pm F}{2A}, \quad F = \left(B^2 - 4AC\right)^{1/2}. \qquad (13.25)$$

As in a uniaxial crystal, the two modes with dispersion relations given by (13.25) are called the *ordinary* and *extraordinary* modes. However, these names are misleading in that neither mode propagates isotropically. The ordinary mode in magnetoionic theory is defined as the mode whose refractive index squared becomes equal to $1 - X = 1 - \omega_p^2 / \omega^2$ for propagation perpendicular to the magnetic field. Plots of the dispersion relations are given in Figures 13.1–13.3. The solutions $n_\pm^2$ are always

Waves in Plasmas

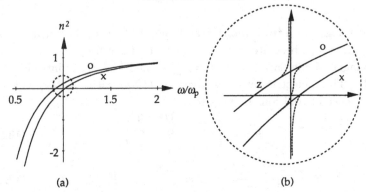

(a)   (b)

Fig. 13.1 Refractive index curves for the magnetoionic waves for $\Omega_e/\omega_p = 0.5$.
(a) For $\theta = 0$ there are two curves plus a vertical line (not shown) at $\omega = \omega_p$.
(b) Circled portion of (a) magnified; the dashed lines is for $\theta \neq 0$.
For $\theta = 0$ the o mode and the z mode join at $\omega = \omega_p$.

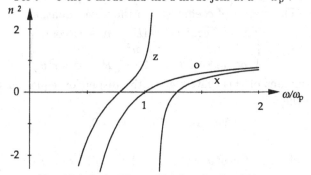

Fig. 13.2 As in Figure 13.1 but for $\theta = 30°$.

real, but they are negative in some ranges of frequency. Negative values correspond to imaginary k for real ω. The waves are then *evanescent*.

The regions of evanescence separate each of the two magnetoionic modes into two branches, so that there are four branches overall. It is convenient to refer to the two higher frequency branches as the *o mode* and the *x mode*, and to the lower frequency branch of the extraordinary mode as the *z mode*. The lower frequency branch of the ordinary mode is called the *whistler mode*; this mode appears in the upper left hand corner of Figure 13.3.

The o mode and the x mode are regarded as magnetically split versions of transverse waves in an isotropic plasma. As for transverse waves described by (13.15), the refractive indices for both modes are less than unity, so that their phase speeds are greater than the speed of light. These modes are of particular importance in radio astronomy because the only radiation that can escape from an astrophysical source is in one

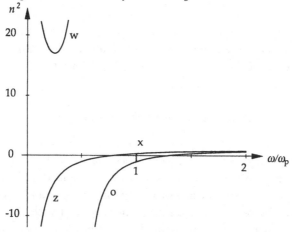

Fig. 13.3 As for Figure 13.1 but plotted on a different scale.
The whistler (w) mode branch is in the upper left hand corner.

or other of these modes. The reason follows by noting that escape to
infinity from a source requires that the radiation propagate along a path
where ω_p and Ω_e decrease away from the source. Qualitatively, as a wave
propagates along such a path it effectively moves to the right along the
refractive index curve on which it started. Waves in the whistler mode
and in the z mode encounter an infinity in refractive index, called a
resonance, where they are strongly absorbed. Such waves cannot escape
themselves, although they may produce escaping radiation in the o mode
or x mode by coupling to these modes in various ways. If, on the other
hand, waves in the o mode or x mode are initially propagating towards
increasing ω_p then they are effectively moving to the left on the refractive
index curve and approach cutoff where the refractive index becomes zero.
Waves are reflected (or refracted through a large angle) as they approach
a cutoff, and after reflection the waves can escape. Hence all radiation
generated in the o mode or the x mode tends to escape to infinity from
an astronomical source.

The z mode and the whistler mode are intrinsically new modes characteristic of the magnetoionic theory. The whistler mode was given its
name from "whistling atmospherics" heard with radio receivers in the
audio range. Whistling atmospherics are known to be due to lightning
flashes, which generate nearly white noise in the radio range, causing excitation of waves in the whistler mode that propagate along the Earth's
field lines from one hemisphere to the other, often reflecting and bounc-
ing back and forth several times. The strong dispersion, that is, the
strong frequency dependence, of the whistler mode causes lower fre-

quencies to become increasingly delayed compared with higher frequencies, leading to the falling tone characteristic of whistling atmospherics. Whistler waves are called *helicon* waves in solid state physics.

13.3 Cutoffs and Resonances

In a frequency range where n_+^2 or n_-^2 is negative the waves are evanescent rather than propagating. Evanescence corresponds to spatial decay in a medium with no transfer of energy to the medium. A simple example is light incident on a perfect mirror. The idealized metal in the mirror may be regarded as a plasma-like medium with the electron density determined by the density of electrons in the conduction band. The corresponding plasma frequency $\omega_p = (e^2 n_e/\varepsilon_0 m_e)^{1/2}$ is at X-ray frequencies, and hence for light one has $\omega \ll \omega_p$. The square of the refractive index for transverse waves $n^2 = k^2 c^2/\omega^2 = 1 - \omega_p^2/\omega^2$ is then large and negative. The implied value of the wavenumber is $k \approx \pm i\omega_p/c$. The waves decay spatially in the mirror as $e^{-\omega_p x/c}$, where x is measured from the surface of the metal. The wave energy penetrates into the metal only to a characteristic distance $\approx c/\omega_p$ called the *skin depth*. All the energy is reflected from an idealized mirror, and no energy is transferred to the metal. In contrast, when damping is included waves decay in space or time or both; waves generated uniformly everywhere initially decay purely in time, and waves generated at a constant point source decay purely in space away from the source. Temporal decay is described by a complex frequency and spatial decay by a complex wavenumber. Mathematically, spatial damping differs from evanescence in that in spatial damping the wavenumber is complex with the imaginary part determined by the antihermitian part of the dielectric tensor, whereas in evanescence the wavenumber is imaginary and is determined entirely by the hermitian part of the dielectric tensor.

The evanescent and real portions of the four branches of the magnetoionic wave modes, as illustrated in Figures 13.1–13.3, are separated by points where the refractive index passes either through zero or through infinity. A zero in refractive index is called a *cutoff*, and an infinity in refractive index is called a *resonance*.

The cutoffs correspond to $n^2 = 0$ and (13.23) implies that all cutoffs occur at $C/A = 0$, and all solutions of interest correspond to $C = 0$. Thus the cutoffs are at

$$P(S^2 - D^2) = 0. \tag{13.26}$$

There are three solutions of (13.26). One corresponds to $P = 0$, which gives $\omega = \omega_p$. This cutoff is in the o mode. There are two positive frequency solutions of $S^2 - D^2 = 0$. These are at $\omega = \omega_x$, and $\omega = \omega_z$, with

$$\omega_x = \tfrac{1}{2}\Omega_e + \tfrac{1}{2}(4\omega_p^2 + \Omega_e^2)^{1/2}, \quad \omega_z = -\tfrac{1}{2}\Omega_e + \tfrac{1}{2}(4\omega_p^2 + \Omega_e^2)^{1/2}, \quad (13.27)$$

and these are the cutoff frequencies in the x mode and the z mode, respectively. Note that none of the cutoff frequencies depends on angle of propagation.

The resonances occur at $n^2 = \infty$; setting $1/n^2 = 0$ in (13.23) implies that resonances occur at $A/C = 0$; all solutions of interest here correspond to $A = 0$. Thus the resonances are at

$$\tan^2 \theta = -\frac{P}{S}. \quad (13.28)$$

In particular, the resonances at $\theta = 0$ are at $P = 0$ and $S = \infty$, which correspond to $\omega = \omega_p$ and $\omega = \Omega_e$, respectively, and the resonances at $\theta = \pi/2$ are at $S = 0$. The solutions of $S = 0$ are called the hybrid frequencies. The only hybrid frequency in the magnetoionic theory is the *upper hybrid frequency*

$$\omega_{UH} = (\omega_p^2 + \Omega_e^2)^{1/2}. \quad (13.29)$$

There is also a lower hybrid frequency associated with the contribution of the ions, and there are ion–ion hybrid frequencies when more than one ionic species is included. In the magnetoionic theory one solves (13.28) explicitly for the resonant frequencies $\omega = \omega_\pm(\theta)$:

$$\omega_\pm^2(\theta) = \tfrac{1}{2}(\omega_p^2 + \Omega_e^2) \pm \tfrac{1}{2}\left[(\omega_p^2 + \Omega_e^2)^2 - 4\omega_p^2\Omega_e^2\cos^2\theta\right]^{1/2}. \quad (13.30)$$

For $\omega_p > \Omega_e$ the resonance at $\omega = \omega_+(\theta)$ is in the z mode and the resonance at $\omega = \omega_-(\theta)$ is in the whistler mode.

Waves propagating through a plasma in which there are gradients in the plasma parameters tend to be refracted into the direction of increasing refractive index and away from the direction of decreasing refractive index. Hence waves tend to refract towards resonances and away from cutoffs. Collisionless damping, due to gyromagnetic absorption by thermal particles, becomes strong at resonances and if a wave encounters a resonance it tends to be absorbed.

13.4 The Polarization of the Magnetoionic Waves

The polarization vector $\mathbf{e}_M(\mathbf{k})$ for any wave mode M in a magnetized

172   *Waves in Plasmas*

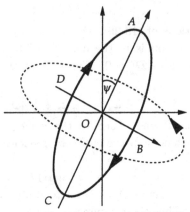

Fig. 13.4 The axial ratio $|T|$ is equal to $|AC|/|BD|$. The solid ellipse
corresponds to $T > 0$. The wave is propagating into the page.
The two ellipses correspond to orthogonal polarizations.

plasma may be expressed in terms of the set of basis vectors

$$\boldsymbol{\kappa} = (\sin\theta, 0, \cos\theta), \quad \mathbf{t} = (\cos\theta, 0, -\sin\theta), \quad \mathbf{a} = (0, 1, 0). \quad (13.31)$$

These are unit vectors along the wave vector $\mathbf{k}$, along the direction perpendicular to $\mathbf{k}$ in the $\mathbf{B}$–$\mathbf{k}$ plane, and along the direction orthogonal to
both $\mathbf{B}$ and $\mathbf{k}$, respectively. The Onsager relations imply that the component of the electric vector along $\mathbf{a}$ is out of phase with the components
in the $\mathbf{B}$–$\mathbf{k}$ plane. It is convenient to write

$$\mathbf{e}_M = \frac{L_M \boldsymbol{\kappa} + T_M \mathbf{t} + \mathrm{i}\mathbf{a}}{(L_M^2 + T_M^2 + 1)^{1/2}}. \quad (13.32)$$

The longitudinal part of the polarization vector is described by L_M and
the transverse part is described by T_M.

The transverse part corresponds to an elliptical polarization, with
$|T_M|$ the *axial ratio* of the polarization ellipse. Here this axial ratio is
chosen to be of the component along $\mathbf{t}$ to the component along $\mathbf{a}$. The
triad of unit vectors (13.31) forms a right hand set, and hence the sign of
T_M determines the handedness of the ellipse, with $T_M > 0$ corresponding to right hand polarization and $T_M < 0$ corresponding to left hand
polarization. The polarization ellipse is illustrated in Figure 13.4.

The polarization vectors for the magnetoionic waves are constructed
from any column of the matrix of cofactors λ_{ij}, corresponding to (13.22).
It is convenient to choose the middle column:

$$\lambda_{i2} = \begin{pmatrix} \mathrm{i}D(P - n^2 \sin^2\theta) \\ -An^2 + PS \\ -\mathrm{i}Dn^2 \sin\theta\cos\theta \end{pmatrix},$$

with A given by (13.24). On dividing each component by $\mathrm{i}(An^2 - PS)$, so that the component along **a** is equal to i, setting $n^2 = n_M^2$, and writing the resulting vector in the form (13.32), one identifies

$$T_M = \frac{DP\cos\theta}{An_M^2 - PS}, \quad L_M = \frac{(P - n_M^2)D\sin\theta}{An_M^2 - PS}. \tag{13.33}$$

On inserting explicit expressions for the refractive indices into (13.33) one finds explicit expressions for the polarization vectors. However, for computational and other purposes it is more convenient to reverse this procedure, and to develop an alternative procedure as follows.

Note first that if one interprets the expression (13.33) for T_M in terms of n_M^2 as a formal relation between T and n^2, then it implies that $1/T$ is a linear function of n^2. It then follows from the fact that n^2 satisfies a quadratic equation that $1/T$ also satisfies a quadratic equation, and hence that T itself satisfies a quadratic equation. This equation is

$$T^2 - \frac{(PS - S^2 + D^2)\sin^2\theta}{PD\cos\theta}T - 1 = 0. \tag{13.34}$$

The alternative procedure is to solve (13.34) for $T = T_M$ and to find n_M^2 and L_M by inverting (13.33) to express n_M^2 and L_M as functions of T_M.

On substituting the explicit forms (13.14) for S, D and P in terms of the magnetoionic parameters X and Y, (13.34) becomes

$$T^2 + \frac{Y\sin^2\theta}{(1 - X)\cos\theta}T - 1 = 0. \tag{13.35}$$

The expressions for n^2 and L as functions of T are

$$n^2 = 1 - \frac{XT}{T - Y\cos\theta} = 1 - \frac{X(1 - X)(1 + YT\cos\theta)}{1 - X - Y^2 + XY^2\cos^2\theta}, \tag{13.36}$$

$$L = \frac{XY\sin\theta}{1 - X}\frac{T}{T - Y\cos\theta} = \frac{XY\sin\theta(1 + YT\cos\theta)}{1 - X - Y^2 + XY^2\cos^2\theta}, \tag{13.37}$$

where the alternative forms follow by use of the quadratic equation (13.35).

Explicit solutions of (13.35) are

$$T = T_\sigma = \frac{Y(1 - X)\cos\theta}{\frac{1}{2}Y^2\sin^2\theta - \sigma\Delta} = \frac{-\frac{1}{2}Y^2\sin^2\theta - \sigma\Delta}{Y(1 - X)\cos\theta}, \tag{13.38}$$

with

$$\Delta^2 = \frac{1}{4}Y^4\sin^4\theta + (1 - X)^2Y^2\cos^2\theta, \tag{13.39}$$

and where $\sigma = \pm$ labels the two modes. The label $\sigma = 1$ corresponds to the ordinary mode and the label $\sigma = -1$ corresponds to the extraordinary mode. The two polarization ellipses are orthogonal:

$$T_+T_- = -1. \tag{13.40}$$

Approximations to the axial ratio follow by considering the ratio of the two terms in (13.39). For $|(1 - X)\cos\theta| \gg \frac{1}{2}Y\sin^2\theta$ one finds

$$T_\sigma \approx -\sigma\frac{\cos\theta}{|\cos\theta|}\frac{1-X}{|1-X|}\left[1 + \frac{\sigma Y\sin^2\theta}{2|(1-X)\cos\theta|} + \cdots\right]. \qquad (13.41)$$

The leading terms in (13.41) correspond to circular polarizations, with the handedness such that the electric vector in the x mode and in the whistler mode rotate in the same sense as that in which electrons gyrate (right hand screw sense relative to **B**) and the electric vector in the o mode and the z mode rotate in the opposite sense. This is called the *quasi-circular limit.* The corresponding approximation to the dispersion relations is, for $Y|\cos\theta| \ll 1$, and $X < 1$,

$$n_\sigma^2 = 1 - X(1-\sigma Y|\cos\theta| + \cdots) = 1 - \frac{\omega_p^2}{\omega^2}(1 - \sigma\frac{\Omega_e|\cos\theta|}{\omega} + \cdots). \qquad (13.42)$$

At high frequencies the magnetoionic waves are nearly circularly polarized except for a small range of angles of propagation close to perpendicular to the magnetic field.

In the opposite limit $|(1 - X)\cos\theta| \ll \frac{1}{2}Y\sin^2\theta$, called the *quasi-planar limit* or *quasi-linear limit,* one has

$$T_o \approx \infty, \quad n_o^2 \approx 1 - X, \quad L_o \approx \frac{XY\sin\theta}{1-X}; \qquad (13.43)$$

$$T_x \approx 0, \quad n_x^2 \approx 1 - \frac{X(1-X)}{1 - X - Y^2 + XY^2\cos^2\theta},$$

$$L_x \approx \frac{XY\sin\theta}{1 - X - Y^2 + XY^2\cos^2\theta}. \qquad (13.44)$$

The transverse parts of the polarization correspond to linear polarizations along **t** and **a** for the ordinary and extraordinary modes respectively.

Exercise Set 13

13.1 Consider a model for Langmuir waves in which the electrons are treated as a compressible gas with pressure $P = n_e m_e V_e^2$ satisfying the adiabatic law $P \propto n_e^\gamma$ where γ is the adiabatic index.

(a) Assuming an equation of fluid motion of the form

$$m_e n_e \frac{d\mathbf{v}}{dt} = -en_e\mathbf{E} - \operatorname{grad} P,$$

show that the implied dispersion relation for Langmuir waves is

$$\omega^2 = \omega_p^2 + \gamma k^2 V_e^2. \qquad (E13.1)$$

Hint: Use a first order perturbation treatment with the zeroth order corresponding to $P = 0$. Assume $\omega \gg \mathbf{k} \cdot \mathbf{v}$.

(b) This model reproduces the correct form (13.1) for $\gamma = 3$. Can this value of γ be justified, or is the model inadequate to describe Langmuir waves?

13.2 Show that there are no solutions of the transverse dispersion equation corresponding to the limit $y_\alpha \ll 1$ for all α in the expression (10.24) for $K^T(\omega, k)$.

(a) Specifically, show that in this limit the dispersion equation for transverse waves reduces to

$$k^2 c^2 / \omega^2 = 1 - 1/k^2 \lambda_D^2. \qquad (E13.2)$$

(b) Show that the solutions of this equation for k^2 are incompatible with the assumption $y_\alpha \ll 1$.

13.3 Show that the magnetoionic wave modes, labeled here as $\pm$, which does not necessarily correspond to $\sigma = \pm$ in (13.38), have the following properties in the special cases indicated:

(a) for $\theta = 0$,

$$n_\pm^2 = \frac{1 - X - Y^2 \mp XY}{1 - Y^2}, \quad T_\pm = \pm 1, \quad L_\pm = 0; \qquad (E13.3)$$

(b) for $\theta = \pi/2$,

$$n_o^2 = 1 - X, \quad T_o = \infty, \quad L_o = \frac{XY}{1 - X};$$
$$\qquad\qquad\qquad\qquad\qquad\qquad\qquad (E13.4)$$
$$n_x^2 = 1 - \frac{X(1 - X)}{1 - X - Y^2}, \quad T_x = 0, \quad L_x = -\frac{XY}{1 - X - Y^2};$$

(c) at a cutoff frequency,

$$n^2 = 0, \quad \text{and} \quad \mathbf{e} = 2^{-1/2}(1, \pm i, 0), \quad \text{or} \quad \mathbf{e} = (0, 0, 1). \qquad (E13.5)$$

Hint: Show that the cutoff in the latter case is at $1 - X = 0$ with $T = \infty$, $L/T = -\cot\theta$.

13.4 The properties of the cold plasma wave modes are derived by starting from the quadratic equation (13.34) for the axial ratio, solving it and inverting (13.33) to find n^2 and L as functions of T.

(a) Show that (13.23), (13.33) and (13.34) imply

$$n^2 = \frac{P}{A}\left(S + \frac{D\cos\theta}{T}\right) = \frac{S^2 - D^2}{S - DT\cos\theta}, \qquad (E13.6)$$

$$L = \frac{\sin\theta}{A}(P - S)T\cos\theta - D$$

$$= \frac{\sin\theta}{P}\frac{(PS - S^2 + D^2)T\cos\theta - PD}{S - DT\cos\theta}. \qquad (E13.7)$$

(b) Substitute the solutions

$$T_\pm = \frac{(PS - S^2 + D^2)\sin^2\theta \pm F}{2PD\cos\theta}, \qquad (E13.8)$$
$$F^2 = (PS - S^2 + D^2)^2\sin^4\theta + 4P^2D^2\cos^2\theta,$$

in (E13.6) and show that (13.25) is reproduced.

13.5 The magnetoionic waves are nearly circularly polarized and nearly linearly polarized in two opposite limits depending on which term dominates in the expression (13.39) for Δ^2.

(a) Show that the angle at which the two contributions are equal is

$$\theta_0 = \arcsin\left(\frac{2^{1/2}}{Y}\left\{\left[(1 - X)^4\right.\right.\right.$$

$$\left.\left.\left. + Y^2(1 - X)^2\right]^{1/2} - (1 - X)^2\right\}^{1/2}\right). \qquad (E13.9)$$

(b) Show that (E13.9) reduces to $\theta_0 \approx \arccos\frac{1}{2}Y$ for $X \ll 1$, $Y \ll 1$.

13.6 Double solutions of the dispersion equation occur when both Λ and its derivative, for example, with respect to n^2, vanish simultaneously.

(a) Show that a double solution of the dispersion equation for the magnetoionic waves occurs at $X = 1$, $\sin\theta = 0$.

(b) Plot n^2 as a function of $X \approx 1$ for small $\sin\theta$.

13.7 Show that the refractive index curves for the magnetoionic modes pass through the points $n^2 = 1$ and $n^2 = 0$ at $\omega = \omega_p$, and determine the polarization vectors at these two points.

13.8 At low frequencies a plasma may be regarded as a magnetized fluid. The linearized forms of the fluid equations, assuming an adiabatic equation of state for the plasma, lead to an MHD wave equation which may written in the form

$$\Gamma_{ij}(\omega, \mathbf{k})\xi_i(\omega, \mathbf{k}) = 0, \qquad (E13.10)$$

where $\boldsymbol{\xi}(\omega, \mathbf{k})$ is the Fourier transform of the fluid displacement, and with

$$\Gamma_{ij}(\omega, \mathbf{k}) = \omega^2 \delta_{ij} - k^2 c_s^2 \kappa_i \kappa_j - k^2 v_A^2 [\kappa_i \kappa_j$$
$$- \cos\theta(\kappa_i b_j + b_i \kappa_j) + \cos^2\theta\, \delta_{ij}], \qquad (E13.11)$$

where θ is the angle between $\boldsymbol{\kappa}$ (in the x–z plane) and $\mathbf{b}$ (along the z axis), c_s is the sound speed, and v_A is the Alfvén speed.

(a)   Show that there are three solutions of the dispersion equation for v_ϕ^2, where v_ϕ is the phase speed:

$$v_\phi^2 = k^2 v_A^2 \cos^2\theta, \quad v_\phi^2 = k^2 v_\pm^2,$$
$$v_\pm^2 = \tfrac{1}{2}(v_A^2 + c_s^2) \pm \tfrac{1}{2}\left[(v_A^2 + c_s^2)^2 - 4v_A^2 c_s^2 \cos^2\theta\right]^{1/2}. \qquad (E13.12)$$

Remark: The first of these solutions is the Alfvén wave mode, which corresponds to shear or torsional waves in the magnetic field, and the other two solutions correspond to the fast and slow magnetoacoustic modes.

(b)   Show that the direction of the fluid displacement in an Alfvén wave is given by

$$\boldsymbol{\xi}_A = (0, 1, 0).$$

(c)   Show that the direction of the fluid displacement in the magneto-acoustic wave is given by

$$\boldsymbol{\xi}_\pm = (\sin\psi_\pm, 0, \cos\psi_\pm) \quad \tan\psi_\pm = \frac{c_s^2 \sin\theta \cos\theta}{v_\pm^2 - v_A^2 - c_s^2 \sin^2\theta}.$$

14

The Polarization of Transverse Waves

Preamble

The general theory used in Chapter 11 to derive the polarization vector for a specific wave mode fails when the dispersion equation has a double solution, corresponding to two degenerate wave modes. The most important example where this occurs is for transverse waves in an isotropic medium. Such waves are regarded as an arbitrary mixture of two degenerate wave modes, and as a consequence it is not possible to describe the polarization of such waves in terms of a single polarization vector in general. The description of the polarization of transverse waves is the question discussed in detail in this chapter.

14.1 The Polarization Tensor

Consider an isotropic medium that is not optically active. As argued in Chapter 10, Λ_{ij} may be separated into longitudinal and transverse parts, and the same argument implies that the matrix of cofactors λ_{ij} may also be separated into longitudinal and transverse parts. Specifically one has

$$\Lambda_{ij}(\omega, \mathbf{k}) = K^{\mathrm{L}}(\omega, k)\kappa_i\kappa_j + \left[K^{\mathrm{T}}(\omega, k) - n^2\right](\delta_{ij} - \kappa_i\kappa_j), \quad (14.1)$$

$$\Lambda(\omega, \mathbf{k}) = K^{\mathrm{L}}(\omega, k)\left[K^{\mathrm{T}}(\omega, k) - n^2\right]^2, \quad (14.2)$$

$$\lambda_{ij}(\omega, \mathbf{k}) = \left[K^{\mathrm{T}}(\omega, k) - n^2\right]^2\kappa_i\kappa_j$$
$$+ K^{\mathrm{L}}(\omega, k)\left[K^{\mathrm{T}}(\omega, k) - n^2\right](\delta_{ij} - \kappa_i\kappa_j). \quad (14.3)$$

The solutions of $K^{\mathrm{L}}(\omega, k) = 0$ are identified in Chapter 10 as corresponding to longitudinal waves, and the solutions of $n^2 = K^{\mathrm{T}}(\omega, k)$ to transverse waves. The proof of the first of these assertions follows by substituting $K^L(\omega, k) = 0$ into (14.3), giving $\lambda_{ij}(\omega, \mathbf{k}) \propto \kappa_i \kappa_j$, and implying that the polarization is longitudinal. However, for transverse waves there is a double solution of the dispersion equation and on substituting $n^2 = K^T(\omega, k)$ into (14.3) one obtains a null result so that the polarization vector is indeterminate. This is as expected for degenerate solutions: no unique polarization vector can be identified. All that can be concluded from the wave equation $\Lambda_{ij}(\omega, \mathbf{k}) A_j(\omega, \mathbf{k}) = 0$ with $n^2 = K^T(\omega, k)$ in (14.1) is that any solution must satisfy $\boldsymbol{\kappa} \cdot \mathbf{A}(\omega, \mathbf{k}) = 0$, that is, the waves are transverse.

The most general description of the polarization of transverse waves is in terms of *polarization tensor*, p_{ij} say. The polarization tensor must be hermitian, and it must have only transverse components:

$$p_{ij} = p_{ji}^*, \qquad \kappa_i p_{ij} = 0 = p_{ij} \kappa_j. \qquad (14.4)$$

A normalization condition needs to be imposed, and we choose to require that the trace be equal to unity:

$$p_{ii} = 1. \qquad (14.5)$$

If one chooses coordinate axes such that one axis is along $\boldsymbol{\kappa}$ then p_{ij} has non-zero components only along the other two directions. It is convenient to adopt a matrix representation in which the row and column of zeros is omitted. Let the basis vectors be $\mathbf{e}^1$, $\mathbf{e}^2$, $\boldsymbol{\kappa}$, and let the 2×2 matrix representation of p_{ij} be denoted by

$$p^{\alpha\beta} = e_i^{*\alpha} e_j^{\beta} p_{ij}, \qquad (14.6)$$

where α, β run over 1, 2.

The most general form of $p^{\alpha\beta}$ is implied by the fact that, apart from normalization factors, there are only four independent 2×2 hermitian matrices. These may be chosen as the unit matrix and the three Pauli matrices. Noting the normalization condition that the trace is equal to unity, the general form is

$$p^{\alpha\beta} = \frac{1}{2} \left[\begin{pmatrix} 1 & 0 \\ 0 & 1 \end{pmatrix} + q \begin{pmatrix} 1 & 0 \\ 0 & -1 \end{pmatrix} + u \begin{pmatrix} 0 & 1 \\ 1 & 0 \end{pmatrix} + v \begin{pmatrix} 0 & -i \\ i & 0 \end{pmatrix} \right], \quad (14.7)$$

where q, u, v are real parameters that describe the state of polarization. Three parameters are required to describe an arbitrary state of transverse polarization (for given direction of wave propagation). In contrast, only two parameters are required to describe an arbitrary polarization vector, for example, the parameters T_M and L_M introduced in (13.32).

The description of the polarization in terms of complex 2×2 matrices is called the *Jones calculus*, cf. §14.4. An alternative description in terms of real four-dimensional matrices is called the *Mueller calculus*. The most familiar form of the Mueller calculus is for the specific intensity of radiation, which is the power per unit area per unit solid angle per unit frequency range. The specific intensity is written as the polarization matrix

$$I^{\alpha\beta} = I\,p^{\alpha\beta} = \frac{1}{2}\begin{pmatrix} I+Q & U-iV \\ U+iV & I-Q \end{pmatrix}, \qquad (14.8)$$

where I, Q, U, V are the *Stokes parameters*. The total intensity is I, and one has

$$Q = qI, \quad U = uI, \quad V = vI. \qquad (14.9)$$

14.2 Interpretation of the Polarization Tensor

An interpretation of the general form of the polarization tensor is developed by considering the two special cases of completely polarized and unpolarized radiation.

The radiation is said to be *completely polarized* if the polarization may be described by a polarization vector. Let this polarization vector have components e_0^1, e_0^2, 0 with respect to the basis $\mathbf{e}^1$, $\mathbf{e}^2$, $\boldsymbol{\kappa}$ used for the polarization tensor. For completely polarized radiation one has

$$p^{\alpha\beta} = e_0^\alpha e_0^{*\beta}. \qquad (14.10)$$

Thus for completely polarized radiation the polarization tensor may be written as the outer product of a vector with itself. In matrix language, the condition for this to be possible is that the matrix be of rank one, that is, the determinant of the 2×2 matrix must vanish. Hence the condition for the radiation to be completely polarized is

$$\det [p^{\alpha\beta}] = \tfrac{1}{4}(1 - q^2 - u^2 - v^2) = 0. \qquad (14.11)$$

The radiation is said to be *unpolarized* if a separation into any pair of orthogonal polarizations gives equal components. The polarization tensor for unpolarized radiation is

$$p^{\alpha\beta} = \frac{1}{2}\begin{pmatrix} 1 & 0 \\ 0 & 1 \end{pmatrix}, \quad q = u = v = 0. \qquad (14.12)$$

In general radiation is partially polarized. The *degree of polarization r* is defined such that $r = 1$ corresponds to completely polarized radiation and $r = 0$ corresponds to unpolarized radiation. In accord with (14.11) and (14.12), one identifies

$$r = (q^2 + u^2 + v^2)^{1/2}. \qquad (14.13)$$

One may make a unique separation of any polarization tensor into an unpolarized part $p_u^{\alpha\beta}$ and a polarized part $p_p^{\alpha\beta}$:

$$p^{\alpha\beta} = p_u^{\alpha\beta} + p_p^{\alpha\beta}, \quad p_u^{\alpha\beta} = \frac{1-r}{2} \begin{pmatrix} 1 & 0 \\ 0 & 1 \end{pmatrix}, \quad p_p^{\alpha\beta} = r e_0^\alpha e_0^{*\beta}, \quad (14.14)$$

where the specific form of the polarization vector e_0^α has yet to be determined.

An arbitrary transverse polarization vector is of the form

$$\mathbf{e} = \gamma \mathbf{e}^1 + \delta \mathbf{e}^2, \quad |\gamma|^2 + |\delta|^2 = 1, \quad (14.15)$$

where γ and δ are complex numbers. Let us show that this corresponds to an elliptical polarization. On making a rotation of the coordinate axes through an angle ψ, one has

$$\mathbf{e}^{1'} = \cos\psi\, \mathbf{e}^1 + \sin\psi\, \mathbf{e}^2, \quad \mathbf{e}^{2'} = -\sin\psi\, \mathbf{e}^1 + \cos\psi\, \mathbf{e}^2. \quad (14.16)$$

Then (14.15) gives

$$\mathbf{e} = (\gamma \cos\psi + \delta \sin\psi)\mathbf{e}^{1'} + (-\gamma \sin\psi + \delta \cos\psi)\mathbf{e}^{2'}. \quad (14.17)$$

It is always possible to choose ψ such that (14.17) is of the form

$$\mathbf{e} = \frac{e^{i\Phi}}{(1+T^2)^{1/2}}\left(T\mathbf{e}^{1'} + i\mathbf{e}^{2'}\right), \quad (14.18)$$

where T is real. Specifically one has

$$\gamma \cos\psi + \delta \sin\psi = \frac{e^{i\Phi}T}{(1+T^2)^{1/2}}, \quad -\gamma \sin\psi + \delta \cos\psi = \frac{ie^{i\Phi}}{(1+T^2)^{1/2}}, \quad (14.19)$$

or

$$\gamma = \frac{e^{i\Phi}}{(1+T^2)^{1/2}}(T \cos\psi - i \sin\psi), \quad \delta = \frac{e^{i\Phi}}{(1+T^2)^{1/2}}(T \sin\psi + i \cos\psi). \quad (14.20)$$

The form (14.18) describes an elliptical polarization which has axial ratio T with respect to the axis $\mathbf{e}^{1'}$. The phase Φ is physically unimportant and reflects the fact that there is an arbitrary phase factor in any polarization vector. The general form of the polarization ellipse for a wave propagating into the page is illustrated in Figure 14.1.

The polarization tensor for the completely polarized radiation described by (14.15) with (14.20) is

$$p^{\alpha\beta} = \begin{pmatrix} \gamma \\ \delta \end{pmatrix} \begin{pmatrix} \gamma^* & \delta^* \end{pmatrix}$$

$$= \frac{1}{1+T^2} \begin{pmatrix} T^2 \cos^2\psi + \sin^2\psi & (T^2-1)\sin\psi\cos\psi - iT \\ (T^2-1)\sin\psi\cos\psi + iT & T^2 \sin^2\psi + \cos^2\psi \end{pmatrix}. \quad (14.21)$$

One identifies the right hand side of (14.21) with the polarization matrix (14.7), and hence identifies q_0, u_0 and v_0 for completely polarized

The Polarization of Transverse Waves

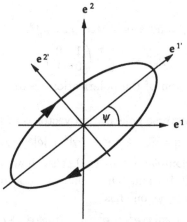

Fig. 14.1 The polarization ellipse for a wave propagating into the page.
The directions $\mathbf{e}^{1'}$ and $\mathbf{e}^{2'}$ are defined by (14.18).

radiation. For partially polarized radiation the polarized part is written as r times this polarization tensor, so that q, u and v are r times q_0, u_0 and v_0, respectively. Thus the parameters q, u, v are identified in terms of the parameters T and ψ:

$$\frac{q}{r} = \left(\frac{T^2 - 1}{T^2 + 1}\right)\cos 2\psi, \quad \frac{u}{r} = \left(\frac{T^2 - 1}{T^2 + 1}\right)\sin 2\psi, \quad \frac{v}{r} = \frac{2T}{T^2 + 1}.$$
(14.22)

The quantities

$$r_1 = (q^2 + u^2)^{1/2} = r\left(\frac{T^2 - 1}{T^2 + 1}\right),$$
(14.23)

$$r_V = v = r\left(\frac{2T}{T^2 + 1}\right),$$
(14.24)

are identified as the *degree of linear polarization* and the *degree of circular polarization*, respectively.

The parameters q/r, u/r, v/r that describe the polarized part of the radiation have a geometric interpretation in terms of the Poincaré sphere. This geometric construction is particularly helpful for understanding the changes in the state of polarization as radiation propagates through a weakly anisotropic medium.

The *Poincaré sphere* is defined as a unit sphere in which the state of polarization is represented by a point on the surface of the sphere. It is conventional to have the north and south poles of the sphere representing right and left hand circular polarizations, respectively. The points on the equator then represent linear polarizations with the plane of polarization rotating through 180° as the representative point moves

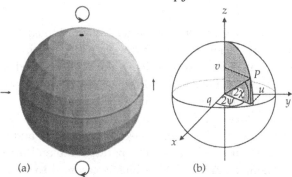

Fig. 14.2 (a) The Poincaré sphere; (b) the parameters ψ, χ, q, u, v for an arbitrary point P on the sphere.

around the equator through $360°$. Let 2χ and 2ψ be the latitude and the longitude, respectively, of a point on the sphere. Then the point on the sphere corresponding to the parameters q/r, u/r, v/r is identified by writing

$$q/r = \cos(2\chi)\cos(2\psi), \quad u/r = \cos(2\chi)\sin(2\psi), \quad v/r = \sin(2\chi).$$
(14.25)

The representative point on the sphere is illustrated in Figure 14.2.

14.3 The Weak Anisotropy Limit

In an isotropic medium transverse waves have two degenerate states of polarization, and in an anisotropic medium the natural modes are non-degenerate and not necessarily transverse. There are many situations where the effect of anisotropy is relatively weak, so that to a first approximation the medium is regarded as isotropic, with the inclusion of the anisotropy breaking the degeneracy of the transverse states of polarization. An anisotropic medium in which the effect of the anisotropy is treated as a perturbation in this sense is said to be *weakly anisotropic*. Radiation incident on or propagating through such a medium splits into components in the two natural modes, and it is implicit in the weak anisotropy assumption that the difference in the ray paths of these two components is unimportant. However, the difference between the refractive indices of the two natural modes is important because it causes these components to get out of phase with each other. When the components are recombined, the phase difference produces radiation with a polarization that is different from that of the incident radiation. Thus a weakly anisotropic medium is regarded as continuously changing the state of polarization of transverse radiation passing through it. Specific examples of this effect include a quarter wave plate, which is a slab of uniaxial crystal that can convert incident linearly polarized radiation into emerging circularly polarized radiation, and Faraday rotation, in which the plane of linear polarization of radiation rotates as the radiation propagates through a medium in a magnetic field. A general method for treating this effect is discussed in this section and in §14.4.

Let the dielectric tensor $K_{ij}(\omega, \mathbf{k})$ be separated into an isotropic part and an anisotropic part, $\Delta K_{ij}(\omega, \mathbf{k})$, which is assumed to be small. To a first approximation, the anisotropic part is neglected, and then the wave equation has two degenerate solutions corresponding to transverse waves, with dispersion relation $n = n_0(\omega)$ say. (The dispersion equation may also have a solution corresponding to longitudinal waves, but such a solution is of no interest here.) On including the anisotropic part $\Delta K_{ij}(\omega, \mathbf{k})$ as a perturbation, the degeneracy is broken. We are interested only in the transverse components and so project the wave equation onto the two transverse directions, denoted by greek superscripts running over 1, 2. The wave equation then becomes

$$\left[\left(n_0^2(\omega) - n^2 \right) \delta^{\alpha\beta} + \Delta K^{\alpha\beta} \right] A^\beta = 0. \tag{14.26}$$

The hermitian part of $\Delta K^{\alpha\beta}$ has zero trace because the trace has been included in the isotropic part of the dielectric tensor. Thus on separating $\Delta K^{\alpha\beta}$ into hermitian and antihermitian parts, one writes the hermitian part in the form

$$\Delta K^{\mathrm{H}\alpha\beta} = \begin{pmatrix} \Delta K_Q^{\mathrm{H}} & \Delta K_U^{\mathrm{H}} - \mathrm{i}\Delta K_V^{\mathrm{H}} \\ \Delta K_U^{\mathrm{H}} + \mathrm{i}\Delta K_V^{\mathrm{H}} & -\Delta K_Q^{\mathrm{H}} \end{pmatrix}, \tag{14.27}$$

and the antihermitian part in the form

$$\Delta K^{\mathrm{A}\alpha\beta} = \begin{pmatrix} \Delta K_I^{\mathrm{A}} + \Delta K_Q^{\mathrm{A}} & \Delta K_U^{\mathrm{A}} - \mathrm{i}\Delta K_V^{\mathrm{A}} \\ \Delta K_U^{\mathrm{A}} + \mathrm{i}\Delta K_V^{\mathrm{A}} & \Delta K_I^{\mathrm{A}} - \Delta K_Q^{\mathrm{A}} \end{pmatrix}. \tag{14.28}$$

All the ΔK_A^{H}, $A = I, Q, U, V$ in (14.27) are then real and all the ΔK_A^{A} in (14.28) are imaginary.

It is possible to orient the axes so that ΔK_U^{H} and ΔK_U^{A} are zero. In a magnetized medium this corresponds to choosing $\mathbf{e}^1$ (or $\mathbf{e}^2$) along the projection of the magnetic field on the plane orthogonal to $\boldsymbol{\kappa}$. Let us suppose that the projection of the magnetic field is along $\mathbf{e}^{1'}$ at an angle ψ to $\mathbf{e}^1$. If one writes, in accordance with (14.22),

$$\Delta K_Q = \Delta K_\perp \cos(2\psi), \quad \Delta K_U = \Delta K_\perp \sin(2\psi), \tag{14.29}$$

then $\Delta K^{\alpha'\beta'}$ relative to the new coordinate system is of the form (14.27) and (14.28) with ΔK_Q replaced by $\Delta K_\perp$ and ΔK_U replaced by zero.

It is straightforward to solve for the properties of the two natural modes in the two-dimensional subspace, cf. (14.26). On neglecting the antihermitian part $\Delta K^{A\alpha\beta}$, the determinant of the matrix of coefficients gives

$$
\begin{aligned}
&\left[n_0^2(\omega) - n^2\right]^2 - (\Delta K^{\mathrm{H}})^2 = 0, \\
&\Delta K^{\mathrm{H}} = [(\Delta K_Q^{\mathrm{H}})^2 + (\Delta K_U^{\mathrm{H}})^2 + (\Delta K_V^{\mathrm{H}})^2]^{1/2},
\end{aligned}
\tag{14.30}
$$

and the matrix of cofactors $\lambda^{\alpha\beta}$ has an hermitian part

$$
\lambda^{\mathrm{H}\alpha\beta} = \begin{pmatrix} n_0^2(\omega) - n^2 - \Delta K_Q^{\mathrm{H}} & -\Delta K_U^{\mathrm{H}} + \mathrm{i}\Delta K_V^{\mathrm{H}} \\ -\Delta K_U^{\mathrm{H}} - \mathrm{i}\Delta K_V^{\mathrm{H}} & n_0^2(\omega) - n^2 + \Delta K_Q^{\mathrm{H}} \end{pmatrix}.
\tag{14.31}
$$

The two solutions of (14.30) are labeled as the $+$ and $-$ modes. Their refractive indices are given by

$$
n^2 = n_\pm^2(\omega, \psi) = n_0^2(\omega) \pm \Delta K^{\mathrm{H}},
\tag{14.32}
$$

and their polarization vectors are given by

$$
e_\pm^\alpha = \frac{(T_\pm \cos\psi - \mathrm{i}\sin\psi)\mathbf{e}^1 + (T_\pm \sin\psi + \mathrm{i}\cos\psi)\mathbf{e}^2}{(1 + T_\pm^2)^{1/2}},
\tag{14.33}
$$

$$
T_+ = \cot\chi, \quad T_- = -\tan\chi.
$$

The parameter χ is as introduced in (14.25) in connection with the Poincaré sphere and appears here on writing, by analogy with (14.25),

$$
\Delta K_Q^{\mathrm{H}} = \Delta K^{\mathrm{H}} \cos(2\chi)\cos(2\psi), \quad \Delta K_U^{\mathrm{H}} = \Delta K^{\mathrm{H}} \cos(2\chi)\sin(2\psi),
$$

$$
\Delta K_V^{\mathrm{H}} = \Delta K^{\mathrm{H}} \sin(2\chi).
\tag{14.34}
$$

One separates a transverse wave with an arbitrary initial polarization into components in the two natural modes. As the wave propagates, the components in the two modes vary in space as $\mathrm{e}^{\mathrm{i}\mathbf{k}_\pm \cdot \mathbf{x}}$. Each component is also damped, but we neglect this for the present. Let s denote distance along the ray path. The relative phase of the component in the $+$ mode to that in the $-$ mode then varies as $\mathrm{e}^{\mathrm{i}\omega(n_+ - n_-)s/c}$. This implies a continually increasing phase difference between the two components. A consequence of such a phase difference is that when the two components are recombined the polarization is different from the initial polarization; this implies a change in the state of polarization which varies with distance of propagation through the medium. There is no widely used generic name for this process. Here the name *generalized Faraday rotation* is used to describe it. *Faraday rotation* is the particular case when the natural modes of the medium are circularly polarized, and then the polarization change involves a rotation of the plane of linear polarization.

14.4 The Transfer Equation in the Jones Calculus

A formal treatment of generalized Faraday rotation, now also including the effect of polarization-dependent damping, is developed as follows. The wave equation (14.26) in the weak anisotropy limit is written in the form

$$\left(n^2 - n_0^2\right)A^\alpha = \Delta K^{\alpha\beta}A^\beta, \tag{14.35}$$

where the argument of $n_0(\omega)$ is now omitted. After multiplying by $i\omega/c$, the coefficient on the left hand side of (14.35) is approximated by $2n_0$ times $i(k - k_0)$ with $k_0 = \omega n_0/c$. On inverting the Fourier transform, the factor $i(k - k_0)$ is replaced by the spatial derivative along the direction of wave propagation. Then (14.35) reduces to

$$\frac{dA^\alpha}{ds} = \frac{i\omega}{2n_0 c}\Delta K^{\alpha\beta}A^\beta. \tag{14.36}$$

In the Jones calculus one is concerned with the transfer of the wave amplitude. The transfer equation (14.36) in the Jones calculus retains information about the phase of the waves, and this is lost when a correlation function is taken below to derive the transfer equation in the Mueller calculus. Thus when one is interested in the phase, it is appropriate to use the Jones calculus. A disadvantage of the Jones calculus is that it does not readily allow one to treat unpolarized or partially polarized radiation. One may treat partially polarized radiation only by taking correlation functions, as is done below, and then the Mueller calculus is more convenient than the Jones calculus.

The right hand side of the transfer equation (14.36) is separated into two parts, by separating $\Delta K^{\alpha\beta}$ into hermitian and antihermitian parts. By using both (14.36) and its complex conjugate, it follows that when only the hermitian part of $\Delta K^{\alpha\beta}$ is retained, the magnitude of the amplitude does not change; that is, $A^{*\alpha}A^\alpha$ remains unchanged. Hence the hermitian part of $\Delta K^{\alpha\beta}$ can cause only a change in the state of polarization of the radiation. The antihermitian part of $\Delta K^{\alpha\beta}$ is associated with damping of the radiation.

The Jones calculus is used to construct the transfer equation for the polarization tensor $I^{\alpha\beta}$. The polarization tensor $I^{\alpha\beta}$ is proportional to the square of the amplitude of the waves; more generally it is defined as the average over phase of the outer product of A^α and $A^{*\beta}$, that is, as the correlation function $\langle A^\alpha A^{*\beta}\rangle$. It then follows from (14.36) and its complex conjugate that the polarization tensor varies according to

$$\frac{dI^{\alpha\beta}}{ds} = \frac{i\omega}{2n_0 c}\left(\Delta K^{\alpha\rho}\delta^{\beta\sigma} - \Delta K^{*\beta\sigma}\delta^{\alpha\rho}\right)I^{\rho\sigma}. \tag{14.37}$$

On separating $\Delta K^{\alpha\beta}$ on the right hand side of (14.37) into its hermitian and antihermitian parts, (14.37) is written in the form

$$\frac{dI^{\alpha\beta}}{ds} = \rho^{\alpha\beta\rho\sigma} I^{\rho\sigma} - \mu^{\alpha\beta\rho\sigma} I^{\rho\sigma}, \qquad (14.38)$$

with

$$\rho^{\alpha\beta\rho\sigma} = \frac{i\omega}{2n_0 c} \left(\Delta K^{\mathrm{H}\alpha\rho} \delta^{\beta\sigma} - \Delta K^{\mathrm{H}\sigma\beta} \delta^{\alpha\rho} \right), \qquad (14.39)$$

$$\mu^{\alpha\beta\rho\sigma} = -\frac{i\omega}{2n_0 c} \left(\Delta K^{\mathrm{A}\alpha\rho} \delta^{\beta\sigma} + \Delta K^{\mathrm{A}\sigma\beta} \delta^{\alpha\rho} \right). \qquad (14.40)$$

The coefficients $\rho^{\alpha\beta\rho\sigma}$ and $\mu^{\alpha\beta\rho\sigma}$ describe the effects of generalized Faraday rotation and of polarization-dependent damping, respectively. Note that they satisfy the following symmetry properties

$$\rho^{\alpha\beta\rho\sigma} = -\rho^{\rho\sigma\alpha\beta}, \quad \mu^{\alpha\beta\rho\sigma} = \mu^{\rho\sigma\alpha\beta}. \qquad (14.41)$$

14.5 The Transfer Equation in the Mueller Calculus

In the Mueller calculus one is concerned with how the Stokes parameters change. It is convenient to write the Stokes parameters as a *Stokes vector* S_A, with $S_1 = I$, $S_2 = Q$, $S_3 = U$, $S_4 = V$. In matrix form this becomes

$$S_A = \begin{pmatrix} I \\ Q \\ U \\ V \end{pmatrix}. \qquad (14.42)$$

A one-to-one correspondence is set up between the components of $I^{\alpha\beta}$ and the components of S_A. This correspondence also allows one to translate (14.38) into a transfer equation in the Mueller calculus. It is convenient to write

$$I^{\alpha\beta} = \tfrac{1}{2}\tau_A^{\alpha\beta} S_A, \quad S_A = \tau_A^{\beta\alpha} I^{\alpha\beta}, \qquad (14.43)$$

with

$$\tau_I^{\alpha\beta} = \begin{pmatrix} 1 & 0 \\ 0 & 1 \end{pmatrix}, \quad \tau_Q^{\alpha\beta} = \begin{pmatrix} 1 & 0 \\ 0 & -1 \end{pmatrix},$$
$$\tau_U^{\alpha\beta} = \begin{pmatrix} 0 & 1 \\ 1 & 0 \end{pmatrix}, \quad \tau_V^{\alpha\beta} = \begin{pmatrix} 0 & -i \\ i & 0 \end{pmatrix}. \qquad (14.44)$$

It is then straightforward to rewrite (14.37) in the Mueller calculus. One obtains

$$\frac{dS_A}{ds} = \rho_{AB} S_B - \mu_{AB} S_B, \qquad (14.45)$$

where the sum over the repeated index $B = 1$–4 is implied. In the matrix representation ρ_{AB} and μ_{AB} are 4×4 matrices which are given by

$$\rho_{AB} = \frac{i\omega}{4n_0 c} \left(\tau_A^{\beta\alpha} \Delta K^{\mathrm{H}\alpha\rho} \tau_B^{\rho\beta} - \tau_B^{\rho\sigma} \Delta K^{\mathrm{H}\sigma\beta} \tau_A^{\beta\rho} \right), \qquad (14.46)$$

$$\mu_{AB} = -\frac{i\omega}{4n_0 c}\left(\tau_A^{\beta\alpha}\Delta K^{A\alpha\rho}\tau_B^{\rho\beta} + \tau_B^{\rho\sigma}\Delta K^{A\sigma\beta}\tau_A^{\beta\rho}\right). \quad (14.47)$$

By inspection these satisfy the symmetry properties

$$\rho_{AB} = -\rho_{BA}, \quad \mu_{AB} = \mu_{BA}. \quad (14.48)$$

The explicit evaluation of the matrices ρ_{AB} and μ_{AB} involves substituting (14.27), and (14.28) into (14.46) and (14.47), and using (14.43). The final expression for ρ_{AB} is of the form

$$\rho_{AB} = \begin{pmatrix} 0 & 0 & 0 & 0 \\ 0 & 0 & -\rho_V & \rho_U \\ 0 & \rho_V & 0 & -\rho_Q \\ 0 & -\rho_U & \rho_Q & 0 \end{pmatrix}. \quad (14.49)$$

Explicit expressions for the parameters in (14.49) are written in terms of the properties of the natural modes using (14.32)–(14.34). On noting that ρ_Q, ρ_U, ρ_V are proportional to ΔK_Q^H, ΔK_U^H ΔK_V^H, one finds:

$$\rho_Q = -\Delta k\,\frac{T_+^2 - 1}{T_+^2 + 1}\cos(2\psi), \quad \rho_U = -\Delta k\,\frac{T_+^2 - 1}{T_+^2 + 1}\sin(2\psi),$$

$$\rho_V = -\Delta k\,\frac{2T_+}{T_+^2 + 1} \qquad , \quad (14.50)$$

with

$$\Delta k = \frac{\omega}{c}(n_+ - n_-) \approx \frac{\Delta K^H \omega}{n_0 c}, \quad (14.51)$$

where ΔK^H, as given by (14.30), is the rate per unit length at which the components in the two natural modes get out of phase.

The absorption matrix in (14.47) is of the form

$$\mu_{AB} = \begin{pmatrix} \mu_I & \mu_Q & \mu_U & \mu_V \\ \mu_Q & \mu_I & 0 & 0 \\ \mu_U & 0 & \mu_I & 0 \\ \mu_V & 0 & 0 & \mu_I \end{pmatrix}. \quad (14.52)$$

Explicit evaluation gives

$$\mu_A = -\frac{i\omega\Delta K_A^A}{n_0 c}, \quad (14.53)$$

where the ΔK_A^A are defined by (14.28). The diagonal terms in (14.52) describe polarization-independent absorption, and the off-diagonal terms are needed to describe polarization-dependent absorption.

The effect of polarization-dependent absorption is readily understood. Consider, for example, a medium with an absorption mechanism that applies only to circular polarization of one handedness, say right hand. Incident unpolarized radiation becomes increasingly left hand polarized

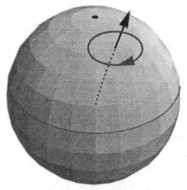

Fig. 14.3 The representative point for radiation in a weakly anisotropic medium rotates about the diagonal joining the points for the two natural modes.

as it propagates through such a medium due to the removal of the right hand polarized component. Similarly initially linearly polarized radiation becomes elliptical in the left hand sense, and the axial ratio T of the polarization ellipse approaches -1 after propagation over an arbitrarily long path.

14.6 Examples of the Transfer of Polarized Radiation

The transfer of polarization through an anisotropic medium has the following geometric interpretation on the Poincaré sphere. Let the initial polarization be described by a point P on the sphere, and let P be at a relative latitude $2\chi'$ and a relative longitude $2\psi'$ to the diagonal D defined by the natural modes. Then as the radiation propagates through the medium, the point representing the polarization rotates about D at constant χ'. The rate at which this rotation occurs is

$$\frac{\mathrm{d}\psi'}{\mathrm{d}z} = -\tfrac{1}{2}\Delta k, \tag{14.54}$$

where $\Delta k = (n_+ - n_-)\omega/c$ is the difference (14.51) in wavenumber between the two natural modes. This rotation is illustrated in Figure 14.3.

Let us apply this general theory to the two examples mentioned at the beginning of §14.3. First, consider a uniaxial crystal. The two modes are linearly polarized, corresponding to $\chi = 0$, $T_+ = 0$, $T_- = \infty$. Consider radiation incident perpendicular to the principal axis and orient the coordinate axes such that $\mathbf{e}^1$ is along the principal axis (which is in the surface of the crystal). This corresponds to setting $\psi = 0$. Then (14.50) implies $\rho_V = 0 = \rho_U$, and the transfer equation (14.45), neglecting any absorption, reduces to

$$\frac{d}{ds}\begin{pmatrix} U \\ V \end{pmatrix} = \rho_Q \begin{pmatrix} 0 & -1 \\ 1 & 0 \end{pmatrix}\begin{pmatrix} U \\ V \end{pmatrix}, \qquad (14.55)$$

with I and Q remaining constant. Linearly polarized radiation with its plane of polarization along the direction at 45° to the principal axis has $Q_0 = 0$, $U_0 = I_0$, $V_0 = 0$, where the subscript 0 refers to the incident values. Then (14.55) implies that its polarization changes periodically from linear to circular and back to linear as a function of distance s along the ray path:

$$U = U_0 \cos(\rho_Q s), \quad V = U_0 \sin(\rho_Q s). \qquad (14.56)$$

In a quarter wave plate the thickness d of the slab of crystal is adjusted so that one has $\rho_Q d = \pi/2$. Using (14.50) this corresponds to a thickness

$$d = \frac{\pi c}{2\omega}\frac{1}{|n_x - n_o|}. \qquad (14.57)$$

A quarter wave plate is used in practice to convert linearly polarized light into circularly polarized light or vice versa.

The other example is Faraday rotation. In this case the natural modes are circularly polarized. The orthogonality of the modes requires $T_+ T_- = -1$ and circular polarization implies a polarization ellipse with axial ratio equal to unity, that is, $|T_\pm| = 1$. Thus circular polarization corresponds to $T_+ = 1 = -T_-$ in (14.50) and hence to $\rho_Q = 0 = \rho_U$. Then the transfer equation (14.44), neglecting any absorption, reduces to

$$\frac{d}{ds}\begin{pmatrix} Q \\ U \end{pmatrix} = \rho_V \begin{pmatrix} 0 & -1 \\ 1 & 0 \end{pmatrix}\begin{pmatrix} Q \\ U \end{pmatrix}, \qquad (14.58)$$

From (14.22) one has $U/Q = \tan(2\psi)$, so that ψ describes the direction of the plane of linear polarization. According to (14.58) this plane rotates at the rate

$$\frac{d\psi}{ds} = \tfrac{1}{2}\rho_V = -\frac{\omega}{2c}(n_R - n_L), \qquad (14.59)$$

where the labeling of the modes is changed so that they correspond to right (R) and left (L) polarizations, with $T_R = 1$, $T_L = -1$.

For a magnetized plasma at frequencies $\omega \gg \omega_p, \Omega_e$, the relevant wave properties are given by (13.41) and (13.42). Using these in (14.47) gives

$$\rho_V = \frac{\omega_p^2 \Omega_e \cos\psi}{2\omega^2 c}. \qquad (14.60)$$

Faraday rotation is important in astronomy, particularly in radio astronomy where sources are often partially linearly polarized, which is a characteristic signature of synchrotron radiation. The amount of Faraday rotation from a given source is determined by measuring the orientation of the plane of polarization as a function of frequency. According to (14.59) with (14.60) this should be proportional to $1/\omega^2$, and this is usually found to be the case. The constant of proportionality then gives a measure of the integral of $n_e B \cos\theta$ along the line of sight. This parameter, called the rotation measure, then provides information on the interstellar medium through which radio waves propagate.

Exercise Set 14

14.1 Show that if one defines a *Stokes vector* S_A as the column vector with I, Q, U, V, that is, with A running over I, Q, U, V, then the polarization tensor $I^{\alpha\beta}$ is written

$$I^{\alpha\beta} = \tfrac{1}{2}\tau_A^{\alpha\beta} S_A, \quad S_A = \tau_A^{*\alpha\beta} I^{\alpha\beta}, \qquad (E14.1)$$

with

$$\tau_I^{\alpha\beta} = \begin{pmatrix} 1 & 0 \\ 0 & 1 \end{pmatrix}, \quad \tau_Q^{\alpha\beta} = \begin{pmatrix} 1 & 0 \\ 0 & -1 \end{pmatrix},$$

$$\tau_U^{\alpha\beta} = \begin{pmatrix} 0 & 1 \\ 1 & 0 \end{pmatrix}, \quad \tau_V^{\alpha\beta} = \begin{pmatrix} 0 & -i \\ i & 0 \end{pmatrix}, \qquad (E14.2)$$

and where the sum over repeated indices A is implied.

14.2 Consider a change in the basis vectors from $\mathbf{e}^1$, $\mathbf{e}^2$ to $\mathbf{e}^{1'}$, $\mathbf{e}^{2'}$, as given by (14.16). The polarization tensor $I^{\alpha\beta}$ transforms into $I^{\alpha'\beta'}$. Show that the Stokes parameters in the new basis, defined by adding primes to the quantities in (14.8), are related to the Stokes parameters in the old basis by

$$I' = I, \quad Q' = Q\cos(2\psi) + U\sin(2\psi),$$

$$U' = -Q\sin(2\psi) + U\cos(2\psi), \quad V' = V.$$

14.3 Suppose that one chooses circular polarizations

$$\mathbf{e}^R = \frac{1}{2^{1/2}}(\mathbf{e}^1 + i\mathbf{e}^2), \quad \mathbf{e}^L = \frac{1}{2^{1/2}}(\mathbf{e}^1 - i\mathbf{e}^2),$$

as basis vectors for the polarization tensor. Show that one has

$$I^{RR} = \tfrac{1}{2}(I + V), \quad I^{LL} = \tfrac{1}{2}(I - V),$$

$$I^{RL} = \tfrac{1}{2}(Q + iU), \quad I^{LR} = \tfrac{1}{2}(Q - iU).$$

14.4 Show that orthogonal transverse polarizations, with axial ratios T and $-1/T$, correspond to points on opposite sides of the Poincaré sphere along a diagonal.

14.5 Two transverse polarizations $\mathbf{e}_1$ and $\mathbf{e}_2$ are represented by points χ_1, ψ_1 and χ_2, ψ_2 on the Poincaré sphere.

(a) Show that one has

$$\mathbf{e}_1 = (\cos\chi_1 \cos\psi_1 - i\sin\chi_1 \sin\psi_1)\mathbf{e}^1$$
$$+ (\cos\chi_1 \sin\psi_1 + i\sin\chi_1 \cos\psi_1)\mathbf{e}^2,$$

$$\mathbf{e}_2 = (\cos\chi_2 \cos\psi_2 - i\sin\chi_2 \sin\psi_2)\mathbf{e}^1$$
$$+ (\cos\chi_2 \sin\psi_2 + i\sin\chi_2 \cos\psi_2)\mathbf{e}^2.$$

(b) Show that these correspond to Stokes vectors

$$(S_1)_A = \begin{pmatrix} I_1 \\ I_1 \cos(2\chi_1)\cos(2\psi_1) \\ I_1 \cos(2\chi_1)\sin(2\psi_1) \\ I_1 \sin(2\chi_1) \end{pmatrix},$$

$$(S_2)_A = \begin{pmatrix} I_2 \\ I_2 \cos(2\chi_2)\cos(2\psi_2) \\ I_2 \cos(2\chi_2)\sin(2\psi_2) \\ I_2 \sin(2\chi_2) \end{pmatrix}.$$

(c) Show that one has

$$|\mathbf{e}_2^* \cdot \mathbf{e}_1|^2 = \tfrac{1}{2}\{1+\cos(2\chi_1)\cos(2\chi_2)\cos[2(\psi_1-\psi_2)] + \sin(2\chi_1)\sin(2\chi_2)\}.$$

14.6 Elliptically polarized radiation with axial ratio T propagating along the z axis is normally incident on a surface of an isotropic inhomogeneous medium that causes the ray to be refracted through $\pi/2$. The ray is assumed to remain in the x–z plane and to emerge normal to another surface of the medium. Assume $|T| > 0$ and that the *major axis* of the polarization ellipse is the axis along $\mathbf{e}^{1'}$ in (14.18). Determine the polarization of the emerging radiation in the following cases:

(a) the major axis of the initial polarization ellipse is along the y axis;
(b) the major axis of the initial polarization ellipse is along the x axis;
(c) the major axis of the initial polarization ellipse is along the line $x = y$.

14.7 Consider the eigenfunctions of the transfer equation in the Jones calculus (14.36).

(a) Show that in the absence of absorption ($\mu^{\alpha\beta\rho\sigma} = 0$) the eigenfunctions correspond to the natural modes of the medium, as given by (14.32).
(b) Find the eigenvalues and eigenfunctions in the presence of absorption for negligible generalized Faraday rotation ($\rho^{\alpha\beta\rho\sigma} = 0$).
(c) Discuss the effect of absorption in the case $\mu_I = (\mu_Q^2+\mu_U^2+\mu_V^2)^{1/2}$. What would occur if one had $(\mu_Q^2 + \mu_U^2 + \mu_V^2)^{1/2} > \mu_I$?

14.8 Find the eigenvalues and eigenfunctions in the Jones calculus for the general case $\rho^{\alpha\beta\rho\sigma} \neq 0$, $\mu^{\alpha\beta\rho\sigma} \neq 0$.

14.9 The following exercise concerns the construction of the eigenvalues and eigenfunctions of the matrix $\rho_{AB} - \mu_{AB}$ in the Mueller calculus, cf. (14.49) and (14.52).

(a)   Show that the eigenvalues of $\rho_{AB} - \mu_{AB}$ are given by the solutions for λ of

$$(\lambda + \mu_I)^4 + (\lambda + \mu_I)^2(\rho_Q^2 + \rho_U^2 + \rho_V^2 - \mu_Q^2 - \mu_U^2 - \mu_V^2)$$
$$- \rho_Q^2\mu_Q^2 - \rho_U^2\mu_U^2 - \rho_V^2\mu_V^2 - 2\rho_Q\rho_U\mu_Q\mu_U$$
$$- 2\rho_Q\rho_V\mu_Q\mu_V - 2\rho_U\rho_V\mu_U\mu_V = 0. \qquad (E14.3)$$

(b)   Construct the matrix of cofactors of the matrix $\rho_{AB} - \mu_{AB} - \lambda\delta_{AB}$.

(c)   Write down the eigenfunctions S_A obtained from the first and from the fourth columns of the matrix constructed in part (b).

(d)   Using (14.43) rewrite the eigenfunctions S_A found in part (c) as polarization tensors $I^{\alpha\beta}$.

14.10 The following problem concerns the interpretation of the eigenvalues and eigenfunctions of the matrix ρ_{AB}. In this problem μ_{AB} is assumed equal to zero.

(a)   Show that the four eigenvalues of ρ_{AB} consist of two that are equal to zero and two that are non-zero and approximately equal to $\pm\Delta k/2$.

(b)   Choose the polarization vectors of the two natural modes of the medium, as determined by (14.33), as the basis for the polarization tensor $I^{\alpha\beta}$ and show that the degenerate eigenfunctions for the eigenvalues equal to zero are chosen to correspond to completely polarized radiation in one or other of the natural modes. Let these eigenvalues correspond to polarization tensors denoted by I^{++} and I^{--}.

(c)   Show that the eigenfunctions corresponding to the non-zero eigenvalues are interpreted as I^{+-} and I^{-+}.

(d)   Find the eigenvectors S_A in the Mueller calculus corresponding to these four polarization tensors.

Remark: The eigenfunctions corresponding to non-zero eigenvalues have zero total intensity and thus do not correspond to natural wave modes.

14.11 The following problem concerns the interpretation of the eigenvalues and eigenfunctions of the matrix μ_{AB}. In this problem ρ_{AB} is assumed equal to zero.

(a)   Show that μ_{AB} has two degenerate eigenvalues that are equal to μ_I, and two non-degenerate eigenvalues that are equal to $\mu_I \pm (\mu_Q^2 + \mu_U^2 + \mu_V^2)^{1/2}$.

(b)   Construct the eigenfunctions corresponding to the non-degenerate eigenvalues and show that they imply that the following combinations damp with absorption coefficients $\mu_I \pm (\mu_Q^2 + \mu_U^2 + \mu_V^2)^{1/2}$:

$$I \pm \frac{\mu_Q Q + \mu_U U + \mu_V V}{(\mu_Q^2 + \mu_U^2 + \mu_V^2)^{1/2}}.$$

(c) Find combinations that damp with absorption coefficient μ_I.

14.12 Show that the antisymmetry property $\rho_{AB} = -\rho_{BA}$ implies that, in the absence of damping, I and $Q^2 + U^2 + V^2$ are conserved.

14.13 Consider a transformation of $\Delta K^{\alpha\beta}$ to $\Delta K^{\alpha'\beta'}$ corresponding to a rotation through an angle ψ. Post multiply $\Delta K^{\alpha\beta}$ by the matrix R and premultiply by the transpose R^{T} of R, with

$$R^{\alpha\alpha'} = \begin{pmatrix} \cos\psi & -\sin\psi \\ \sin\psi & \cos\psi \end{pmatrix}.$$

Show that for

$$\Delta K_Q = \Delta K_\perp \cos(2\psi), \quad \Delta K_U = \Delta K_\perp \sin(2\psi),$$

one has

$$\Delta K'_Q = \Delta K_\perp, \quad \Delta K'_U = 0, \quad \Delta K'_V = \Delta K_V.$$

Remark: Thus one may eliminate the U component by rotating the coordinate axes, and this corresponds to orienting one axis along the major axis of the polarization ellipse of one of the natural modes.

15

Energetics and Damping of Waves

Preamble

Once the amplitude of a wave is defined, the electric and magnetic energies in the waves are calculated in terms of this amplitude. However, the total energy in the waves cannot be identified in any simple way in general. The damping of waves is used to identify the total energy by relating the damping to the dissipative part of the response tensor in two ways. One way involves calculating the work done by the dissipative process and equating this to the energy lost by the waves. The other way involves including damping in terms of an imaginary part of the frequency (§11.4). The equivalence of the two ways of treating damping provides an explicit expression for the total energy in the waves. A semiclassical description of a distribution of waves is useful for both formal and practical purposes; the semiclassical description is based on regarding the waves as a collection of wave quanta.

15.1 The Electric and Magnetic Energies in Waves

In any physical theory, energy is defined in terms of its mechanical equivalent. In §15.2 this is achieved by calculating the work done by an arbitrary dissipative process and equating it to the energy lost by the waves. An important preliminary step is to define the amplitude of the waves and to calculate the electric energy in waves in Fourier space by using the fact that the electric energy density in coordinate space is given by $\frac{1}{2}\varepsilon_0|\mathbf{E}|^2$. The magnetic energy in waves is calculated in an analogous way.

A solution of the wave equation defines the dispersion relation $\omega = \omega_M(\mathbf{k})$ and the polarization vector $\mathbf{e}_M(\mathbf{k})$ for a wave mode M. The magnitude of the vector potential $\mathbf{A}_M(\omega, \mathbf{k})$ remains arbitrary, and its level is determined by the amplitude of the waves. To be specific, let us define the *amplitude* $a_M(\mathbf{k})$ of waves in the mode M by writing

$$\mathbf{A}_M(\omega, \mathbf{k}) = a_M(\mathbf{k})\mathbf{e}_M(\mathbf{k})2\pi\delta(\omega - \omega_M(\mathbf{k})). \qquad (15.1)$$

Negative frequencies contain the same information as positive frequencies, and the definition (15.1) of the amplitude includes both positive and negative frequencies. This is made explicit by noting that under the transformation $\mathbf{k} \to -\mathbf{k}$ one has

$$\omega_M(\mathbf{k}) \to \omega_M(-\mathbf{k}) = -\omega_M(\mathbf{k}), \quad \mathbf{e}_M(\mathbf{k}) \to \mathbf{e}_M(-\mathbf{k}) = \mathbf{e}_M^*(\mathbf{k}). \qquad (15.2)$$

The reality condition $\mathbf{A}(-\omega, -\mathbf{k}) = \mathbf{A}^*(\omega, \mathbf{k})$ then requires $a_M(-\mathbf{k}) = a_M^*(\mathbf{k})$. The inclusion of negative frequencies in (15.1) is made explicit by writing (15.1) in the form

$$\mathbf{A}_M(\omega, \mathbf{k}) = a_M(\mathbf{k})\mathbf{e}_M(\mathbf{k})2\pi\delta(\omega - |\omega_M(\mathbf{k})|)$$
$$+ a_M^*(-\mathbf{k})\mathbf{e}_M^*(-\mathbf{k})2\pi\delta(\omega + |\omega_M(-\mathbf{k})|). \qquad (15.3)$$

A more detailed argument leading to (15.3) is as follows. Suppose one writes the positive-frequency and negative-frequency solutions separately as $\omega = \omega_{M+}(\mathbf{k}) > 0$ and $\omega = \omega_{M-}(\mathbf{k}) < 0$. Then the right hand side of (15.1) is replaced by the sum of two terms, one with $M \to M+$ and the other with $M \to M-$. It is assumed that the transformation $\mathbf{k} \to -\mathbf{k}$ corresponds to interchanging the roles of the $+$ and $-$ solutions. This corresponds to $\omega_{M\pm}(\mathbf{k}) = -|\omega_{M\mp}(-\mathbf{k})|$, and to $\mathbf{e}_{M\pm}(\mathbf{k}) = \mathbf{e}_{M\mp}^*(-\mathbf{k})$. Finally, one uses these relations to write the $-$ solution in terms of the $+$ solution, and omits the $+$ label.

The electric energy density in the electromagnetic field is identified as $\frac{1}{2}\varepsilon_0|\mathbf{E}(t, \mathbf{x})|^2$, cf. (1.7). On integrating over an arbitrarily large volume V and averaging over an arbitrarily long time T, the time-averaged electric energy density is expressed in term of Fourier transforms using the power theorem (4.11):

$$\lim_{TV\to\infty} \frac{1}{TV} \int_{-T/2}^{T/2} dt \int d^3\mathbf{x}\, \tfrac{1}{2}\varepsilon_0|\mathbf{E}(t, \mathbf{x})|^2$$
$$= \lim_{TV\to\infty} \frac{1}{TV} \int \frac{d\omega d^3\mathbf{k}}{(2\pi)^4}\, \tfrac{1}{2}\varepsilon_0|i\omega\mathbf{A}(\omega, \mathbf{k})|^2, \qquad (15.4)$$

where the relation (5.39), *viz.*, $\mathbf{E}(\omega, \mathbf{k}) = i\omega\mathbf{A}(\omega, \mathbf{k})$, of the electric field to the vector potential in the temporal gauge is inserted.

It is convenient to define $W_M^E(\mathbf{k})$ to be the electric energy in the mode M per unit volume of phase space. Specifically, one writes the electric energy in the range $d^3x d^3k/(2\pi)^3$ in the waves in the mode M as $W_M^E(\mathbf{k})d^3x d^3k/(2\pi)^3$. The electric energy is identified as V times the electric energy density (15.4), and this is equated to $W_M^E(\mathbf{k})$ integrated over $d^3x d^3k/(2\pi)^3$. (The integral over $d^3\mathbf{x}$ gives the volume V so that these powers of V cancel.) The field $\mathbf{A}(\omega, \mathbf{k})$ in (15.4) is identified with the form (15.1) for waves in the mode M. The resulting square of the δ-function that appears in (15.4) is rewritten using (4.37), viz., $[2\pi\delta(\omega)]^2 = T2\pi\delta(\omega)$, and factors of T cancel. The positive and negative frequencies contribute equally, so that one obtains the following explicit expression for the *electric energy* in waves in the mode M:

$$W_M^E(\mathbf{k}) = \varepsilon_0 |\omega_M(\mathbf{k})a_M(\mathbf{k})|^2/V. \tag{15.5}$$

The *magnetic energy* $W_M^M(\mathbf{k})$ in the waves is derived in a similar way. The magnetic energy density is $|\mathbf{B}(t,\mathbf{x})|^2/2\mu_0$, cf. (1.7). Repeating the derivation of (15.5) using (5.5), viz., $\mathbf{B}(\omega, \mathbf{k}) = i\mathbf{k} \times \mathbf{A}(\omega, \mathbf{k})$, one obtains

$$W_M^M(\mathbf{k}) = \frac{|\mathbf{k}|^2 c^2}{|\omega_M(\mathbf{k})|^2} \left[1 - |\boldsymbol{\kappa} \cdot \mathbf{e}_M(\mathbf{k})|^2\right] W_M^E(\mathbf{k}). \tag{15.6}$$

The total energy $W_M(\mathbf{k})$ in the waves may be decomposed into electric and magnetic parts, plus a part associated with the energy in forced particle motions $W_M^P(\mathbf{k})$:

$$W_M(\mathbf{k}) = W_M^E(\mathbf{k}) + W_M^M(\mathbf{k}) + W_M^P(\mathbf{k}). \tag{15.7}$$

Two undefined quantities $W_M(\mathbf{k})$ and $W_M^P(\mathbf{k})$ are contained in (15.7). For sufficiently simple media, the model of the medium used to calculate the response tensor may also be used to calculate $W_M^P(\mathbf{k})$ directly, and then (15.7) defines $W_M(\mathbf{k})$. However, this is not possible in general because all one is given in the general case is the form of the response tensor. One needs to find a way of identifying $W_M(\mathbf{k})$ in terms of the response tensor. Once this has been done, (15.7) defines $W_M^P(\mathbf{k})$.

15.2 The Work Done by a Dissipative Process

The interpretation of (1.7) as the the continuity equation for electromagnetic energy implies that $-\mathbf{J} \cdot \mathbf{E}$ is the rate per unit volume at which work is done on the electromagnetic field by the current $\mathbf{J}$. In principle, the current $\mathbf{J}$ may be separated into two parts: a non-dissipative part that does no work on average (where the average is over an arbitrarily long time), and a dissipative part that causes the waves to damp or grow. If one averages $-\mathbf{J} \cdot \mathbf{E}$ over an arbitrarily long time, then only the dissipative part of the current can contribute. This allows one to identify the dissipative part of the current, as follows.

The time-averaged power transferred to the electromagnetic field per unit volume is evaluated by analogy with (15.4):

$$-\lim_{TV\to\infty}\frac{1}{TV}\int_{-T/2}^{T/2}dt\int d^3x\,\mathbf{E}(t,\mathbf{x})\cdot\mathbf{J}(t,\mathbf{x})$$

$$=-\lim_{TV\to\infty}\frac{1}{TV}\int\frac{d\omega d^3k}{(2\pi)^4}\,\mathbf{E}^*(\omega,\mathbf{k})\cdot\mathbf{J}(\omega,\mathbf{k})$$

$$=\lim_{TV\to\infty}\frac{1}{TV}\int\frac{d\omega d^3k}{(2\pi)^4}\,i\omega\mathbf{A}^*(\omega,\mathbf{k})\cdot\mathbf{J}(\omega,\mathbf{k}),\quad(15.8)$$

We identify $\mathbf{J}$ as the induced current in the medium, and insert the relation (6.16), *viz.*,

$$J_i(\omega,\mathbf{k})=\alpha_{ij}(\omega,\mathbf{k})A_j(\omega,\mathbf{k}).\tag{15.9}$$

The integral in (15.8) is real, and hence, without loss of generality, the integrand on the right hand side is replaced by half the sum of itself and its complex conjugate. After inserting (15.9), this procedure shows that only the antihermitian part of the response tensor, cf. (7.4), contributes to the integral. Thus (15.8) gives

$$-\lim_{TV\to\infty}\frac{1}{TV}\int_{-T/2}^{T/2}dt\int d^3x\,\mathbf{E}(t,\mathbf{x})\cdot\mathbf{J}(t,\mathbf{x})$$

$$=\lim_{TV\to\infty}\frac{1}{TV}\int\frac{d\omega d^3k}{(2\pi)^4}\,i\omega A_i^*(\omega,\mathbf{k})\alpha_{ij}^A(\omega,\mathbf{k})A_j(\omega,\mathbf{k}).$$

$$(15.10)$$

The vector potential $\mathbf{A}(\omega,\mathbf{k})$ in (15.10) is identified with that given by (15.1) for waves in the mode M. The power transferred to waves in the mode M in the range $d^3x d^3k/(2\pi)^3$ is identified as $Q_M(\mathbf{k})d^3x d^3k/(2\pi)^3$. A derivation similar to that of (15.5) starting from (15.10) gives

$$Q_M(\mathbf{k})=2i\omega_M(\mathbf{k})|a_M(\mathbf{k})|^2 e_{Mi}^*(\mathbf{k})e_{Mj}(\mathbf{k})\alpha_{ij}^A(\omega_M(\mathbf{k}),\mathbf{k})/V.\tag{15.11}$$

Damping of the waves implies a power loss by the waves that is proportional to the energy in the waves. Thus the power transferred to the waves in a damping process is written in the form

$$Q_M(\mathbf{k})=-\gamma_M(\mathbf{k})W_M(\mathbf{k}),\tag{15.12}$$

where the quantity $\gamma_M(\mathbf{k})$ is called the *absorption coefficient* for waves in the mode M. Let us also define the *ratio of electric to total energy* $R_M(\mathbf{k})$ in the waves in the mode M by writing

$$R_M(\mathbf{k})=\frac{W_M^E(\mathbf{k})}{W_M(\mathbf{k})}.\tag{15.13}$$

Then using (15.5), (15.11) and (15.12) one obtains an explicit expression for the absorption coefficient:

$$\gamma_M(\mathbf{k})=-2i\omega_M(\mathbf{k})R_M(\mathbf{k})e_{Mi}^*(\mathbf{k})e_{Mj}(\mathbf{k})K_{ij}^A(\omega_M(\mathbf{k}),\mathbf{k}),\tag{15.14}$$

where the antihermitian part of $\alpha_{ij}(\omega, \mathbf{k})$ is replaced by the antihermitian part of $K_{ij}(\omega, \mathbf{k})$ using the relation, cf. (6.17),

$$\mu_0 \alpha_{ij}^A(\omega, \mathbf{k}) = \omega^2 K_{ij}^A(\omega, \mathbf{k})/c^2. \tag{15.15}$$

The expression (15.14) with (15.15) is formally identical to (11.22). However, these two expressions are derived by two independent arguments. Their equivalence implies the expression (11.23) for $R_M(\mathbf{k})$. Let us repeat the arguments given in §11.4 to emphasize this point, and generalize the derivation to introduce another point, specifically, the relation between temporal and spatial damping.

15.3 Temporal and Spatial Damping

In a theory based on Fourier transforms, temporal damping is described in terms of an imaginary part of the frequency ω and spatial damping is described in terms of an imaginary part of the wavevector $\mathbf{k}$. The wave amplitude varies as $e^{-i(\omega t - \mathbf{k} \cdot \mathbf{x})}$, and adding imaginary parts $i\omega_I$ and $i\mathbf{k}_I$ to ω and $\mathbf{k}$, respectively, causes the wave amplitude to vary secularly as $e^{\omega_I t - \mathbf{k}_I \cdot \mathbf{x}}$. The wave energy varies as the square of the wave amplitude, implying that it varies secularly according to

$$\frac{\partial W_M(\mathbf{k})}{\partial t} = 2\omega_I W_M(\mathbf{k}), \quad \frac{\partial W_M(\mathbf{k})}{\partial \mathbf{x}} = -2\mathbf{k}_I W_M(\mathbf{k}). \tag{15.16}$$

The variation of $W_M(\mathbf{k})$ on t and $\mathbf{x}$ is assumed to be slow. The spatial variation must be slow or weak in the sense of geometric optics. In simple cases this means that the distribution of waves must vary slowly over a wavelength. Such a slow variation is then compatible with a separation of spatial variations into two scale lengths. One is the scale length over which rapid variations occur and one describes these rapid variations in terms of Fourier transforms. The other scale length is that of the slow variations in the distribution of the waves. Similarly, the slow variation in time requires a separation into two time scales, one characterized by the wave frequency and the other being much slower. (Such a separation of time scales is evident in the simple example of beating between two musical tones of similar frequencies where one hears the slow rise and fall superimposed on the tone.)

In (15.16), the imaginary part of the frequency describes temporal damping and the imaginary part of the wavevector describes spatial damping. These two processes are not independent. Waves that are uniformly excited everywhere remain homogeneous in space, and any damping involves a purely temporal reduction in their energy density.

In contrast, waves that are excited by a point source radiating uniformly
for an arbitrarily long time reach a steady state in which the damping
involves a purely spatial decay of the wave energy away from the point
source.

Damping is included in the dispersion equation by including the imaginary parts of ω and $\mathbf{k}$, and by also including the antihermitian part of
the response tensor. On expanding the determinant of the coefficients
in the wave equation to first order in the dissipative processes, one generalizes (11.20) to

$$\det\left[\Lambda^{\mathrm{H}}_{ij}(\omega+i\omega_{\mathrm{I}},\mathbf{k}+i\mathbf{k}_{\mathrm{I}})+K^{\mathrm{A}}_{ij}(\omega+i\omega_{\mathrm{I}},\mathbf{k}+i\mathbf{k}_{\mathrm{I}})\right]=\Lambda(\omega,\mathbf{k})$$

$$+i(\omega_{\mathrm{I}}\frac{\partial}{\partial\omega}+\mathbf{k}_{\mathrm{I}}\cdot\frac{\partial}{\partial\mathbf{k}})\Lambda(\omega,\mathbf{k})+\lambda_{ji}(\omega,\mathbf{k})K^{\mathrm{A}}_{ij}(\omega,\mathbf{k})+\cdots.$$

$$(15.17)$$

As in §11.4, one solves the dispersion equation by setting the left hand
side of (15.17) equal to zero in two steps. The initial step is to neglect the
dissipative terms, so that only the term $\Lambda(\omega,\mathbf{k})$ remains. The resulting
zeroth order dispersion equation is $\Lambda(\omega,\mathbf{k})=0$, which is the same as
(11.10). The next step is to include the first order terms in the dissipative
processes, giving

$$(\omega_{\mathrm{I}}\frac{\partial}{\partial\omega}+\mathbf{k}_{\mathrm{I}}\cdot\frac{\partial}{\partial\mathbf{k}})\Lambda(\omega,\mathbf{k})=i\lambda_{ji}(\omega,\mathbf{k})K^{\mathrm{A}}_{ij}(\omega,\mathbf{k}),\qquad(15.18)$$

which generalizes (11.21) to allow either temporal or spatial damping.
Then in place of (11.21) one obtains

$$\omega_{\mathrm{I}}-\mathbf{k}_{\mathrm{I}}\cdot\mathbf{v}_{gM}(\mathbf{k})=i\left[\frac{\lambda_{ss}(\omega,\mathbf{k})}{\partial\Lambda(\omega,\mathbf{k})/\partial\omega}\right]e^*_{Mi}(\mathbf{k})e_{Mj}(\mathbf{k})K^{\mathrm{A}}_{ij}(\omega,\mathbf{k}),\quad(15.19)$$

which applies only for $\omega=\omega_M(\mathbf{k})$. The *group velocity* for waves in the
mode M is defined by

$$\mathbf{v}_{gM}(\mathbf{k})=\frac{\partial\omega_M(\mathbf{k})}{\partial\mathbf{k}}=-\left[\frac{\partial\Lambda(\omega,\mathbf{k})/\partial\mathbf{k}}{\partial\Lambda(\omega,\mathbf{k})/\partial\omega}\right]\Bigg|_{\omega=\omega_M(\mathbf{k})},\qquad(15.20)$$

where the latter relation follows from the chain rule for partial differentiation.

Two important results follow from (15.19). First, if the damping is
purely temporal, then the definition of the absorption coefficient $\gamma_M(\mathbf{k})$
implies that it is equal to $-2\omega_{\mathrm{I}}$, and hence (15.19), after multiplication
by -2, is identified with (15.14). This leads to the identification (11.23),
viz.,

$$R_M(\mathbf{k})=\left[\frac{\lambda_{ss}(\omega,\mathbf{k})}{\omega\partial\Lambda(\omega,\mathbf{k})/\partial\omega}\right]\Bigg|_{\omega=\omega_M(\mathbf{k})}.\qquad(15.21)$$

When combined with (15.13), the derivation of (15.21) may be regarded as a formal identification of the total energy $W_M(\mathbf{k})$ in waves in the mode M.

The other important result implied by the derivation of (15.19) is that *the velocity of energy propagation* of waves in the mode M is the group velocity for waves in the mode M. One identifies the velocity of energy propagation formally as the ratio of the time-derivative to the spatial gradient of the energy in a wave packet. Then (15.18) and (15.20) imply that this is indeed equal to the group velocity.

The form (15.21) for the ratio of the electric to the total energy $R_M(\mathbf{k})$ is useful for formal purposes, but is inconvenient for calculating $R_M(\mathbf{k})$ for specific wave modes. A more useful alternative form is

$$R_M(\mathbf{k}) = \left\{ \frac{\omega}{\partial[\omega^2 K_M(\omega, \mathbf{k})]/\partial\omega} \right\} \bigg|_{\omega = \omega_M(\mathbf{k})}, \qquad (15.22)$$

with $K_M(\omega, \mathbf{k}) = e^*_{Mi}(\mathbf{k})e_{Mj}(\mathbf{k})K^{\mathrm{H}}_{ij}(\omega, \mathbf{k})$. An alternative that applies only if the medium is not spatially dispersive is

$$R_M(\mathbf{k}) = \frac{1}{[1 - |\boldsymbol{\kappa} \cdot \mathbf{e}_M(\mathbf{k})|^2]\, 2n_M(\omega, \boldsymbol{\kappa})\partial[\omega n_M(\omega, \boldsymbol{\kappa})]/\partial\omega}, \qquad (15.23)$$

where the form $n = kc/\omega = n_M(\omega, \boldsymbol{\kappa})$ is used for the dispersion relation.

15.4 The Energy Flux in Waves

Three other quantities associated with waves are the energy flux, the momentum density and the stress tensor. Explicit expressions for these quantities are derived as follows.

The energy flux is identified by manipulating (15.17) into the form of a continuity equation for energy. To achieve this, multiply (15.17) by $-2W^{\mathrm{E}}_M(\mathbf{k})\omega/\lambda_{ss}(\omega, \mathbf{k})$, with $\omega = \omega_M(\mathbf{k})$ and with $W^{\mathrm{E}}_M(\mathbf{k})$ given by (15.5), and use (15.12), (15.14), (15.15) and (15.20). Thus one obtains

$$\frac{\partial}{\partial t} W_M(\mathbf{k}) + \operatorname{div} \mathbf{F}_M(\mathbf{k}) = -\gamma_M(\mathbf{k})W_M(\mathbf{k}), \qquad (15.24)$$

and hence one identifies the energy flux as

$$\mathbf{F}_M(\mathbf{k}) = v_{\mathrm{g}M}(\mathbf{k})\, W_M(\mathbf{k})$$

$$= -\varepsilon_0\omega^3_M(\mathbf{k})|a_M(\mathbf{k})|^2 \left[\frac{1}{\lambda_{ss}(\omega, \mathbf{k})} \frac{\partial}{\partial \mathbf{k}} \Lambda(\omega, \mathbf{k}) \right] \bigg|_{\omega = \omega_M(\mathbf{k})}, \qquad (15.25)$$

where the group velocity is defined by (15.20).

Before interpreting (15.25) it is helpful to write it in an alternative way. Specifically, the derivative of $\Lambda(\omega, \mathbf{k})$ in (15.25) is evaluated as follows. On taking a differential of (11.24) and using (11.25) one obtains

$$d[\Lambda(\omega,\mathbf{k})] = \lambda_{ji}(\omega,\mathbf{k})\,d[\Lambda_{ij}(\omega,\mathbf{k})]. \tag{15.26}$$

The derivative with respect to $\mathbf{k}$ of $\Lambda_{ij}(\omega,\mathbf{k})$ follows directly from (11.5):

$$\frac{\partial}{\partial k_r}\Lambda_{ij}(\omega,\mathbf{k}) = \frac{c^2}{\omega^2}(k_i\delta_{jr} + k_j\delta_{ir} - 2k_r\delta_{ij}) + \frac{\partial}{\partial k_r}K_{ij}(\omega,\mathbf{k}). \tag{15.27}$$

Then on using (11.18) one has

$$\left[\frac{1}{\lambda_{ss}(\omega,\mathbf{k})}\frac{\partial}{\partial k_r}\Lambda(\omega,\mathbf{k})\right] \tag{15.28}$$

$$= e_{Mi}^*(\mathbf{k})e_{Mj}(\mathbf{k})\left[\frac{c^2}{\omega^2}(k_i\delta_{jr} + k_j\delta_{ir} - 2k_r\delta_{ij}) + \frac{\partial}{\partial k_r}K_{ij}(\omega,\mathbf{k})\right],$$

which is to be evaluated at $\omega = \omega_M(\mathbf{k})$. On substituting (15.28) into (15.25), the energy flux is separated into two parts:

$$\mathbf{F}_M(\mathbf{k}) = \mathbf{F}_M^{\mathrm{EM}}(\mathbf{k}) + \mathbf{F}_M^{\mathrm{P}}(\mathbf{k}), \tag{15.29}$$

with

$$\mathbf{F}_M^{\mathrm{EM}}(\mathbf{k}) = \frac{\omega_M(\mathbf{k})|a_M(\mathbf{k})|^2}{\mu_0}\,2\mathrm{Re}\,[\mathbf{k} - \mathbf{e}_M(\mathbf{k})\,\mathbf{e}_M^*(\mathbf{k})\cdot\mathbf{k}], \tag{15.30}$$

$$\mathbf{F}_M^{\mathrm{P}}(\mathbf{k}) = -\frac{\omega_M(\mathbf{k})|a_M(\mathbf{k})|^2}{\mu_0}\,e_{Mi}^*(\mathbf{k})e_{Mj}(\mathbf{k})\left[\frac{\partial K_{ij}(\omega,\mathbf{k})}{\partial\mathbf{k}}\right]\Bigg|_{\omega\,=\,\omega_M(\mathbf{k})} \tag{15.31}$$

The terms (15.30) and (15.31) are interpreted as the electromagnetic energy flux and the energy flux associated with forced particle motions, respectively. The basis for this interpretation is that the flux (15.30) is that associated with the Poynting vector $\mathbf{E}\times\mathbf{B}/\mu_0$. The contribution to the Poynting vector from waves in the mode M in the range $d^3\mathbf{k}/(2\pi)^3$ is calculated by repeating the steps in the derivation of (15.5), starting with $\mathbf{E}\times\mathbf{B}/\mu_0$ in place of $\frac{1}{2}\varepsilon_0|\mathbf{E}|^2$ in (15.4). The calculation reproduces (15.30) confirming the interpretation of this term as the energy flux associated with the electromagnetic field. The other part of the flux (15.31) is non-zero only in a medium which is spatially dispersive. Thus one concludes that the energy flux is given by the Poynting vector only in media which are not spatially dispersive.

The momentum density $\mathbf{P}_M(\mathbf{k})$ and the stress tensor $[\mathbf{T}_M(\mathbf{k})]_{ij}$ in the waves is derived by considering the equation for momentum balance (3.5) in place of the energy equation (1.7) in the analysis given in the derivations of (15.12) and (15.24). One finds

$$\mathbf{P}_M(\mathbf{k}) = \frac{\mathbf{k}}{\omega_M(\mathbf{k})}\,W_M(\mathbf{k}), \tag{15.32}$$

$$[\mathbf{T}_M(\mathbf{k})]_{ij} = [\mathbf{v}_{gM}(\mathbf{k})]_i\,[\mathbf{P}_M(\mathbf{k})]_j. \tag{15.33}$$

In place of the continuity of energy (15.24), the continuity equation for momentum is

$$\frac{\partial}{\partial t}\left[\mathbf{P}_M(\mathbf{k})\right]_j + \frac{\partial}{\partial x_r}\left[\mathbf{T}_M(\mathbf{k})\right]_{rj} = -\gamma_M(\mathbf{k})\left[\mathbf{P}_M(\mathbf{k})\right]_j. \qquad (15.34)$$

The derivative of the stress tensor is non-zero only if the distribution of waves is spatially inhomogeneous. A slow variation of $\mathbf{P}_M(\mathbf{k}; t, \mathbf{x})$ and $[\mathbf{T}_M(\mathbf{k}; t, \mathbf{x})]_{ij}$ on t and $\mathbf{x}$ is implicit in (15.34).

15.5 Semiclassical Description of Waves

A *semiclassical theory* in electromagnetism is a theory in which the particles are treated quantum mechanically and the waves are treated classically. For some purposes it is convenient to use a semiclassical formalism even when no intrinsically quantum effects are included. In particular, use of detailed balance (which relies on quantum mechanical concepts) in the form of the Einstein coefficients allows one to relate absorption to emission, cf. §16.4, and this procedure is simpler and more physically transparent than relating absorption to emission in a purely classical way.

The basic idea in the semiclassical formalism used here is to regard a distribution of waves as a collection of wave quanta. Each wave quantum in the mode M has an energy $\hbar\omega_M(\mathbf{k})$ and momentum $\hbar\mathbf{k}$, where $\hbar$ is Planck's constant divided by 2π. The distribution of wave quanta is described by their occupation number $N_M(\mathbf{k})$, which is defined such that the number of wave quanta in the range $d^3\mathbf{x}d^3\mathbf{k}/(2\pi)^3$ of phase space is $N_M(\mathbf{k})d^3\mathbf{x}d^3\mathbf{k}/(2\pi)^3$. The energy in this range is then $\hbar\omega_M(\mathbf{k})N_M(\mathbf{k})d^3\mathbf{x}d^3\mathbf{k}/(2\pi)^3$. However this energy is also identified as $W_M(\mathbf{k})d^3\mathbf{x}d^3\mathbf{k}/(2\pi)^3$. This equivalence leads to the identification

$$N_M(\mathbf{k}) = \frac{W_M(\mathbf{k})}{\hbar\omega_M(\mathbf{k})}. \qquad (15.35)$$

A powerful tool that fits in naturally with the semiclassical formalism is the Hamiltonian description of the propagation of rays. Hamilton's equations of motion for a particle or other mechanical system are familiar in classical mechanics, and are the basis for modern quantum mechanics and statistical mechanics. Hamilton's equations for a wave system are written down by analogy with their mechanical counterpart. For a time-independent system the Hamiltonian function is the energy expressed as a function of position and momentum. With this and the foregoing identifications of the energy and momentum of a wave quantum, Hamilton's equations for the wave quantum are

$$\frac{d\mathbf{x}}{dt} = \frac{\partial}{\partial \mathbf{k}}\omega_M(\mathbf{k}), \qquad \frac{d\mathbf{k}}{dt} = -\frac{\partial}{\partial \mathbf{x}}\omega_M(\mathbf{k}). \qquad (15.36a, b)$$

The frequency $\omega_M(\mathbf{k})$ in $(15.36a, b)$ is assumed to be a slowly varying function of $\mathbf{x}$ but this is not shown explicitly. Examples of such slow spatial variations are the gradient in concentration in the center of an optical fiber that causes light to be confined to the fiber, and the spatial variation of the electron plasma frequency in the ionosphere. The point is that the spatial variation is in one or more of the physical properties on which $\omega_M(\mathbf{k})$ depends.

The interpretation of $(15.36a, b)$ requires some care because difficulties arise if the parameter t is interpreted literally as the time. The equations may be thought of as describing a set of wave quanta propagating through the spatially varying system from a time-independent source to a time-independent sink. The parameter t is regarded as a parametrized distance along the ray path. Let s denote distance along the actual (curved) ray path. Then $(15.36a)$ implies that the tangent to the ray path is along the direction of the group velocity and that $\dot{s}$, where the dot denotes differentiation with respect to t, is equal to the group speed $|\mathbf{v}_{gM}|$. The parameter t is then identified as $t = \int ds/|\mathbf{v}_{gM}|$, where the integral is along the ray path. Equation $(15.36b)$ describes the variation of $\mathbf{k}$ along the ray path. A simple implication of this equation is that the components of $\mathbf{k}$ that are orthogonal to the direction of the gradient $-\operatorname{grad}\omega_M$ are constants. Let Ψ be the angle between $\mathbf{k}$ and this gradient. Then $(15.36b)$ implies

$$k \sin \Psi = \text{constant}, \qquad (15.37)$$

which is *Snell's law*. The more familiar form of Snell's law in terms of the refractive index follows by changing the independent variable from ω to the refractive index n; on multiplying by c/ω, (15.37) reduces to the more familiar form $n \sin \Psi = \text{constant}$.

It is important to note that the angle Ψ in Snell's law refers to the direction of $\mathbf{k}$ (the wave-normal direction) and not to the direction of $\mathbf{v}_g$ (the ray direction). This greatly complicates the determination of the ray path in an inhomogeneous anisotropic medium, where the wave-normal and ray directions are different; one first determines how the wave-normal direction changes in accord with Snell's law, and then uses the wave properties to find the changes in the group velocity and hence in the ray direction.

An alternative method for determining the ray path involves the use

of Fermat's principle, but as might be expected this is also much more
cumbersome to use in an anisotropic system than in an isotropic system. *Fermat's principle* is the variational principle that the actual ray
path between two points is an extremum in time. It is formulated in the
context of Lagrangian rather than Hamiltonian mechanics so that the in-
dependent variables are $\mathbf{x}$ and $\dot{\mathbf{x}}$, rather than $\mathbf{x}$ and $\mathbf{k}$ in the Hamiltonian
approach. The variational principle is

$$\delta \int_{t_1}^{t_2} dt\, L_M(\mathbf{x}, \dot{\mathbf{x}}) = 0, \tag{15.38}$$

where the two points are labeled 1 and 2, and where t is the parametrized
distance along the ray path that plays the role of time in (15.36). Also
in (15.38), δ denotes a variation in the path and $L_M(\mathbf{x}, \dot{\mathbf{x}})$ plays the role
of the Lagrangian. On reversing the construction of a Hamiltonian from
a Lagrangian, one obtains

$$L_M(\mathbf{x}, \dot{\mathbf{x}}) = \mathbf{k} \cdot \dot{\mathbf{x}} - \omega_M(\mathbf{k}), \tag{15.39}$$

where $\mathbf{k}$ on the right hand side is to be regarded as an implicit function
of $\mathbf{x}$ and $\dot{\mathbf{x}}$. In a time-independent system the frequency in (15.39) does
not contribute in (15.38) and the interpretation of t as a parametrized
distance along the ray path allows one to rewrite (15.38), after multiplying by c/ω, in the form

$$\delta \int_{s_1}^{s_2} ds\, M = 0, \tag{15.40}$$

where the *ray refractive index* is defined by

$$M = \frac{c\mathbf{k} \cdot \dot{\mathbf{x}}}{\omega |\dot{\mathbf{x}}|} = n(\omega, \theta) \cos(\theta - \theta_r). \tag{15.41}$$

The relation on the right of (15.41) applies for a uniaxial system (the
label M for the mode is suppressed) where θ_r is the angle between the
ray direction and the principal axis.

In the absence of any source or sink for the waves, such as emission,
absorption or scattering, the occupation number is a constant along the
ray. Formally, this is a consequence of Liouville's theorem applied to
the system of waves that satisfy (15.36). Physically, it expresses the fact
that in the absence of sources and sinks the occupation number $N_M(\mathbf{k})$
of the wave quanta is a constant along a ray. The implied conservation
law is

$$\frac{dN_M(\mathbf{k})}{dt} = \left[\frac{\partial}{\partial t} + \frac{d\mathbf{x}}{dt} \cdot \frac{\partial}{\partial \mathbf{x}} + \frac{d\mathbf{k}}{dt} \cdot \frac{\partial}{\partial \mathbf{k}}\right] N_M(\mathbf{k}) = 0, \tag{15.42}$$

with $d\mathbf{x}/dt$ and $d\mathbf{k}/dt$ given by (15.36a, b). In the presence of sources and sinks for the waves appropriate source terms are included on the right hand side of (15.42).

Suppose that the system is also a slowly varying function of time (for example, a partially ionized plasma whose plasma frequency is increasing slowly with time due to a slow increase in the degree of ionization) as well as of position. Then $\omega_M(\mathbf{k})$ in (15.36) denotes $\omega_M(\mathbf{k}; t, \mathbf{x})$, with the slow time and space variations implicit in (15.36). There is then a further equation,

$$\frac{d\omega}{dt} = \frac{\partial}{\partial t}\omega_M(\mathbf{k}), \tag{15.43}$$

that describes how the wave frequency varies slowly with time due to the time dependence of the medium. Again care is required in interpreting the parameter t in (15.43). On the left t is to be regarded as a parametrized distance along the ray path, and the implicit t dependence on the right involves the actual temporal variation of the parameters of the system.

Exercise Set 15

15.1 The ratio of electric to total energy for transverse waves is evaluated using (15.22) with $K_M(\omega, \mathbf{k})$ interpreted as the transverse part $K^T(\omega, k)$ of the dielectric tensor.

(a)   Show that for a non-dispersive dielectric with K^T a constant equal to n^2 with $\partial n/\partial \omega = 0$, the electric, magnetic and particle contributions to the wave energy are in the ratios

$$W_T^E : W_T^M : W_T^P = \frac{1}{2n^2} : \frac{1}{2} : \frac{n^2 - 1}{2n^2}. \tag{E15.1}$$

(b)   Show that for transverse waves in an isotropic plasma with $n^2 = K^T(\omega) = 1 - \omega_p^2/\omega^2$ one has

$$W_T^E : W_T^M : W_T^P = \frac{1}{2} : \frac{1}{2}\left(1 - \frac{\omega_p^2}{\omega^2}\right) : \frac{\omega_p^2}{2\omega^2}. \tag{E15.2}$$

(c)   Show that for Langmuir waves, for which $K_M(\omega, \mathbf{k})$ in (15.22) is interpreted as $K^L(\omega, k)$, the approximate form of (10.23) used in deriving the dispersion relation (13.1) implies, cf. (13.7),

$$R_L(\mathbf{k}) = \frac{\omega_p^2}{2\omega_L^2(k)}. \tag{E15.3}$$

(d)   Show that for Langmuir waves one has ($\omega^2 = \omega_L^2(k)$)

$$W_T^E : W_T^M : W_T^P = \frac{\omega_p^2}{2\omega^2} : 0 : 1 - \frac{\omega_p^2}{2\omega^2}. \tag{E15.4}$$

15.2 Collisional damping of high-frequency waves is treated as a frictional effect in the magnetoionic theory.

(a)   Show that the antihermitian part of the dielectric tensor (10.9) in the magnetoionic theory is obtained from the hermitian part by

$$K_{ij}^A(\omega) = i\frac{\nu_e}{\omega}\left\{\frac{\partial}{\partial\omega}\left[\omega K_{ij}^H(\omega)\right] - \delta_{ij}\right\}. \tag{E15.5}$$

(b)   Show that in the absence of an ambient magnetic field the antihermitian part $(E15.5)$ implies the collisional damping rate (13.19) for transverse waves.

(b)   Show that for magnetoionic waves the absorption coefficient implied by $(E15.5)$ may be written in the form

$$\gamma_M^c(\mathbf{k}) = 2\nu_e \frac{W_M^P(\mathbf{k})}{W_M(\mathbf{k})}. \tag{E15.6}$$

Remark: One interprets $(E15.6)$ as follows. The ratio $W_M^P(\mathbf{k})/W_M(\mathbf{k})$ is the fraction of the energy in the waves that is in forced motions of the electrons. Each time an electron has a collision, that part of the coherent motion of the electron associated with this forced motion in the waves is randomized, and so is lost to the waves. Hence the total energy in the waves reduces at the rate determined by twice the imaginary part of the frequency ($2\nu_e$ here) times this ratio.

15.3 The following exercise concerns the derivation of the alternative forms (15.22) and (15.23) for the ratio of electric to total energy from the form (15.21).

(a) Use the differential relation (15.26) for $\Lambda(\omega, \mathbf{k}) = 0$ together with (11.18) to show that (15.21) may be written

$$R_M(\mathbf{k}) = \left\{ \frac{1}{e^*_{Mi}(\mathbf{k})e_{Mj}(\mathbf{k})\partial[\omega\Lambda_{ij}(\omega,\mathbf{k})]/\partial\omega} \right\}\Bigg|_{\omega\,=\,\omega_M(\mathbf{k})}.$$

(b) Use the definition $K_M(\omega, \mathbf{k}) = e^*_{Mi}(\mathbf{k})e_{Mj}(\mathbf{k})K^{\mathrm{H}}_{ij}(\omega, \mathbf{k})$ and the wave equation to complete the derivation of (15.22) by showing

$$e^*_{Mi}(\mathbf{k})e_{Mj}(\mathbf{k})\frac{\partial[\omega\Lambda_{ij}(\omega,\mathbf{k})]}{\partial\omega} = \frac{1}{\omega}\frac{\partial[\omega^2 K_M(\omega,\mathbf{k})]}{\partial\omega}.$$

(c) Complete the derivation of (15.23) by using the wave equation to show

$$K_M(\omega, \mathbf{k}) = n^2_M(\omega, \boldsymbol{\kappa})|\boldsymbol{\kappa} \times \mathbf{e}_M(\mathbf{k})|^2.$$

15.4 There is a quantity $\mathbf{D}^* \cdot \mathbf{E}$ which may be identified as the wave energy in a non-dispersive medium, but which appears to have no physical significance in general.

(a) Use (i) the wave equation, (ii) the definitions of the electric field $\mathbf{E}_M(\mathbf{k})$ and the magnetic field $\mathbf{B}_M(\mathbf{k})$ in waves in the mode M in terms of the wave amplitude (15.1), and (iii) the definition $D_{Mi} = \varepsilon_0 K_{ij}(\omega_M(\mathbf{k}), \mathbf{k})E_{Mi}(\mathbf{k})$ to show

$$\mathbf{E}^*_M(\mathbf{k}) \cdot \mathbf{D}_M(\mathbf{k}) = |\mathbf{B}_M(\mathbf{k})|^2/\mu_0. \qquad (E15.7)$$

(b) Consider a non-dispersive medium, in which K_M does not depend on ω. Use the form (15.22) for the ratio of electric to total energy and the expression (15.5) for the electric energy to show

$$W_M(\mathbf{k}) = \mathbf{E}^*_M(\mathbf{k}) \cdot \mathbf{D}_M(\mathbf{k}). \qquad (E15.8)$$

15.5 Consider an isotropic medium in which the response is described in terms of a dielectric constant $\varepsilon(\omega)$ and a magnetic response function $\mu(\omega)$, so that it is described by the dielectric tensor (6.18). Such a medium is spatially dispersive with a quadratic dependence on k.

(a)   Use the form (15.22) to show that the energy in transverse waves in such a medium may be expressed in the form

$$W_T(\mathbf{k}) = |\mathbf{E}_T(\mathbf{k})|^2 \frac{\partial[\omega^2\varepsilon(\omega)]}{\partial\omega} + |\mathbf{B}_T(\mathbf{k})|^2 \frac{\partial[\omega^2\mu^{-1}(\omega)]}{\partial\omega}. \quad (E15.9)$$

(b)   What is the corresponding form for the energy in waves in an arbitrary mode M in an anisotropic magnetoelectric medium with response described by (6.20)?

15.6 The energy flux in waves may be identified from the Poynting vector in the form $\mathbf{E}\times\mathbf{B}/\mu_0$ only in non-spatially dispersive media. However, if the Poynting vector is defined by $\mathbf{E}\times\mathbf{H}$ then this identification extends to a somewhat wider class of media.

(a)   Show that for the medium considered in part (a) of Exercise 15.5 that the energy flux is given by the more general form of the Poynting vector.

(b)   Show that this result also applies to anisotropic, magnetoelectric media, as described in part (b) of Exercise 15.5.

(c)   Consider the proposition that the energy flux is always given by the Poynting vector defined in this way. Show this proposition to be untenable by considering the case of longitudinal waves in a spatially dispersive isotropic medium.

15.7 Landau damping is treated using the antihermitian part of the response tensor (10.23):

$$\operatorname{Im} K^L(\omega, k) = \left(\frac{\pi}{2}\right)^{1/2} \frac{\omega\omega_p^2}{k^3 V_e^3} e^{-\omega^2/2k^2 V_e^2}. \quad (E15.10)$$

(a)   Using (E15.10), derive the absorption coefficient (13.6) for Landau damping of Langmuir waves.

(b)   Using (E15.10) and its counterpart for ions, derive the absorption coefficient (13.13) for Landau damping of ion sound waves.

15.8 The fact that the group velocity is equal to the ratio of the Poynting vector to the wave energy in a non-spatially dispersive medium has some subtle consequences.

(a)   The ratio of the Poynting vector to the wave energy follows from (15.30) and (15.5). For a non-spatially dispersive medium this implies

$$\mathbf{v}_{Mg} = \frac{2kc}{\omega_M(\mathbf{k})} R_M(\mathbf{k}) \operatorname{Re}\left[\boldsymbol{\kappa} - \mathbf{e}_M^*(\mathbf{k})\boldsymbol{\kappa}\cdot\mathbf{e}_M(\mathbf{k})\right]. \quad (E15.11)$$

(b) Evaluate (E15.11) for the extraordinary mode in a uniaxial crystal, compare the result with the form (12.29) and hence infer the identity

$$\frac{1}{n_x(\omega,\theta)}\frac{\partial n_x(\omega,\theta)}{\partial\theta} = \frac{\boldsymbol{\kappa}\cdot\mathbf{e}_x(\omega,\theta)\boldsymbol{\kappa}\cdot\mathbf{e}_x(\omega,\theta)}{1-|\boldsymbol{\theta}\cdot\mathbf{e}_x(\omega,\theta)|^2}. \qquad (E15.12)$$

(c) Repeat part (a) for waves in a cold plasma using the form (13.32) for the polarization vector, and hence infer the identity

$$\frac{1}{n_M(\omega,\theta)}\frac{\partial n_M(\omega,\theta)}{\partial\theta} = \frac{L_M(\omega,\theta)T_M(\omega,\theta)}{1+T_M^2(\omega,\theta)}. \qquad (E15.13)$$

Remark: The foregoing procedure also leads directly to the identification of the ratio of electric to total energy in the form (15.23).

PART FOUR

Theory of Emission Processes

The emission of waves is treated by solving the inhomogeneous wave equation and deriving an emission formula. The emission formula derived here may be used to describe the emission of waves in an arbitrary medium by an arbitrary source, described by an extraneous current. The current corresponding to electric and magnetic dipoles and electric quadrupoles is discussed in detail, and the electric dipole case is used to derive the Larmor formula, which describes emission by an accelerated nonrelativistic charge *in vacuo*. The more conventional treatment of emission based on the Lienard–Wierchert potentials is then developed. The back reaction of the radiating system to the emission of radiation may be taken into account in two different ways, depending on the context: by the use of quasi-linear theory, and through a radiation reaction force. These two procedures are discussed critically here.

16

The Emission Formula

Preamble

The energy radiated is treated here assuming that the source is described in terms of an extraneous current. The energy radiated is identified as minus the work done by this current against the electric field that it generates. The first step is to identify and solve the homogeneous wave equation to find the field generated by the extraneous current. After identifying the energy radiated, it is convenient to introduce a semiclassical formalism to derive a kinetic equation that describes both emission and absorption by a statistical distribution of radiating particles.

16.1 The Photon Propagator

To understand the basic idea behind the treatment of emission here, consider the equation (1.7) for continuity of electromagnetic energy:

$$\frac{\partial}{\partial t}\left(\tfrac{1}{2}\varepsilon_0|\mathbf{E}|^2 + \tfrac{1}{2}|\mathbf{B}|^2/\mu_0\right) + \mathrm{div}\,(\mathbf{E}\times\mathbf{B}/\mu_0) = -\mathbf{J}\cdot\mathbf{E}. \qquad (16.1)$$

The terms on the left are interpreted as the rate of change of the energy density in the electromagnetic field and the divergence of the energy flux in the electromagnetic field, respectively. The term on the right hand side is the rate per unit volume at which work is done by a current on the electromagnetic field. This interpretation is straightforward *in vacuo*, and one treats emission either by evaluating the left hand side, as in the traditional approach outlined above, or by evaluating the right hand side, as in the approach adopted here. However, in the presence of a medium the interpretation of (16.1) is not straightforward.

In a medium, the current $\mathbf{J}$ is separated into an induced part $\mathbf{J}_{\mathrm{ind}}$ and an extraneous part $\mathbf{J}_{\text{ext}}$ by writing $\mathbf{J} = \mathbf{J}_{\text{ind}} + \mathbf{J}_{\text{ext}}$. The time-reversible part of $\mathbf{J}_{\text{ind}}$ is transferred from the right hand side to the left hand side because it is part of the self-consistent field in the medium. This can be achieved in a simple way only after Fourier transforming, and then (16.1) is not the appropriate energy continuity equation. However, the inclusion of a medium causes no particular difficulty in evaluating the right hand side of (16.1). One simply replaces $\mathbf{J}$ by $\mathbf{J}_{\text{ext}}$. Then on integrating the resulting source term on the right hand side of (16.1) over an arbitrarily large volume V and an arbitrarily long time T one obtains the energy transferred from the source to the self-consistent electromagnetic field.

In treating any specific emission process, it is assumed that the extraneous current is given. Examples of extraneous currents are given in Chapter 17. The first step in treating emission is to find the electromagnetic field generated by this extraneous current. This involves solving the inhomogeneous wave equation (11.4), *viz.*,

$$\Lambda_{ij}(\omega, \mathbf{k})A_j(\omega, \mathbf{k}) = -\frac{\mu_0 c^2}{\omega^2} J_{\text{ext}\,i}(\omega, \mathbf{k}), \qquad (16.2)$$

for the the vector potential $\mathbf{A}(\omega, \mathbf{k})$ generated by the extraneous current. In treating the emission it is assumed that dissipative processes are ignored, so that $\Lambda_{ij}(\omega, \mathbf{k})$ is to be interpreted as hermitian. The antihermitian part is neglected because it is included separately when treating damping of the waves. Specifically, in a general transfer equation one has an emission term and an absorption term, and one treats emission and absorption (or damping) as separate processes in deriving the emission and absorption coefficients.

The solution of (16.2) is of the form

$$A_i(\omega, \mathbf{k}) = -\frac{\mu_0 c^2}{\omega^2} D_{ij}(\omega, \mathbf{k})J_{\text{ext}\,j}(\omega, \mathbf{k}), \qquad (16.3)$$

where $D_{ij}(\omega, \mathbf{k})$ is the appropriate Greens function for the wave equation. In other contexts, specifically in a quantum treatment, the Greens function $D_{ij}(\omega, \mathbf{k})$ is called the *photon propagator*, and this name is used here. The photon propagator is found almost trivially by using the definition (11.16), *viz.*,

$$\Lambda_{ij}(\omega, \mathbf{k})\lambda_{jl}(\omega, \mathbf{k}) = \delta_{il}\Lambda(\omega, \mathbf{k}). \qquad (16.4)$$

One contracts (16.2) with $\lambda_{il}(\omega, \mathbf{k})$ and then (16.4) implies

$$D_{ij}(\omega, \mathbf{k}) = \frac{\lambda_{ij}(\omega, \mathbf{k})}{\Lambda(\omega, \mathbf{k})}. \qquad (16.5)$$

Only the hermitian part of the response tensor is included here, and so

$D_{ij}(\omega, \mathbf{k})$ is hermitian. However, an antihermitian part appears when one imposes the causal condition. The photon propagator has poles at the zeros of the dispersion equation $\Lambda(\omega, \mathbf{k}) = 0$. Near the pole corresponding to the dispersion relation for waves in the mode M, that is, for $\omega \approx \omega_M(\mathbf{k})$, one approximates $\Lambda(\omega, \mathbf{k})$ by

$$\Lambda(\omega, \mathbf{k}) \approx \frac{\partial \Lambda(\omega, \mathbf{k})}{\partial \omega} \left[\omega - \omega_M(\mathbf{k}) + \mathrm{i}0 \right], \qquad (16.6)$$

where the causal condition has been imposed by including $+\mathrm{i}0$. On evaluating (16.5) with (16.6) for each mode using the Plemelj formula (4.32), the antihermitian part reduces to

$$D_{ij}^{\mathrm{A}}(\omega, \mathbf{k}) = \sum_M -\mathrm{i}\pi \omega_M(\mathbf{k}) R_M(\mathbf{k}) \, e_{Mi}(\mathbf{k}) e_{Mj}^*(\mathbf{k}) \, \delta\big(\omega - \omega_M(\mathbf{k})\big), \quad (16.7)$$

where (11.18) and (15.21) are used, and where the sum is over all wave modes. This antihermitian part (16.7) is referred to here as the *resonant part* of the propagator.

When only the hermitian part of the propagator is included in (16.3), the resulting field is the inductive field. The resonant part of the propagator in (16.3) gives the radiative part of the field in each wave mode.

16.2 The Emission Formula

As discussed above, the source term for energy in waves in a medium is the term $-\mathbf{J} \cdot \mathbf{E}$ in (16.1) with $\mathbf{J}$ replaced by the extraneous current $\mathbf{J}_{\mathrm{ext}}$. The energy radiated is calculated by integrating this source term over the normalization time T and normalization volume V. The resulting quantity is interpreted as the energy added to the field in the waves due to the work done by the extraneous current against the electric field that it generates. The energy radiated is the quantity of interest in an impulsive radiation process. For continuous radiation processes the power radiated is the relevant quantity; in this case the energy radiated is proportional to the normalization time T, and the power radiated is identified simply by dividing by T.

Let $U_M(\mathbf{k}) \mathrm{d}^3\mathbf{k}/(2\pi)^3$ be the energy added to waves in the mode M in the range $\mathrm{d}^3\mathbf{k}/(2\pi)^3$. On integrating $U_M(\mathbf{k})$ over $\mathbf{k}$-space and summing over all wave modes, the basic identification made in deriving the emission formula is

$$-\int_{-T/2}^{T/2} \mathrm{d}t \int \mathrm{d}^3\mathbf{x} \, \mathbf{J}_{\mathrm{ext}}(t, \mathbf{x}) \cdot \mathbf{E}(t, \mathbf{x}) = \sum_M \int \frac{\mathrm{d}^3\mathbf{k}}{(2\pi)^3} \, U_M(\mathbf{k}). \quad (16.8)$$

The left hand side of (16.8) is evaluated using the power theorem (4.11), and then identifying $\mathbf{E}(\omega, \mathbf{k})$ as $i\omega \mathbf{A}(\omega, \mathbf{k})$, with $\mathbf{A}(\omega, \mathbf{k})$ identified as the solution (16.3) of the inhomogeneous wave equation with the extraneous current as the source term. This gives

$$-\int_{-T/2}^{T/2} dt \int d^3x\, \mathbf{J}_{ext}(t, \mathbf{x}) \cdot \mathbf{E}(t, \mathbf{x}) = -\int \frac{d\omega d^3k}{(2\pi)^4}\, \mathbf{J}_{ext}^*(\omega, \mathbf{k}) \cdot \mathbf{E}(\omega, \mathbf{k})$$

$$= \int \frac{d\omega d^3k}{(2\pi)^4}\, \frac{i}{\varepsilon_0 \omega}\, J_{ext\,i}^*(\omega, \mathbf{k}) D_{ij}(\omega, \mathbf{k}) J_{ext\,j}(\omega, \mathbf{k}), \qquad (16.9)$$

where $\mu_0 c^2$ is rewritten as $1/\varepsilon_0$. Only the antihermitian part D_{ij}^A of D_{ij} contributes in (16.9). This is seen by first noting that the integral must be real, due to the left hand side being real; then, due to the factor i in the integrand, the hermitian part of D_{ij} gives a purely imaginary contribution, and the antihermitian part of D_{ij} gives the real contribution. The antihermitian part D_{ij}^A is given by (16.7). On inserting (16.7) into (16.9), the integral over ω is performed over the δ-function. Note that, as in the treatment of damping in §15.2, negative frequencies are implicitly contained in (16.7) and make a contribution equal in magnitude to the contribution from the positive frequencies and that this leads to an additional factor of 2 in the result, cf. (15.2) and (15.11).

The resulting expression for the energy radiated $U_M(\mathbf{k})$ in waves in the mode is

$$U_M(\mathbf{k}) = \frac{R_M(\mathbf{k})}{\varepsilon_0}\, \left| \mathbf{e}_M^*(\mathbf{k}) \cdot \mathbf{J}_{ext}\big(\omega_M(\mathbf{k}), \mathbf{k}\big) \right|^2. \qquad (16.10)$$

The result (16.10) is referred to here as the *emission formula*. It gives the energy radiated in the range $d^3k/(2\pi)^3$ in the mode M due to an arbitrary extraneous current $\mathbf{J}_{ext}$.

The emission formula (16.10) needs to be modified for transverse waves because of the two degenerate states of transverse polarization. All the information on the polarization of the emitted radiation is retained by replacing (16.10) by a polarization tensor that describes the emission:

$$U^{\alpha\beta}(\mathbf{k}) = \frac{R_T(k)}{\varepsilon_0}\, e_i^{*\alpha} e_j^{\beta} J_{ext\,i}\big(\omega_T(k), \mathbf{k}\big) J_{ext\,j}^*\big(\omega_T(k), \mathbf{k}\big), \qquad (16.11)$$

where α, β take on the values 1 or 2, as in (14.6). The total energy radiated is found by summing over the two states of polarization, which corresponds to taking the trace of the polarization tensor. This sum is performed in a formal way as follows. By definition, $\mathbf{e}^1$, $\mathbf{e}^2$, and $\boldsymbol{\kappa}$ form an orthonormal set of vectors. Hence a representation of the unit tensor is

$$e_i^{*1} e_j^1 + e_i^{*2} e_j^2 + \kappa_i \kappa_j = \delta_{ij}. \qquad (16.12)$$

It follows that the sum over states of polarization reduces to

$$e_i^{*\alpha} e_j^{\alpha} = \delta_{ij} - \kappa_i \kappa_j, \tag{16.13}$$

where the sum of α over 1 and 2 is implied. Then the trace of (16.11) gives, for the energy radiated summed over states of polarization,

$$U^{\alpha\alpha}(\mathbf{k}) = \frac{R_T(k)}{\varepsilon_0} \left| \boldsymbol{\kappa} \times \mathbf{J}_{\text{ext}}(\omega_T(k), \mathbf{k}) \right|^2. \tag{16.14}$$

The emission formula in the form (16.10) or (16.14) gives the energy radiated per unit volume of $\mathbf{k}$-space. It is often more appropriate to consider the energy radiated per unit frequency and per unit solid angle (about the wave normal direction). Let $U_M(\omega, \boldsymbol{\kappa})$ be the energy radiated per unit frequency and per unit solid angle. Then one has

$$\int \frac{d^3\mathbf{k}}{(2\pi)^3} U_M(\mathbf{k}) = \int_0^\infty d\omega \int d^2\Omega \, U_M(\omega, \boldsymbol{\kappa}), \tag{16.15}$$

where $d^2\Omega$ denotes an element of solid angle about the direction $\boldsymbol{\kappa}$. The relation between $U_M(\omega, \boldsymbol{\kappa})$ and $U_M(\mathbf{k})$ follows from (16.15) by changing the independent variables from $\mathbf{k}$, with ω the dependent variable, to the independent variables ω, $\boldsymbol{\kappa}$, with $n = kc/\omega$ the dependent variable. One has

$$\int \frac{d^3\mathbf{k}}{(2\pi)^3} = \int_0^\infty \frac{d\omega\, \omega^2}{(2\pi c)^3} \int d^2\Omega \, n_M^2(\omega, \boldsymbol{\kappa}) \frac{\partial[\omega n_M(\omega, \boldsymbol{\kappa})]}{\partial \omega}. \tag{16.16}$$

The derivative of the refractive index that appears in (16.16) cancels with a corresponding factor in the expression (15.23) for the ratio of electric to total energy. Hence the energy radiated in the mode M per unit frequency and per unit solid angle is given by

$$U_M(\omega, \boldsymbol{\kappa}) = \frac{\omega^2 n_M}{\varepsilon_0 (2\pi c)^3} \frac{|\mathbf{e}_M^* \cdot \mathbf{J}|^2}{1 - |\mathbf{e}^* \cdot \boldsymbol{\kappa}|^2}, \tag{16.17}$$

where the arguments $\mathbf{k}$ or $(\omega, \boldsymbol{\kappa})$ are omitted for simplicity in writing.

The polarization tensor $U^{\alpha\beta}(\omega, \boldsymbol{\kappa})$ for the energy radiated per unit frequency and per unit solid angle in transverse waves follows by analogy with (16.14). After summing over the two states of polarization and integrating over frequency and solid angle, the energy radiated in transverse waves reduces to

$$U^{\alpha\alpha}(\omega, \boldsymbol{\kappa}) = \frac{\omega^2 n(\omega)}{16\pi^3 \varepsilon_0 c^3} \left| \boldsymbol{\kappa} \times \mathbf{J}_{\text{ext}}(\omega, \boldsymbol{\kappa}) \right|^2. \tag{16.18}$$

The emission formula in the form (16.10), (16.14), (16.17) or (16.18) is used to treat continuous or periodic emission as follows. In such cases the current is proportional to a δ-function or to a sum of such terms, and the square contains a power of the normalization time T. However, in such cases it is the power radiated $P_M = U_M/T$ rather than the energy radiated that is of physical interest. Let us suppose that the current is of the form

$$\mathbf{J}_{ext}(\omega, \mathbf{k}) = \sum_s \mathbf{J}^{(s)}(\mathbf{k}) \, 2\pi\delta\big(\omega - \omega_s(\mathbf{k})\big), \qquad (16.19)$$

where s labels some set of natural frequencies $\omega_s(\mathbf{k})$, which may depend on $\mathbf{k}$ in general. On substituting (16.19) into (16.10), the squares of the δ-functions are rewritten using (4.37) in the form

$$\big[2\pi\delta\big(\omega - \omega_s(\mathbf{k})\big)\big]^2 = T \, 2\pi\delta\big(\omega - \omega_s(\mathbf{k})\big), \qquad (16.20)$$

where T is the normalization time. The average power radiated $P_M(\mathbf{k})$ is identified as

$$P_M(\mathbf{k}) = \lim_{T \to \infty} \frac{U_M(\mathbf{k})}{T}, \qquad (16.21)$$

so that the powers of T cancel. It then follows that the power radiated, analogous to (16.10) for the energy radiated, is given by

$$P_M(\mathbf{k}) = \sum_s \frac{R_M(\mathbf{k})}{\varepsilon_0} \left|\mathbf{e}_M^*(\mathbf{k}) \cdot \mathbf{J}^{(s)}(\mathbf{k})\right|^2 2\pi\delta\big(\omega_M(\mathbf{k}) - \omega_s(\mathbf{k})\big). \qquad (16.22)$$

Expressions for the power radiated analogous to the expressions (16.14), (16.17) and (16.18) follow in a similar way. When expressed in terms of the variables used in (16.17), the power radiated per unit frequency and per unit solid angle is referred to as the *emissivity*. The emissivity in the mode M is

$$\eta_M(\omega, \boldsymbol{\kappa}) = \frac{\omega^2 n_M}{\varepsilon_0(2\pi c)^3} \frac{|\mathbf{e}_M^* \cdot \mathbf{J}|^2}{1 - |\mathbf{e}^* \cdot \boldsymbol{\kappa}|^2} 2\pi\delta(\omega - \omega_s), \qquad (16.23)$$

where arguments are omitted for simplicity in writing.

16.3 The Einstein Coefficients

As remarked in §15.5, a semiclassical theory is one in which the waves are treated classically and the particles are treated quantum mechanically. A semiclassical treatment is required for many important applications, such as emission by atoms and molecules, where the use of quantum mechanics is essential. Another reason for introducing the semiclassical treatment here is to allow one to introduce the Einstein coefficients to relate emission to absorption. There is a subtle but important difference between a quantum mechanical theory and a classical theory of emission. This arises from the fact that energy is not automatically conserved in a classical theory. The energy (and momentum and angular momentum) in the radiation field must come from the radiating system, but in a classical theory the back reaction on the particles needs to be taken into account separately. For classical emission *in vacuo* the back reaction is introduced through a radiation reaction force (§19.5) whose sole purpose is to ensure that energy and momentum are conserved. However, this procedure cannot be applied, at least not without major modification, to emission in an arbitrary medium. On the other hand, in quantum mechanics the conservation of energy on a microscopic level is built into the theory.

In a classical theory absorption and emission are usually treated independently. In §15.2 absorption is treated in terms of the dissipative part of the response tensor and in §16.2 emission is treated in a quite a different way. There is necessarily a relation between emission and absorption by thermal particles implied by the second law of thermodynamics. Classically, this is expressed in terms of *Kirchhoff's law* which requires that for a Maxwellian distribution of particles at a given temperature, when emission and absorption are in balance the resulting wave spectrum must be a thermal spectrum at the same temperature. This relates the emission and absorption coefficients for a thermal distribution of particles. In a quantum approach the second law of thermodynamics is imposed on a microscopic level where it leads to *detailed balance* between the emission and absorption processes. In radiation theory, detailed balance is expressed in terms of the Einstein coefficients.

Consider a quantum mechanical system that has quantum states labeled q, q', and so on. Let the energy of state q be denoted ε_q. Quantum mechanics requires that in a transition from state q to state q' the electromagnetic field be given a quantum of energy $\hbar\omega$ which is determined by energy conservation:

$$\hbar\omega = \hbar\omega_{qq'}, \quad \hbar\omega_{qq'} := \varepsilon_q - \varepsilon_{q'}. \tag{16.24}$$

Let $\hat{\mathbf{J}}(\mathbf{x})$ be the current operator for our quantum mechanical system. The time-independent part of the matrix element of this operator between the states q and q' is denoted $\mathbf{J}_{qq'}(\mathbf{x})$. The matrix element varies with time as $e^{-i(\varepsilon_q - \varepsilon_{q'})t/\hbar}$ due to the wave function for the initial state evolving as $e^{-i\varepsilon_q t/\hbar}$, and the complex conjugate of the wave function for the final state evolving as $e^{i\varepsilon_{q'} t/\hbar}$. Hence the Fourier transform of the matrix element of the current is of the form

$$\mathbf{J}(\omega, \mathbf{k}) = \mathbf{J}_{qq'}(\mathbf{k}) \, 2\pi\delta(\omega - \omega_{qq'}), \tag{16.25}$$

which is the appropriate source term to be inserted in the emission formula (16.10). This current involves a δ-function and its square appears when inserted into (16.10).

On writing the current (16.25) in the form (16.19), the power radiated

(16.22) reduces to

$$[P_M(\mathbf{k})]_{qq'} = \frac{R_M(\mathbf{k})}{\varepsilon_0} \left| \mathbf{e}_M^*(\mathbf{k}) \cdot \mathbf{J}_{qq'}(\mathbf{k}) \right|^2 2\pi\delta\big(\omega_M(\mathbf{k}) - \omega_{qq'}\big). \quad (16.26)$$

In a quantum mechanical treatment the power emitted (16.26) is attributed to the emission of wave quanta. Each wave quantum has an energy $\hbar\omega_M(\mathbf{k})$. It is convenient to introduce the *probability of spontaneous emission* by writing

$$w_{Mqq'}(\mathbf{k}) := \frac{[P_M(\mathbf{k})]_{qq'}}{\hbar\omega_M(\mathbf{k})}. \quad (16.27)$$

As so defined, the probability is such that $w_{Mqq'}(\mathbf{k})\mathrm{d}^3\mathbf{k}/(2\pi)^3$ is the probability per unit time that a photon in the mode M is emitted in the range $\mathrm{d}^3\mathbf{k}/(2\pi)^3$ with the quantum system making a transition from state q to q'.

A form of Einstein coefficients is derived as follows. The quantum mechanical version of a thermal distribution of waves is the *Planck distribution*

$$N_M(\mathbf{k}) = \frac{1}{e^{\hbar\omega_M(\mathbf{k})/\Theta} - 1}, \quad (16.28)$$

where Θ is the temperature times Boltzmann's constant. Part of Einstein's argument was based on the fact that the transition rate for spontaneous emission does not depend on $N_M(\mathbf{k})$ whereas the transition rate for absorption is proportional to $N_M(\mathbf{k})$. To satisfy detailed balance it is necessary to introduce an additional form of emission, called *stimulated emission*, that is also proportional to $N_M(\mathbf{k})$. Collectively, stimulated emission and true absorption are called the *induced processes*.

Consider emission by a particle in which the particle has a transition from a state with quantum numbers q to a state with quantum numbers q'. Absorption is due to the inverse transition from q' to q. The probability of spontaneous emission is defined by (16.27). Let us also introduce probabilities $w_{Mqq'}^{\text{stim}}(\mathbf{k})\,N_M(\mathbf{k})$ and $w_{Mqq'}^{\text{abs}}(\mathbf{k})\,N_M(\mathbf{k})$ for stimulated emission and true absorption, respectively. Let n_q be the occupation number for the state q. A thermal distribution corresponds to $n_q \propto e^{-\varepsilon_q/\Theta}$. Energy conservation at a microscopic level implies $\varepsilon_q = \varepsilon_{q'} + \hbar\omega_M(\mathbf{k})$, and hence

$$n_{q'} = n_q e^{\hbar\omega_M(\mathbf{k})/\Theta}. \quad (16.29)$$

Detailed balance requires that in thermal equilibrium the rate of transitions $q \to q'$ be in balance with the rate of transitions $q' \to q$. This implies

$$\big[w_{Mqq'}(\mathbf{k}) + w_{Mqq'}^{\text{stim}}(\mathbf{k})\,N_M(\mathbf{k})\big]\,n_q = w_{Mqq'}^{\text{abs}}(\mathbf{k})\,N_M(\mathbf{k})\,n_{q'}. \quad (16.30)$$

On inserting (16.28) and (16.29), and requiring that the result hold for all Θ, one finds

$$w^{\text{stim}}_{M q q'}(\mathbf{k}) = w^{\text{abs}}_{M q q'}(\mathbf{k}) = w_{M q q'}(\mathbf{k}), \qquad (16.31)$$

which is one form of the *Einstein coefficients*. Although the relations (16.31) are derived by appealing to a thermal distribution, the result is independent of the form of the distribution of particles.

16.4 The Kinetic Equation for the Waves

A transfer equation for the waves in the semiclassical formalism is derived as follows. Each emission of a wave quantum increases the occupation number $N_M(\mathbf{k})$ by unity, and each absorption of a wave quantum decreases the occupation number $N_M(\mathbf{k})$ by unity. The quantity $w_{M q q'}(\mathbf{k})d^3\mathbf{k}/(2\pi)^3$ is the probability per unit time that one atom or molecule emits a wave quantum in the range $d^3\mathbf{k}/(2\pi)^3$. The occupation number $N_M(\mathbf{k})$ of the wave quanta is defined as the density in phase space, and an element of phase space corresponds to a range $d^3\mathbf{x}d^3\mathbf{k}/(2\pi)^3$ which, after integration, contains a power of the normalization volume V. As a consequence, for a single atom or molecule, the rate of increase of $N_M(\mathbf{k})$ due to spontaneous emission $w_{M q q'}(\mathbf{k})/V$ also involves a power of V. The difference between the sum of spontaneous and stimulated emissions and of true absorptions by a collection of atoms or molecules with occupation number n_q is given by:

$$\frac{dN_M(\mathbf{k})}{dt} = \frac{1}{V} \sum_{q q'} w_{M q q'}(\mathbf{k}) \left\{ \left[1 + N_M(\mathbf{k}) \right] n_q - N_M(\mathbf{k}) n_{q'} \right\}, \quad (16.32)$$

where the sum is over all pairs of states for which the transition is allowed.

The unit term in the factor $[1 + N_M(\mathbf{k})]$ on the right hand side of (16.32) describes the effect of spontaneous emission by the distribution of particles with occupation number n_q, and the term proportional to $N_M(\mathbf{k})$ describes the effect of stimulated emission. The two terms on the right hand side of (16.32) that are proportional to $N_M(\mathbf{k})$ are combined to describe *net absorption*. Thus, if one rewrites (16.32) in the form

$$\frac{dN_M(\mathbf{k})}{dt} = \alpha_M(\mathbf{k}) - \gamma_M(\mathbf{k})N_M(\mathbf{k}) \qquad (16.33)$$

$$\begin{pmatrix} \alpha_M(\mathbf{k}) \\ \gamma_M(\mathbf{k}) \end{pmatrix} = \frac{1}{V} \sum_{q q'} w_{M q q'}(\mathbf{k}) \begin{pmatrix} n_q \\ -(n_q - n_{q'}) \end{pmatrix} \qquad (16.34)$$

then the two terms on the right hand side of (16.33) describe the effects of spontaneous emission and net absorption, respectively.

Net absorption can be negative, leading to maser action. It follows from (16.34) that negative absorption requires $n_q > n_{q'}$, that is, that the population of the higher energy state exceed that of the lower energy state. However, this is a necessary but not sufficient condition for maser action.

A deficiency in classical electromagnetic theory is that conservation of energy is not taken into account explicitly. For emission *in vacuo* a radiation reaction force is introduced to overcome this difficulty, cf. §19.5. In the semiclassical approach this deficiency is overcome trivially: energy (and momentum where relevant) is conserved on a microscopic level. A kinetic equation that describes the reaction of the particles to the emission and absorption of radiation is derived by considering two transitions, one from q to q', and the other from q'' to q. The rate of change in the occupation number n_q is determined by the difference between the gains due to emission $q'' \to q$ and absorption $q' \to q$, and losses due to absorption $q \to q''$ and emission $q \to q'$. The net rate of change gives

$$\frac{dn_q}{dt} = \frac{1}{V} \sum_{q''} w_{Mq''q}(\mathbf{k}) \big\{ \left[1 + N_M(\mathbf{k}) \right] n_{q''} - N_M(\mathbf{k}) \, n_q \big\}$$

$$- \frac{1}{V} \sum_{q'} w_{Mqq'}(\mathbf{k}) \big\{ \left[1 + N_M(\mathbf{k}) \right] n_q - N_M(\mathbf{k}) \, n_{q'} \big\}. \quad (16.35)$$

The terms of the right hand side of (16.35) that are independent of $N_M(\mathbf{k})$ describe the back reaction to spontaneous emission, and the terms proportional to $N_M(\mathbf{k})$ describe the back reaction of the distribution of particles to the induced processes.

Exercise Set 16

16.1 Consider emission of longitudinal waves. Show that, with ρ_{ext} the extraneous charge density, the emission formula (16.10) implies

$$U_M(\mathbf{k}) = \frac{R_M(\mathbf{k})\omega_M^2(\mathbf{k})}{|\mathbf{k}|^2\varepsilon_0} \left|\rho_{\text{ext}}\left(\omega_M(\mathbf{k}), \mathbf{k}\right)\right|^2. \qquad (E16.1)$$

16.2 A classical system with a natural frequency of oscillation ω_0 is immersed in a plasma with plasma frequency ω_p slightly less than ω_0. The current associated with the system is $\mathbf{J}_0 2\pi\delta(\omega - \omega_0)$.

(a)   Show that the power radiated in transverse waves is

$$P_T = \frac{\omega_0^2 n(\omega_0)|\mathbf{J}_0|^2}{3\pi\varepsilon_0 c^3}. \qquad (E16.2)$$

(b)   Derive a formula analogous to (16.26) for the power radiated in Langmuir waves assuming $\omega_0 - \omega_p \ll \omega_p$, and hence that the wave properties are given by $\omega_L(k) = \omega_p + 3k^2 V_e^2/2\omega_p$, $R_L(k) = \frac{1}{2}$.

(c)   Hence show that the power radiated in Langmuir waves is

$$P_L = \frac{[\omega_p^3(\omega_0 - \omega_p)]^{1/2}|\mathbf{J}_0|^2}{9(6)^{1/2}\pi\varepsilon_0 V_e^3}. \qquad (E16.3)$$

16.3 The polarization of emitted transverse waves is uniquely determined by the requirements that $\mathbf{e}$ be orthogonal to $\boldsymbol{\kappa}$ and that the energy or power radiated be non-zero only when $\mathbf{e}$ has a non-zero projection along the direction of $\mathbf{J}_{\text{ext}}(\omega, \mathbf{k})$.

(a)   Show that the polarization is along $\mathbf{k} \times [\mathbf{k} \times \mathbf{J}_{\text{ext}}(\omega, \mathbf{k})]$.

(b)   What are the physical consequences on the polarization of the emitted radiation if the components of $\mathbf{J}_{\text{ext}}(\omega, \mathbf{k})$: (i) all have the same phase, and (ii) have different relative phases?

16.4 The emission formula (16.10) is derived with the source term for energy identified from the energy equation (1.7). An analogous formula for the emission of momentum is derived by starting from the source term in the momentum equation (3.5).

(a)   Starting with the source term

$$\rho_{\text{ext}}(t, \mathbf{x})\mathbf{E}(t, \mathbf{x}) + \mathbf{J}_{\text{ext}}(t, \mathbf{x}) \times \mathbf{B}(t, \mathbf{x}),$$

repeat the analysis leading to (16.9) and (16.10). In particular, show that the ratio of the momentum emitted to the energy emitted in the range $d^3\mathbf{k}/(2\pi)^3$ is equal to $\mathbf{k}/\omega$.

16.5 The fluctuations in the electromagnetic field are described in terms of the autocorrelation function of the potential $\mathbf{A}$; the level of such fluctuations is related to the correlation function for the extraneous current. The autocorrelation function is denoted

$$\langle A_i A_j \rangle (\tau, \boldsymbol{\xi}) = \langle A_i(t, \mathbf{x}) A_j(t + \tau, \mathbf{x} + \boldsymbol{\xi}) \rangle, \qquad (E16.4)$$

where the angular brackets on the right hand side denote an ensemble average. In practice the ensemble average is replaced by an average over a time T and a volume V.

(a) By expressing the Fourier transform $\mathbf{A}(\omega, \mathbf{k})$ in terms of $\mathbf{A}(t, \mathbf{x})$ using the Fourier integral theorem, and using (E16.1), show

$$\langle A_i(\omega, \mathbf{k}) A_j(\omega', \mathbf{k}') \rangle = \langle A_i A_j \rangle (\omega, \mathbf{k}) 2\pi \delta(\omega + \omega') (2\pi)^3 \delta^3(\mathbf{k} + \mathbf{k}').$$
$$(E16.5)$$

(b) Assuming that $\mathbf{A}(\omega, \mathbf{k})$ is the field generated by the extraneous current, and hence is given by (16.3), show that the autocorrelation functions of $\mathbf{A}$ and $\mathbf{J}_{\text{ext}}$ are related by

$$\langle A_i A_j \rangle (\omega, \mathbf{k}) = \frac{\lambda_{ir}(\omega, \mathbf{k}) \lambda_{js}^*(\omega, \mathbf{k})}{|\varepsilon_0 \omega^2 \Lambda(\omega, \mathbf{k})|^2} \langle J_r J_s \rangle_{\text{ext}}(\omega, \mathbf{k}). \qquad (E16.6)$$

(c) Show that on making the separation

$$\Lambda(\omega, \mathbf{k}) = \text{Re}\,[\Lambda(\omega, \mathbf{k})] + i\text{Im}\,[\Lambda(\omega, \mathbf{k})],$$

and assuming that the imaginary part is small, one has

$$\frac{1}{|\Lambda(\omega, \mathbf{k})|^2} = \frac{\pi \delta[\Lambda(\omega, \mathbf{k})]}{|\text{Im}\,\Lambda(\omega, \mathbf{k})|}. \qquad (E16.7)$$

Hint: Use the Plemelj formula (4.32) with $\omega \to \text{Re}\,\Lambda$, $\eta \to \text{Im}\,\Lambda$.

(d) Using (15.1), (15.5) and (15.13), show that for waves in the mode M one has

$$\langle A_i A_j \rangle_M (\omega, \mathbf{k}) = \frac{V R_M(\mathbf{k}) W_M(\mathbf{k})}{\varepsilon_0 \omega_M^2(\mathbf{k})} e_{Mi}(\mathbf{k}) e_{Mj}^*(\mathbf{k})\, 2\pi \delta\big(\omega - \omega_M(\mathbf{k})\big).$$
$$(E16.8)$$

(e) Show that (E16.6)–(E16.8), with (11.18), (15.13), (15.17) and (15.21), imply

$$W_M(\mathbf{k}) = \frac{R_M(\mathbf{k})}{V \varepsilon_0 \gamma_M(\mathbf{k})} e_{Mr}^*(\mathbf{k}) e_{Ms}(\mathbf{k}) \langle J_r J_s \rangle_{\text{ext}}(\omega_M(\mathbf{k}), \mathbf{k}).$$
$$(E16.9)$$

16.6 *The fluctuation–dissipation theorem* gives the steady state level of fluctuations when spontaneous emission and dissipation are in balance in a thermal plasma. Let the temperature (in energy units) be T_e.

(a) Show that a sufficient condition for (E16.9), with $W_M(\mathbf{k}) = T_e/V$

for a thermal distribution of waves, to apply for every wave mode in the plasma is

$$\langle J_i J_j \rangle_{\text{ext}}(\omega, \mathbf{k}) = -2i\varepsilon_0 T_e \omega K_{ij}^{\text{A}}(\omega, \mathbf{k}). \qquad (E16.10)$$

(b)   Hence, using $(E16.8)$, find the autocorrelation function for the vector potential for waves in the mode M in a thermal plasma.

16.7 Generalize (16.34) and (16.35) to the cases where the particles are treated quantum mechanically. This involves assuming that a transition $q \to q'$ occurs at a rate proportional to $n_q(1 \pm n_{q'})$, rather than to n_q for classical particles, with the upper sign for bosons and the lower sign for fermions.

17

Emission by Multipoles

Preamble

A localized distribution of charge and current, such as an atom or molecule, may be expanded in multipole moments. Provided that this expansion converges rapidly, the total emission by the system may be separated into its multipole components, that is, into an electric dipole term, a magnetic dipole term, an electric quadrupole term, and so on.

17.1 The Multipole Source Terms

The multipole expansion introduced in §3.2 is to be regarded formally as an expansion in a ratio of two lengths. The smaller of the two lengths is a characteristic scale length associated with the charge and current system. The other length is the distance to the observer or, more relevant here, the wavelength of radiation. The multipole expansion converges rapidly only if this ratio is small.

In the emission formula (16.10) the source term appears as the Fourier transform of an extraneous current. This Fourier transform is related to the multipole expansion by writing

$$\mathbf{J}(\omega, \mathbf{k}) = \int dt\, e^{i\omega t} \int d^3x\, e^{-i\mathbf{k}\cdot\mathbf{x}}\, \mathbf{J}(t, \mathbf{x})$$

$$= \int dt\, e^{i\omega t} \int d^3x\, [1 - i\mathbf{k}\cdot\mathbf{x} - \tfrac{1}{2}(\mathbf{k}\cdot\mathbf{x})^2 + \cdots]\, \mathbf{J}(t, \mathbf{x}). \quad (17.1)$$

One rewrites each of the terms in (17.1) by introducing the multipole quantities defined by (3.8) and (3.11). Then, using (3.14), the leading term corresponding to the unit term inside the square brackets in

(17.1) is expressed as the time-derivative of the electric dipole moment, and, using (3.15), the next term corresponding to the $-\mathrm{i}\mathbf{k}\cdot\mathbf{x}$ inside the square brackets in (17.1) is separated into its symmetric and antisymmetric parts and rewritten in terms of the time-derivative of the electric quadrupole moment and of the magnetic dipole moment, respectively. Thus (17.1) reduces to

$$J_i(\omega,\mathbf{k}) = \int \mathrm{d}t\, \mathrm{e}^{\mathrm{i}\omega t} \left\{ \frac{\partial}{\partial t}\, d_i(t) + \mathrm{i}[\mathbf{k}\times\boldsymbol{\mu}(t)]_i - \mathrm{i}\tfrac{1}{2}k_s \frac{\partial}{\partial t}\, q_{is}(t) + \cdots \right\}$$
$$= -\mathrm{i}\omega d_i(\omega) + \mathrm{i}[\mathbf{k}\times\boldsymbol{\mu}(\omega)]_i - \tfrac{1}{2}\omega k_s q_{is}(\omega) + \cdots . \qquad (17.2)$$

The terms shown explicitly on the right hand side of (17.2) are identified as the source terms for electric dipole emission, magnetic dipole emission and electric quadrupole emission, respectively.

On inserting the multipole source terms from (17.2) into the emission formula (16.10), one obtains the following expression for electric dipole emission, magnetic dipole emission and electric quadrupole emission, respectively:

$$U_M(\mathbf{k}) = \frac{R_M(\mathbf{k})}{\varepsilon_0} \left| \omega_M(\mathbf{k})\, \mathbf{e}_M^*(\mathbf{k})\cdot \mathbf{d}\big(\omega_M(\mathbf{k})\big) \right|^2, \qquad (17.3)$$

$$U_M(\mathbf{k}) = \frac{R_M(\mathbf{k})}{\varepsilon_0} \left| \mathbf{e}_M^*(\mathbf{k})\cdot \mathbf{k}\times\boldsymbol{\mu}\big(\omega_M(\mathbf{k})\big) \right|^2, \qquad (17.4)$$

$$U_M(\mathbf{k}) = \frac{R_M(\mathbf{k})}{4\varepsilon_0} \left| \omega_M(\mathbf{k})\, e_{Mi}^*(\mathbf{k}) k_j q_{ij}\big(\omega_M(\mathbf{k})\big) \right|^2. \qquad (17.5)$$

The formulas give the energy $U_M(\mathbf{k})\mathrm{d}^3\mathbf{k}/(2\pi)^3$ radiated in the mode M in the range $\mathrm{d}^3\mathbf{k}/(2\pi)^3$ due to these three lowest multipoles.

It is conventional to treat emission by different multipole moments of the same system as independent processes. Strictly this is not correct because the total source term is a sum over the individual currents from all the multipole components. The cross terms between the different multipole terms interfere when (17.2) is inserted into (16.10). However, in practice one is usually interested only in the emission due to the lowest multipole moment, and interference with higher order moments is necessarily only a small correction when the multipole expansion converges rapidly. Also, it may be shown, cf. Exercise 19.4, that in the total power radiated in transverse waves, the contribution of the cross terms averages to zero. In a quantum mechanical application, the assumption that interference between the currents due to different multipoles is unimportant is justified for a different reason. In this case there are different selection rules for different multipole transitions. Thus if a transition from one quantum state of an atom or molecule is allowed as an electric quadrupole transition, then it is usually not allowed as a magnetic dipole transition.

A subtle point arises concerning the traces of the quadrupole and higher moments. As discussed in §3.3, when one defines the multipole moments as cartesian tensors, the trace of the electric quadrupole moment and the traces of all higher order moments do not contribute to the electromagnetic field *in vacuo*. As a consequence it is conventional to redefine the cartesian tensors to be traceless. Specifically, as in (3.9), one replaces q_{ij} by $d_{ij} = 3q_{ij} - q_{ss}\delta_{ij}$ and one removes the traces of the higher order moments in a similar way. In (17.5) one has

$$e_{Mi}^*(\mathbf{k})k_j q_{ij}(\omega) = \tfrac{1}{3}\left[e_{Mi}^*(\mathbf{k})k_j d_{ij}(\omega) + \mathbf{k}\cdot\mathbf{e}_M^*(\mathbf{k})q_{ss}(\omega)\right], \qquad (17.6)$$

from which it is clear that the trace q_{ss} does not contribute for transverse waves (with $\mathbf{e}\cdot\mathbf{k} = 0$). On the other hand, the trace does contribute to the emission of any waves for which $\mathbf{e}_M(\mathbf{k})$ has a component along $\mathbf{k}$. Thus the neglect of the trace is only justified for the emission of waves that are strictly transverse. In particular, the use of the spherical tensor forms of the multipole moments, cf. §3.5, which exclude all traces, is incomplete when applied to a system radiating in an arbitrary medium.

The simplest example of a case where the conventional multipole expansion is inadequate is a spherically symmetric charge distribution $\rho(r,t)$ which is changing as a function of time. Then the only non-zero multipole moment (in the conventional sense with traces removed) is the zeroth moment, that is, the total charge. As the charge does not change with time, any time variation that does not affect the spherical symmetry leads to no change in the electromagnetic field (outside the charge distribution) and to no radiation of transverse waves. Thus a spherically symmetric oscillating charged sphere does not radiate *in vacuo* or in a medium where the only waves are transverse. However, such a spherically symmetric charge distribution can radiate waves which are not strictly transverse, that is, waves such as Langmuir waves in an isotropic plasma, extraordinary waves in a uniaxial crystal or magnetoionic waves in a magnetized electron plasma.

It follows that to treat the multipole expansion for emission in a medium where the waves are not strictly transverse, one needs to retain the traces of the multipole moments. This is done in two different ways. One way is to adopt the definitions of the cartesian multipole moments given by (3.8) and (3.11) and retain all the traces. Alternatively, one adopts the conventional definition of the electric and magnetic multipole moments as being traceless and introduces a further set of moments that take the traces into account. There are relatively few applications where this needs to be done explicitly.

17.2 Multipole Emission in Transverse Waves

We concentrate on multipole emission of transverse waves *in vacuo* or in an isotropic dielectric. The angular distribution of the radiation and its polarization are characteristic features of each multipole term.

Consider first the angular distribution of the emitted radiation without reference to the polarization (except that it is assumed transverse). The total power radiated is found by summing over the two states of transverse polarization. This is achieved using (16.13), leading to the expression (16.14) for the total energy radiated $U^{\alpha\alpha}(\mathbf{k})d^3\mathbf{k}/(2\pi)^3$ in the range $d^3\mathbf{k}/(2\pi)^3$. Let us introduce the energy radiated per unit frequency $U_{\text{rad}}(\omega)$ by defining $U_{\text{rad}}(\omega)d\omega$ to be the energy radiated in the frequency range $d\omega$. By definition the total energy radiated is

$$E_{\text{rad}} = \int \frac{d^3\mathbf{k}}{(2\pi)^3} U^{\alpha\alpha}(\mathbf{k}) = \int_0^\infty d\omega \, U_{\text{rad}}(\omega). \tag{17.7}$$

Now using $k = n(\omega)\omega/c$ one finds

$$U_{\text{rad}}(\omega) = \frac{[\omega n(\omega)]^2}{(2\pi c)^3} \frac{\partial[\omega n(\omega)]}{\partial\omega} \int d^2\Omega \, U^{\alpha\alpha}(\mathbf{k}), \tag{17.8}$$

where the integral is over 4π steradians. In the absence of spatial dispersion transverse waves have a dispersion relation given by $n^2(\omega) = K^{\text{T}}(\omega)$ and then (15.22) implies that the ratio of electric to total energy is given by $R(\omega) = 1/\{2n(\omega)\partial[\omega n(\omega)]/\partial\omega\}$. When (17.3)–(17.5) are inserted into (17.8) the factors $\partial[\omega n(\omega)]/\partial\omega$ cancel. Let us introduce the unit vector $\boldsymbol{\kappa}$ along $\mathbf{k}$, and replace the integral over solid angle by multiplication by 4π and an average over all directions of $\boldsymbol{\kappa}$. Then for electric dipole emission, magnetic dipole emission and electric quadrupole emission, respectively, one finds

$$U_{\text{rad}}(\omega) = \frac{n(\omega)\,\omega^4}{4\pi^2\varepsilon_0 c^3} \langle |\boldsymbol{\kappa} \times \mathbf{d}(\omega)|^2 \rangle, \tag{17.9}$$

$$U_{\text{rad}}(\omega) = \frac{n^3(\omega)\,\omega^4}{4\pi^2\varepsilon_0 c^5} \langle |\boldsymbol{\kappa} \times \{\boldsymbol{\kappa} \times \boldsymbol{\mu}(\omega)\}|^2 \rangle, \tag{17.10}$$

$$U_{\text{rad}}(\omega) = \frac{n^3(\omega)\,\omega^6}{144\pi^2\varepsilon_0 c^5} \langle \kappa_r\kappa_s(\delta_{ij} - \kappa_i\kappa_j) \rangle d_{ir}(\omega)d_{js}^*(\omega), \tag{17.11}$$

where the angular brackets denote this average. In (17.11) it is noted that the traces of q_{ir} and q_{js} do not contribute, and these moments are

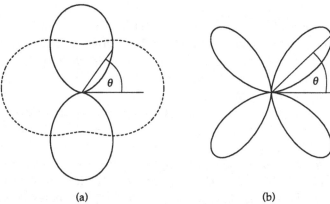

(a)   (b)

Fig. 17.1 (a) Dipolar radiation patterns: solid curve for oscillation
along the axis, dashed curve for rotation about the axis.
(b) A quadrupolar pattern for oscillation along the axis.

each replaced by one third times the corresponding traceless moments
in accord with (17.6).

Before averaging over the direction of emission $\boldsymbol{\kappa}$, let us discuss the
angular pattern of the emitted radiation. For electric dipole emission,
the angular dependence is determined by the factor $|\boldsymbol{\kappa} \times \mathbf{d}(\omega)|^2$. If $\mathbf{d}(\omega)$
is real the angular dependence is simple. One then chooses spherical
polar coordinates with the polar axis along $\mathbf{d}(\omega)$, and chooses θ to be
the angle between the direction $\boldsymbol{\kappa}$ and $\mathbf{d}(\omega)$. It follows that for $\mathbf{d}(\omega)$
real, there is a $\sin^2 \theta$ pattern about the direction $\mathbf{d}(\omega)$, as illustrated
in Figure 17.1. However, $\mathbf{d}(\omega)$ need not be real in general. Another
simple example is when $\mathbf{d}(\omega)$ corresponds to a system rotating about
the z axis, which is the polar axis here. In this case the angular pattern
of the emission varies with θ as $1 + \cos^2 \theta$, cf. Exercise 17.1. This raises
the question of what is the most general form of the angular pattern
for dipole emission. One general statement that can be made is that
the angular distribution may be expressed as a sum of three terms each
proportional to the modulus squared of a sum (over $-l < m < l$) of the
form $d_l^m Y_{lm}(\theta, \phi)$ with $l = 1$, cf. §3.5.

The angular pattern for quadrupolar emission implied by (17.11) is
that of $|\boldsymbol{\kappa} \times \mathbf{C}|^2$ with $C_i = d_{ij}(\omega)\kappa_j$. Suppose that $d_{ij}(\omega)$ is real. On
diagonalizing it the sum of the diagonal terms is zero. In the simplest
non-trivial case two of the diagonal components are equal, with the value
D say, so that the third is equal to $-2D$. Let this third component correspond to the 3-axis. One is free to orient the axes so that $\boldsymbol{\kappa}$ is in the 1-
3 plane, with $\boldsymbol{\kappa} = (\sin\theta, 0, \cos\theta)$. Then one has $\mathbf{C} = D(\sin\theta, 0, -2\cos\theta)$

and hence $|\boldsymbol{\kappa} \times \mathbf{C}|^2 = 9D^2 \sin^2 \theta \cos^2 \theta$. Thus in this simplest case, the quadrupolar radiation has a $\sin^2 \theta \cos^2 \theta$ pattern, which is illustrated in Figure 17.1. A somewhat more general case is where no two of the diagonal components of the quadrupole tensor are equal after diagonalization. Let the three components be D_1, D_2, $-(D_1 + D_2)$, which need not be real in general. Then with $\boldsymbol{\kappa} = (\sin \theta \cos \phi, \sin \theta \sin \phi, \cos \theta)$ one has $\mathbf{C} = (D_1 \sin \theta \cos \phi, D_2 \sin \theta \sin \phi, -(D_1 + D_2) \cos \theta)$, and $|\boldsymbol{\kappa} \times \mathbf{C}|^2 = \left(|D_1 + 2D_2|^2 \sin^2 \phi + |2D_1 + D_2|^2 \cos^2 \phi\right) \sin^2 \theta \cos^2 \theta + |D_1 - D_2|^2 \sin^4 \theta \cos^2 \phi \sin^2 \phi$. This angular pattern reduces to a $\sin^2 \theta \cos^2 \theta$ pattern only for $D_1 = D_2$. In the general case, $d_{ij}(\omega)$ is not real, and all that can be said is that the angular pattern may be expressed in terms of the modulus squared of a sum (over $-l < m < l$) of the form $d_l^m Y_{lm}(\theta, \phi)$ with $l = 2$.

Now let us evaluate the averages over the angle of emission to determine the total energy radiated per unit frequency. The averages may be performed by setting up coordinate axes, with $\boldsymbol{\kappa}$ in the general polar form introduced above, and then integrating over $0 < \phi < 2\pi$ and $-1 < \cos \theta < 1$. An alternative procedure is used here based on a more abstract argument. First consider the average of $\boldsymbol{\kappa}$; this average must be a vector because $\boldsymbol{\kappa}$ itself is a vector. Any such average vector would define a preferred direction in space, and because there is no such preferred direction this average must be zero. Next consider the average of $\kappa_i \kappa_j$. The average must be a second rank tensor and any such tensor in the absence of a preferred direction must be proportional to the unit tensor δ_{ij}. Let the constant of proportionality be A: $\langle \kappa_i \kappa_j \rangle = A \delta_{ij}$. On taking the trace and using $\kappa_i \kappa_i = |\boldsymbol{\kappa}|^2 = 1$ and $\delta_{ii} = 3$, one finds $A = 1/3$. On extending the argument, the average over any product of an odd number of factors $\boldsymbol{\kappa}$ gives zero, and that over any product of an even number of factors $\boldsymbol{\kappa}$ is expressed in terms of sums of products of the unit tensor, cf. Exercise 2.5. The relevant averages here are

$$\langle \kappa_i \kappa_j \rangle = \tfrac{1}{3} \delta_{ij}, \quad \langle \kappa_i \kappa_j \kappa_r \kappa_s \rangle = \tfrac{1}{15} (\delta_{ij} \delta_{rs} + \delta_{ir} \delta_{js} + \delta_{is} \delta_{jr}). \quad (17.12)$$

Explicit expressions for the energy radiated per unit frequency in electric dipole, magnetic dipole and electric quadrupole emission then follow by using (17.12) in (17.9)–(17.11). Thus one finds

$$U_{\text{rad}}(\omega) = \frac{n(\omega) \, \omega^4}{6\pi^2 \varepsilon_0 c^3} \, |\mathbf{d}(\omega)|^2, \quad (17.13)$$

$$U_{\text{rad}}(\omega) = \frac{\mu_0 n^3(\omega) \, \omega^4}{6\pi^2 c^3} \, |\boldsymbol{\mu}(\omega)|^2, \quad (17.14)$$

$$U_{\mathrm{rad}}(\omega) = \frac{n^3(\omega)\,\omega^6}{720\pi^2\varepsilon_0 c^5}\, d_{ij}(\omega)d_{ij}^*(\omega). \tag{17.15}$$

In the quadrupole case the average over angles involves the specific identity $\langle \kappa_r \kappa_s (\delta_{ij} - \kappa_i \kappa_j)\rangle d_{ir}d_{js}^* = d_{ij}d_{ij}^*/5$.

A characteristic feature of multipole emission in an isotropic dielectric is its dependence on the refractive index $n(\omega)$. According to (17.13)–(17.15), electric dipole emission is proportional to the first power of $n(\omega)$ and magnetic dipole and electric quadrupole emission are proportional to $n^3(\omega)$. More generally, for radiation in an isotropic dielectric, emission due to a 2^l-electric multipole ($l = 1$ for dipole, $l = 2$ for quadrupole, $l = 3$ for octupole, and so on) is proportional to $n^{2l-1}(\omega)$, and emission due to a 2^l-magnetic multipole is proportional to $n^{2l+1}(\omega)$.

17.3 Multipole Emission *in Vacuo*

The most familiar examples of multipole emission are for emission of transverse waves *in vacuo*, in which case there are simple formulas for the total power radiated. Using such formulas and the expressions derived above for the frequency dependence and the angular dependence of the emission one can identify all relevant properties of the emission in a simple and direct way.

For emission *in vacuo* one sets $n(\omega) = 1$ in (17.13)–(17.15). The total energy radiated due to any specific multipole emission follows by inserting the relevant one of (17.13)–(17.15) (or their generalization to higher multipoles) into (17.7). The remaining integral over frequency in (17.7) is evaluated using the power theorem (4.11) for temporal Fourier transforms. One uses (4.11) to rewrite the energy radiated as given by the integral over ω of the energy radiated per unit frequency in terms of the integral over time t of the power radiated $P(t)$. This gives

$$E_{\mathrm{rad}} = \int_0^\infty \mathrm{d}\omega\, U_{\mathrm{rad}}(\omega) = \int_{-\infty}^\infty \mathrm{d}t\, P(t). \tag{17.16}$$

The integral over frequency is extended to the complete range $-\infty < \omega < \infty$ by noting that the integrand is an even function of ω and hence the integral over $-\infty < \omega < \infty$ is twice that over $0 < \omega < \infty$. The power theorem is used in the form, for the electric dipole term for example,

$$\int_{-\infty}^\infty \frac{\mathrm{d}\omega}{2\pi}\, \omega^4 |\mathbf{d}(\omega)|^2 = \int_{-\infty}^\infty \mathrm{d}t\, |\ddot{\mathbf{d}}(t)|^2, \tag{17.17}$$

where a dot denotes differentiation with respect to time, and where $(-\mathrm{i}\omega)^2 \mathbf{d}(\omega)$ is the temporal Fourier transform of $\ddot{\mathbf{d}}(t)$.

Formally, it is not possible to identify $P(t)$ from (17.16). After using (17.17), (17.16) relates two integrals over time and to identify $P(t)$ one needs to equate the integrands. Except under very special circumstances, the equality of two integrals does not imply the equality of their integrands. However, if the emission is continuous or periodic then in an average sense one can identify $P(t)$ by equating the integrands. Ignoring the formal difficulty of equating the integrands, on doing so, for electric dipole emission, magnetic dipole emission and electric quadrupole emission, respectively, one finds

$$P(t) = \frac{|\dddot{\mathbf{d}}(t)|^2}{6\pi\varepsilon_0 c^3}, \tag{17.18}$$

$$P(t) = \frac{\mu_0 |\dddot{\boldsymbol{\mu}}(t)|^2}{6\pi c^3}, \tag{17.19}$$

$$P(t) = \frac{\dddot{d}_{ij}(t)\,\dddot{d}_{ij}(t)}{720\pi\varepsilon_0 c^5}. \tag{17.20}$$

It is clear from this derivation that any inference that the power emitted $P(t)$ at time t is localized in time to the quantities on the right hand sides of (17.18)–(17.20) should be viewed with considerable suspicion. This point is discussed further in §19.5.

An important application of formula (17.18) for the electric dipole case is to the treatment of emission by charged particles using the Larmor formula, as discussed in Chapter 18. Here, by way of illustration, we consider examples of multipole emission by classical systems in which the temporal variation is due to vibration or rotation.

Consider a charged system that is vibrating harmonically at a frequency ω_0. Specifically, let an arbitrary point in the system oscillate along the z axis about its mean position proportional to $[1 + \alpha\cos(\omega_0 t)]$, where α describes the amplitude of the oscillation.

The electric dipole emission of such a system is calculated as follows. Only the z component of the electric dipole moment of the system varies with time, and its variable part is proportional to $\alpha\cos(\omega_0 t)$. One inserts this term into (17.18), carries out the differentiations with respect to time and averages over one cycle of the vibration to find the mean power emitted. This gives

$$\bar{P} = \frac{\alpha^2 \omega_0^4 |d_z^{(0)}|^2}{12\pi\varepsilon_0 c^3}, \tag{17.21}$$

where $\mathbf{d}^{(0)}$ denotes the unperturbed value of the electric dipole moment. The frequency of the radiation follows by Fourier transforming $d_z^{(0)}\alpha\cos(\omega_0 t)$. The temporal Fourier transform of $\cos(\omega_0 t)$ is $\pi[\delta(\omega -$

$\omega_0) + \delta(\omega + \omega_0)]$. It follows that the emission is monochromatic at frequency $\omega = \omega_0$. The angular distribution is as $\sin^2 \theta$ about the z axis, that is, about the axis of vibration. The dipole radiation is linearly polarized along the projection of the axis of vibration onto the transverse plane.

The electric quadrupole emission of such an idealized vibrating system may be discussed in an analogous way. The quadrupole components $d_{xz} = d_{zx}, d_{yz} = d_{zy}$ vary proportional to $[1 + \alpha \cos(\omega_0 t)]$ and d_{zz} varies proportional to $[1 + \alpha \cos(\omega_0 t)]^2$. It is a straightforward exercise to evaluate the triple time-derivatives of these components and to substitute them into (17.20). After averaging over time one finds

$$\bar{P} = \frac{\alpha^2 \omega_0^6}{1440\pi\varepsilon_0 c^5} [(d_{xz}^{(0)})^2 + (d_{yz}^{(0)})^2 + 4(d_{zz}^{(0)})^2(1 + 4\alpha^2)]. \tag{17.22}$$

The frequency spectrum of the emission is determined by Fourier transforming the oscillating part of the quadrupole moment. There are contributions from $d_{xz} = d_{zx}, d_{yz} = d_{zy}$ and d_{zz} that vary at the fundamental frequency $\omega = \omega_0$ and a contribution from d_{zz} that varies at the second harmonic frequency $\omega = 2\omega_0$. The powers radiated at these two frequencies correspond to the terms in (17.22) that are proportional to α^2 and α^4, respectively. The angular pattern of the second harmonic emission in this case is proportional to $\sin^2 \theta \cos^2 \theta$.

Now let us consider a rotating system. Let Ω be the angular velocity of rotation. The time-derivative of a rotating vector field is evaluated using the algorithm $d/dt \rightarrow \Omega \times$. Specifically, for rotating electric and magnetic dipole moments one has

$$\ddot{\mathbf{d}}(t) = \Omega \times [\Omega \times \mathbf{d}], \quad \ddot{\boldsymbol{\mu}}(t) = \Omega \times [\Omega \times \boldsymbol{\mu}]. \tag{17.23}$$

Then in (17.18) one has $|\ddot{\mathbf{d}}(t)|^2 = \Omega^4 |\mathbf{d}|^2 \sin^2 \chi$, and in (17.19) one has $|\ddot{\boldsymbol{\mu}}(t)|^2 = \Omega^4 |\boldsymbol{\mu}|^2 \sin^2 \chi$, where in each case χ is the angle between the dipole axis and the rotation axis. The powers emitted for rotating electric and magnetic dipoles, respectively, are

$$P = \frac{\Omega^4 |\mathbf{d}|^2 \sin^2 \chi}{6\pi\varepsilon_0 c^3}, \tag{17.24}$$

$$P = \frac{\mu_0 \Omega^4 |\boldsymbol{\mu}|^2 \sin^2 \chi}{6\pi c^3}. \tag{17.25}$$

The emission is at the frequency of rotation $\omega = \Omega$ in both cases. For a rotating system $\mathbf{d}(\omega)$ or $\boldsymbol{\mu}(\omega)$ is complex with the x and y components out of phase, and the z component equal to zero, where Ω is along the z axis. That is, one has $\mathbf{d}(\omega) \propto (1, i, 0)$. The angular distribution is proportional to $|\boldsymbol{\kappa} \times \mathbf{d}(\omega)|^2 \propto 1 + \cos^2 \theta$, where $\theta = 0$ is the axis of rotation. The polarization of the radiation is elliptical in general; it is along the direction $\kappa \times [\kappa \times \mathbf{d}(\omega)]$. This corresponds to a polarization vector

$$\mathbf{e} = \frac{(\cos^2\theta, \mathrm{i}, -\sin\theta\cos\theta)}{(1+\cos^2\theta)^{1/2}}. \tag{17.26}$$

which implies an axial ratio $T = \cos\theta$ for the polarization ellipse, cf. (13.31) and (13.32).

17.4 Quantum Mechanical Transitions

An expansion in multipoles is usually appropriate when considering emission by atoms and molecules, in which case one needs to describe the radiating system using quantum mechanics. The multipole expansion usually converges rapidly, so that the dominant emission is due to the electric dipole transition, provided it is allowed.

For a quantum system whose states are eigenvalues of angular momentum and of parity, there are selection rules that are different for different multipole transitions. The rapid convergence of the multipole expansion in most applications means that transition rates for electric dipole transitions tend to be much greater than the transition rates for higher multipole transitions. Transitions that satisfy the selection rules for an electric dipole transition are said to be *allowed*, and transitions that do not satisfy the selection rules for electric dipole transitions are said to be *forbidden*. Forbidden transitions may still occur due to magnetic dipole, electric quadrupole or other higher order multipole terms. Line emission due to forbidden transitions can be important in some applications, notably in astrophysics. For example, the emission spectra of low density nebulas, such as planetary nebulas, can be dominated by forbidden lines. Also, one of the most important emission lines in radio astronomy is the 21 cm line due to a spin-flip transition of hydrogen in its ground state, and this is a magnetic dipole transition.

The emission by a quantum mechanical system is treated semiclassically in §16.3. The power radiated $[P_M(\mathbf{k})]_{qq'}$ in the mode M due to a transition from the state q to the state q' is given by (16.26) in terms of the matrix element $\mathbf{J}_{qq'}(\mathbf{k})$ of the current. The current is expanded in multipole moments as in (17.2). The expressions for the power radiated in electric dipole emission, magnetic dipole emission or electric quadrupole emission follow by analogy with (17.10)–(17.12), respectively:

$$[P_M(\mathbf{k})]_{qq'} = \frac{R_M(\mathbf{k})}{\varepsilon_0} \left| \omega_M(\mathbf{k}) e_M^*(\mathbf{k}) \cdot (\mathbf{d})_{qq'} \right|^2$$
$$\times 2\pi\delta\big(\omega_M(\mathbf{k}) - \omega_{qq'}\big), \qquad (17.27)$$

$$[P_M(\mathbf{k})]_{qq'} = \frac{R_M(\mathbf{k})}{\varepsilon_0} \left| e_M^*(\mathbf{k}) \cdot \mathbf{k} \times (\boldsymbol{\mu})_{qq'} \right|^2$$
$$\times 2\pi\delta\big(\omega_M(\mathbf{k}) - \omega_{qq'}\big), \qquad (17.28)$$

$$[P_M(\mathbf{k})]_{qq'} = \frac{R_M(\mathbf{k})}{4\varepsilon_0} \left| \omega_M(\mathbf{k}) e_{Mi}^*(\mathbf{k}) k_j (q_{qq'})_{ij} \right|^2$$
$$\times 2\pi\delta\big(\omega_M(\mathbf{k}) - \omega_{qq'}\big). \qquad (17.29)$$

The emission in the quantum case is in a single line at the transition frequency $\omega = \omega_{qq'}$.

Emission and absorption of radiation by atoms and molecules are usually discussed only in terms of transverse waves. For the specific case of emission of transverse waves in an isotropic medium one evaluates the power radiated explicitly, that is, one evaluates the integral of $[P(\mathbf{k})]_{qq'}$ over $d^3\mathbf{k}/(2\pi)^3$ explicitly. Let the resulting expression for the power radiated *in vacuo* due to a transition $q \to q'$ be $P_{qq'}$. The derivation is closely analogous to that of (17.13)–(17.16), and it gives

$$P_{qq'} = \frac{n(\omega_{qq'})\,\omega_{qq'}^4}{6\pi^2\varepsilon_0 c^3}\,|\mathbf{d}_{qq'}|^2, \qquad (17.30)$$

$$P_{qq'} = \frac{\mu_0 n^3(\omega_{qq'})\,\omega_{qq'}^4}{6\pi^2 c^3}\,|\boldsymbol{\mu}_{qq'}|^2, \qquad (17.31)$$

$$P_{qq'} = \frac{n^3(\omega_{qq'})\,\omega_{qq'}^6}{720\pi^2\varepsilon_0 c^5}\,(d_{qq'})_{ij}(d_{qq'})_{ij}^*. \qquad (17.32)$$

The energy emitted in the transition $q \to q'$ is $\hbar\omega_{qq'}$, and it is convenient to write the power radiated as the product of a transition rate per unit time $\Gamma_{qq'}$ times the energy released in the transition. Thus the *transition rate* is identified by writing

$$P_{qq'} = \hbar\omega_{qq'}\Gamma_{qq'}. \qquad (17.33)$$

The transition rates for electric dipole emission, magnetic dipole emission and electric quadrupole transitions due to the emission of transverse waves then follow from (17.30)–(17.32):

$$\Gamma_{qq'} = \frac{n(\omega_{qq'})\,\omega_{qq'}^3}{6\pi^2\varepsilon_0\hbar c^3}\,|\mathbf{d}_{qq'}|^2, \qquad (17.34)$$

$$\Gamma_{qq'} = \frac{\mu_0 n^3(\omega_{qq'})\,\omega_{qq'}^3}{6\pi^2\hbar c^3}\,|\boldsymbol{\mu}_{qq'}|^2, \qquad (17.35)$$

$$\Gamma_{qq'} = \frac{n^3 (\omega_{qq'}) \, \omega_{qq'}^5}{720\pi^2 \varepsilon_0 \hbar c^5} (d_{qq'})_{ij} (d_{qq'})_{ij}^*. \tag{17.36}$$

The *lifetime* of an excited state of a quantum system is defined as the inverse of the transition rate from the excited state to lower energy states. Thus the lifetime for the state q' is found by performing the sum $\sum_q \Gamma_{qq'}$ over all lower energy states q and taking the inverse of this sum. Often one transition dominates, in that the rate for it is much faster than the rate of all other transitions, and then the lifetime is approximately the inverse of this dominant transition rate.

One point that is not often recognized is that the transition rate and hence the lifetime of an atom or a molecule in an excited state depends on the properties of the ambient medium. This is seen in (17.34)–(17.36) through the dependence on the refractive index. Thus, for example, if an atom that emits in the optical range were embedded in a diamond crystal (refractive index near 2), then in principle the lifetime of an excited state that decays due to an electric dipole transition would be reduced by a factor of 2 from its value *in vacuo*. Note, however, that this example is highly idealized because of the neglect of any microscopic effect of the carbon atoms in the diamond on the atomic structure of the radiating system; in practice the structure of a test atom would be affected by the surrounding atoms in a crystal. However, this does not affect the point being made: the transition rate for a given transition in an atom depends on the macroscopic properties of the ambient medium in which it is embedded. Another example of this effect is for a transition at a frequency lower than the plasma frequency in an isotropic plasma; such emission simply cannot occur, so that the transition rate is zero. More generally, for a radiating system in a medium, if the refractive index is less than unity the transition rate is lower than *in vacuo* and the presence of the medium is said to suppress emission, and if the refractive index is greater than unity the transition rate is higher than *in vacuo* and the presence of the medium is said to enhance emission.

The modification of the transition rate from its value *in vacuo* is attributed to a combination of two effects. One is the phase space available for emission. The phase space available is determined by the integral over $d^3\mathbf{k}$, and the smaller the value of $|\mathbf{k}|$ for fixed ω, the less phase space is available to the emission, and the lower the transition rate. The other effect is the ratio of electric to total energy $R_T(\omega)$. Only the electric field is involved in the coupling to the current, and the lower the ratio $R_T(\omega)$, the weaker is the emission for a given current. For a non-dispersive medium the phase space factor is proportional to n^3 and

$R_T(\omega)$ is proportional to n^{-2}, and for an isotropic plasma the phase space effect is proportional to $n(\omega)$ and $R_T(\omega)$ is independent of $n(\omega)$.

A further point relating to emission by atoms and molecules that is rarely discussed is that it can occur in a medium where the waves are not transverse. In general, emission by atoms and molecules can occur into any mode in an ambient medium. For example, in an isotropic plasma, molecules that have a transition frequency just above the plasma frequency could emit and absorb Langmuir waves due to this particular transition, and in a magnetized plasma the magnetoionic waves are not strictly transverse. One treats emission into such modes using (17.27)–(17.29). Note, however, that if the waves have a longitudinal component the trace of the quadrupole moment contributes in (17.29). In principle, the dependence on the trace of the quadrupole moment allows transitions that are not allowed for emission of transverse waves.

17.5 Two Derivations of the Absorption Coefficient

To every emission process there is a corresponding absorption process. In a semiclassical approach this is obvious: as discussed in §16.3 the Einstein coefficients are used to relate absorption to emission. The result is not so obvious in the classical case, because emission and absorption are treated in quite different ways. Here we show that the classical and semiclassical methods of treating absorption are equivalent for electric dipole radiation by a quantum system.

Using the Einstein coefficients, the expression obtained for the absorption coefficient is given by (16.34), *viz.*,

$$\gamma_M(\mathbf{k}) = -\frac{1}{V} \sum_{qq'} w_{Mqq'}(\mathbf{k})(n_q - n_{q'}). \qquad (17.37)$$

Let us apply this to the case of electric dipole emission. The relevant probability is given by inserting (17.27) into (16.27). One obtains

$$w_{Mqq'}(\mathbf{k}) = \frac{R_M(\mathbf{k})\omega_M(\mathbf{k})}{\hbar\varepsilon_0} \left| \mathbf{e}_M^*(\mathbf{k}) \cdot (\mathbf{d})_{qq'} \right|^2 2\pi\delta\big(\omega_M(\mathbf{k}) - \omega_{qq'}\big). \qquad (17.38)$$

The explicit value of the absorption coefficient due to a collection of absorbing dipoles implied by (17.37) and (17.38) is

$$\gamma_M(\mathbf{k}) = -\sum_{qq'} \frac{(n_q - n_{q'})}{V} \frac{R_M(\mathbf{k})\omega_M(\mathbf{k})}{\hbar\varepsilon_0}$$
$$\times \left| \mathbf{e}_M^*(\mathbf{k}) \cdot (\mathbf{d})_{qq'} \right|^2 2\pi\delta\big(\omega_M(\mathbf{k}) - \omega_{qq'}\big). \qquad (17.39)$$

Classically, the absorption coefficient is derived in quite a different way. As discussed in §11.4 and §15.2, it is related to the antihermitian part of the response tensor. The response tensor of relevance here is for a system of dipoles, and this is written down in terms of the polarizability (9.30). The polarization is given explicitly by (9.31), which may be written

$$P_i(\omega) = \sum_{q,q'} \frac{(n_q - n_{q'})}{V\hbar} \frac{(d_i)_{q'q}(d_j)_{qq'}}{\omega - \omega_{qq'}} E_j(\omega). \tag{17.40}$$

On writing (17.40) in the form $P_i(\omega) = \varepsilon_0 \chi_{ij}(\omega) E_j(\omega)$, one identifies the susceptibility tensor $\chi_{ij}(\omega)$. The antihermitian part is given by imposing the causal condition $1/(\omega - \omega_{qq'}) \to 1/(\omega - \omega_{qq'} + i0)$ and using the Plemelj formula (4.32) to rewrite the resonant denominator. The semiresidue term gives the antihermitian part

$$K_{ij}^A(\omega) = \chi_{ij}^A(\omega) = -\frac{i\pi}{\varepsilon_0 V\hbar} \sum_{q,q'} (n_q - n_{q'})\, (d_i)_{qq'}(d_j)_{q'q} \delta(\omega - \omega_{qq'}). \tag{17.41}$$

On inserting (17.41) into (11.22) or (15.14), one finds that the result reproduces (17.39). Thus the two procedures for treating absorption are equivalent in this case.

The equivalence of these two ways of treating absorption in this particular case of dipole emission by a quantum system generalizes to all emission and absorption processes. In one sense this is an obvious result in that it is clearly necessary for the internal consistency of the theory. However, from another point of view it is somewhat surprising. In the quantum mechanical approach, emission and absorption are related by an argument that relies on the second law of thermodynamics. In the classical approach absorption and emission are treated quite differently, with absorption deduced by imposing the causal condition on the response tensor and emission treated by the methods described in §16.2 or §18.2. Thus the equivalence of the two methods of treating absorption reflects an underlying consistency between two of the fundamental concepts of physics, specifically, between causality and the second law of thermodynamics.

The second law of thermodynamics is also used in the classical context to argue that to every emission process there is a corresponding absorption process. However, the argument is less powerful than in the quantum case because it allows one to relate particular emission and absorption formulas only for thermal particles. The argument is as follows. Suppose that a particle were able to emit by a process to which there is no corresponding absorption process. Then, by constructing a system composed of an arbitrarily large number of such particles, one could make the system emit arbitrarily brightly. Let the particles have a thermal distribution at a temperature T_0. Now consider another system at temperature $T_1 > T_0$, and suppose that this system can absorb radiation in the frequency range emitted by the first system. By including a sufficient number of emitting particles in the first system one could make that radiation incident on the second system brighter than black body radiation at temperature T_1. Then energy would flow from the first system to the second, that is, from a cooler body to a hotter body. To avoid the implied inconsistency with the second law of thermodynamics no such emission process can exist.

The foregoing argument is the basis of Kirchhoff's law. The essential requirement is that a thermal distribution of particles never emits radiation that is brighter than that of a black body at the given temperature. From the classical limit of the Planck distribution (16.28), a thermal distribution of radiation at temperature Θ (in energy units) is

$$T_M(\mathbf{k}) := \hbar\omega_M(\mathbf{k})N_M(\mathbf{k}) = \Theta, \tag{17.42}$$

where $T_M(\mathbf{k})$ is the effective temperature of the radiation. A statement of Kirchhoff's law is that when emission and absorption are in balance for a thermal distribution of particles the wave distribution is thermal at the same temperature. In terms of the emission and absorption coefficients in (16.34), this requires

$$\hbar\omega_M(\mathbf{k})\alpha_M(\mathbf{k}) = \gamma_M(\mathbf{k})\Theta \tag{17.43}$$

For transverse waves in an isotropic medium it is conventional to describe the level of radiation in terms of the specific intensity, and to express Kirchhoff's law in terms of the specific intensity for thermal radiation. The *specific intensity* or *brightness of radiation*, written as $I_M(\omega, \boldsymbol{\kappa})$ for an arbitrary mode M, is defined as the power per unit frequency and per unit solid angle crossing unit area normal to the direction of propagation. To convert the photon occupation $N_T(\mathbf{k})$, which is the number density of photons per unit range of the integral over $d^3\mathbf{k}/(2\pi)^3$, into the specific intensity, one multiplies by three factors: (i) the energy $\hbar\omega$ per photon, (ii) the group speed at which the energy propagates and (iii) the factor in (16.16) relating the integrals over $d^3\mathbf{k}/(2\pi)^3$ and over $d\omega d^2\boldsymbol{\Omega}$. Thus one finds

$$I_M(\omega, \boldsymbol{\kappa}) = \frac{\hbar\omega^3 n_M^2 N_M}{(2\pi)^3 c^2}, \tag{17.44}$$

where the implicit arguments of n_M are to be written as functions of the independent variables ω, $\boldsymbol{\kappa}$. For a thermal distribution of transverse waves (17.44) implies the *Rayleigh–Jeans* distribution

$$I(\omega, \boldsymbol{\kappa}) = \frac{2\omega^2 [n(\omega)]^2 \Theta}{(2\pi)^3 c^2},\qquad (17.45)$$

where an extra factor of 2 arises from summing over the two states of polarization.

The transfer of radiation through a medium that is weakly inhomogeneous and time-independent has a particularly simple form in terms of the occupation number: N_M is a constant along a ray. Note, however, that $\mathbf{k}$ changes along a ray in accord with Snell's law, which is such as to ensure that $\omega_M(\mathbf{k})$ remains constant. The transfer equation for the specific intensity may be derived from this fact, but it is relatively cumbersome in the general case. This transfer simplifies considerably for transverse in an isotropic dielectric: $I(\omega, \boldsymbol{\kappa})/[n(\omega)]^2$ is a constant along a ray, as suggested by (17.44). Thus, if s denotes distance along the curved ray path, then the transfer equation is

$$[n(\omega)]^2 \frac{\mathrm{d}}{\mathrm{d}s} \left\{ \frac{I(\omega, \boldsymbol{\kappa})}{[n(\omega)]^2} \right\} = 0 \qquad (17.46)$$

in the absence of sources or sinks. The result (17.46) does not apply in the case of an anisotropic medium because the element of solid angle changes as a result of refraction and this effect is not included in (17.46)

Exercise Set 17

17.1 The angular patterns of the radiation by a linearly oscillating dipole and by a rotating dipole vary as $\sin^2\theta$ and $1 + \cos^2\theta$, respectively. In general the variation of a dipole consists of a linear oscillation and a rotation. The coordinate axes and the initial phase may be chosen such that $d_x(\omega)$ and $d_z(\omega)$ are real and $d_y(\omega)$ is imaginary. Thus one may write $\mathbf{d}(\omega) \propto (\alpha, i, \beta)$, where α and β are real constants.

(a)   Determine the angular pattern of the dipole emission in this general case.

(b)   Argue that the case $\beta = 1$ corresponds to the linear oscillation being along the axis about which the rotation occurs (the z axis here).

17.2 The polarization of dipole emission is determined in terms of $\mathbf{d}(\omega)$ by noting that the energy emitted is proportional to $|\mathbf{e}^* \cdot \mathbf{d}(\omega)|^2$ and hence the polarization is along the projection of $\mathbf{d}(\omega)$ on the plane orthogonal to the direction $\boldsymbol{\kappa}$ of emission. Determine the polarization in the following cases as a function of the angle θ between $\boldsymbol{\kappa}$ and the z axis.

(a)   The system is oscillating such that $\mathbf{d}(\omega)$ is real and along the z axis.

(b)   The system is rotating about the z axis such that $\mathbf{d}(\omega)$ is proportional to $(1, \pm i, 0)$, where the $\pm$ signs refer to the right and left hand senses of rotation, respectively.

(c)   The dipole moment is of the form $\mathbf{d}(\omega) \propto (\alpha, i, \beta)$, as in Exercise 17.1.

17.3 Fill in the details of the derivation of (17.22). Specifically, write $d_{xz} = d_{zx} = d_{xz}^{(0)}[1 + \alpha\cos(\omega_0 t)]$, $d_{yz} = d_{zy} = d_{yz}^{(0)}[1 + \alpha\cos(\omega_0 t)]$, $d_{zz} = d_{zz}^{(0)}[1 + \alpha\cos(\omega_0 t)]^2$ and derive (17.25) in two ways.

(a)   First evaluate the triple derivative with respect to time and use (17.20) to derive (17.22).

(b)   Fourier transform in time to find $d_{ij}(\omega)$, and use (17.15) and (4.37) to evaluate the power radiated per unit frequency. Hence show that the power radiated at the fundamental is proportional to α^2 and the power radiated at the second harmonic is proportional to α^4.

(c)   Integrate the result obtained in part (b) over frequency to rederive (17.22).

17.4 Pulsars are rotating neutron stars with very strong surface magnetic fields. The observed slowing down of rotation of pulsars is attributed to magnetic dipole radiation.

The Crab pulsar rotates with a period of 33 ms and is estimated to radiate a power 10^{31} W. Assuming that the magnetic moment is orthogonal to the rotation axis, estimate the surface magnetic field at the pole of the neutron star, assuming a stellar radius of 10 km.

17.5 The hydrogen atom has a binding energy of about 13.6 eV. The classical lifetime of the hydrogen atom is finite due to the electron losing energy as it radiates and so spiraling into the nucleus. In the following problem the mass of the nucleus is assumed infinite and the orbit of the electron is assumed circular.

(a) Use the virial theorem, which is that the kinetic energy of the bound electron is equal in magnitude to the binding energy and that the potential energy is twice the binding energy in magnitude, to find how the radius of the orbit of the electron changes as a function of the energy radiated.

(b) Calculate the power radiated assuming electric dipole radiation due to an electron moving in a classical circular orbit.

(c) Show that the electron spirals into the nucleus in a finite time, and estimate this time.

17.6 Consider an atom that has a magnetic dipole transition from an excited state to the ground state. It follows from (17.36) that the rate of a magnetic dipole transition in a medium with refractive index n is n^3 times the rate of transition *in vacuo*. Are the following statements true or false? Give reasons for each answer.

(a) In a plasma with $n^2 = 1 - \omega_p^2/\omega^2$ the transition rate is less than *in vacuo* because the presence of the plasma modifies the frequency of transition.

(b) As in part (a), but with the explanation that the transition rate is less than *in vacuo* because the plasma absorbs some of the energy.

(c) A transition with a natural frequency ω_0 is forbidden in a plasma with $\omega_p > \omega_0$.

(d) In water or glass with $n > 1$, the transition rate is greater than *in vacuo* because the presence of the medium reduces the dimensions of the system so causing it to radiate faster.

17.7 A spherical drop of liquid of radius R has a surface charge with surface charge density $\sigma(\theta)$, where θ is the polar angle about an axis of symmetry, which is along the unit vector $\hat{\mathbf{z}}$.

(a) By writing the charge density in the form $\rho(\mathbf{x}) = \sigma(\theta)\delta(r - R)$, show that the first three moments of the distribution give

$$Q = 2\pi R^2 \int_{-1}^{1} d\cos\theta\, \sigma(\theta), \quad \mathbf{d} = \hat{\mathbf{z}}\, 2\pi R^2 \int_{-1}^{1} d\cos\theta\, \cos\theta\, \sigma(\theta),$$

$$d_{ij} = (3\hat{z}_i\hat{z}_j - \delta_{ij})\, 2\pi R^2 \int_{-1}^{1} d\cos\theta\, \tfrac{1}{2}(3\cos^2\theta - 1)\sigma(\theta),$$

$$(E17.1)$$

(b) By expanding in Legendre polynomials $P_l(\cos\theta)$, show that Q, $\mathbf{d}$ and d_{ij} in $(E17.1)$ depend only σ_0, σ_1 and σ_2, respectively, in

$$\sigma(\theta) = \sum_{l=1}^{\infty} \sigma_l P_l(\cos\theta). \qquad (E17.2)$$

(c) Find the power radiated *in vacuo* due to each of these first three moments when the radius R is made to oscillate (at fixed Q) with small amplitude $\delta R \ll R$ at a frequency ω_0.

17.8 Two electrons are repelled by the Coulomb force between them. Let the origin be at the center of mass with one electron at $\mathbf{x}/2$ and the other at $-\mathbf{x}/2$. The equation of motion of the first electron is

$$\frac{m}{2}\ddot{\mathbf{x}} = \frac{e^2}{4\pi\varepsilon_0}\frac{\mathbf{x}}{r^3}, \qquad (E17.3)$$

with $r = |\mathbf{x}|$ and where $m/2$ is the reduced mass of the electron.

(a) Show that if the origin is chosen at the center of mass, then the first three electric moments of this system are $Q = -2e$, $\mathbf{d} = 0$, $d_{ij} = -\tfrac{1}{2}e(3x_i x_j - r^2\delta_{ij})$.

(b) What are the values of these three moments if the origin is chosen at the position of one of the electrons? Why is this not a sensible choice of origin?

(c) Show that the triple time-derivative of the electric quadrupole moment reduces to

$$\dddot{d}_{ij} = -\frac{e^3}{2\pi\varepsilon_0 m r^4}\left[6r(\dot{x}_i x_j + x_i \dot{x}_j) - 9\dot{r}x_i x_j - \dot{r}r^2\delta_{ij}\right]. \quad (E17.4)$$

(d) Using $(E17.4)$ in (17.20) show that the power emitted in electric quadrupole radiation is

$$P = \frac{e^6}{480\pi^3\varepsilon_0^3 m^2 c^5}\frac{v^2 + 11v_\phi^2}{r^4}, \qquad (E17.5)$$

where $v^2 = \dot{r}^2 + v_\phi^2$ denotes a separation into the radial and azimuthal components of the velocity in the orbital plane.

17.9 The following exercise concerns the evaluation of the absorption coefficient (17.39) for a thermal distribution of dipoles. It is assumed that the quantum numbers q contain a number n labeling an energy eigenvalue E_n and that the remaining quantum numbers determine a statistical weight g_n, corresponding to g_n degenerate states. The occupation number n_n of the state is

$$n_n = \frac{g_n e^{-E_n/\Theta}}{\Phi(\Theta)}, \quad \Phi(\Theta) = \sum_n g_n e^{-E_n/\Theta}. \qquad (E17.6)$$

where $\Phi(\Theta)$ is the partition function.

(a) Show that the absorption coefficient (17.39) for the transition $n' \to n$ (the inverse of the emission $n \to n'$) with $g_{n'} = g_n$ reduces to

$$\gamma_M(\mathbf{k}) = \frac{g_n e^{-E_n/\Theta}[e^{\hbar\omega_M(\mathbf{k})/\Theta} - 1]}{\Phi(\Theta)V} \frac{R_M(\mathbf{k})\omega_M(\mathbf{k})}{\hbar\varepsilon_0}$$

$$\times \left|\mathbf{e}_M^*(\mathbf{k}) \cdot (\mathbf{d})_{nn'}\right|^2 2\pi\delta\big(\omega_M(\mathbf{k}) - \omega_{nn'}\big). \qquad (E17.7)$$

(b) Evaluate $(E17.7)$ explicitly for a simple harmonic oscillator with natural frequency ω_0 and hence energy eigenvalues

$$E_n = (n + \tfrac{1}{2})\hbar\omega_0, \quad g_n = 1. \qquad (E17.8)$$

17.10 Maser action by a simple harmonic oscillator, whether treated classically or quantum mechanically, is not possible. Let $w(n, \omega) \propto \delta(\omega - \omega_0)$ be the probability of a transition $n \to n - 1$ and let n_n be the occupation number of the nth level.

(a) Show that the absorption coefficient in the classical limit is

$$\gamma(\omega) = -\int dn\, w(n, \omega) \frac{dF(n)}{dn}, \qquad (E17.9)$$

where $\int dn\, F(n)$ is the number of oscillators.

(b) By partially integrating in $(E17.9)$ show that $\gamma(\omega)$ cannot be negative for any continuous, normalizable function $F(n)$.

(c) Show that if an *anharmonic* correction term $\Delta\omega(n)$ is included in the δ-function by making the replacement $\delta(\omega - \omega_0) \to \delta(\omega - \omega_0 + \Delta\omega_n)$, then negative absorption (implying maser action) is possible when $F(n)$ is an increasing function of n over a finite range.

18

The Larmor Formula

Preamble

The Larmor formula describes emission of radiation *in vacuo* by an accelerated non-relativistic charged particle (charge q, mass m). The Larmor formula is derived here and is then applied to some simple radiation processes, specifically, emission by a charge in simple harmonic motion, a charge in circular motion, Thomson scattering and bremsstrahlung. Further details of these and other radiation processes are discussed in Part Five.

18.1 The Current Associated with a Moving Particle

As a particle moves either freely or under the influence of a net force, its position changes as a function of time. The path of the particle is referred to as its orbit or its trajectory. Mathematically, the *orbit* of the particle is defined by an equation that describes its position as a function of time. Thus the orbit is described by an equation of the form

$$\mathbf{x} = \mathbf{X}(t). \tag{18.1}$$

The orbit is found by solving the equation of motion of the particle with given initial position and velocity. Let the initial position (at $t = 0$) be $\mathbf{x}_0$ and the initial velocity be $\mathbf{v}_0$. Then $\mathbf{X}(t)$ is an implicit function of $\mathbf{x}_0$ and $\mathbf{v}_0$. The instantaneous velocity of a particle is its velocity at time t. Once the orbit is known, the instantaneous velocity is found by differentiation. Thus the *instantaneous velocity* is given by

$$\mathbf{v}(t) = \dot{\mathbf{X}}(t) = \frac{\mathrm{d}}{\mathrm{d}t}\,\mathbf{X}(t). \tag{18.2}$$

The instantaneous acceleration $\mathbf{a}(t) = \ddot{\mathbf{X}}(t)$ is obtained by a further differentiation.

The electromagnetic fields due to a moving charge are found by solving Maxwell's equations with the source terms identified as the charge and current densities associated with the moving charge. The charge is assumed to be a point particle. Its charge density is then zero except at the point $\mathbf{x} = \mathbf{X}(t)$ where the particle is located and there the charge density is infinite. Hence the charge density is proportional to $\delta(\mathbf{x} - \mathbf{X}(t))$. The total charge is assumed to be q, and so the charge density integrated over all space must be equal to q. The current density is equal to the charge density times the instantaneous velocity. Thus one identifies

$$\rho(t, \mathbf{x}) = q\, \delta^3(\mathbf{x} - \mathbf{X}(t)), \quad \mathbf{J}(t, \mathbf{x}) = q\mathbf{v}(t)\, \delta^3(\mathbf{x} - \mathbf{X}(t)), \qquad (18.3)$$

as the charge and current densities, respectively, for a moving charge.

When treating radiation, the source term is the Fourier transform of the current density. With the current density given by (18.3), the source term for emission by a moving charge is

$$\mathbf{J}(\omega, \mathbf{k}) = q \int_{-\infty}^{\infty} dt\, \mathbf{v}(t)\, e^{i(\omega t - \mathbf{k} \cdot \mathbf{X}(t))}. \qquad (18.4)$$

This source term may be evaluated explicitly once the orbit is known.

To treat emission by the particle, the current (18.4) is to be identified as the extraneous current in the emission formula (16.10). Here we consider only emission of transverse waves by non-relativistic particles. This case may be treated using the (electric) dipole approximation, which effectively corresponds to omitting the term $\mathbf{k} \cdot \mathbf{X}(t)$ in the exponent in (18.4). In the present context, the dipole approximation is equivalent to neglecting relativistic corrections. To see this, first note that the multipole expansion of the current involves an expansion in powers of k, cf. (17.1), and that for transverse waves in an isotropic dielectric one has $k = n(\omega)/c$. The leading term in this expansion is the electric dipole term. The first correction is smaller than the dipole term by some dimensionless factor proportional to k; one-dimensionally this factor is written as a numerical factor times $kv/\omega = n(\omega)v/c$, where v is a yet to be identified speed. In practice this v is characteristic of the motion of the particle and, provided that the motion remains non-relativistic, one has $v/c \ll 1$. It follows that for emission of transverse waves the dipole approximation is expected to be a reasonable approximation provided that the motion of the particle is non-relativistic.

18.2 Derivation of the Larmor Formula

The Larmor formula gives the power radiated *in vacuo* by a charged particle due to its accelerated motion. The Larmor formula is derived in the dipole approximation, and so it applies only to emission by particles whose motion is non-relativistic.

The electric dipole approximation to the current (18.4) follows by inspection of (17.2). Retaining only the leading term in (17.2), the dipole approximation to (18.4) gives

$$-\mathrm{i}\omega\mathbf{d}(\omega) = q \int_{-\infty}^{\infty} \mathrm{d}t\, \mathbf{v}(t)\, \mathrm{e}^{\mathrm{i}\omega t} = -\mathrm{i}q \int_{-\infty}^{\infty} \mathrm{d}t\, \omega\mathbf{X}(t)\, \mathrm{e}^{\mathrm{i}\omega t}. \qquad (18.5)$$

It then follows that the temporal Fourier transform of the electric dipole moment for a moving charge is

$$\mathbf{d}(\omega) = q\mathbf{X}(\omega). \qquad (18.6)$$

One could deduce (18.6) more directly by noting that the dipole moment is $\mathbf{d}(t) = q\mathbf{X}(t)$, and then taking the temporal Fourier transform of this relation.

Let us first consider the energy radiated per unit frequency in an isotropic dielectric. Applying (17.13) to the present case gives

$$U_{\mathrm{rad}}(\omega) = \frac{q^2\, n(\omega)\, \omega^4\, |\mathbf{X}(\omega)|^2}{6\pi^2\varepsilon_0 c^3}. \qquad (18.7)$$

For a non-relativistic particle subject to a force $\mathbf{F}(t)$, the temporal Fourier transform of the equation of motion

$$m\,\mathbf{a}(t) = \mathbf{F}(t), \quad \mathbf{a}(t) = \frac{\mathrm{d}}{\mathrm{d}t}\,\mathbf{v}(t) = \frac{\mathrm{d}^2}{\mathrm{d}t^2}\,\mathbf{X}(t), \qquad (18.8)$$

allows one to replace (18.7) by

$$U_{\mathrm{rad}}(\omega) = \frac{q^2\, n(\omega)\, |\mathbf{F}(\omega)|^2}{6\pi^2\varepsilon_0 m^2 c^3}, \qquad (18.9)$$

which gives the energy radiated (in the electric dipole approximation) per unit frequency in terms of the temporal Fourier transform of the force accelerating the charged particle. The angular distribution of the emitted radiation is determined by the angular dependence of $|\boldsymbol{\kappa} \times \mathbf{F}(\omega)|^2$. The energy radiated (per unit frequency here) is the appropriate quantity to consider for a particle subjected to an impulsive force, or more specifically to a force whose effective action is restricted to a finite time. For a particle whose motion is periodic, which corresponds to a force that persists indefinitely, the Fourier transform $\mathbf{F}(\omega)$ is proportional to a δ-function, for example, to $\delta(\omega - \omega_0)$ where ω_0 is the natural frequency of the system. When such an $\mathbf{F}(\omega)$ is inserted into (18.9) the square of the

δ-function is rewritten using (4.37). The factor T of the normalization time is removed by considering the power radiated per unit frequency in place of the energy radiated per unit frequency.

Now consider emission *in vacuo*. The power radiated in electric dipole emission is given by (17.18), and applying this to the present case gives

$$P(t) = \frac{q^2 |\mathbf{a}(t)|^2}{6\pi\varepsilon_0 c^3}, \qquad (18.10)$$

which is the *Larmor formula*. The acceleration in (18.10) may be rewritten in terms of the force acting on the particle using (18.8), that is, by writing $\mathbf{a}(t) = \mathbf{F}(t)/m$. Unlike (18.9), which needs to be modified to apply to periodic motions, the Larmor formula (18.10) applies to any non-relativistic charged particle irrespective of the form of its motion or of the force acting on it.

The derivation of (17.18), on which (18.10) is based, involves the dubious step discussed following (17.16) and (17.17). Specifically, the energy radiated is written as the integral of $P(t)$ over t, and in effect an identity between two infinite integrals over t is used to infer the equality of the integrands. Thus, strictly all that may be said with confidence is that the integral over $-\infty < t < \infty$ of the left hand side of (18.10) is equal to the corresponding integral over the right hand side. The main point is that the derivation of (18.10) does not allow one to draw any inferences concerning the actual temporal relation between the acceleration and the emission of the radiation. Indeed, if one assumes that the power $P(t)$ is lost by the particle at time t, then the radiation reaction force that needs to be introduced to account for this energy loss leads to conceptual difficulties. This point is discussed in §19.5 in connection with the radiation reaction force.

The Larmor formula is valid only for non-relativistic motion *in vacuo*. However, in principle this is not as restrictive as might at first be supposed. The reason is that one is free to make a Lorentz transformation to the inertial frame in which the particle is instantaneously at rest. In this frame the Larmor formula is valid and so may be used to calculate the power radiated. The power radiated is a Lorentz invariant (it is the ratio of the time components of two 4-vectors) and hence is equal to the power radiated in any other frame, cf. §18.4.

18.3 Applications of the Larmor Formula

The Larmor formula is used to find the power radiated *in vacuo* by an accelerated charge under a variety of different conditions. The following are some illustrative examples.

Linear Harmonic Motion

Suppose that the particle is subjected to an elastic (or other) force that causes it to execute simple harmonic motion at frequency ω_0. Let the motion be along the x axis with an amplitude x_0. The x component of the orbit is then of the form

$$X(t) = x_0 \cos(\omega_0 t), \quad a(t) = \ddot{X}(t) = -\omega_0^2 x_0 \cos(\omega_0 t). \tag{18.11}$$

On inserting the acceleration from (18.11) into (18.10), the resulting harmonic variation of $P(t)$ has no physical significance; this is because only the integral of $P(t)$ over t is well defined. On averaging $P(t)$ over time, and denoting the result by $\bar{P}$, one obtains

$$\bar{P} = \frac{q^2 \omega_0^4 x_0^2}{12\pi\varepsilon_0 c^3}. \tag{18.12}$$

Now consider the frequency spectrum. The temporal Fourier transform of (18.11) gives

$$X(\omega) = \pi x_0 \left[\delta(\omega - \omega_0) + \delta(\omega + \omega_0) \right]. \tag{18.13}$$

On inserting (18.13) into (18.7), the squares of the δ-functions are rewritten using (4.27) in the form

$$[\delta(\omega \pm \omega_0)]^2 = (T/2\pi) \, \delta(\omega \pm \omega_0),$$

and the factor T is incorporated into the power radiated per unit frequency $P(\omega) = U_{\text{rad}}(\omega)/T$. Thus one finds

$$P(\omega) = \frac{q^2 n(\omega_0) \omega_0^4 x_0^2}{12\pi\varepsilon_0 c^3} \, \delta(\omega - \omega_0). \tag{18.14}$$

where only the positive frequency contribution has relevance. The emission is all at the natural frequency ω_0 of oscillation of the system. On integrating (18.14) over all $\omega > 0$ for a vacuum ($n(\omega) = 1$) one rederives (18.12); the power radiated is modified by the multiplicative factor $n(\omega_0)$ in the presence of a medium.

The polarization of the emission is along the projection of the x axis on the plane orthogonal to the direction $\boldsymbol{\kappa}$ of emission.

Circular Motion: Cyclotron Emission

Consider a particle in circular motion in the x–y plane. Let the radius of the circle be R_0, and the frequency of motion be ω_0. The orbit is described by

$$X(t) = R_0 \cos(\omega_0 t), \quad Y(t) = R_0 \sin(\omega_0 t), \tag{18.15}$$

where the particle is assumed to be at the point $(R_0, 0)$ at $t = 0$. The power radiated is given by (18.12) with x_0 replaced by R_0.

An important example is the emission due to a particle in a magnetostatic field **B**. The equation of motion for a non-relativistic particle is

$$m\frac{d}{dt}\mathbf{v}(t) = q\mathbf{v}(t) \times \mathbf{B}. \qquad (18.16)$$

The motion is circular if the particle has no component of velocity along the field lines. If $v_\perp$ is its speed perpendicular to the field lines, then one has

$$R_0 = \frac{v_\perp}{\Omega_0}, \quad \Omega_0 = \frac{|q|B}{m}, \qquad (18.17)$$

which are its *radius of gyration* and its *gyrofrequency* or *cyclotron frequency*, respectively. The mean power radiated is

$$\bar{P} = \frac{q^4 B^2 v_\perp^2}{12\pi\varepsilon_0 m^2 c^3}. \qquad (18.18)$$

The emission is referred to as *cyclotron emission*. As for any classical system executing simple harmonic motion (circular motion is simple harmonic motion), the frequency radiated is monochromatic at the natural frequency $\omega = \Omega_0$.

The angular distribution of the radiation and its polarization are those characteristic of circular motion, cf. the discussion following (17.25). Thus the angular pattern varies as $1 + \cos^2\theta$, where $\theta = 0$ is the direction of the magnetic field. Similarly, from the discussion following (17.25), the polarization is elliptical with axial ratio $T = \pm\cos\theta$; the plus sign applies for negative charges which gyrate in a right hand sense relative to **B** and the minus sign applies for positive charges.

Thomson Scattering

Next consider a charged particle whose motion is perturbed by the field of a test electromagnetic wave *in vacuo*. The test wave is assumed monochromatic with frequency ω_0. Let the electric field in the wave be of the form

$$\mathbf{E}(t, \mathbf{x}) = \mathbf{E}_0 \cos(\omega_0 t - \mathbf{k}_0 \cdot \mathbf{x}), \qquad (18.19)$$

with $|\mathbf{k}_0| = \omega_0/c$. In the non-relativistic approximation, the magnetic force on the particle is negligible, and hence the acceleration is approximated by $\mathbf{a}(t) = q\mathbf{E}(t, \mathbf{x})/m$. The mean power radiated implied by (18.10) is then

$$\bar{P} = \frac{q^4 |\mathbf{E}_0|^2}{12\pi\varepsilon_0 m^2 c^3}. \qquad (18.20)$$

The average electrical energy density in the test wave field described by (18.19) is $\frac{1}{4}\varepsilon_0|\mathbf{E}_0|^2$, and the ratio of electric to total energy for an electromagnetic wave *in vacuo* is $\frac{1}{2}$. Hence the average energy density W_0 in the test wave is $W_0 = \frac{1}{2}\varepsilon_0|\mathbf{E}_0|^2$. On using this relation, (18.20) becomes

$$\bar{P} = \frac{8\pi}{3}\left(\frac{q^2}{4\pi\varepsilon_0 mc^2}\right)^2 cW_0. \tag{18.21}$$

The power radiated (18.21) is interpreted as that produced in scattered radiation due to scattering of the test wave by the particle. For an electron this scattering of one electromagnetic wave into another electromagnetic wave is called *Thomson scattering*. In the approximation made here, the frequency of the scattered wave is equal to ω_0.

For electrons, (18.21) is written in the form $\bar{P} = \sigma_\mathrm{T} cW_0$, where

$$\sigma_\mathrm{T} = \frac{8\pi}{3}\left(\frac{e^2}{4\pi\varepsilon_0 m_e c^2}\right)^2 = \frac{8\pi r_0^2}{3} \tag{18.22}$$

is the *Thomson cross-section*. The quantity $r_0 = e^2/4\pi\varepsilon_0 m_e c^2$ is the *classical radius of the electron*.

Bremsstrahlung

Consider an electron moving near an ion with charge Ze situated at the origin. Let us ignore corrections of order the ratio of the mass of the electron to the mass of the ion, so that the ion is regarded as fixed. The equation of motion of the electron is then

$$m_e\ddot{\mathbf{X}}(t) = -\frac{Ze^2\mathbf{X}(t)}{4\pi\varepsilon_0|\mathbf{X}(t)|^3}. \tag{18.23}$$

On inserting $\ddot{\mathbf{X}}(t) = \mathbf{a}(t)$ in the Larmor formula (18.10), one obtains

$$P(t) = \frac{2Z^2}{3}\left(\frac{e^2}{4\pi\varepsilon_0}\right)^3 \frac{1}{|\mathbf{X}(t)|^4 m_e^2 c^3}. \tag{18.24}$$

The emission described by (18.24) is called *bremsstrahlung*.

The power radiated is of little interest in a single encounter between an electron and an ion. Each encounter is regarded as an impulsive event, and the quantity of interest is the energy radiated in such an event. The energy radiated in a single encounter is given by integrating (18.24) over all time. For this purpose it is convenient to write $|\mathbf{X}(t)| = r(t)$, where $r(t)$ is the radial distance of the electron from the ion, and to write the integral as one over $r(t)$:

$$E_\mathrm{rad} = \int_{-\infty}^{\infty} dt\, P(t) = 2\int_{r_\mathrm{min}}^{\infty} \frac{dr}{\dot{r}}\frac{2Z^2}{3}\left(\frac{e^2}{4\pi\varepsilon_0}\right)^3 \frac{1}{r^4 m_e^2 c^3}, \tag{18.25}$$

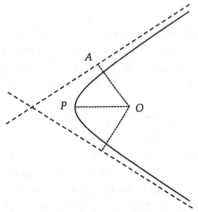

Fig. 18.1 A hyperbolic orbit of an electron for an ion at the origin O;
$OP = r_{\min}$, $OA = b$, and the dashed lines are asymptotes.

where $r_{\min}$ is the distance of closest approach. The integral in (18.25) may be performed exactly. The angular momentum and the energy of the electron are conserved. Let v_0 be the speed of the electron at infinity, and b be the impact parameter; the orbit is illustrated in Figure 18.1. Then these conservation laws become

$$m_e r^2 \dot\theta = m_e b v_0, \tag{18.26}$$

$$\tfrac{1}{2} m_e (\dot r^2 + r^2 \dot\theta^2) - \frac{Ze^2}{4\pi\varepsilon_0 r} = \tfrac{1}{2} m_e v_0^2, \tag{18.27}$$

respectively. After eliminating $\dot\theta$ between (18.26) and (18.27), one solves for $\dot r$ as a function of r and substitute the result into (18.25). At the distance of closest approach $r = r_{\min}$ one has $\dot r = 0$, and (18.26) and (18.27) allow one to determine $r_{\min}$ in terms of the other parameters. The integral gives

$$E_{\text{rad}} = \frac{2Z^2}{3m_e^2 c^3 v_0 b^3} \left(\frac{e^2}{4\pi\varepsilon_0}\right)^3$$
$$\times \left\{ 3\frac{b_0}{b} + \left(1 + 3\frac{b_0^2}{b^2}\right) \left[\frac{\pi}{2} + \arctan\left(\frac{b_0}{b}\right)\right] \right\}, \tag{18.28}$$

with

$$b_0 := \frac{e^2}{4\pi\varepsilon_0} \frac{Z}{m_e v_0^2}. \tag{18.29}$$

Distant encounters correspond to $b \gg b_0$ (when one has $b \approx r_{\min}$). In this case (18.28) gives

$$E_{\text{rad}} \approx \frac{\pi Z^2}{3m_e^2 c^3 v_0 b^3} \left(\frac{e^2}{4\pi\varepsilon_0}\right)^3. \tag{18.30}$$

The form of the energy spectrum is inferred from the fact that the encounter is impulsive with a characteristic time scale of b/v_0. One expects the energy emitted per unit frequency to be approximately independent of ω for $\omega \ll v_0/b$, to fall off for $\omega \approx v_0/b$, and to be negligible for $\omega \gg v_0/b$.

18.4 Lorentz Transformation of the Larmor Formula

Although strictly the Larmor formula applies only in the rest frame of the radiating particle, it may be generalized to an arbitrary frame by applying a Lorentz transformation to it. By making a Lorentz transformation one obtains a formula for the power emitted by a particle whose motion may be relativistic.

The underlying physical principle used in this argument is the special theory of relativity, which requires that the laws of physics have the same form in all inertial frames. A Lorentz transformation is a transformation from one inertial frame to another, and so the special theory of relativity requires that the laws of physics be invariant under a Lorentz transformation. A theory that is written in such a way that it trivially satisfies this requirement is said to be *manifestly covariant*. Our initial objective here is to write the Larmor formula in a manifestly covariant form. Once this is done, the generalization of the Larmor formula to the emission by a particle with arbitrary velocity **v** follows by making a Lorentz transformation from the rest frame (the frame in which the particle is instantaneously at rest at the retarded time) to the frame moving with velocity −**v**.

One writes a formula in a manifestly covariant form by introducing a 4-tensor notation. The tensor notation introduced in Chapter 2 is for tensors in three dimensions, called 3-tensors. In 3-tensor notation, the basic transformation is that of the components of the position vector x_i of a point relative to one set of coordinate axes into the components $x_{i'}$ of the same point relative to a rotated set of coordinate axes. That is, the group of transformations that defines the tensorial character of 3-tensors is rotations in three dimensions. The group of transformations that defines the tensorial character of 4-tensors is that of Lorentz transformations, which may be regarded as rotations in four dimensions with the fourth dimension being time. In introducing 4-tensor notation it is convenient to introduce time as the 0th rather than the 4th component. (An older form of 4-tensor notation has time an imaginary 4th coordinate, and modern notation has time as a real 0th coordinate.) Let the

4-tensor indices be lower case greek symbols (usually μ or ν) which run over 0, 1, 2, 3. The basic 4-vector describes an *event* that occurs at time t at point $\mathbf{x}$. The 4-vector for an event is written as $x^\mu = [ct, \mathbf{x}]$, where the entries inside the square brackets denote the time and space components of the 4-vector, respectively. Relevant examples of 4-vectors are

an event	$x^\mu = [ct, \mathbf{x}]$,	
4-velocity	$u^\mu = [\gamma c, \gamma \mathbf{v}]$,	
4-momentum	$p^\mu = [\varepsilon/c, \mathbf{p}]$,	
wave 4-vector	$k^\mu = [\omega/c, \mathbf{k}]$,	(18.31)
4-current density	$J^\mu = [\rho c, \mathbf{J}]$,	
4-potential	$A^\mu = [\phi/c, \mathbf{A}]$,	
4-force	$F^\mu = [\gamma \mathbf{v} \cdot \mathbf{F}/c, \gamma \mathbf{F}]$.	

where $\gamma = (1 - v^2/c^2)^{-1/2}$ is the Lorentz factor, $\varepsilon = \gamma m c^2$ is the energy and $\mathbf{p} = \gamma m \mathbf{v}$ is the 3-momentum of a particle with rest mass m.

A quantity that is unchanged under a Lorentz transformation is called an *invariant*. In 4-tensor notation, the invariant formed from any two 4-vectors plays a role analogous to the scalar product of two 3-vectors in 3-tensor notation. The invariant formed from the 4-vectors $a^\mu = [a^0, \mathbf{a}]$ and $b^\mu = [b^0, \mathbf{b}]$ is $a^0 b^0 - \mathbf{a} \cdot \mathbf{b}$. To take account of the minus sign one defines the *contravariant* components a^μ with raised index and the *covariant* components $a_\mu = [a^0, -\mathbf{a}]$ with lowered index for each 4-vector. Covariant components transform like the 4-gradient:

$$\text{4-gradient} \qquad \partial_\mu = \partial/\partial x^\mu = [\partial/\partial ct, \partial/\partial \mathbf{x}]. \qquad (18.32)$$

One modifies the tensor rules in §2.4 by requiring that a pair of *dummy* 4-tensor indices consist of one raised and one lowered index. The invariant formed from $a^\mu = [a^0, \mathbf{a}]$ and $b^\mu = [b^0, \mathbf{b}]$ is

$$a^\mu b_\mu = a_\mu b^\mu = a^0 b^0 - \mathbf{a} \cdot \mathbf{b}. \qquad (18.33)$$

An invariant of particular importance here is the *proper time* τ, which is interpreted as the actual time according to an observer moving along an arbitrary orbit. The orbit (18.1) is rewritten in the covariant form as a function of the proper time:

$$x^\mu = X^\mu(\tau). \qquad (18.34)$$

The space components of (18.34) correspond to (18.1). The time component $X^0(\tau)$ is such that the invariant $dx^\mu dx_\mu = c^2 d\tau^2$ formed from the differential of (18.34) defines an element $d\tau$ of proper time. This gives

$$d\tau = dt/\gamma, \qquad (18.35)$$

where γ is regarded as a function of t in a specific inertial frame, and integrating (18.35) along the orbit gives the proper time. The 4-velocity and the 4-acceleration are given by differentiating with respect to proper time:

$$u^\mu(\tau) = dX^\mu(\tau)/d\tau, \quad a^\mu(\tau) = d^2X^\mu(\tau)/d\tau^2. \qquad (18.36)$$

The covariant form of Newton's equation of motion is

$$ma^\mu(\tau) = F^\mu. \qquad (18.37)$$

A Lorentz transformation of a 4-vector b^μ from an inertial frame K to another inertial frame K' is written

$$b^{\mu'} = L^{\mu'}{}_\mu b^\mu. \qquad (18.38)$$

The Lorentz transformation of relevance here is from the rest frame of a particle, which is the inertial frame in which it is instantaneously at rest, to the laboratory frame in which the particle has 3-velocity **v**. It is convenient to write $\mathbf{v} = c\boldsymbol{\beta}$, $\gamma = (1 - \beta^2)^{-1/2}$ and to consider a *boost* in which K' is moving along the 3 axis of K, with the coordinate axes in K' and K parallel to each other and with their origins coincident at $t = t' = 0$. (An arbitrary Lorentz transformation may be decomposed into a rotation of the coordinate axes and a boost, and here we consider only a boost along the 3 axis.) The transformation matrix is

$$L^{\mu'}{}_\mu = \begin{pmatrix} \gamma & 0 & 0 & -\gamma\beta \\ 0 & 1 & 0 & 0 \\ 0 & 0 & 1 & 0 \\ -\gamma\beta & 0 & 0 & \gamma \end{pmatrix}. \qquad (18.39)$$

Now let us return to the Larmor formula (18.10). In the rest frame we have $\mathbf{v} = 0$, $\gamma = 1$, and $\mathbf{a} = \dot{\mathbf{v}} = c\dot{\boldsymbol{\beta}}$, which imply $u^\mu = [c, \mathbf{0}]$, $a^\mu = [0, c\dot{\boldsymbol{\beta}}]$ and $a^\mu a_\mu = -c^2\dot{\beta}^2$. Hence (18.10) is written in the form

$$P = -\frac{q^2 a^\mu a_\mu}{6\pi\varepsilon_0 c^3}, \qquad (18.40)$$

in the rest frame. The left hand side of (18.40) is the ratio of the time components of two vectors (the energy radiated and the time) and hence is an invariant, and the right hand side is in invariant form. Hence the special theory of relativity implies that (18.40) applies in all inertial frames. It is therefore the desired generalization of the Larmor formula.

In the laboratory frame (18.40) is rewritten in 3-vector notation by noting the relations

$$a^\mu = c\gamma^4[\boldsymbol{\beta} \cdot \dot{\boldsymbol{\beta}}, \boldsymbol{\beta} \cdot \dot{\boldsymbol{\beta}}\boldsymbol{\beta} + (1 - \beta^2)\dot{\boldsymbol{\beta}}],$$
$$a^\mu a_\mu = -c^2\gamma^6\left[\dot{\boldsymbol{\beta}}^2 - |\boldsymbol{\beta} \times \dot{\boldsymbol{\beta}}|^2\right]. \tag{18.41}$$

The generalization of the Larmor formula is therefore

$$P = \frac{q^2}{6\pi\varepsilon_0 c}\frac{\dot{\boldsymbol{\beta}}^2 - |\boldsymbol{\beta} \times \dot{\boldsymbol{\beta}}|^2}{(1 - \beta^2)^3}. \tag{18.42}$$

In the rest frame $\boldsymbol{\beta} = 0$ (18.42) reduces to (18.10), as required. The acceleration is a somewhat artificial quantity for a relativistic particle and it is more relevant to write (18.42) in terms of the force $\mathbf{F}$ acting on the particle. This gives

$$P = \frac{q^2}{6\pi\varepsilon_0 c}\frac{\gamma^2(|\mathbf{F}|^2 - |\boldsymbol{\beta} \cdot \mathbf{F}|^2)}{m^2 c^2}. \tag{18.43}$$

Exercise Set 18

18.1 A non-relativistic charge q moves in simple harmonic motion along the z axis. The orbit is described by

$$Z(t) = Z_0 \cos(\omega_0 t), \qquad (E18.1)$$

where Z_0 and ω_0 are constants. Write $\beta = Z_0 \omega_0 / c$.

(a) Show that the instantaneous power radiated per unit solid angle is

$$P(t) = \frac{q^2 Z_0^2 \omega_0^4}{(4\pi)^2 \varepsilon_0 c^3} \frac{\sin^2 \theta \cos^2(\omega_0 t)}{[1 + \beta \cos \theta \sin(\omega_0 t)]^5}. \qquad (E18.2)$$

(b) Show that the time-averaged power radiated per unit solid angle is given by

$$\bar{P} = \frac{q^2 Z_0^2 \omega_0^4}{8(4\pi)^2 \varepsilon_0 c^3} \frac{\sin^2 \theta (4 + \cos^2 \theta)}{(1 - \beta^2 \cos^2 \theta)^{7/2}}. \qquad (E18.3)$$

18.2 A non-relativistic particle (charge q, mass m) with kinetic energy E at infinity makes a head-on collision with a fixed central potential field which goes to zero at large radial distance. The interaction is repulsive and is described by a potential $V(r)$ which exceeds E at small r.

(a) Show that the total energy radiated *in vacuo* is

$$E_{\text{rad}} = \frac{4q^2}{12\pi\varepsilon_0 m^2 c^3} \left(\frac{m}{2}\right)^{1/2} \int_R^\infty dr \, \frac{|dV(r)/dr|^2}{[V(R) - V(r)]^{1/2}}, \qquad (E18.4)$$

where $r = R$ is the distance of closest approach, and $V(R) = E$.

(b) Suppose that the central potential is a Coulomb potential $V(r) = Qq/4\pi\varepsilon_0 r$. Show that

$$E_{\text{rad}} = \frac{8}{45} \frac{qmv^5}{Qc^3}, \quad v = \left(\frac{2E}{m}\right)^{1/2}. \qquad (E18.5)$$

18.3 Fill in the steps in the derivation of (18.28) from (18.25) using (18.26) and (18.27).

18.4 A charge q moves in a Keplerian orbit in the x–y plane. The semimajor radius of the orbit is R, the eccentricity is e and the period is $2\pi/\omega_0$. The orbit is described in terms of an expansion in Bessel functions by $Z(t)/R = 0$ and

$$X(t)/R = -3e/2 + \sum_{n=1}^\infty (2/n) J_n'(ne) \cos(n\omega_0 t),$$
$$\qquad (E18.6)$$
$$Y(t)/R = e^{-1}(1 - e^2)^{1/2} \sum_{n=1}^\infty (2/n) J_n(ne) \sin(n\omega_0 t).$$

(a)   Use the Larmor formula to find the time-averaged power P radiated *in vacuo*.

(b)   Show that on summing the series found in part (a), one obtains

$$P = \frac{q^2 R^2 \omega_0^4}{12\pi\varepsilon_0 c^3} \frac{2+e^2}{(1-e^2)^{5/2}}. \qquad (E18.7)$$

Hint: The sums involved are the Kapteyn series (for $0 \le x < 1$)

$$\sum_{n=1}^{\infty} n^2 J_n^2(nx) = \frac{x^2(4+x^2)}{16(1-x^2)^{7/2}}, \qquad (E18.8)$$

$$\sum_{n=1}^{\infty} n^2 J_n'^2(nx) = \frac{(4+3x^2)}{16(1-x^2)^{5/2}}. \qquad (E18.9)$$

18.5 Using (18.26) and (18.27) determine the radial distance $r = r_{\min}$ of closest approach in terms of the parameters v_0 and b.

18.6 Consider the identification of the 4-force (18.31).

(a)   Show that the 4-velocity u^μ satisfies $u^\mu u_\mu = c^2$.

(b)   Hence, argue that the equation of motion in the form (18.37) or $dp^\mu/d\tau = F^\mu$ requires that the 4-force satisfy $u_\mu F^\mu = 0$.

(c)   Show that the space components of the equation of motion, the definition (18.35) of the proper time and the identity derived in part (b) imply $F^\mu = [\gamma \mathbf{v} \cdot \mathbf{F}/c, \gamma \mathbf{F}]$, where $\mathbf{F}$ is the 3-force.

18.7 The following exercise is concerned with the relation between the force and the acceleration of a relativistic particle, as they appear in (18.42) and (18.43).

(a)   On writing the 4-velocity in the form $u^\mu = c\gamma[1, \boldsymbol{\beta}]$, show that the 4-acceleration $a^\mu = \gamma du^\mu/dt$ reduces to

$$a^\mu = c\gamma^4[\boldsymbol{\beta} \cdot \dot{\boldsymbol{\beta}}, \boldsymbol{\beta}\boldsymbol{\beta} \cdot \dot{\boldsymbol{\beta}} + (1-\beta^2)\dot{\boldsymbol{\beta}}]. \qquad (E18.10)$$

(b)   Show that ($E18.10$) implies

$$a^\mu a_\mu = -c^2\gamma^6\left(\dot{\boldsymbol{\beta}}^2 - |\boldsymbol{\beta} \times \dot{\boldsymbol{\beta}}|^2\right). \qquad (E18.11)$$

(c)   On writing Newton's equation of motion in the form $\dot{\mathbf{p}} = \mathbf{F}$ with $\mathbf{p} = mc\gamma\boldsymbol{\beta}$, show that one has

$$|\mathbf{F}|^2 - |\boldsymbol{\beta} \cdot \mathbf{F}|^2 = m^2 c^2 \gamma^4\left(\dot{\boldsymbol{\beta}}^2 - |\boldsymbol{\beta} \times \dot{\boldsymbol{\beta}}|^2\right). \qquad (E18.12)$$

19

Alternative Treatment of Emission Processes

Preamble

The conventional treatment of the emission of radiation *in vacuo* by a particle involves calculating the electric and magnetic fields due to the particle, finding the Poynting vector, and then identifying the power radiated as the integral of the radial component of the Poynting vector over an arbitrarily large sphere centered on the origin. This procedure is carried out here. The standard technique for including the back reaction to the radiation through a radiation reaction force is then discussed.

19.1 The Lienard–Wierchert Potentials

The following procedure may be used to find the electromagnetic field due to a moving charge *in vacuo* as a function of space and time. First, the potentials in the Lorentz gauge are found by solving d'Alembert's equation. Next, the resulting solution takes a simple form when the origin is chosen appropriately, and the resulting expressions are referred to as the Lienard–Wierchert potentials. The electric and magnetic fields are then found using (1.8) and (1.9). The details are as follows.

In the absence of an ambient medium, the electromagnetic potentials in the Lorentz gauge satisfy d'Alembert's equation, cf. (1.20) and (1.21). The appropriate solution requires choosing a Greens function, and the appropriate choice here is the causal Greens function given by (5.33). The causal function is appropriate because of the causal relation between the sources and the fields. Thus the fields for an arbitrary source are given by

$$\phi(t,\mathbf{x}) = \frac{1}{4\pi\varepsilon_0} \int dt' d^3x' \frac{\rho(t',\mathbf{x}')\,\delta(t-t'-|\mathbf{x}-\mathbf{x}'|/c)}{|\mathbf{x}-\mathbf{x}'|}, \quad (19.1)$$

$$\mathbf{A}(t,\mathbf{x}) = \frac{\mu_0}{4\pi} \int dt' d^3x' \frac{\mathbf{J}(t',\mathbf{x}')\,\delta(t-t'-|\mathbf{x}-\mathbf{x}'|/c)}{|\mathbf{x}-\mathbf{x}'|}. \quad (19.2)$$

To find the potentials for a moving charge, the source terms (18.3) are inserted into (19.1) and (19.2). The integral over d^3x' is performed over the δ-function in (18.3), giving

$$\phi(t,\mathbf{x}) = \frac{q}{4\pi\varepsilon_0} \int dt' \frac{\delta(t-t'-|\mathbf{x}-\mathbf{X}(t')|/c)}{|\mathbf{x}-\mathbf{X}(t')|}, \quad (19.3)$$

$$\mathbf{A}(t,\mathbf{x}) = \frac{\mu_0 q}{4\pi} \int dt' \frac{\dot{\mathbf{X}}(t')\,\delta(t-t'-|\mathbf{x}-\mathbf{X}(t')|/c)}{|\mathbf{x}-\mathbf{X}(t')|}. \quad (19.4)$$

The integral over t' is then performed over the remaining δ-functions. The integral is of the form

$$\int dt'\,\delta(t-t'-|\mathbf{x}-\mathbf{X}(t')|/c) = \frac{|\mathbf{x}-\mathbf{X}(t')|}{\left| |\mathbf{x}-\mathbf{X}(t')| - [\mathbf{x}-\mathbf{X}(t')]\cdot\dot{\mathbf{X}}(t')/c \right|}. \quad (19.5)$$

which is to be evaluated at the *retarded time* $t' = t_r$ determined by

$$t - t_r - |\mathbf{x}-\mathbf{X}(t_r)|/c = 0. \quad (19.6)$$

The resulting expressions for the potentials are

$$\phi(t,\mathbf{x}) = \frac{q}{4\pi\varepsilon_0} \frac{1}{\left| |\mathbf{x}-\mathbf{X}(t')| - [\mathbf{x}-\mathbf{X}(t')]\cdot\dot{\mathbf{X}}(t')/c \right|}, \quad (19.7)$$

$$\mathbf{A}(t,\mathbf{x}) = \frac{\mu_0 q}{4\pi} \frac{\dot{\mathbf{X}}(t')}{\left| |\mathbf{x}-\mathbf{X}(t')| - [\mathbf{x}-\mathbf{X}(t')]\cdot\dot{\mathbf{X}}(t')/c \right|}, \quad (19.8)$$

evaluated at $t' = t_r$.

The explicit forms (19.7) and (19.8) are relatively cumbersome. Considerable algebraic simplification occurs if one chooses the origin to be at the position of the particle at the retarded time. This corresponds to choosing the coordinate axes to satisfy

$$\mathbf{X}(t') = 0. \quad (19.9)$$

Let r be the radial distance from this origin. Then (19.6) simplifies to give the retarded time in the form

$$t_r = t - \frac{r}{c}. \quad (19.10)$$

Further, if one writes $\mathbf{v}$ for $\dot{\mathbf{X}}(t')$ in this coordinate system, then (19.7) and (19.8) reduce to

$$\phi(t, \mathbf{x}) = \frac{q}{4\pi\varepsilon_0(r - \mathbf{x} \cdot \mathbf{v}/c)}, \tag{19.11}$$

$$\mathbf{A}(t, \mathbf{x}) = \frac{\mu_0 q\, \mathbf{v}}{4\pi(r - \mathbf{x} \cdot \mathbf{v}/c)}, \tag{19.12}$$

which are the *Lienard–Wierchert potentials*. They give the potentials in the Lorentz gauge when the particle is at the origin at the retarded time.

Let us consider the potentials (19.11) and (19.12) in some simple cases. The simplest case is for a charge at rest ($\mathbf{v} = 0$). In this case (19.11) gives $\phi = q/4\pi\varepsilon_0 r$, which is just the Coulomb field. Another simple case is for a charge in constant rectilinear motion ($\mathbf{v} = $ constant). Let us start with the potentials relative to a fixed origin using (19.7) and (19.8). The orbit is written $\mathbf{X}(t) = \mathbf{x}_0 + \mathbf{v}t$. Then, according to (19.6), the retarded time is found by solving $c(t - t_r) = |\mathbf{x} - \mathbf{x}_0 - \mathbf{v}t_r|$. Some simplification results by choosing the origin such that $\mathbf{x}_0$ is zero, but even in this simple case one needs to solve a quadratic equation to find t_r as a function of t and $\mathbf{x}$, cf. Exercise 19.1. As a result the explicit expressions resulting from (19.7) and (19.8) involve quite cumbersome expressions in the denominator. The algebraic simplification achieved in (19.11) and (19.12) results from choosing the origin to be at the point $\mathbf{x} = \mathbf{x}_0 + \mathbf{v}t_r$. The position of this moving origin is found explicitly after the expression for t_r is found. Thus, using (19.11) and (19.12) in place of (19.7) and (19.8) transfers the algebraic complexity from the explicit evaluation of the denominators in (19.7) and (19.8) into the implicit location of the moving origin in (19.11) and (19.12).

It should be emphasized that the mathematical simplicity introduced by assuming the origin to be at the position of the moving charge at the retarded time is offset by the considerable increase in the mathematical complexity of taking the derivatives of the potentials to find the fields. The point is that the partial derivatives $\partial/\partial t$ and $\partial/\partial \mathbf{x}$ are defined relative to a fixed origin, and their evaluation for potentials expressed in terms of a moving origin requires considerable care.

19.2 Calculation of the Poynting Vector

The electric and magnetic fields are derived from the potentials (19.11) and (19.12) using (1.8) and (1.9), *viz.*,

$$\mathbf{E} = -\operatorname{grad}\phi - \partial\mathbf{A}/\partial t, \quad \mathbf{B} = \operatorname{curl}\mathbf{A}. \tag{19.13}$$

However, the evaluation of the derivatives when (19.11) and (19.12) are inserted into (19.13) cannot be carried out directly because the origin at the position of the particle at the retarded time is moving and accelerating relative to an inertial frame. The important point is that the moving origin depends on the retarded time t_r which is an implicit function of both position and time. Hence to evaluate the partial derivatives one needs to have expressions for the derivatives of t_r. It is helpful to use a fixed coordinate system to evaluate these derivatives, and then to revert to the moving coordinate system. Starting from (19.6) and differentiating with respect to time then transforming to the moving coordinate system (19.9), and repeating the calculation for the spatial derivative of (19.6), one obtains

$$\frac{\partial t_r}{\partial t} = \frac{1}{1 - \mathbf{x} \cdot \mathbf{v}/rc}, \quad \mathrm{grad}\, t_r = -\frac{\mathbf{x}}{c(r - \mathbf{x} \cdot \mathbf{v}/c)}. \tag{19.14}$$

To evaluate the fields (19.13) one needs to find the temporal and spatial derivatives of the denominator in (19.11) and (19.12). The denominator $r - \mathbf{x} \cdot \mathbf{v}/c$ is not an explicit function of time and so the temporal derivative is equal to $\partial t_r/\partial t$ times the partial derivative with respect to t_r. There are two contributions to the spatial derivative, an explicit one and an implicit one through the dependence of t_r on position. Thus the derivatives required are

$$\frac{\partial}{\partial t_r} \left(r - \frac{\mathbf{x} \cdot \mathbf{v}}{c} \right) = -\frac{\mathbf{x} \cdot \mathbf{v}}{r} + \frac{v^2}{c} - \frac{\mathbf{x} \cdot \mathbf{a}}{c},$$

$$\mathrm{grad} \left(r - \frac{\mathbf{x} \cdot \mathbf{v}}{c} \right) = \frac{\mathbf{x}}{r} - \frac{\mathbf{v}}{c} + (\mathrm{grad}\, t_r) \frac{\partial}{\partial t_r} \left(r - \frac{\mathbf{x} \cdot \mathbf{v}}{c} \right), \tag{19.15}$$

where $\mathbf{a}$ is the acceleration.

The evaluation of the fields then involves some straightforward but lengthy algebra. The final expression for the electric field is

$$\mathbf{E} = \frac{q \left[(\mathbf{x} - r\mathbf{v}/c)(1 - v^2/c^2 + \mathbf{x} \cdot \mathbf{a}/c^2) - (\mathbf{a}r/c^2)(r - \mathbf{x} \cdot \mathbf{v}/c) \right]}{4\pi\varepsilon_0 (r - \mathbf{x} \cdot \mathbf{v}/c)^3}. \tag{19.16}$$

The magnetic field is found to be related to the electric field by

$$\mathbf{B} = \frac{\mathbf{x} \times \mathbf{E}}{rc}. \tag{19.17}$$

These fields are expressed as functions of position relative to the moving origin. The velocity $\mathbf{v}$ and the acceleration $\mathbf{a}$ are relative to a fixed (but arbitrary) origin; they are to be evaluated at the retarded time.

Inspection of the form of the electric field (19.16) shows that it may be separated into a part that varies with r as $1/r^2$, and a part that varies as $1/r$. The part that varies as $1/r^2$ does not involve the acceleration, and is interpreted as the *inductive* part of the field. The part that varies as $1/r$ is proportional to the acceleration and is interpreted as the *radiative* part of the field. The relation (19.17) implies that the magnetic field may be separated into analogous inductive and radiative parts.

The Poynting vector gives the energy flux in electromagnetic radiation. From (19.16) and (19.17) one finds

$$\mathbf{F}_{EM} = \frac{\mathbf{E} \times \mathbf{B}}{\mu_0} = c\varepsilon_0 \left(|\mathbf{E}|^2 \mathbf{n} - \mathbf{n} \cdot \mathbf{E}\mathbf{E} \right), \qquad (19.18)$$

with $\mathbf{n} = \mathbf{x}/r$ a unit vector in the radial direction (away from the origin which is at the position of the particle at the retarded time). The inductive parts of the field give rise to a contribution to the Poynting vector that varies with r as $1/r^4$, and the radiative parts give a contribution that varies as $1/r^2$; there are also cross terms that vary as $1/r^3$. The inductive terms are associated with a flow of electromagnetic energy that remains localized near the particle. The radiative terms imply an irreversible outflow of energy from the particle to infinity. The radiative terms give

$$\mathbf{F}_{EMrad} = \frac{q^2}{16\pi^2 \varepsilon_0 c^3} \frac{|(\mathbf{x} - r\mathbf{v}/c)\,\mathbf{x} \cdot \mathbf{a} - \mathbf{a}r\,(r - \mathbf{x} \cdot \mathbf{v}/c)|^2}{(r - \mathbf{x} \cdot \mathbf{v}/c)^6} \mathbf{n}. \qquad (19.19)$$

19.3 Alternative Derivation of the Larmor Formula

The power radiated by a particle is calculated from the rate energy flows to infinity. Consider a large sphere centered on the moving origin. The energy flowing to infinity is given by integrating the radial component of the Poynting vector over the surface of the sphere, and allowing the radius of the sphere to tend to infinity.

The Larmor formula (18.10) gives the power radiated *in vacuo* by a non-relativistic charge. For a non-relativistic particle the expression (19.19) for the radiative part of the Poynting vector reduces to

$$\mathbf{F}_{EMrad} = \frac{q^2}{16\pi^2 \varepsilon_0 r^6 c^3} |\mathbf{x}\mathbf{x} \cdot \mathbf{a} - \mathbf{a}r^2|^2\,\mathbf{n} = \frac{q^2}{16\pi^2 \varepsilon_0 r^2 c^3} |\mathbf{n} \times (\mathbf{n} \times \mathbf{a})|^2\,\mathbf{n},$$
$$(19.20)$$

with $\mathbf{n} = \mathbf{x}/r$. The power radiated is identified by multiplying the magnitude of $\mathbf{F}_{EMrad}$ by $4\pi r^2$ and averaging over all directions. Thus one rederives the Larmor formula

$$P_{rad} = \frac{q^2}{6\pi\varepsilon_0 c^3} |\mathbf{a}|^2, \qquad (19.21)$$

where the average over directions has been performed using

$$\langle |\mathbf{n} \times (\mathbf{n} \times \mathbf{a})|^2 \rangle = \tfrac{2}{3} |\mathbf{a}|^2. \qquad (19.22)$$

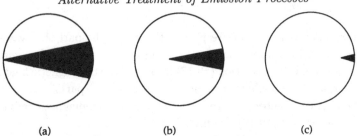

(a)   (b)   (c)

Fig. 19.1 Radiation confined to a forward cone (shown in black) for a particle which (a) enters a sphere from the left, (b) propagates through the sphere, and (c) leaves the sphere on the right.

The Larmor formula is derived here in two different ways. The procedure used in §18.2 is based on the emission formula which is derived from the work done by the current. The procedure used in deriving (19.21) is based on a calculation of the escaping electromagnetic radiation. The evaluation of the power radiated by the two procedures is equivalent under the assumptions made in deriving the Larmor formula. However, when the procedures are applied to treat emission under more general conditions, they lead to different results.

It is helpful to make a distinction between the power radiated and the power received by a fixed distant observer. Consider the radiation emitted (per unit solid angle) in the direction of a fixed distant observer by a source *in vacuo*. The emission formula gives the *power radiated*, and the Poynting vector gives the *power received* by the observer. That these are not equal in general is seen by considering the example illustrated in Figure 19.1. The point is that the power radiated and the power escaping from a fixed volume are the same only if the electromagnetic energy contained in the volume is a constant, and as illustrated in Figure 19.1 this proviso is not satisfied even in some simple cases.

One idea underlying the example illustrated in Figure 19.1 relates to the fact that the emission from a highly relativistic particle is strongly concentrated in a forward cone about the direction of motion of the particle. To be specific consider synchrotron radiation, in which case the radiation emitted by the particle forms a highly anisotropic radiation pattern that is independent of time in a coordinate system fixed to the center of gyration of the particle. The power radiated by the particle is then equal to the power escaping across the surface of a sphere *comoving* with the particle. Now consider a fixed sphere such that the observer is at a point on its surface. The particle remains inside the sphere for only a finite time, and during that time the electromagnetic energy inside the sphere changes due to the motion of the radiation pattern with the particle. As a consequence the power in radiation crossing the surface of the sphere during the time in which the particle is inside the sphere exceeds the power radiated by the particle. This is seen in a more formal way by considering the continuity equation for the electromagnetic energy, cf. (1.7),

$$\frac{\partial}{\partial t} W_{\text{EM}} + \text{div}\, \mathbf{F}_{\text{EM}} = -\mathbf{J} \cdot \mathbf{E}. \tag{19.23}$$

On integrating over a fixed volume, the first term gives the change in the energy in the radiation within the volume, the second term gives the escaping power and the term on the right hand side gives the power radiated. Clearly the escaping power and the power radiated are equal only if the energy in the radiation within the volume remains constant.

19.4 Generalizations of the Larmor Formula

The foregoing point is illustrated further by considering an alternative derivation of the generalization (18.42) or (18.43) of the Larmor formula. Specifically consider the power radiated by a particle and the power received by a fixed distant observer using the expression (19.19) for the Poynting vector. Writing $\boldsymbol{\beta} = \mathbf{v}/c$ and $\dot{\boldsymbol{\beta}} = \mathbf{a}/c$, the power received is

$$P_{\text{rec}} = \frac{q^2}{4\pi\varepsilon_0 c} \left\langle \frac{|\mathbf{n} \times [(\mathbf{n} - \boldsymbol{\beta}) \times \dot{\boldsymbol{\beta}}]|^2}{(1 - \mathbf{n} \cdot \boldsymbol{\beta})^6} \right\rangle, \tag{19.24}$$

and the power radiated is

$$P_{\text{rad}} = \frac{q^2}{4\pi\varepsilon_0 c} \left\langle \frac{|\mathbf{n} \times [(\mathbf{n} - \boldsymbol{\beta}) \times \dot{\boldsymbol{\beta}}]|^2}{(1 - \mathbf{n} \cdot \boldsymbol{\beta})^5} \right\rangle. \tag{19.25}$$

The difference between these two quantities is attributed to time dilation: the energy emitted in unit time by the moving source is received in a shorter time due to the motion of the source towards the observer.

The power radiated for a particle with arbitrary speed is calculated in §18.4 using a Lorentz transformation. An alternative way of deriving the result is to perform the average over directions implied by the angular brackets in (19.25) explicitly. The calculation proceeds as follows.

Let the angle between $\boldsymbol{\beta}$ and $\dot{\boldsymbol{\beta}}$ be ψ, and let the angles between $\mathbf{n}$ and $\boldsymbol{\beta}$ and $\dot{\boldsymbol{\beta}}$ be θ and θ', respectively:

$$\boldsymbol{\beta} \cdot \dot{\boldsymbol{\beta}} = \beta\dot{\beta}\cos\psi, \quad \mathbf{n} \cdot \boldsymbol{\beta} = \beta\cos\theta, \quad \mathbf{n} \cdot \dot{\boldsymbol{\beta}} = \dot{\beta}\cos\theta'. \tag{19.26}$$

The azimuthal angle ϕ is defined such the the following identity holds:

$$\cos\theta' = \cos\theta\cos\psi + \sin\theta\sin\psi\cos\phi. \tag{19.27}$$

Then (19.19), after integration over the arbitrarily large sphere, gives

$$P_{\text{rad}} = \frac{q^2 \dot{\beta}^2}{(4\pi)^2 \varepsilon_0 c} \int_0^{2\pi} \mathrm{d}\phi \int_{-1}^{1} \mathrm{d}\cos\theta \, f(\theta, \theta', \psi; \beta),$$

$$f(\theta, \theta', \psi; \beta) = (1 - \beta\cos\theta)^{-5} \big[(1 - \beta\cos\theta)^2 \tag{19.28}$$
$$- 2\cos\theta'(1 - \beta\cos\theta)(\cos\theta' - \beta\cos\psi)$$
$$+ \cos^2\theta'(1 - 2\beta\cos\theta + \beta^2).$$

The remainder of the calculation is outlined in Exercise 19.4. The result is

$$P_{\text{rad}} = \frac{2q^2 \dot{\beta}^2}{12\pi\varepsilon_0 c} \frac{(1 - \beta^2\sin^2\psi)}{(1 - \beta^2)^3}, \tag{19.29}$$

which reproduces (18.42).

The Larmor formula applies to emission by a non-relativistic particle *in vacuo*, and (19.29) is a generalization that relaxes the assumption that the particle is non-relativistic. There is some interest in considering a generalization that relaxes the assumption that the emission is *in vacuo*.

Let us consider how the foregoing method based on calculating the fields and the Poynting vector for emission by a particle might be generalized to emission in a medium. This requires replacing (19.3) and (19.4) by the solutions for the potentials in the presence of the medium. In the temporal gauge the relevant result is given by solving (11.4) for the vector potential, as in (16.3) with (16.5), and inverting the Fourier transform. Thus one replaces (19.3) and (19.4) by $\phi(t, \mathbf{x}) = 0$ and

$$A_i(t, \mathbf{x}) = \mu_0 q c^2 \int \mathrm{d}t' \, \dot{X}_j(t') \int \frac{\mathrm{d}\omega \mathrm{d}^3\mathbf{k}}{(2\pi)^4} \frac{\lambda_{ij}(\omega, \mathbf{k})}{\omega^2 \Lambda(\omega, \mathbf{k})}$$
$$\times \, \mathrm{e}^{-i\omega(t-t')+i\mathbf{k}\cdot[\mathbf{x}-\mathbf{X}(t')]}. \tag{19.30}$$

The inductive part of the field is given by the non-resonant part of the integrand, and the radiative part of the field is given by replacing $\lambda_{ij}(\omega, \mathbf{k})/\Lambda(\omega, \mathbf{k})$ by $D_{ij}^A(\omega, \mathbf{k})$ as given by (16.7). The resulting expression for the radiation field in the mode M is

$$\mathbf{A}_{M\text{rad}}(t, \mathbf{x}) = -\frac{i\mu_0 q c^2}{2} \int \mathrm{d}t' \int \frac{\mathrm{d}^3\mathbf{k}}{(2\pi)^3} \frac{R_M(\mathbf{k})}{\omega_M(\mathbf{k})} \mathbf{e}_M(\mathbf{k}) \mathbf{e}_M^*(\mathbf{k}) \cdot \dot{\mathbf{X}}(t')$$
$$\times \, \mathrm{e}^{-i\omega_M(\mathbf{k})(t-t')+i\mathbf{k}\cdot[\mathbf{x}-\mathbf{X}(t')]}. \tag{19.31}$$

The negative frequency parts are implicit in (19.31) through the relations (15.2). The calculation of the Poynting vector using (19.31) is then straightforward but tedious.

However an important point in principle is that the energy flux in radiation is *not* given by the Poynting vector in general. It is shown in §15.4 that the energy flux is given by the Poynting vector only in media that are not spatially dispersive. Hence the generalization of the foregoing method is inadequate to treat emission in a spatially dispersive medium. The method used in §18.2 does not suffer from this deficiency.

Thus one concludes that the method based on the calculation of the Poynting vector is considerably more tedious to use than the method adopted in Chapter 18 for a medium that is not spatially dispersive, and that the method is not valid for a medium that is spatially dispersive.

19.5 Radiation Reaction

One of the unsatisfactory features of classical electromagnetic theory is that it does not conserve energy and momentum; this is because the back reaction of the emission on the radiating system is not automatically taken into account. One way of complementing classical electromagnetic theory to overcome this difficulty is to introduce a *radiation reaction force*. The idea is to postulate the existence of a force, $\mathbf{F}_{react}(t)$ say, such that minus the rate work is done by the particle against this force is equal to the power radiated. Thus we require

$$P(t) = -\mathbf{v}(t) \cdot \mathbf{F}_{react}(t). \tag{19.32}$$

On integrating the Larmor formula (19.21) over time, one partially integrates to find

$$\int dt\, P(t) = \frac{q^2}{6\pi\varepsilon_0 c^3} \int dt\, |\dot{\mathbf{v}}(t)|^2$$

$$= -\frac{q^2}{6\pi\varepsilon_0 c^3} \int dt\, \mathbf{v}(t) \cdot \ddot{\mathbf{v}}(t). \tag{19.33}$$

Comparison with (19.32) then leads to the identification

$$\mathbf{F}_{react}(t) = \frac{q^2}{6\pi\varepsilon_0 c^3} \ddot{\mathbf{v}}(t). \tag{19.34}$$

However, although introducing this force overcomes one difficulty in classical electromagnetic theory, it introduces other difficulties. Specifically, if the radiation reaction force is interpreted as a real force it leads to runaway solutions and to an acausal effect.

The equation of motion in the presence of an external force, $\mathbf{F}_0(t)$ say, and of the radiation reaction force is called the *Abraham–Lorentz equation of motion*:

$$m\left[\dot{\mathbf{v}}(t) - \tau_0\ddot{\mathbf{v}}(t)\right] = \mathbf{F}_0(t), \quad \tau_0 := \frac{q^2}{6\pi\varepsilon_0 c^3}. \tag{19.35}$$

In the absence of an external force, (19.35) has the physically acceptable solution $\dot{\mathbf{v}}(t) = 0$, corresponding to constant rectilinear motion.

However, it also has the physically unacceptable *runaway* solution

$$\dot{\mathbf{v}}(t) \propto e^{t/\tau_0}. \tag{19.36}$$

One way of excluding this solution is to replace the differential equation (19.35) by the integro-differential equation

$$m\dot{\mathbf{v}}(t) = \int_0^\infty dx\, e^{-x}\, \mathbf{F}_0(t + \tau_0 x). \tag{19.37}$$

The runaway solution then does not occur, but (19.37) requires the acausal effect called *preacceleration*. The point is that (19.37) implies that the acceleration at time t is due to the force $\mathbf{F}_0$ at times $t + \tau_0 x$ later than t. Thus the particle is required to respond to the force before the force has acted. This preacceleration is of no practical significance because it occurs on a time scale that is intrinsically unmeasurable. (The characteristic time scale τ_0 is the light travel time over the classical radius of the particle, and this is much shorter than the light travel time over a Compton wavelength, so that the time scale is unresolvable due to quantum effects.)

These conceptual difficulties associated with the radiation reaction are attributed to the fact that in the derivation of the radiation reaction force from (19.33) one is not justified in omitting the integral over time and equating the integrands. All that the theory requires is that the integrals be equal, and it is incorrect to infer any causal relations by equating the integrands. Similarly, the derivation of the Larmor formula (18.10) is based on equating integrands, cf. (17.17) and the remarks following (17.20). One should not infer that $P(t)$ is due to $\mathbf{a}(t)$ at the same time t. The power radiated is a meaningful concept only over a finite time.

For specific radiation mechanisms it is possible to treat the radiation reaction in quite a different way. This involves using the kinetic equation (16.35). Examples are Cerenkov emission, cf. §20.6, and for gyromagnetic emission, cf. §23.5.

Exercise Set 19

19.1 Consider the electromagnetic field at a point described by polar coordinates r, θ due to a charge q in constant rectilinear motion at velocity βc along the polar axis. Assume that the particle is at the origin $r = 0$ at $t = 0$.

(a) Show that the retarded time is given by the smaller of the solutions of the quadratic equation

$$(t - t_r)^2 = r^2/c^2 - 2t_r(r/c)\beta \cos\theta + \beta^2 t_r^2. \qquad (E19.1)$$

(b) Show that the electric and magnetic fields of a charge q in constant rectilinear motion at velocity βc are given by

$$\mathbf{E} = \frac{q}{4\pi\varepsilon_0\gamma^2} \frac{\mathbf{x} - r\boldsymbol{\beta}}{(r - \mathbf{x}\cdot\boldsymbol{\beta})^3},$$

$$\mathbf{B} = \frac{q\mu_0 c}{4\pi\gamma^2} \frac{r\boldsymbol{\beta}\times\mathbf{x}}{(r - \mathbf{x}\cdot\boldsymbol{\beta})^3}, \qquad (E19.2)$$

with $\gamma = (1 - \beta^2)^{-1/2}$.

(c) In the non-relativistic limit the first of $(E19.2)$ gives the Coulomb field. Give a physical interpretation of the non-relativistic limit of the other relation in terms of the Biot–Savart law.

19.2 A Lorentz transformation of the electric and magnetic fields from an unprimed frame K to a primed frame K' moving a velocity $\mathbf{v}$ relative to K is

$$\mathbf{E}' = \gamma(\mathbf{E} + \boldsymbol{\beta}\times\mathbf{B}) - \frac{\gamma^2}{\gamma + 1}\boldsymbol{\beta}(\boldsymbol{\beta}\cdot\mathbf{E}),$$

$$\mathbf{B}' = \gamma(\mathbf{B} - \boldsymbol{\beta}\times\mathbf{E}/c^2) - \frac{\gamma^2}{\gamma + 1}\boldsymbol{\beta}(\boldsymbol{\beta}\cdot\mathbf{B}). \qquad (E19.3)$$

Identify the fields $(E19.2)$ with the primed fields in $(E19.3)$, apply the transformation to find the unprimed fields, and interpret the result.

19.3 Derive the radiation fields *in vacuo* by expanding in multipole moments and neglectingt all fields that fall off faster than $1/r$.

(a) Consider the expansion of fields (1.26) and (1.27) in powers of $1/r$ and show that the leading terms correspond to $(\mathbf{n} = \mathbf{x}/r)$

$$\phi(t, \mathbf{x}) = \frac{1}{4\pi\varepsilon_0 r}\int d^3x' \left[1 + \frac{\mathbf{n}\cdot\mathbf{x}'}{c}\frac{\partial}{\partial t}\right.$$

$$\left. + \left(\frac{\mathbf{n}\cdot\mathbf{x}'}{c}\frac{\partial}{\partial t}\right)^2 + \cdots\right]\rho, \qquad (E19.4)$$

$$\mathbf{A}(t,\mathbf{x}) = \frac{\mu_0}{4\pi r} \int d^3x' \left[1 + \frac{\mathbf{n}\cdot\mathbf{x}'}{c}\frac{\partial}{\partial t} \right.$$

$$\left. + \left(\frac{\mathbf{n}\cdot\mathbf{x}'}{c}\frac{\partial}{\partial t}\right)^2 + \cdots \right]\mathbf{J}, \qquad (E19.5)$$

where the arguments $(t - r/c, \mathbf{x})$ of ρ and $\mathbf{J}$ are omitted.

(b) Using the definitions of the multipole moments and the results (3.14) and (3.15) derived in §3.2, show that $(E19.4)$ and $(E19.5)$ imply, respectively,

$$\phi(t,\mathbf{x}) = \frac{1}{4\pi\varepsilon_0 rc}\left[n_s \dot{d}_s + \tfrac{1}{2c}\ddot{q}_{ij}n_i n_j + \cdots \right], \qquad (E19.6)$$

$$A_i(t,\mathbf{x}) = \frac{\mu_0}{4\pi r}\left[\dot{d}_i - \tfrac{1}{c}\epsilon_{ijk}n_j\dot{m}_k \right.$$

$$\left. + \tfrac{1}{2c}\ddot{q}_{ij}n_j + \cdots \right]. \qquad (E19.7)$$

where a dot denotes a time-derivative and where the moments are implicitly functions of $t - r/c$.

(c) Show that to leading order in the expansion in $1/r$ the functional dependence on $t - r/c$ implies

$$\operatorname{grad}\phi = -\frac{\mathbf{n}}{c}\dot{\phi}, \quad \operatorname{curl}\mathbf{A} = -\frac{\mathbf{n}}{c}\times\dot{\mathbf{A}}. \qquad (E19.8)$$

(d) Hence show that (1.8) and (1.9) imply

$$E_i(t,\mathbf{x}) = \frac{1}{4\pi\varepsilon_0 c^2}\left[\ddot{d}_i + \tfrac{1}{c}\epsilon_{ijk}n_j\ddot{m}_k \right.$$

$$\left. + \tfrac{1}{6c}(n_a n_b \delta_{ci} - \delta_{ai}\delta_{bc})\dddot{d}_{ab}n_c \cdots \right], \quad (E19.9)$$

$$\mathbf{B}(t,\mathbf{x}) = (1/c)\mathbf{n}\times\mathbf{E}(t,\mathbf{x}). \qquad (E19.10)$$

19.4 Rederive the formulas (17.18), (17.19) and (17.20) by starting from $(E19.9)$ and $(E19.10)$ as follows.

(a) Construct $\mathbf{n}\cdot\mathbf{F}_{\text{EM}}$ using (19.18) and identify its integral over the surface of an arbitrarily large sphere as the power lost to radiation.

(b) Perform the integral over solid angle by writing it as 4π times the average over directions, and evaluate this average using (17.12).

(c) Show that the average of each of the cross terms between the three multipole components gives zero.

(d) Show that the remaining terms reproduce (17.18), (17.19) and (17.20).

19.5 Show that the length of the pulse of radiation emitted in the direction $\mathbf{n}$ in unit time by a particle with velocity $\boldsymbol{\beta}c$ is $(1 - \mathbf{n}\cdot\boldsymbol{\beta})c$. Hence show that the power received by a fixed distant observer exceeds the power radiated in the direction $\mathbf{n}$ by the inverse of the factor $(1 - \mathbf{n} \cdot \boldsymbol{\beta})$, thereby justifying the ratio of (19.24) to (19.25).

19.6 Complete the derivation of (19.29) from (19.28) as follows.

(a) Show that the ϕ-integral reduces to

$$\int_0^{2\pi} d\phi\, f(\theta, \theta', \psi; \beta) = 2\pi \big[(1 - \beta \cos\theta)^2 - (1 - \beta^2)g(\theta, \psi)$$
$$+ 2\beta(1 - \beta \cos\theta)\cos\theta \cos^2\psi\big],$$
$$g(\theta, \psi) = \cos^2\theta \cos^2\psi + \tfrac{1}{2}\sin^2\theta \sin^2\psi. \qquad (E19.11)$$

(b) Show that the integral over $\cos\theta$ reduces to a sum of terms of the form

$$I_n = \int_{-1}^{1} d\cos\theta\, \frac{1}{(1 - \beta \cos\theta)^n},$$
$$I_3 = \frac{2}{(1 - \beta^2)^2}, \quad I_4 = \frac{2(3 + \beta^2)}{3(1 - \beta^2)^3}, \quad I_5 = \frac{2(1 + \beta^2)}{(1 - \beta^2)^4}. \qquad (E19.12)$$

19.7 Show that (19.29) may be rewritten in terms of the applied force $\mathbf{F}$ in the form

$$P_{\text{rad}} = \frac{q^2 \gamma^3}{6\pi\varepsilon_0 m^2 c^3}\big[|\mathbf{F}|^2 - |\boldsymbol{\beta} \cdot \mathbf{F}|^2\big]. \qquad (E19.13)$$

19.8 Consider a particle whose acceleration is parallel to its velocity at the retarded time, with both along the polar axis $\theta = 0$.

(a) Show that the maximum intensity in the observed emission is at an angle θ_0 given by

$$\cos\theta_0 = \frac{1}{24}\big[(1 + 24\beta^2)^{1/2} - 1\big].$$

(b) At what angle is the power radiated a maximum?

19.9 In the absence of the radiation reaction force a particle executes simple harmonic motion at frequency ω_0. Consider the effect of radiation reaction on such an oscillator that is oscillating freely at time $t = 0$.

(a) Show that the equation of motion is

$$\ddot{x} = -\omega_0^2 x + \frac{q^2}{6\pi\varepsilon_0 mc^3}\dddot{x}. \qquad (E19.14)$$

(b) Show that if the radiation reaction term is evaluated assuming free oscillations, then $(E19.14)$ implies that the frequency of the damped oscillator is of the form $\omega = \omega_0 - i\Gamma/2$ for small Γ, with

$$\Gamma = \frac{q^2 \omega_0^2}{6\pi\varepsilon_0 mc^3}. \qquad (E19.15)$$

(c)   Show that for small Γ the initial energy W_0 in the oscillations decays proportional to $e^{-\Gamma t}$, and show that the initial power loss ΓW_0 balances the power radiated by the oscillator, as given by (18.12), so that the total energy is conserved.

(d)   Show that the power radiated per unit frequency when the radiation reaction is included is related to the power radiated $\bar{P}$ in the absence of the radiation reaction, as given by (18.12), by

$$P(\omega) = \frac{\bar{P}\Gamma/2}{\pi\left[(\omega - \omega_0)^2 + (\Gamma/2)^2\right]}. \qquad (E19.16)$$

(e)   Plot the shape of $P(\omega)$ over the range $|\omega - \omega_0| < 3\Gamma/2$.

19.10 Derive the radiation reaction force in terms of the multipole expansion as follows.

(a)   Identify the momentum flux in the radiation field as $\mathbf{E} \times \mathbf{B}/\mu_0 c$ with $\mathbf{E}$ and $\mathbf{B}$ given by (E19.9) and (E19.10), respectively, and evaluate the rate at which momentum is lost by integrating the momentum flux over the surface of an arbitrarily large sphere.

(b)   Evaluate the integral over solid angle by replacing it by 4π times the average over directions and evaluate this average using (17.12).

(c)   Identify the radiation reaction force by appealing to conservation of momentum, and hence identify

$$\left[\mathbf{F}_{\mathrm{react}}\right]_i = -\frac{1}{960\pi^3 \varepsilon_0^2}\left(\dddot{d}_{ij}\ddot{d}_j + 10\epsilon_{ijk}\dddot{d}_j\ddot{\mu}_k\right). \qquad (E19.17)$$

Remark: Although the cross terms between different multipoles do not contribute to the power emitted in transverse waves, the cross terms play an essential role in the radiation reaction.

PART FIVE

Specific Emission Processes

Specific emission and absorption processes may be classified in several different ways. The presentation here is based on the nature of the extraneous current that acts as the source term for the emission. The simplest emission processes are "direct" emission in which the current is due solely to the motion of the emitting particle. The specific direct emission processes discussed here are (i) Cerenkov emission, which occurs for a particle in constant rectilinear motion at greater than the phase speed of the emitted wave, (ii) bremsstrahlung, which results from the accelerated motion of an electron due to the influence of the Coulomb field of an ion, and (iii) gyromagnetic emission, which is due to the accelerated motion of a particle in a magnetostatic field. To each of these processes there is a corresponding absorption process, and negative absorption is possible for both the Cerenkov and gyromagnetic processes. Another class of emission processes corresponds to scattering of waves by particles. The unscattered wave perturbs the orbit of the particle, and the emission due to this perturbed motion corresponds to the generation of the scattered radiation. Non-linear plasma currents need to be taken into account in treating scattering in a plasma. The non-linear currents allow another class of emission processes that are attributed to wave–wave interactions, which include a variety of processes such as frequency doubling in non-linear optics and radiation from turbulent plasmas.

20

Cerenkov Emission

Preamble

Cerenkov emission is due to a particle in constant rectilinear motion at a speed greater than the phase speed of the emitted radiation. It was discovered by accident in the early 1930s when Cerenkov observed a blue glow from the water near a radioactive source. Cerenkov emission is used in the laboratory in Cerenkov detectors in which fast particles are detected through the light they emit as they pass through a chamber filled with a dielectric material. Cerenkov emission occurs naturally in the atmosphere due to cosmic ray showers, which consist of secondary electrons and positrons generated by a very energetic cosmic ray. The absorption process corresponding to Cerenkov emission is called Landau damping in plasma physics. Under relatively mild conditions Landau damping can be negative, and some important plasma instabilities involve negative Landau damping of longitudinal waves. Appearance, disappearance and transition radiations may all be treated in a similar way to Cerenkov radiation.

20.1 A Charge in Constant Rectilinear Motion

The first step in treating Cerenkov emission is to identify the appropriate extraneous current. This is then inserted in the emission formula (16.10).

The extraneous current (18.4) for a charge q involves the orbit of the particle. For a particle in constant rectilinear motion at a velocity $\mathbf{v}$ the orbit is described by

$$\mathbf{X}(t) = \mathbf{x}_0 + \mathbf{v}t, \tag{20.1}$$

where $\mathbf{x}_0$ is the position of the particle at the time $t = 0$. The Fourier transform of the particle current (18.4) then reduces to

$$\mathbf{J}(\omega, \mathbf{k}) = q\mathbf{v}\, 2\pi\delta(\omega - \mathbf{k} \cdot \mathbf{v})\, e^{-i\mathbf{k}\cdot\mathbf{x}_0}. \qquad (20.2)$$

The current (20.2) contains a δ-function, and so is of the form (16.19). Hence it is appropriate to use the form (16.22) which gives the power radiated $P_M(\mathbf{k})d^3\mathbf{k}/(2\pi)^3$ in the mode M in the range $d^3\mathbf{k}/(2\pi)^3$. One finds

$$P_M(\mathbf{k}) = \frac{q^2 R_M(\mathbf{k})}{\varepsilon_0}\, |\mathbf{e}_M^*(\mathbf{k}) \cdot \mathbf{v}|^2\, 2\pi\delta(\omega_M(\mathbf{k}) - \mathbf{k} \cdot \mathbf{v}). \qquad (20.3)$$

It is convenient to use the semiclassical description of the emission introduced in §16.3. Here it is only the notation that is quantum mechanical, with all the detailed calculations being classical. It might be remarked that the use of quantum mechanical ideas and notation to describe "purely classical" processes is sometimes unjustifiably viewed with suspicion. It should be emphasized that the modern-day theory of electromagnetism is quantum electrodynamics (QED for short) and classical electrodynamics is to be viewed as an approximation to QED. Of course there are many processes for which classical theory is entirely adequate (these are the "purely classical" processes), but nevertheless the best available modern theory for all electromagnetic processes is one based on QED. In the case of Cerenkov emission the quantum mechanical theory was developed (by Ginzburg in 1940) within two years of the first classical theory, and so in this case there is a long tradition (primarily in the Russian literature) of quantum and classical ideas being used interchangeably.

In the semiclassical description one interprets the emission in terms of wave quanta with energy $\hbar\omega_M(\mathbf{k})$ and momentum $\hbar\mathbf{k}$. The *probability of emission* $w_M(\mathbf{p}, \mathbf{k})$ is defined by (16.27). The introduction of quantum mechanical notation for a free particle (which is a particle subject to no force) is trivial: the momentum $\mathbf{p}$ is a continuous quantum number and the energy $\varepsilon = (m^2c^4 + p^2c^2)^{1/2}$ is a continuous function of $\mathbf{p}$. Thus the quantum numbers q that describe the initial state in (16.27) are identified as the components of $\mathbf{p}$. Conservation of momentum requires that after emission the particle have momentum $\mathbf{p} - \hbar\mathbf{k}$. Hence the quantum numbers q' that describe the final state in (16.27) are identified as the components of $\mathbf{p} - \hbar\mathbf{k}$. It is convenient to replace the labels qq' on the probability by a functional dependence on the initial state $\mathbf{p}$. The fact that the final state depends on $\mathbf{p} - \hbar\mathbf{k}$ is implicit. That is, let $w_M(\mathbf{p}, \mathbf{k})d^3\mathbf{k}/(2\pi)^3$ be the probability per unit time of emission of a wave quantum in the mode M in the range $d^3k/(2\pi)^3$ by a particle with momentum $\mathbf{p} \rightarrow \mathbf{p} - \hbar\mathbf{k}$. By definition the probability is then given by, cf. (16.27),

$$P_M(\mathbf{k}) = \hbar\omega_M(\mathbf{k}) w_M(\mathbf{p}, \mathbf{k}). \tag{20.4}$$

On inserting (20.3) into (20.4) one obtains

$$w_M(\mathbf{p}, \mathbf{k}) = \frac{q^2 R_M(\mathbf{k})}{\varepsilon_0 \hbar\omega_M(\mathbf{k})} |\mathbf{e}_M^*(\mathbf{k}) \cdot \mathbf{v}|^2 \, 2\pi\delta(\omega_M(\mathbf{k}) - \mathbf{k} \cdot \mathbf{v}). \tag{20.5}$$

The expression (20.5) is the basis for the quantitative treatment of Cerenkov emission in the following discussion.

20.2 The Cerenkov Condition

An essential requirement for Cerenkov emission to occur is that there exist waves such that the argument of the δ-function in (20.5) vanishes. The condition implied by the δ-function in (20.2) is

$$\omega - \mathbf{k} \cdot \mathbf{v} = 0, \tag{20.6}$$

which is called the *Cerenkov condition*. There is a purely classical interpretation of (20.6), and there is also a quantum mechanical interpretation.

The classical interpretation of the Cerenkov condition is based on the idea of a wave-particle resonance. For a system with a natural frequency ω_0 a resonance occurs at $\omega = \omega_0$, as discussed in §9.3. If a particle is moving through a medium then two inertial frames are relevant: the laboratory frame where the medium is at rest and where the particle has velocity $\mathbf{v}$, and the rest frame of the particle where the particle is at rest and the medium is moving with velocity $-\mathbf{v}$. Consider a wave that has a frequency ω and wave vector $\mathbf{k}$ in the laboratory frame. By applying a Lorentz transformation from the laboratory frame to the rest frame of the particle one obtains the corresponding quantities ω', $\mathbf{k}'$ that describe the wave in the rest frame. One finds

$$\omega' = \gamma(\omega - \mathbf{k} \cdot \mathbf{v}), \tag{20.7}$$

where γ is the Lorentz factor of the particle. A resonance that occurs at $\omega' = \omega_0$ in the rest frame of the particle then occurs at

$$\omega - \mathbf{k} \cdot \mathbf{v} = \omega_0/\gamma \tag{20.8}$$

in the laboratory frame. It follows by comparison of (20.8) and (20.6) that the Cerenkov condition corresponds to a resonance at zero frequency in the rest frame of the particle.

The emission and absorption of Cerenkov radiation may be understood by noting that in the rest frame of the particle the fields of the wave appear to be static. Spontaneous Cerenkov emission may be understood by considering the spatial Fourier components of the electric field associated with the particle in its rest frame. The purely spatial Fourier components associated with the electric field of the particle at rest in its rest frame become Fourier components with both an ω and a $\mathbf{k}$ after Lorentz transforming to the laboratory frame. Specifically, they become components that satisfy the Cerenkov emission $\omega - \mathbf{k} \cdot \mathbf{v} = 0$ in the laboratory frame. Provided there exist waves whose dispersion relation is compatible with this condition, Cerenkov emission of these waves occurs spontaneously, effectively because the shielding field of the emitting particle in its rest frame includes a radiating component when transformed to the laboratory frame.

Landau damping or wave growth may be understood by considering the effect of the wave field on a particle in its rest frame. When the Cerenkov condition is satisfied by a wave in the laboratory frame, that is, the phase velocity of the wave is equal to the resonant velocity, then the wave fields in the rest frame are static fields. The implied static electric field in the rest frame freely accelerates the resonant particles, which are initially at rest. There is an efficient exchange of energy between the wave and the particles. Particles that are traveling slightly more slowly than the wave tend to be accelerated into exact resonance, and particles traveling slightly faster than the wave tend to be slowed down into exact resonance. The energy flow is from the waves to the particles if the number of particles decreases with increasing velocity at the resonant velocity, and this corresponds to Landau damping of the waves. Conversely, if the number of particles increases with increasing velocity at the resonant velocity then the energy flow is from the particles to the waves and the waves grow.

The Cerenkov condition can be satisfied only for waves with phase speed less than the speed of light, which corresponds to a refractive index greater than unity. This is seen by writing the Cerenkov condition (20.6) in the form

$$1 - n\boldsymbol{\kappa} \cdot \boldsymbol{\beta} = 1 - n\beta\cos\chi = 0, \tag{20.9}$$

with $\boldsymbol{\beta} := \mathbf{v}/c$, and where χ is the angle between $\mathbf{k}$ and $\mathbf{v}$. It follows from $\beta < 1$ and $\cos\chi \leq 1$ that (20.9) can be satisfied only for $n \geq 1/\beta > 1$. Put another way, the emission occurs only for particles with $\beta \geq 1/n$. Emission at the threshold $\beta = 1/n$ is at $\chi = 0$ and the angle χ increases

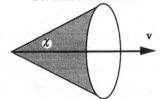

Fig. 20.1 The Cerenkov cone half angle χ.

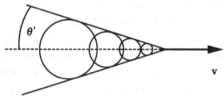

Fig. 20.2 Cerenkov emission forms a bow shock with half-angle $\theta' = \pi/2 - \chi$.

with the speed of particles to a maximum value $\chi_{\max} = \arccos(1/n)$ for highly relativistic particles. Thus Cerenkov emission is confined to the surface of a cone, with the cone half-angle χ determined by the speed of the electron. This emission pattern is illustrated in Figure 20.1.

The Cerenkov emission pattern may also be viewed as analogous to the bow wave of a ship or a supersonic aircraft, as illustrated in Figure 20.2. The half-angle of the bow wave is $\theta' = \pi/2 - \chi = \arcsin(1/\beta n)$.

There is quite a different interpretation of the Cerenkov condition from a quantum mechanical viewpoint. In this case the condition is interpreted as a consequence of conservation of energy and momentum on a microscopic scale. The semiclassical interpretation of emission is as a process in which a particle emits a quantum of energy. This quantum has energy $\hbar\omega$ and momentum $\hbar\mathbf{k}$. Let the initial energy and momentum of the particle be ε, $\mathbf{p}$ and let the corresponding quantities after emission of the wave quantum be ε', $\mathbf{p}'$. Conservation of energy and momentum then require

$$\varepsilon' = \varepsilon - \hbar\omega, \quad \mathbf{p}' = \mathbf{p} - \hbar\mathbf{k}. \tag{20.10}$$

The relativistically correct formula for the energy is

$$\varepsilon = \left(m^2c^4 + \mathbf{p}^2c^2\right)^{1/2}, \quad \varepsilon' = \left(m^2c^4 + \mathbf{p}'^2c^2\right)^{1/2}. \tag{20.11}$$

On inserting $\mathbf{p}' = \mathbf{p} - \hbar\mathbf{k}$ into the expression for ε' in (20.11) one expands it in powers of $\hbar$ to obtain the classical counterpart. This gives

$$\omega = \mathbf{k} \cdot \mathbf{v} - \frac{\hbar}{2\varepsilon}\left[k^2c^2 - (\mathbf{k} \cdot \mathbf{v})^2\right] + \cdots. \tag{20.12}$$

The lowest order terms reproduce the Cerenkov condition (20.6). Thus the Cerenkov condition is interpreted as the requirement that energy conservation be compatible with momentum conservation on a microscopic level. It is only when both energy and momentum are conserved that the particle can emit a wave quantum. The next order term in $\hbar$ in (20.12) is interpreted in terms of the quantum recoil of the emitting particle.

The quantum recoil is needed below (§20.6) to derive a purely classical equation that describes the effect of the back reaction of the emission on the emitting particle. Consequently, it is relevant to include it in a generalized expression for the probability. The first order quantum recoil is included in the probability (20.5) simply by replacing the δ-function appropriately. Specifically, on writing

$$w_M(\mathbf{p}, \mathbf{k}) = \frac{q^2 R_M(\mathbf{k})}{\varepsilon_0 \omega_M(\mathbf{k})} |\mathbf{e}_M^*(\mathbf{k}) \cdot \mathbf{v}|^2 \, 2\pi\delta(\varepsilon' - \varepsilon + \hbar\omega_M(\mathbf{k})), \qquad (20.13)$$

and inserting the expansion used in deriving (20.12), the result reproduces the result of a fully quantum mechanical calculation to first order in $\hbar$. (To second order in $\hbar$ the quantum mechanical result depends on the spin of the particle and is not reproduced by extending the expansion of (20.13) to second order in $\hbar$.)

20.3 The Power Radiated in Transverse Waves

The power radiated in Cerenkov emission follows from (20.3) by multiplying by $d^3k/(2\pi)^3$ and integrating over $\mathbf{k}$-space. The power radiated is evaluated explicitly here for transverse waves in an isotropic dielectric.

It is convenient to change the variables of integration to ω and the solid angle about the direction of $\mathbf{k}$, as is done in (16.16). The ratio of electric to total energy is given explicitly by (15.23). Thus one obtains

$$P = \frac{q^2}{2(2\pi)^2 \varepsilon_0 c} \int_0^{2\pi} d\phi \int_{-1}^1 d\cos\chi \int_0^\infty d\omega\, \omega^2 n(\omega)$$
$$\times |\mathbf{e} \cdot \boldsymbol{\beta}|^2 \, \delta(\omega - \omega n(\omega)\beta\cos\chi), \qquad (20.14)$$

where the integral over solid angle is written in polar coordinates with the polar axis along $\mathbf{v} = \boldsymbol{\beta} c$.

The polarization of Cerenkov emission is determined by replacing (20.14) by a polarization tensor, as in (16.11). One finds that the radiation is completely polarized (the determinant of the polarization tensor is zero). The polarization is linear along the direction of the projection of $\mathbf{v}$ on the plane orthogonal to $\boldsymbol{\kappa}$. The total power radiated is found by taking the sum over the two states of polarization, which reduces to making the replacement

$$|\mathbf{e} \cdot \boldsymbol{\beta}|^2 \to |\boldsymbol{\kappa} \times \boldsymbol{\beta}|^2 = \beta^2 \sin^2\chi. \qquad (20.15)$$

On making this replacement in (20.14), the ϕ-integral is trivial and the $\cos\chi$-integral is performed over the δ-function. Once the $\cos\chi$-integral is performed, one replaces $\cos\chi$ by $1/n(\omega)\beta$ in the integrand. Furthermore, the remaining integral over ω is restricted to the range of ω for which the condition $\cos\chi < 1$ is satisfied, that is, the range of ω for which one has $n(\omega)\beta > 1$. Thus one obtains

$$P = \frac{q^2\beta}{4\pi\varepsilon_0 c} \int\limits_{n(\omega)\beta > 1} d\omega\,\omega \left(1 - \frac{1}{n^2(\omega)\beta^2}\right). \qquad (20.16)$$

The formula (20.16) was first derived by Tamm and Frank in 1937, just a few years after Cerenkov discovered the phenomenon experimentally.

In familiar dielectrics, such as air, glass and water, $n(\omega)$ is a slowly varying function of ω at optical frequencies. (The refractive index is $n(\omega) = [K(\omega)]^{1/2}$ with $K(\omega)$ given by the Cauchy formula (9.18) and so is nearly constant with a small dispersive term that increases proportional to ω^2.) With the refractive index assumed approximately independent of ω, (20.16) implies a power per unit frequency that increases with frequency. As a consequence, the power emitted in blue light is greater than the power emitted in red light. This accounts for the characteristic blue color of Cerenkov emission, evident in some photographs of nuclear reactors whose cores are surrounded by water.

20.4 The Power Radiated in Langmuir Waves

An important application of Cerenkov emission and Landau damping is to longitudinal waves in a plasma. The wave–particle interaction at the Cerenkov resonance is the dominant energy transfer process between particles and longitudinal waves in an isotropic collisionless plasma.

Cerenkov emission of Langmuir waves in an isotropic thermal plasma is treated using (20.3). One modifies (20.3) as follows: (i) replace the label M by L, (ii) note that the polarization is longitudinal $\mathbf{e}_M(\mathbf{k}) \to \boldsymbol{\kappa}$, (iii) insert the expression $(E15.3)$ for the ratio of electric to total energy, and (iv) insert the dispersion relation (13.1) for Langmuir waves. Prior to the final step one has

$$P_{\mathrm{L}}(\mathbf{k}) = \frac{q^2\omega_{\mathrm{p}}^2}{2\varepsilon_0\omega_{\mathrm{L}}^2(\mathbf{k})} |\boldsymbol{\kappa}\cdot\mathbf{v}|^2\, 2\pi\delta(\omega_{\mathrm{L}}(\mathbf{k}) - \mathbf{k}\cdot\mathbf{v}). \qquad (20.17)$$

The power radiated in Langmuir waves is given by integrating (20.17) over $\mathbf{k}$-space. Using spherical polar coordinates in $\mathbf{k}$-space with the polar axis along $\mathbf{v}$, the ϕ-integral is trivial and the $\cos\chi$-integral is performed over the δ-function, as in the derivation of (20.16). The condition $\cos\chi <$ 1 implies $\omega_{\mathrm{L}}(\mathbf{k}) < kv$, or $v > v_\phi$ where $v_\phi = \omega_{\mathrm{L}}(\mathbf{k})/k$ is the phase speed of the waves. Thus the power radiated reduces to

$$P = \frac{q^2\omega_{\mathrm{p}}^2}{4\pi\varepsilon_0 v} \int\limits_{v > v_\phi} \frac{\mathrm{d}k}{k}. \tag{20.18}$$

The remaining integral gives a logarithm which is cut off at its lower limit when $k = \omega_{\mathrm{L}}(k)/v$, and which needs to be cut off at some maximum value k_{max} of k. There is a natural maximum due to the fact that Langmuir waves become heavily damped for sufficiently large k and cease to exist for $k\lambda_{\mathrm{D}} \gtrsim 1$. (Specifically Langmuir waves exist only where $\phi(z)$ in Figure 10.1 exceeds unity.) For particles with $v \gg V_{\mathrm{e}}$, the value of the logarithm is insensitive to k_{max}, and it is convenient to choose $k_{\mathrm{max}}\lambda_{\mathrm{D}} = 1$. Then (20.18) gives

$$P = \frac{q^2\omega_{\mathrm{p}}^2}{4\pi\varepsilon_0 v} \ln \frac{v}{V_{\mathrm{e}}}. \tag{20.19}$$

The power (20.19) is that emitted due to spontaneous emission of Langmuir waves. The power emitted is balanced by an energy loss by the particles. Hence the emitting particle loses energy due to Cerenkov emission at the rate $\dot{\varepsilon} = -P$ with P given by (20.19). An interesting generalization of (20.19) is to the total energy loss by an electron. As noted above, Cerenkov emission may be interpreted in terms of the spatial Fourier transform of the Coulomb field of the particle having components that satisfy the Cerenkov condition. These $\mathbf{k}$ values correspond to $k < 1/\lambda_{\mathrm{D}}$. On the other hand, the $\mathbf{k}$ values that correspond to $k > 1/\lambda_{\mathrm{D}}$ are associated with the collective interaction with other particles in the medium. If one simply extends the k-integral in (20.18) to $k > 1/\lambda_{\mathrm{D}}$ one includes the effect of the energy loss of the particle due to collective interactions with the ambient particles. These losses are attributed to "collisions". The collisional loss of energy is given by

$$\frac{\mathrm{d}\varepsilon}{\mathrm{d}t} = -\frac{q^2\omega_{\mathrm{p}}^2}{4\pi\varepsilon_0 v} \ln \Lambda_{\mathrm{c}}, \tag{20.20}$$

where $\ln \Lambda_c$ is the Coulomb logarithm. The Coulomb logarithm may be written in terms of the ratio of two wavenumbers at which the integral is cut off, and the actual value of $\ln \Lambda_c$ is insensitive to the detailed choices of these cutoffs. The conventional choice is to cut off the integral at $k = 1/\lambda_{\mathrm{D}}$ and at $k = 1/b_0$ where b_0 is the impact parameter at which the deflection of an electron by an ion is through an angle $\pi/2$. The Coulomb logarithm is a special case of the Gaunt factor, which is discussed in detail in §21.4. The energy loss may also be extended to apply to ionization losses by a fast particle in neutral matter. In fact essentially the same formula (20.20) applies to "ionization losses" by a fast particle in neutral matter. The underlying reason for this is that for a fast particle whose energy is much greater than the binding energy of an electron in an atom, bound electrons look like free electrons. The fast particle loses energy by scattering the bound electrons which become free electrons leaving the ambient atom ionized. As a consequence, there is a close analogy between the collisional losses of a particle in a fully ionized plasma and the ionization losses of a fast particle in neutral matter.

20.5 Cerenkov Absorption (Landau Damping)

So far we have considered only spontaneous emission by a single particle. Emission and absorption by a distribution of particles are described by the emission and absorption coefficients in the transfer equation (16.32). The occupation number n_q in (16.32) is replaced by V times the classical distribution function $f(\mathbf{p})$, and the sum over q is replaced by the integral over momentum space. As in (10.35) the normalization is such that $\int \mathrm{d}^3\mathbf{p}\, f(\mathbf{p})$ is equal to the number density of the particles. Then (16.32) translates into

$$\frac{\mathrm{d}N_M(\mathbf{k})}{\mathrm{d}t} = \int \mathrm{d}^3\mathbf{p}\, w_M(\mathbf{p},\mathbf{k})\big\{\big[1 + N_M(\mathbf{k})\big]\,f(\mathbf{p}) - N_M(\mathbf{k})\,f(\mathbf{p} - \hbar\mathbf{k})\big\}.$$
(20.21)

In the classical limit one makes the expansion

$$f(\mathbf{p} - \hbar\mathbf{k}) = \left[1 - \hbar\mathbf{k}\cdot\frac{\partial}{\partial\mathbf{p}} + \frac{1}{2}\left(\hbar\mathbf{k}\cdot\frac{\partial}{\partial\mathbf{p}}\right)^2 + \cdots\right]f(\mathbf{p}),$$
(20.22)

in (20.21) and retains only those terms that do not vanish in the limit $\hbar \to 0$. On multiplying (20.21) by $\hbar\omega_M(\mathbf{k})$ and using (15.35) one rewrites the resulting transfer equation in terms of $W_M(\mathbf{k})$ rather than $N_M(\mathbf{k})$. Thus one obtains, cf. (16.33),

$$\frac{\mathrm{d}W_M(\mathbf{k})}{\mathrm{d}t} = S_M(\mathbf{k}) - \gamma_M(\mathbf{k})W_M(\mathbf{k}),$$
(20.23)

with, for the emission coefficient,

$$S_M(\mathbf{k}) = \int \mathrm{d}^3\mathbf{p}\, P_M(\mathbf{k})\,f(\mathbf{p}) = \int \mathrm{d}^3\mathbf{p}\,\hbar\omega_M(\mathbf{k})\,w_M(\mathbf{p},\mathbf{k})\,f(\mathbf{p}),$$
(20.24)

and, for the absorption coefficient,

$$\gamma_M(\mathbf{k}) = -\int \mathrm{d}^3\mathbf{p}\, w_M(\mathbf{p},\mathbf{k})\,\hbar\mathbf{k}\cdot\frac{\partial f(\mathbf{p})}{\partial\mathbf{p}}.$$
(20.25)

The emission coefficient (20.24) gives the power radiated in waves in the mode M in unit range of $d^3\mathbf{k}/(2\pi)^3$ due to spontaneous emission. On inserting the probability (20.5) for Cerenkov emission, (20.24) gives

$$S_M(\mathbf{k}) = \int d^3\mathbf{p} \, \frac{q^2 R_M(\mathbf{k})}{\varepsilon_0} \, |\mathbf{e}_M^*(\mathbf{k}) \cdot \mathbf{v}|^2 \, 2\pi\delta\big(\omega_M(\mathbf{k}) - \mathbf{k}\cdot\mathbf{v}\big) f(\mathbf{p}). \quad (20.26)$$

As discussed in §17.5 the absorption coefficient $\gamma_M(\mathbf{k})$ is derived in two different methods which should be equivalent. One method is to use the expression (11.22) or (15.14) which gives $\gamma_M(\mathbf{k})$ in terms of the antihermitian part of the dielectric tensor. The other method is to use the semiclassical approach, as in the derivation of (20.25). Let us show that the two methods give the same result for Cerenkov emission. The relevant dielectric tensor is that calculated using kinetic theory for an unmagnetized plasma, as given by (10.38). The antihermitian part follows by imposing the causal condition on (10.38) and retaining only the semiresidue at the pole at $\omega - \mathbf{k}\cdot\mathbf{v} = 0$. Proceeding in this way the expression obtained is

$$K_{ij}^{\mathrm{A}}(\omega, \mathbf{k}) = \sum_\alpha -\frac{i\pi q_\alpha^2}{\varepsilon_0 \omega^2} \int d^3\mathbf{p} \, v_i v_j \, \delta(\omega - \mathbf{k}\cdot\mathbf{v}) \, \mathbf{k}\cdot\frac{\partial f_\alpha(\mathbf{p})}{\partial \mathbf{p}}, \quad (20.27)$$

where the sum is over all species of particle. The absorption coefficient for waves in an arbitrary mode M then follows from (15.14):

$$\gamma_M(\mathbf{k}) = \sum_\alpha -\frac{2\pi q_\alpha^2 R_M(\mathbf{k})}{\varepsilon_0 \omega_M(\mathbf{k})} \int d^3\mathbf{p} \, |\mathbf{e}_M(\mathbf{k})\cdot\mathbf{v}|^2 \, \delta(\omega_M(\mathbf{k}) - \mathbf{k}\cdot\mathbf{v}) \, \mathbf{k}\cdot\frac{\partial f_\alpha(\mathbf{p})}{\partial \mathbf{p}}.$$
$$(20.28)$$

The classical result (20.28) is to be compared with the semiclassical expression (20.25) with the probability given by (20.5). The contribution of only one species of particle is included in (20.5), and after summing over all species of particle, (20.25) with (20.5) reproduces (20.28). As expected, the two methods are equivalent.

An important example of Cerenkov absorption is Landau damping of Langmuir waves. The absorption coefficient is given by assuming a thermal distribution of particles in (20.28), and inserting the properties appropriate for Langmuir waves. The resulting absorption coefficient (for electrons) is given by (13.6). When one evaluates the absorption coefficient it is found that the damping at a given phase velocity ω/k is dominated by particles with $v \approx \omega/k$. This corresponds to absorption occurring predominantly at small angles $\chi = \arccos(\omega/kv)$. In contrast, spontaneous emission does not favor small angles particularly, and emission at a given ω/k has a significant contribution from all particles with $v > \omega/k$.

The absorption coefficient (20.25) or (20.28) for Cerenkov emission can

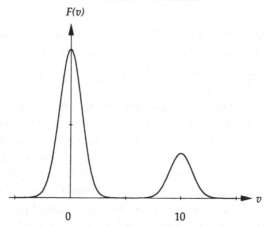

Fig. 20.3 A bump-in-tail distribution $F(v)$.
Growth is restricted to the range $dF(v)/dv > 0$.

be negative, implying that the waves can grow. Negative absorption corresponds to maser action, and the growth of the waves is attributed to an inverted population in which the number of particles in the upper state $\mathbf{p}$ exceeds the number in the lower state $\mathbf{p} - \hbar\mathbf{k}$. According to (20.28), a necessary condition for negative absorption is that $\mathbf{k} \cdot \partial f(\mathbf{p})/\partial \mathbf{p}$ be positive, which requires that $f(\mathbf{p})$ be an increasing function of the component of $\mathbf{p}$ along the direction of $\mathbf{k}$. The distribution of particles must also be anisotropic to lead to negative Cerenkov absorption; specifically, it is shown that an isotropic distribution with $\partial f(p)/\partial p > 0$ over any finite range of p does not cause negative absorption, cf. Exercise 20.8. A distribution that does cause negative absorption is said to be *unstable* to the growth of the waves, and the growth of the waves is referred to as an *instability*.

An example of a non-relativistic distribution that is unstable is the *bump-in-tail* distribution illustrated in Figure 20.3. Wave growth is restricted to the range of phase velocities where the one-dimensional velocity distribution $F(v)$ is an increasing function of v. The growth rate is identified here as being equal to minus the absorption coefficient. The growth rate for Langmuir waves at phase velocity v_ϕ in the one-dimensional case is given by

$$\frac{dW(v_\phi)}{dt} = -\gamma_{\mathrm{L}}(v_\phi)W(v_\phi), \quad \gamma_{\mathrm{L}}(v_\phi) = -\frac{\pi\omega_{\mathrm{p}}}{n_{\mathrm{e}}}v_\phi^2\frac{dF(v_\phi)}{dv_\phi}, \quad (20.29)$$

where the non-relativistic approximation is made with the normalization changed to

$$\int_{-\infty}^{\infty} dv \, F(v) = n_1, \qquad (20.30)$$

where n_1 is the number density of the fast particles.

Growth of Langmuir waves is one example of a wide class of plasma microinstabilities. Another important example is the growth of low frequency waves in current-driven instabilities, where the instability is due to the streaming motion of the electrons relative to the ions, cf. Exercise 20.9. As a result of such instabilities the plasma is said to become *microturbulent*, where *microturbulence* refers to the resulting spectrum of waves in the plasma. Microturbulent plasmas can give rise to non-thermal radiation, greatly enhanced scattering of fast particles and anomalous conductivity.

20.6 Quasi-linear Relaxation

As discussed in §17.5, the semiclassical approach enables one to describe the back reaction of the distribution of particles to the emission and absorption of radiation. The back reaction to spontaneous emission is interpreted as a radiation reaction, as described in §19.5 for emission *in vacuo*. The back reaction to the induced processes is often referred to as quasi-linear relaxation.

The kinetic equation (16.35) applied to classical particles emitting and absorbing through the Cerenkov process is evaluated by making the identification of the quantum numbers $q \to \mathbf{p}$, $q' \to \mathbf{p} - \hbar\mathbf{k}$, $q'' \to \mathbf{p} + \hbar\mathbf{k}$, $n_q \to V f(\mathbf{p})$ and $\sum_q \to \int d^3\mathbf{p}$. Then on making the expansion (20.22) for both $f(\mathbf{p} \pm \hbar\mathbf{k})$, retaining terms up to order $\hbar^2$ in this case, one obtains the kinetic equation

$$\frac{df(\mathbf{p})}{dt} = \int \frac{d^3\mathbf{k}}{(2\pi)^3} \, \hbar\mathbf{k} \cdot \frac{\partial}{\partial \mathbf{p}} \left\{ w_M(\mathbf{p}, \mathbf{k}) \left[f(\mathbf{p}) + N_M(\mathbf{k}) \, \hbar\mathbf{k} \cdot \frac{\partial f(\mathbf{p})}{\partial \mathbf{p}} \right] \right\}.$$
$$(20.31)$$

One may rewrite (20.31) in the form

$$\frac{df(\mathbf{p})}{dt} = \frac{\partial}{\partial p_i} \left[A_i(\mathbf{p}) f(\mathbf{p}) + D_{ij}(\mathbf{p}) \frac{\partial f(\mathbf{p})}{\partial p_i} \right], \qquad (20.32)$$

with

$$\begin{pmatrix} A_i(\mathbf{p}) \\ D_{ij}(\mathbf{p}) \end{pmatrix} = \int \frac{d^3\mathbf{k}}{(2\pi)^3} \, w_M(\mathbf{p}, \mathbf{k}) \begin{pmatrix} \hbar k_i \\ \hbar^2 k_i k_j N_M(\mathbf{k}) \end{pmatrix}. \qquad (20.33)$$

The term independent of $N_M(\mathbf{k})$ in (20.31), which is the term involving $A_i(\mathbf{p})$ in (20.32), describes the back reaction to spontaneous emission. This term is attributed to a radiation reaction force. The radiation reaction force is identified as follows. First consider the rate of change of the mean momentum (the mean is denoted by angular brackets), which is identified as the mean radiation reaction force acting on the particles. Retaining only the term that describes spontaneous emission, one has

$$\frac{d\langle \mathbf{p} \rangle}{dt} = \frac{1}{\int d^3\mathbf{p}\, f(\mathbf{p})} \int d^3\mathbf{p}\, \mathbf{p}\, \frac{df(\mathbf{p})}{dt}, \qquad (20.34)$$

On inserting (20.32) into (20.34), and partially integrating over $\mathbf{p}$ one defines a radiation reaction force $\langle d\mathbf{p}/dt \rangle$ by writing

$$\int d^3\mathbf{p}\, \mathbf{p}\, \frac{df(\mathbf{p})}{dt} = \int d^3\mathbf{p}\, \left\langle \frac{d\mathbf{p}}{dt} \right\rangle f(\mathbf{p}). \qquad (20.35)$$

The term that describes spontaneous emission gives $\langle d\mathbf{p}/dt \rangle = -\mathbf{A}(\mathbf{p})$, that is,

$$\left\langle \frac{d\mathbf{p}}{dt} \right\rangle = -\int \frac{d^3\mathbf{k}}{(2\pi)^3}\, \hbar\mathbf{k}\, w_M(\mathbf{p},\mathbf{k}). \qquad (20.36)$$

The radiation reaction force (20.36) has an obvious interpretation. Each emitted wave quantum carries off momentum $\hbar\mathbf{k}$. Hence, the right hand side of (20.36) is identified as minus the rate that momentum is carried off by the emitted waves. Momentum conservation requires that this momentum be provided by the emitting particle. Hence, the reaction force (20.36) on the particle follows directly from momentum conservation.

It would be of interest to compare the reaction force derived using quasi-linear theory and the conventional radiation reaction force (19.34). However, this is not possible in the present case. This is because (20.36) applies only to Cerenkov emission, which involves emission in a medium by a charge in constant rectilinear motion, and (19.34) applies only to emission by an accelerated charge *in vacuo*. However, it is possible to compare the two reaction forces in the case of gyromagnetic emission, as discussed in §22.5.

The reaction of the distribution of particles to the induced processes is described by the term in (20.32) that involves $D_{ij}(\mathbf{p})$. This term describes a diffusion of particles in momentum space. This is in contrast to the case of spontaneous emission, when the reaction is described in terms of the reaction force (20.36) so that the momentum of a particle changes in a systematic way as it emits. The effect of the induced processes on a group of particles in a small region of momentum space is to cause them to diffuse in such a way that the volume of momentum space which they fill expands with time. This is illustrated schematically in

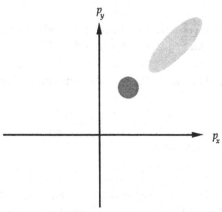

Fig. 20.4 An idealized example of diffusion in momentum space, with diffusion stronger along the radial than the polar direction.

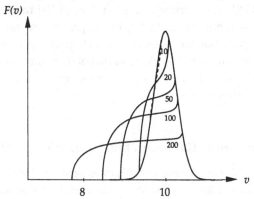

Fig. 20.5 Quasi-linear evolution of the beam particles in Figure 20.3; labels refer to time in units of the initial growth time.

Figure 20.4. As indicated in Figure 20.4, the mean momentum of this idealized group of particles changes in a systematic way. This systematic change is described by a radiation reaction force $\langle d\mathbf{p}/dt \rangle$, but this describes only one aspect of the effect of the diffusion. The energy flow can be either from the waves to the particles or from the particles to the waves. The rate at which the reaction force does work is

$$\langle d\varepsilon/dt \rangle = \mathbf{v} \cdot \langle d\mathbf{p}/dt \rangle , \qquad (20.37)$$

which is positive when energy flows from the waves to the particles at a given $\mathbf{p}$.

One specific application is the back reaction of the particles to the bump-in-tail instability. Quite generally, the back reaction tends to reduce the gradient in momentum space that is driving the instability. In particular, a positive gradient is reduced towards the *plateau distribution* $\partial f(\mathbf{p})/\partial \mathbf{p} = 0$. An example of this effect is developed by reducing the diffusive term in (20.32) to a one-dimensional form using the same notation as in (20.29) and (20.30). As outlined in Exercise 20.10, one finds

$$\frac{\mathrm{d}F(v)}{\mathrm{d}t} = \frac{\partial}{\partial v} D(v) \frac{\partial F(v)}{\partial v}, \tag{20.38}$$

with the diffusion coefficient given by

$$D(v) = \frac{\pi \omega_p}{n_e m_e} v W(v), \tag{20.39}$$

where $\int \mathrm{d}v\, W(v)$ is the energy density in the waves. The evolution of the coupled system of Langmuir waves and electrons in an idealized one-dimensional beam instability may be described by a numerical solution of (20.29) and (20.38). (For formal reasons it is essential to include the effect of spontaneous emission in (20.29) but in practice rounding errors in the numerical calculation simulate the important effects of spontaneous emission.) The evolution of the electron distribution involves formation of a plateau and its gradual extension to lower velocities, as illustrated schematically in Figure 20.5.

20.7 Appearance Radiation and Transition Radiation

When a charge suddenly appears or disappears it produces appearance radiation or disappearance radiation, respectively. Transition radiation is associated with the passage of a charge across the interface between two different media.

Appearance Radiation

Charge is always conserved, but nevertheless a charge can effectively appear or disappear, as in the sudden appearance of an electron when a neutron β decays. Consider a charge that appears at $t = 0$ at some point $\mathbf{x} = \mathbf{x}_0$ and thereafter propagates at constant velocity $\mathbf{v}$. The Fourier transform of the current due to a charge in constant rectilinear motion is given by (20.2). This current may be separated into its contributions from the motion at $t < 0$ and the motion at $t > 0$ using the step function $H(t)$, which is equal to zero for $t < 0$ and to unity for $t > 0$. Let $\mathbf{J}^+(\omega, \mathbf{k})$ denote the current for $t > 0$. Then the convolution theorem and the Fourier transform (4.26) with (4.27) of $H(t)$ applied to

$\mathbf{J}^+(t, \mathbf{x}) = H(t)\mathbf{J}(t, \mathbf{x})$ give

$$\mathbf{J}^+(\omega, \mathbf{k}) = \frac{i}{2\pi} \int_{-\infty}^{\infty} d\omega' \, \frac{\mathbf{J}(\omega', \mathbf{k})}{\omega - \omega' + i0}. \tag{20.40}$$

On inserting the current (20.2) for a particle in constant rectilinear motion, (20.40) gives

$$\mathbf{J}^+(\omega, \mathbf{k}) = \frac{iq\mathbf{v}e^{-i\mathbf{k}\cdot\mathbf{x}_0}}{\omega - \mathbf{k}\cdot\mathbf{v} + i0}, \tag{20.41}$$

which current is identified as the source term for appearance radiation. The corresponding current for disappearance radiation is the complex conjugate of (20.41).

Here we concentrate on emission *in vacuo*, when one has $\omega - \mathbf{k}\cdot\mathbf{v} > 0$. On applying the Plemelj formula (4.32) to (20.41), only the principal value part contributes. On inserting the principal part of (20.41) into the emission formula (16.10), one obtains

$$U_M(\mathbf{k}) = \frac{q^2 R_M(\mathbf{k})}{\varepsilon_0} \frac{|\mathbf{e}_M^*(\mathbf{k}) \cdot \mathbf{v}|^2}{(\omega_M(\mathbf{k}) - \mathbf{k}\cdot\mathbf{v})^2}, \tag{20.42}$$

which is interpreted as the energy radiated into waves in the mode M in the range $d^3\mathbf{k}/(2\pi)^3$ due to appearance or disappearance radiation. For transverse waves *in vacuo* one sets $R(\mathbf{k}) = \frac{1}{2}$, $\omega = kc$, $\mathbf{k} = k\boldsymbol{\kappa}$, and sums over the two transverse states of polarization using (16.13). On integrating over $\mathbf{k}$-space, the energy radiated becomes

$$U_{\text{rad}} = \frac{q^2}{2\varepsilon_0} \int \frac{d^3\mathbf{k}}{(2\pi)^3} \frac{|\boldsymbol{\kappa} \times \mathbf{v}|^2}{\omega^2(1 - \boldsymbol{\kappa}\cdot\mathbf{v}/c)^2}. \tag{20.43}$$

The integral is evaluated in spherical polar coordinates by writing $\mathbf{k}\cdot\mathbf{v} = kv\cos\chi$. On introducing the energy radiated per unit frequency $U_{\text{rad}}(\omega)$,

$$E_{\text{rad}} = \int_0^{\infty} d\omega \, U_{\text{rad}}(\omega), \tag{20.44}$$

one finds

$$\begin{aligned} U_{\text{rad}}(\omega) &= \frac{q^2}{2(2\pi c)^3 \varepsilon_0} \int_0^{2\pi} d\phi \int_{-1}^{1} d\cos\chi \, \frac{\sin^2\chi}{(1 - \beta\cos\chi)^2} \\ &= \frac{q^2}{4\pi^2 c\varepsilon_0} \left[\frac{1}{\beta} \ln\left(\frac{1+\beta}{1-\beta}\right) - 2 \right], \end{aligned} \tag{20.45}$$

with $\beta = v/c$. For non-relativistic particles (20.45) simplifies to

$$U_{\text{rad}}(\omega) = \frac{q^2}{4\pi\varepsilon_0} \frac{2v^2}{3\pi c^3}. \tag{20.46}$$

In β decay the emitted electrons are usually mildly to highly relativistic, and the general form (20.45) is then appropriate. One minor simplification is to write the logarithmic factor as $2\ln[\gamma(1 + \beta)]$, where $\gamma = (1 - \beta^2)^{-1/2}$ is the Lorentz factor of the particle.

In estimating the energy in appearance radiation one needs to cut the ω-integral off at some maximum frequency. From a quantum mechanical viewpoint it is obvious that no emitted photon can carry off more energy than the particle has initially. Thus one requires $\hbar\omega < (\gamma - 1)mc^2$. On cutting the ω-integral off at $\omega = (\gamma - 1)mc^2/\hbar$, the energy radiated reduces to

$$E_{rad} = \frac{q^2 mc(\gamma - 1)}{2\pi^2 \varepsilon_0 \hbar \beta} \left\{ \ln[\gamma(1 + \beta)] - \beta \right\}. \qquad (20.47)$$

A treatment based on QED is needed to improve on the estimate (20.47).

An interesting feature of β decay is that the emission due to the magnetic moment of the electron can be more important than that due to its charge. The current due to a magnetic moment $\boldsymbol{\mu}$ in constant rectilinear motion is written down in (E20.1). The emission due to the sudden appearance of a magnetic moment is treated by repeating the analysis of the preceding paragraph. In place of (27.4) one finds

$$E_{rad} = \frac{1}{2\varepsilon_0 c^2} \int \frac{d^3\mathbf{k}}{(2\pi)^3} \frac{|\boldsymbol{\kappa} \times \boldsymbol{\mu}|^2}{(1 - \boldsymbol{\kappa} \cdot \mathbf{v}/c)^2}. \qquad (20.48)$$

The integral over angles is somewhat more lengthy in this case. Some of the details are outlined in Exercise 20.6. The magnetic moment of the electron is $|\boldsymbol{\mu}| = e\hbar/2mc$. One finds that the ratio of the energy radiated due to the magnetic moment and due to the charge is proportional to $(\hbar\omega/mc^2)^2$. It follows that emission due to the magnetic moment dominates at high frequency for relativistic electrons.

Transition Radiation

Transition radiation may be interpreted in terms of appearance radiation and disappearance radiation, associated with the sudden changes in the electromagnetic field as the charge crosses the interface. The simplest example of transition radiation is when a charged particle is incident on a perfectly conducting metal surface. The electromagnetic field due to the moving charge vanishes inside the metal. The boundary conditions on the electromagnetic fields require that the tangential component of the electric field $\mathbf{E}$ and the normal component of the magnetic induction $\mathbf{B}$ be continuous across the surface, and hence they must vanish at the surface. These boundary conditions enable one to determine the electromagnetic field due to the charge in the half-space outside the metal surface. Transition radiation is attributed to the sudden disappearance of this electromagnetic field when the charge crosses the metal surface.

If the metal is a foil then there is also a burst of appearance radiation as the charge emerges into the half-space on the other side of the foil.

Consider an electron normally incident on a metal surface. The electromagnetic field due to the electron in the half-space in which it is located may be regarded as being composed of two fields. One is the electromagnetic field generated directly by the electron. The other is the field reflected in the metal surface, and this may be regarded as being generated by an image charge on the other side of the surface. The *method of images* requires that the image charge be chosen so that the boundary conditions are satisfied. Consider the electric field due to an image charge which is of opposite sign to that of the electron and which is located inside the metal at an equal distance from the surface along the same normal line as the electron. At the metal surface, the tangential component of the electric field due to this image charge is equal and opposite to that of the electron. The image charge is moving towards the metal surface like a mirror image of the electron, and hence the electric currents associated with the motion of the image and of the electron are equal, and equal to $-e\mathbf{v}$. The sum of the magnetic fields due to these two currents is non-zero at the metal surface, but it has zero normal component.

Thus, using the method of images, the electromagnetic field in the half-space in which the electron is located is regarded as the sum of two fields, these being that due to the electron itself, and that due to the image charge. The electron and its image are propagating directly towards each other and disappear simultaneously when they meet at the surface of the metal. Transition radiation is the disappearance radiation when the electron and its image charge disappear from the half-space.

If the current due to a charge q and its image charge persisted for all time, it would have the value

$$\mathbf{J}(\omega,\mathbf{k}) = 2\pi q \mathbf{v} e^{-i\mathbf{k}\cdot\mathbf{x}_0}\left[\delta(\omega - \mathbf{k}\cdot\mathbf{v}) + \delta(\omega + \mathbf{k}\cdot\mathbf{v})\right], \qquad (20.49)$$

where $\mathbf{x}_0$ is the point on the metal surfaces at which the charge and its image meet at $t = 0$. The actual current, assuming that the charge and its image disappear at $t = 0$, follows by replacing the δ-functions as in (20.41). This current is

$$\mathbf{J}^-(\omega,\mathbf{k}) = -iq\mathbf{v} e^{-i\mathbf{k}\cdot\mathbf{x}_0}\left(\frac{1}{\omega - \mathbf{k}\cdot\mathbf{v} - i0} + \frac{1}{\omega + \mathbf{k}\cdot\mathbf{v} - i0}\right). \qquad (20.50)$$

The principal value part of the current (20.50) is the source term for the transition radiation.

For a non-relativistic electron one may make the dipole approximation to (20.50). This corresponds to $\mathbf{d}(\omega) = 2q\mathbf{v}/\omega$. The energy radiated per unit frequency then follows from (16.18), except that this energy is to be multiplied by a factor of $\frac{1}{2}$ to take account of the emission of energy in only one half-space. The energy per unit frequency in disappearance radiation for a non-relativistic charge incident normally on the surface of a perfectly conducting metal is

$$U_{\mathrm{rad}}(\omega) = \frac{\omega^2 |\mathbf{d}(\omega)|^2}{12\pi^2 \varepsilon_0 c^3} = \frac{q^2 |\mathbf{v}|^2}{3\pi^2 \varepsilon_0 c^3}. \tag{20.51}$$

Exercise Set 20

20.1 Estimate the minimum energy for an electron to emit Cerenkov radiation in air, assuming the refractive index of air to be $n = 1 + 3 \times 10^{-4}$. Express your answer in electron volts ($1\,\text{eV} = 1.60 \times 10^{-19}\,\text{J}$).

20.2 Show that the blue color of Cerenkov emission is predicted by (20.16).

(a) Estimate the ratio of the power emitted at the blue end to that at the red end of the visible spectrum by an electron produced in β decay (energy several MeV) passing through water ($n = 1.33$). (These assumptions imply $\beta \approx 1$, $n(\omega)\beta \gg 1$.) Specifically, separate the visible range into the blue end covering the range $4 \times 10^{-7}\,\text{m} < \lambda < 5.5 \times 10^{-7}\,\text{m}$ and the red end covering the range $5.5 \times 10^{-7}\,\text{m} < \lambda < 7 \times 10^{-7}\,\text{m}$, where λ is the free space wavelength.

(b) Water is weakly dispersive and the dispersion may be described by a term of the Cauchy form, cf. (9.18). Does inclusion of dispersion enhance or diminish the blue excess for $n(\omega)\beta \gg 1$?

(c) Relax the assumption $n(\omega)\beta \gg 1$ and consider Cerenkov emission by an electron near threshold ($n(\omega)\beta \approx 1$). Would the emission be bluer or less blue than in case (b)?

20.3 A particle with speed βc is moving through a dielectric with refractive index n parallel to a plane surface between the dielectric and a vacuum.

(a) Using Snell's law, show that the emitted Cerenkov radiation observed in the vacuum is on the surface of a cone with half-angle θ_0 determined by $\sin\theta_0 = (\beta^2 n^2 - 1)^{1/2}/\beta$.

(b) What happens in the case where the motion is perpendicular to the interface? Does total internal reflection occur?

20.4 The Cerenkov condition applies to the emission of waves by particles in a variety of contexts.

(a) A supersonic aircraft emits sound waves as a sonic boom. Estimate the half-angle of the bow shock for the following speeds: (i) Mach 1.1, (ii) Mach 1.5 (iii) Mach 3.

(b) A ship generates a bow wave of surface water waves which have a dispersion relation $\omega = (kg)^{1/2}$, where g is the acceleration due to gravity. In a specific case the dominant wavelets within the bow shock are observed to have a wavelength of about a meter and the

half-angle of the bow shock is estimated to be 40°. Estimate the
speed of the ship.

(c)   The two modes of an optically active medium have refractive in-
dices $n_\pm = n_0 \pm \Delta n$. Show that the half-angles $\theta_\pm = \theta_0 \pm \Delta \theta$ of
the Cerenkov cones for the two modes satisfy $\Delta \theta \approx (\Delta n/n_0) \cot \theta_0$,
where θ_0 is the Cerenkov angle in the approximation $\Delta n = 0$.

(d)   Ion sound waves have dispersion relation $\omega = k v_s/(1 + k^2 \lambda_D^2)^{1/2}$.
Plot the Cerenkov angle as a function of $k^2 \lambda_D^2$ for an ion with speed
v equal to twice the ion sound speed v_s.

20.5 Consider Cerenkov emission due to a fast particle with speed βc
passing through a uniaxial crystal.

(a)   Show that the threshold condition for emission into the ordinary
mode is $K_\perp(\omega)\beta^2 > 1$.

(b)   Show that for a particle propagating along the principal axis the
threshold condition for emission into the extraordinary mode is the
same as for the ordinary mode.

20.6 Consider Cerenkov emission by a neutron, which has no charge but
has a magnetic dipole moment $\boldsymbol{\mu}$.

(a)   Show that the current associated with a neutron in constant rectilinear motion at velocity $\mathbf{v}$ is

$$\mathbf{J}(\omega, \mathbf{k}) = i\mathbf{k} \times \boldsymbol{\mu}\, 2\pi\delta(\omega - \mathbf{k} \cdot \mathbf{v}). \qquad (E20.1)$$

(b)   Show that the power radiated in transverse waves in an isotropic
dielectric is given by, where the angle between $\boldsymbol{\mu}$ and $\mathbf{v}$ is ζ,

$$P = \frac{\mu^2}{4\pi\varepsilon_0 c^5 \beta} \int\limits_{n(\omega)\beta > 1} d\omega\, \omega^3 n^2(\omega)$$

$$\times \left[1 - \tfrac{1}{2}\sin^2 \zeta - \frac{1}{n^2(\omega)\beta^2}(1 - \tfrac{3}{2}\sin^2 \zeta) \right]. \qquad (E20.2)$$

Hint: Choose polar angles χ, ϕ to perform the integral. Let χ be the
angle between $\mathbf{k}$ and $\mathbf{v}$, and then if χ' is the angle between $\mathbf{k}$ and $\boldsymbol{\mu}$,
choose ϕ to be the angle such that

$$\cos\chi' = \cos\chi\cos\zeta + \sin\chi\sin\zeta\cos\phi.$$

(c)   Estimate to within a factor of 10 the ratio of the power radiated
per unit frequency in the optical range in Cerenkov emission by a
relativistic neutron and a proton with the same speed.

20.7 The following exercise outlines a proof that Kirchhoff's law is satisfied for Langmuir waves in a thermal plasma.

(a)   Evaluate the emission coefficient (20.24) for Langmuir waves for the particular case of a Maxwellian distribution of electrons at temperature $m_e V_e^2$.

(b)   Show that the result corresponds to $S_L(\mathbf{k}) = \gamma_L(\mathbf{k}) W_L(\mathbf{k})$ with $\gamma_L(\mathbf{k})$ given by (13.6) and with $W_L(\mathbf{k}) = m_e V_e^2$ in accord with (17.42) for a thermal distribution of waves.

20.8 The following exercise outlines the proof that maser emission due to the inverse of Cerenkov emission is not possible for particles with an isotropic distribution $f(p)$ even when $\partial f(p)/\partial p$ is positive over any finite range of p.

(a)   Apply the expression (20.28) for the absorption coefficient to Langmuir waves and an isotropic distribution of particles. Evaluate the angular part of the integral over momentum space explicitly in spherical polar coordinates by carrying out the $\cos\chi$-integral over the δ-function, with $\mathbf{k} \cdot \mathbf{v} = kv \cos\chi$ and

$$\mathbf{k} \cdot \frac{\partial f(p)}{\partial \mathbf{p}} = k \cos\chi \, \frac{\partial f(p)}{\partial p}.$$

(b)   Partially integrate the resulting expression over p and show that the absorption coefficient is strictly positive.

(c)   The foregoing argument does not alter the fact that for given $\mathbf{k}$ and $\mathbf{p}$ the contribution to the absorption is negative for $\mathbf{k} \cdot \partial f(p)/\partial \mathbf{p} > 0$. Discuss how this fact is compatible with the net absorption being positive for an isotropic distribution of particles.

20.9 A current-carrying plasma can be unstable to the generation of low-frequency waves when the relative drift velocity between the electrons and the ions exceeds an appropriate threshold. The following exercise illustrates this effect in a plasma where the electrons are sufficiently hot compared to the ions to allow ion sound waves to be weakly damped.

(a)   Apply (20.28) to the absorption of ion sound waves by thermal electrons and ions, and show that for $v_s \ll V_e$ and $k\lambda_D \ll 1$ the resulting expression for the absorption coefficient reproduces (13.13).

(b)   Repeat the calculation assuming that the electrons are drifting with a velocity $\mathbf{v}_d$ relative to the ions and show that the contribution of the electrons to the absorption is negative for $\mathbf{k} \cdot \mathbf{v}_d > \omega_s(\mathbf{k})$.

(c)   Show that for $k\lambda_D \ll 1$ the electronic contribution in part (b) reduces to

$$\gamma_s(\mathbf{k}) \approx (\pi/2)^{1/2}(v_s/V_e)(kv_s - \mathbf{k} \cdot \mathbf{v}_d). \tag{E20.3}$$

20.10 The one-dimensional equations for a beam instability are derived from the three-dimensional equations as follows.

(a) Consider a distribution of non-relativistic electrons that have a net streaming motion along the z axis. Define the one-dimensional distribution by writing $\mathbf{p} = m_e\mathbf{v}$ and $F(v_z) = m_e \int dp_x dp_y\, F(\mathbf{p})$.

(b) Evaluate the absorption coefficient (20.28) for Langmuir waves that are directed along the z axis. Show that the result reproduces (20.29).

(c) Similarly, writing $k_z = \omega/v_\phi$, with $\int dv_\phi W(v_\phi)$ the energy density in the Langmuir waves, show that for Langmuir waves along the streaming direction the diffusive term in (20.33) reduces to the one-dimensional form (20.38).

20.11 Discuss the possibility of negative absorption of Cerenkov emission in air due to a cosmic-ray shower.

(a) Show that the absorption coefficient averaged over the two states of transverse polarization for a one-dimensional distribution of particles with distribution function $F(p)$ and number density $n_1 = \int dp\, F(p)$ reduces to

$$\gamma(\omega) = -\frac{\pi e^2}{2\varepsilon_0 \omega}\left(1 - \frac{1}{n^2\beta^2}\right)v^2\frac{\partial p}{\partial v}\frac{\partial F(p)}{\partial p}. \tag{E20.4}$$

(b) Argue that in a cosmic-ray shower the absence of particles with small p requires $dF(p)/dp > 0$ below some peak p_{max}, and hence argue that the absorption should be negative for a range of angles and determine this range in terms of n and p_{max}.

(c) Negative absorption of Cerenkov emission does not appear to be important in practice. Suggest a reason why this might be the case.

20.12 Consider quasi-linear relaxation of a thermal distribution of Langmuir waves at one temperature interacting with thermal electrons at a different temperature. Apply (20.31) to discuss this case as follows.

(a) After making the non-relativistic approximation in (20.31), derive a formula for the rate of change of the temperature by multiplying (20.31) by $\frac{1}{2}mv^2$, integrating over momentum space and dividing by the the number density of particles.

(b)   Assume that distribution of particles is thermal at a temperature T_1 and that the distribution of waves is thermal at a temperature T_2. Let $\gamma_{\mathrm{L}}(\omega)$ be the absorption coefficient, as given by (13.6) with $m_{\mathrm{e}}V_{\mathrm{e}}^2 \to T_1$ and evaluate the right hand side of the resulting equation using Kirchhoff's law. Hence show that one has

$$\frac{\mathrm{d}T_1}{\mathrm{d}t} = \alpha\gamma_{\mathrm{L}}(\omega)\big[T_1 - T_2\big]. \qquad (E20.5)$$

(c)   Show that α in $(E20.5)$ is of the form $4\pi/n_{\mathrm{e}}\bar{\lambda}^3$ and interpret $\bar{\lambda}$.

(d)   Give a physical interpretation of $(E20.5)$ for (i) $T_1 > T_2$ and (ii) $T_1 < T_2$.

20.13 Evaluate the energy radiated *in vacuo* due to transition radiation at a perfectly conducting metal surface for a particle of arbitrary energy and show that the generalization of (20.51) is

$$\begin{aligned}
U_{\mathrm{rad}}(\omega) &= \frac{q^2\beta^2}{2\pi^2\varepsilon_0 c} \int_0^1 \mathrm{d}\cos\chi \, \frac{\sin^2\chi}{(1 - \beta^2\cos^2\chi)^2} \\
&= \frac{q^2}{4\pi^2\varepsilon_0 c} \left[\frac{1+\beta^2}{2\beta} \ln\left(\frac{1+\beta}{1-\beta}\right) - 1\right],
\end{aligned} \qquad (E20.6)$$

with $\beta = |\mathbf{v}|/c$.

21

Bremsstrahlung

Preamble

"Bremsstrahlung" is used both as the generic name to describe emission due to an accelerated charged particle and as the specific name for the emission when this acceleration is due to the Coulomb field of another particle. Here we are concerned with bremsstrahlung in the latter more restrictive sense. Bremsstrahlung can result in the emission of waves in any wave mode in a plasma, but most interest is in the emission of transverse waves. Emission of Langmuir waves is of less interest because, unlike transverse waves in a plasma, Langmuir waves are generated efficiently by the Cerenkov process. The absorption process corresponding to bremsstrahlung is called collisional damping or, in some astrophysical literature, free–free absorption.

21.1 Qualitative Discussion of Bremsstrahlung

Bremsstrahlung due to Coulomb interactions between electrons and ions is an important emission process in a wide variety of plasmas. We mention only three general applications. (i) For laboratory plasmas and many space plasmas, bremsstrahlung is the basic thermal emission process at radio frequencies. Radio frequency emission results from distant electron–ion encounters in which the motion of the electron is perturbed only slightly by the Coulomb field of the ion. (ii) High-frequency photons, with an energy comparable with the initial energy of the electron, can result from a close encounter between an electron and an ion. So-called non-thermal bremsstrahlung due to energetic electrons is an important source of X-rays from a plasma. In fusion plasmas runaway electrons can be a problem due to their production of hard X-rays, and hard X-ray emission from solar flares is due to energetic (10–20 keV) electrons propagating down magnetic field lines from the site of the flare in the corona into the denser regions of the solar atmosphere. (iii) Even in very tenuous plasmas where interactions between electrons and ions are infrequent, bremsstrahlung and collisional damping can play an important role as a source or sink of photons which otherwise experience only Thomson scattering in their passage through the plasma.

304Coulomb interactions in a plasma are referred to as *collisions*. This terminology is clearly appropriate for close encounters when the interaction is obviously of a binary nature. For distant encounters it is not necessarily obvious that one is justified in treating collisions in terms of binary interactions. Consider the effect of the Coulomb fields of surrounding particles on a test charge. Each Coulomb field falls off with distance r from the test charge as r^{-2}. However the number of field particles between r and $r + dr$ increases as $r^2 dr$, so that the field at the test charge appears to contain equal contributions from field particles at all distances. In other words a single test charge appears to interact simultaneously with all other charges in the plasma. The procedure for treating this situation is well established in plasma physics and in other fields such as solid state physics. One introduces a mean field for which one solves in a self-consistent way. In a thermal plasma the mean field for a test charge at rest at the origin modifies the Coulomb field $\phi(\mathbf{x}) = q/4\pi\varepsilon_0 r$ by introducing *Debye shielding* so that the field becomes

$$\phi(\mathbf{x}) = \frac{q\,e^{-r/\lambda_D}}{4\pi\varepsilon_0 r}, \tag{21.1}$$

where λ_D is the Debye length. As a consequence of the exponential cutoff in the potential (21.1) at the Debye length, the number of field particles with which an individual test charge effectively interacts at any one time is restricted to the number of particles within a Debye sphere, where a *Debye sphere* is a sphere of radius λ_D. In practice this is a large number of particles. The *Coulomb logarithm* is effectively the natural logarithm of this number, and for typical plasmas of interest in the laboratory and in astrophysics the Coulomb logarithm varies between approximately 10 and 20.

It follows from these remarks that bremsstrahlung due to distant encounters involves many particles influencing a single charge at one time. The standard method for treating this problem is to regard the effect of all distant encounters as perturbations on the orbit of the test charge.

The unperturbed orbit is taken to be constant rectilinear motion, and the effect of the Coulomb interaction with each field particle is treated as a separate perturbation.

Bremsstrahlung due to electron–electron encounters is usually negligible in comparison with that due to electron–ion encounters. The reason is that electric dipole emission cannot occur in electron–electron collisions. According to (17.18), the power in electric dipole emission is proportional to the second time-derivative of the electric dipole moment. In an electron–electron encounter this second derivative is equal to $-e(\ddot{\mathbf{x}}_1 + \ddot{\mathbf{x}}_2)$, where $\mathbf{x}_1$ and $\mathbf{x}_2$ are the position vectors of the two electrons. However, Newton's third law implies that the accelerations of the two electrons are equal and opposite, so that there is no dipole emission. Electric dipole emission is allowed if either the charges or the masses of the particles are different. Consequently, only electron–ion encounters need be considered when treating bremsstrahlung in a non-relativistic plasma.

Bremsstrahlung needs to be treated in a variety of special and limiting cases. The main emphasis in the discussion here is on the simplest limiting case of a distant encounter treated classically using non-relativistic perturbation theory. More general classical treatments involve assuming that the orbit is hyperbolic for a non-relativistic electron, or including relativistic effects in the perturbation approach. For some purposes it is essential to include quantum effects. An obvious case when quantum effects are important is when the energy of the emitted photon is comparable with the initial energy of the electron, and this is an important case in practice. A less obvious case where a quantum effect is significant is when the de Broglie wavelength of the electron is comparable with its impact parameter. These and other special cases of bremsstrahlung are considered both for individual electrons and also for a thermal distribution of electrons. The average over a Maxwellian distribution leads to different results in the different limits. Thus there is a confusing variety of special cases that needs to be considered to provide an adequate description of bremsstrahlung. Fortunately, there is a relatively simple way of describing bremsstrahlung in general: one derives a formula that describes the emission by a single electron, and notes that all the subtle details appear in a single factor, which is a logarithmic factor in the simplest case. More generally this factor is identified as the Gaunt factor and its detailed evaluation is regarded as a separate complementary problem. In the following treatment of bremsstrahlung we use the simplest theory (the classical straight line approximation) to derive an

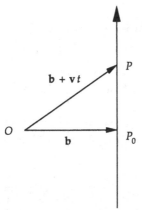

Fig. 21.1 The orbit of the electron in the straight line approximation.

expression for the power radiated per unit frequency $P(\omega)$ by a single electron, and then we summarize generalizations of it by quoting the relevant forms of the Gaunt factor that result from alternative detailed calculations.

21.2 The Straight Line Approximation

The *straight line approximation* is a perturbation treatment in which the zeroth order orbit of the electron is assumed to be constant rectilinear motion. The effect of the Coulomb field of an ion on the electron is taken into account as a perturbation. The equation of motion is given by (18.23), *viz.*,

$$m_e \ddot{\mathbf{X}}(t) = -\frac{Ze^2 \mathbf{X}(t)}{4\pi\varepsilon_0 |\mathbf{X}(t)|^3}. \tag{21.2}$$

The unperturbed orbit is assumed to be

$$\mathbf{X}^{(0)}(t) = \mathbf{b} + \mathbf{v}t, \tag{21.3}$$

where the velocity $\mathbf{v}$ is assumed constant in this approximation. The orbit is illustrated in Figure 21.1. The displacement $\mathbf{b}$ is the position vector of the electron relative to the ion at the point of closest approach. In the present case this corresponds to the impact parameter for the collision. In the more general case when the orbit is hyperbolic, the impact parameter is as shown in Figure 18.1.

The first order force is found by substituting the zeroth order orbit (21.3) into the force on the right hand side of (21.2). This gives

$$\mathbf{F}^{(1)}(t) = -\frac{Ze^2 \mathbf{X}^{(0)}(t)}{4\pi\varepsilon_0 |\mathbf{X}^{(0)}(t)|^3}. \tag{21.4}$$

This force implies a first order acceleration

$$\mathbf{a}^{(1)}(t) = \mathbf{F}^{(1)}(t)/m_e, \tag{21.5}$$

and (21.5) is solved to find the first order corrections to the orbit. However, one need not solve the first order equation of motion (21.5) explicitly.

The power radiated in transverse waves *in vacuo* in the straight line approximation follows by substituting (21.5) into the Larmor formula (18.10). This gives

$$P(t) = \frac{e^2}{6\pi\varepsilon_0 c^3} \left(\frac{Ze^2}{4\pi\varepsilon_0 m_e |\mathbf{X}^{(0)}(t)|^2} \right)^2. \tag{21.6}$$

The energy radiated in a single encounter is found by integrating (21.6) over t. The integral gives

$$E_{\text{rad}} = \int_{-\infty}^{\infty} dt\, P(t) = \frac{2Z^2 e^6}{3(4\pi\varepsilon_0)^3 m_e^2 c^3} \int_{-\infty}^{\infty} \frac{dt}{(b^2 + v^2 t^2)^2}$$

$$= \frac{\pi Z^2 e^6}{3(4\pi\varepsilon_0)^3 m_e^2 c^3 v b^3}. \tag{21.7}$$

The result (21.7) also follows from (18.28) in the limit of large impact parameters $b \gg b_0$.

The frequency spectrum of bremsstrahlung is of particular importance in applications. The energy emitted per unit frequency in transverse waves is given by (18.9), which becomes

$$U_{\text{rad}}(\omega) = \frac{e^2 n(\omega)}{6\pi^2 \varepsilon_0 m_e^2 c^3} |\mathbf{F}^{(1)}(\omega)|^2. \tag{21.8}$$

The single power of the refractive index appearing in (21.8) is characteristic of electric dipole emission. The temporal Fourier transform of the force is

$$\mathbf{F}^{(1)}(\omega) = -\frac{Ze^2}{4\pi\varepsilon_0} \int_{-\infty}^{\infty} dt\, e^{i\omega t} \frac{\mathbf{b} + \mathbf{v}t}{(b^2 + v^2 t^2)^{3/2}}. \tag{21.9}$$

The integral in (21.9) is evaluated in terms of Macdonald functions $K_\nu(z)$, using their integral representations

$$K_\nu(xz) = \frac{\Gamma(\nu + \frac{1}{2})(2z)^\nu}{2x^\nu \Gamma(\frac{1}{2})} \int_{-\infty}^{\infty} dt\, \frac{e^{\pm ixt}}{(t^2 + z^2)^{\nu + 1/2}}. \tag{21.10}$$

These functions satisfy the recursion formulas

$$K_{\nu-1}(z) + K_{\nu+1}(z) = -2K_\nu'(z), \tag{21.11}$$

$$K_{\nu-1}(z) - K_{\nu+1}(z) = -2\frac{\nu}{z} K_\nu(z). \tag{21.12}$$

Expansion in z for integral $\nu = n$ and for small z gives the following leading terms:

$$K_n(z) \approx \begin{cases} \ln(2/\Gamma z) & \text{for } n = 0, \\ 1/z^n & \text{for } n \neq 0. \end{cases} \qquad (21.13)$$

with $\Gamma = 1.7811\ldots$, $\ln\Gamma = 0.5722\ldots$. The asymptotic expansions give

$$K_\nu(z) \approx \left(\frac{\pi}{2z}\right)^{1/2} e^{-z}. \qquad (21.14)$$

Using the relations (21.10) and (21.11), with $K_{-\nu}(z) = K_\nu(z)$, the integral in (21.9) may be evaluated explicitly:

$$\mathbf{F}^{(1)}(\omega) = -\frac{2Ze^2\omega}{4\pi\varepsilon_0 v^2} \left[\frac{\mathbf{b}}{b} K_1\left(\frac{b\omega}{v}\right) - i\frac{\mathbf{v}}{v} K_0\left(\frac{b\omega}{v}\right)\right]. \qquad (21.15)$$

Then (21.8) reduces to

$$U_{\text{rad}}(\omega) = \frac{8}{3\pi} \frac{n(\omega)Z^2 e^6 \omega^2}{(4\pi\varepsilon_0)^3 m_e^2 c^3 v^4} \left[K_1^2\left(\frac{b\omega}{v}\right) + K_0^2\left(\frac{b\omega}{v}\right)\right]. \qquad (21.16)$$

Using the approximations (21.13) and (21.14), (21.16) has the following approximate forms

$$U_{\text{rad}}(\omega) = \frac{8}{3\pi} \frac{n(\omega)Z^2 e^6 \omega^2}{(4\pi\varepsilon_0)^3 m_e^2 c^3 v^4} \begin{cases} v^2/b^2\omega^2 & \text{for } b\omega/v \ll 1, \\ \pi(v/2b\omega)\,e^{-2b\omega/v} & \text{for } b\omega/v \gg 1. \end{cases} \qquad (21.17)$$

It follows that the energy emitted per unit frequency is approximately independent of frequency for $b\omega/v \ll 1$, and falls off rapidly with frequency for $b\omega/v \gg 1$. This type of spectrum is characteristic of emission due to an impulsive event of duration $\Delta t = b/v$: the frequency spectrum is white noise for $\omega \ll (\Delta t)^{-1}$ and falls off rapidly for $\omega \gg (\Delta t)^{-1}$.

21.3 Continuous Emission due to Distant Encounters

The energy radiated per unit frequency (21.16) is for a single distant encounter with impact parameter b. In practice the emission from a single electron due to a single distant encounter cannot be observed. This is because of the large number of ions within the Debye sphere of the electron, and the fact that the electron is influenced simultaneously by all these ions. This does not invalidate the approach based on a single encounter because the bremsstrahlung resulting from interactions with different ions is independent in the sense that there is no interference due to the random distribution and motion of the ions. However, the simultaneous interaction with many ions implies that each electron has a quasi-continuous emission due to the effect of interactions with all the ions within its Debye sphere.

The power emitted by an electron as it experiences distant encounters

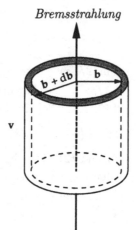

Fig. 21.2 An electron has encounters per unit time with impact parameter
between b and $b + db$ with all ions in the cylindrical surface shown.

with many ions along its path is found as follows. First, let us suppose
that there are ions of various species, with the ith species having charge
$Z_i e$ and number density n_i. The number of encounters that an electron
with speed v has per unit time with ions of species i with impact parameters between b and $b + db$ is $2\pi db\, b\, n_i v$, cf. Figure 21.2. Hence the
power radiated is

$$P(\omega) = \sum_i 2\pi n_i v \int_0^\infty db\, b\, U(\omega), \qquad (21.18)$$

where the sum is over all species of ion. On substituting the expression
(21.16) for $U(\omega)$ into (21.18), the integral diverges at $b = 0$. Let us cut
the integral off at $b = b_1$. The integral is evaluated using

$$\int_{z_1}^\infty dz\, z \left[K_0^2(z) + K_1^2(z) \right] = z_1 K_0(z_1) K_1(z_1). \qquad (21.19)$$

Thus (21.18) reduces to

$$P(\omega) = \sum_i \frac{16}{3} \frac{n(\omega) Z_i^2 n_i e^6}{(4\pi\varepsilon_0)^3 m_e^2 c^3 v} \frac{b_1 \omega}{v} K_0\left(\frac{b_1 \omega}{v}\right) K_1\left(\frac{b_1 \omega}{v}\right). \qquad (21.20)$$

In the limits of low and high frequencies, (21.20) gives

$$P(\omega) = \sum_i \frac{16}{3} \frac{n(\omega) Z_i^2 n_i e^6}{(4\pi\varepsilon_0)^3 m_e^2 c^3 v} \begin{cases} \ln\left(\dfrac{2}{\Gamma} \dfrac{v}{b_1 \omega}\right) & \text{for } \omega \ll v/b_1, \\[2mm] \dfrac{\pi}{2} e^{-2b_1 \omega/v} & \text{for } \omega \gg v/b_1. \end{cases} \qquad (21.21)$$

The power emitted (21.21) diverges at low frequencies. This is surprising
in view of the fact that the energy emitted in each encounter (21.17)
is independent of frequency at low frequencies. The divergence in the
power is attributed to the number of distant encounters diverging as

b diverges. This divergence is removed by taking Debye shielding into account. The assumed Coulomb field of an ion is a valid approximation to the actual field of an ion only for distances less than about a Debye length, cf. (21.1). Encounters with $b \geq \lambda_D$ are ineffective due to the Debye shielding, and it is appropriate to restrict the integral over b in (21.18) to the range $b_1 < b < b_2$ with $b_2 = \lambda_D$. If one cuts the integral off in this way, then (21.20) is replaced by

$$P(\omega) = \sum_i \frac{16}{3} \frac{n(\omega)Z_i^2 n_i e^6}{(4\pi\varepsilon_0)^3 m_e^2 c^3 v}$$

$$\times \left[\frac{b_1\omega}{v} K_0\left(\frac{b_1\omega}{v}\right) K_1\left(\frac{b_1\omega}{v}\right) - \frac{b_2\omega}{v} K_0\left(\frac{b_2\omega}{v}\right) K_1\left(\frac{b_2\omega}{v}\right) \right]. \tag{21.22}$$

The low-frequency limit in (21.21) is then modified to

$$P(\omega) = \sum_i \frac{16}{3} \frac{n(\omega)Z_i^2 n_i e^6}{(4\pi\varepsilon_0)^3 m_e^2 c^3 v} \ln\left(\frac{2}{\Gamma} \frac{b_2}{b_1} \right). \tag{21.23}$$

The high-frequency limit in (21.21) is unaffected by this change for $b_2 \gg b_1$. The logarithmic factor in (21.23) is related to the Gaunt factor introduced below.

We chose to describe the emission in terms of $P(\omega)$, which is the power per unit frequency per electron. For formal purposes it is sometimes preferable to describe the emission in terms of the emissivity $\eta(\omega)$ per electron. The emissivity is the power per unit frequency and per unit solid angle. The net emission due to encounters with all the ions within the Debye sphere is isotropic, and hence the emissivity is given by dividing by 4π steradians. That is, one has $\eta(\omega) = P(\omega)/4\pi$.

21.4 The Gaunt Factor

As mentioned above, the treatment of bremsstrahlung involves approximations of various kinds, and it is convenient to absorb all the approximations into a single factor, identified as the Gaunt factor $G(v, \omega)$. We now generalize (21.23) by regarding it as the general form of the power radiated in bremsstrahlung per thermal electron by replacing the logarithmic factor by the Gaunt factor. Estimations of the Gaunt factor are outlined here in three general contexts: (i) classical calculations, (ii) quantum calculations and (iii) averages over a thermal distribution. The discussion is restricted to the non-relativistic limit.

Classical Expression for the Gaunt Factor

The expression for the Gaunt factor implied by the calculation based on (21.22) and (21.23) leads to the form

$$\frac{\pi}{\sqrt{3}}\, G(v,\omega) = \ln\left(\frac{2}{\Gamma}\frac{b_2}{b_1}\right). \qquad (21.24)$$

The inclusion of the factor $\pi/\sqrt{3}$ is an historical convention. The Gaunt factor (21.24) is insensitive to factors of order unity in the argument of the logarithm, due to the fact that the logarithmic factor itself is of order 10–20 in plasmas of practical interest. As a consequence there is no point in attempting to identify the cutoffs with high precision. An appropriate choice for the maximum impact parameter for a fast electron is taken to be $b_2 = v/\omega$, corresponding to the value at which the frequency spectrum (21.21) starts to fall off exponentially. The minimum impact parameter is identified as that for which a deflection through $\pi/2$ occurs, when the straight line approximation clearly breaks down. This corresponds to $b_1 = b_0$, where b_0 is given by (18.29). These choices give

$$\frac{\pi}{\sqrt{3}}\, G(v,\omega) = \ln\left(\frac{2}{\Gamma}\frac{m_e v^3}{Z_i^2 e^2 \omega/4\pi\varepsilon_0}\right), \qquad (21.25)$$

which is the classical form for the Gaunt factor for a single electron with speed v.

Quantum Mechanical Forms for the Gaunt Factor

At small impact parameters quantum effects can be important and the de Broglie wavelength $\hbar/p$ of the electron needs to be taken into account. The minimum impact parameter b_1 is the greater of the de Broglie wavelength and the impact parameter b_0 for deflection through $\pi/2$. Thus, for $\hbar/p > b_0$, the Gaunt factor (21.25) is replaced by

$$\frac{\pi}{\sqrt{3}}\, G(v,\omega) = \ln\left(\frac{2m_e v^2}{\hbar\omega}\right). \qquad (21.26)$$

At high frequencies, a quantum treatment is essential when the energy of the emitted photon is comparable with the energy of the electron. Energy is not automatically conserved in classical theory, and this is a serious weakness when the energy of the electron changes significantly, that is, when the final speed v' is significantly different from v. Conservation of energy implies

$$\tfrac{1}{2}m_e v'^2 = \tfrac{1}{2}m_e v^2 - \hbar\omega. \qquad (21.27)$$

A quantum calculation in the Born approximation gives

$$\frac{\pi}{\sqrt{3}} G(v,\omega) = \ln\left(\frac{v + v'}{v - v'}\right). \tag{21.28}$$

In the limit $\hbar\omega \ll \frac{1}{2}m_e v^2$, (21.28) reduces to (21.26).

A more general quantum calculation involves using the exact wavefunctions for an electron in the Coulomb field of an ion. The wave function involves a hypergeometric function $_2F_1(i\nu'; i\nu; 1, \xi)$ which we write as $F(\xi)$, with

$$\nu = \frac{Z_i e^2}{4\pi\varepsilon_0 \hbar v}, \quad \nu' = \frac{Z_i e^2}{4\pi\varepsilon_0 \hbar v'}, \quad \xi = \frac{-4vv'}{(v - v')^2}. \tag{21.29}$$

The general expression for the Gaunt factor is

$$\frac{\pi}{\sqrt{3}} G(v,\omega) = \frac{\pi^2}{(1 - e^{-2\pi\nu'})(e^{2\pi\nu} - 1)}\left(-\xi\frac{\mathrm{d}}{\mathrm{d}\xi}|F(\xi)|^2\right). \tag{21.30}$$

The Born approximation (21.28) corresponds to the limit $Z_i\alpha \ll \nu, \nu' \ll 1$, with $\alpha = e^2/4\pi\varepsilon_0\hbar c \approx 1/137$ the fine structure constant, in which case one has $F(\xi) \approx 1$, $F'(\xi) \approx (\nu\nu')\ln(1 - \xi)$. A more general case is for $Z_i\alpha \ll \nu \ll 1$ with ν' not necessarily small. This allows for the case where the electron loses most of its energy to the photon so that v' is much less than v. In this case one has

$$-\xi \approx 4v/v', \quad F(\xi) \approx 1, \quad F'(\xi) \approx -\nu\nu'.$$

The implied approximation to the Gaunt factor is

$$\frac{\pi}{\sqrt{3}} G(v,\omega) = \frac{4\pi\nu}{(1 - e^{-2\pi\nu'})}. \tag{21.31}$$

In the limit $\nu' \ll 1$ (21.31) reproduces the result of the Born approximation (21.30) for $v' \ll v$.

An important special case of (21.31) is the limit $v' \to 0$. This corresponds to $\nu' \to \infty$ in (21.31) in which case the denominator is equal to unity. The Gaunt factor approaches a constant value for $v' = 0$, which is the case where the electron loses all its initial energy to a single photon. The case $\hbar\omega = \frac{1}{2}mv^2$ separates free–free and free–bound transitions. Emission at $\hbar\omega > \frac{1}{2}mv^2$ is possible due to free–bound transitions in which the initially free electron becomes a bound electron. The relatively high efficiency of emission at $\hbar\omega \approx \frac{1}{2}mv^2$ may be attributed to the high density of bound states in the energy range separating bound and free electrons. In practice, X-ray emission from energetic electrons in a plasma is dominated by emission of photons with energy comparable to that of the electrons. For example, the hard X-ray spectrum observed in a solar flare is a direct signature of the energy spectrum of the precipitating electrons, because the X-rays with energy $\hbar\omega$ are due primarily to electrons with the same energy $\frac{1}{2}mv^2 = \hbar\omega$.

The Gaunt Factor for Thermal Electrons

The bremsstrahlung emission from a thermal plasma is found by integrating the power radiated per electron over a thermal distribution of electrons. The relevant integral in the low-frequency limit, according to (21.23), involves the average over $1/v$. The average over a Maxwellian distribution of $1/v$ is given by

$$\left\langle \frac{1}{v} \right\rangle = \int \frac{d^3\mathbf{v}}{(2\pi)^{3/2}V_e^3} \frac{1}{v} e^{-v^2/2V_e^2} = \left(\frac{2}{\pi}\right)^{1/2} \frac{1}{V_e}, \qquad (21.32)$$

where V_e is the thermal speed of electrons. It is convenient to replace the average over the velocity-dependent factor in (21.32) in terms of the *thermal Gaunt factor* $\bar{G}(V_e, \omega)$. Thus, in the classical case, we write

$$\left\langle \frac{1}{v} \left[\frac{b_1\omega}{v} K_0\left(\frac{b_1\omega}{v}\right) K_1\left(\frac{b_1\omega}{v}\right) - \frac{b_2\omega}{v} K_0\left(\frac{b_2\omega}{v}\right) K_1\left(\frac{b_2\omega}{v}\right) \right] \right\rangle$$

$$= \left(\frac{2}{\pi}\right)^{1/2} \frac{1}{V_e} \frac{\pi}{\sqrt{3}} \bar{G}(V_e, \omega). \qquad (21.33)$$

The power radiated per thermal electron is then given by

$$P(\omega) = \sum_i \frac{16}{3} \frac{n(\omega)Z_i^2 n_i e^6}{(4\pi\varepsilon_0)^3 m_e^2 c^3} \left(\frac{2}{\pi}\right)^{1/2} \frac{1}{V_e} \frac{\pi}{\sqrt{3}} \bar{G}(V_e, \omega). \qquad (21.34)$$

The Gaunt factor for a classical thermal plasma is expressed in terms of the Coulomb logarithm $\ln\Lambda_c$:

$$\frac{\pi}{\sqrt{3}} \bar{G}(V_e, \omega) = \ln\Lambda_c - \ln\left(\frac{\Gamma}{2}\frac{\omega}{\omega_p}\right). \qquad (21.35)$$

The classical expression for the Coulomb logarithm leads to the following numerical value:

$$\ln\Lambda_c = 16.0 - \tfrac{1}{2}\ln(n_e) + \tfrac{3}{2}\ln(T_e) - \ln(Z_i), \qquad (21.36)$$

with n_e is the electron number density per cubic meter, T_e is the electron temperature in kelvin, and $Z_i e$ is the ionic charge.

The Gaunt factor at high frequencies in a thermal plasma is

$$\frac{\pi}{\sqrt{3}} \bar{G}(V_e, \omega) = K_0(\hbar\omega/m_e V_e^2) e^{-\hbar\omega/m_e V_e^2}$$

$$\approx \begin{cases} \ln(2m_e V_e^2/\Gamma\hbar\omega) & \text{for } \hbar\omega \ll m_e V_e^2, \\ (\pi m_e V_e^2/2\hbar\omega)^{1/2} e^{-\hbar\omega/m_e V_e^2} & \text{for } \hbar\omega \gg m_e V_e^2. \end{cases}$$

$$\qquad (21.37)$$

21.5 Collisional Damping

The absorption process corresponding to bremsstrahlung is collisional damping or free–free absorption. One procedure for treating collisional damping is to include the effect of collisions in the response tensor, as outlined in Exercise 6.3. Two other procedures are discussed here: one is based on using Kirchhoff's law and applies only to damping by thermal particles; the other is based on detailed balance and is used to treat collisional damping by non-thermal electrons.

The detailed discussion here is restricted to transverse waves. (Collisional damping of Langmuir waves also occurs, cf. (13.9), and is of interest when Landau damping is weak or absent.) Landau damping is absent for waves with phase speed greater than the speed of light, as is the case for transverse waves in an unmagnetized plasma. Hence collisional damping is usually the dominant absorption process for transverse waves in an unmagnetized plasma.

The power radiated in bremsstrahlung per unit frequency and per unit volume by a thermal plasma is used to find the absorption coefficient for a thermal plasma by appealing to Kirchhoff's law. For any isotropic emission process due to thermal particles, let $d\eta(\omega)/dV$ be the *emissivity per unit volume*, where the emissivity is the power emitted per unit frequency and per unit solid angle. Let $\gamma(\omega)$ be the absorption coefficient per unit frequency. The relation between these implied by Kirchhoff's law in an isotropic dielectric is

$$\gamma(\omega) = d\eta(\omega)/dV \left[\frac{(2\pi c)^3}{2n(\omega)\omega^2 m_e V_e^2} \right], \qquad (21.38)$$

where a factor of 2 arises due to averaging over the states of polarization in treating absorption, compared with summing over the two states of polarization in treating emission. For bremsstrahlung the emissivity is identified as $d\eta(\omega)/dV = n_e P(\omega)/4\pi$, with $P(\omega)$ given by (21.34). Then (21.38) gives the absorption coefficient for collisional damping, which may be written in the form (13.19), *viz.*,

$$\gamma_c(\omega) = \frac{\omega_p^2}{\omega^2} \nu(\omega), \qquad (21.39)$$

with the *collision frequency* identified as

$$\nu(\omega) = \sum_i \frac{1}{3} \left(\frac{2}{\pi} \right)^{1/2} \frac{4\pi Z_i^2 n_i e^4}{(4\pi\varepsilon_0)^2 m_e^2 V_e^3} \frac{\pi}{\sqrt{3}} \bar{G}(V_e, \omega). \qquad (21.40)$$

Thus the appeal to Kirchhoff's law, coupled with the detailed calculation of thermal bremsstrahlung, reproduces the result (13.19) obtained by including the frictional effect of collisions in deriving the response antihermitian part of the response tensor. Furthermore, it allows one to identify the expression (21.40) for the collision frequency.

An average probability per unit time of emission of bremsstrahlung by an electron, $\bar{w}(p, k)$, is obtained from the power emitted (21.23) with

(21.24). The average is over directions of emission, and is justified by regarding the probability as an average probability per electron for an isotropic distribution of electrons. To construct the probability from $P(\omega)$, divide by three factors: the energy $\hbar\omega$ per photon, 4π steradians, and the factor $k^2(\mathrm{d}k/\mathrm{d}\omega)/(2\pi)^3$ relating the appropriate ranges of integration. This leads to an *average emission probability for bremsstrahlung*

$$\bar{w}(p,k) = \sum_i \frac{32\pi^2}{3\hbar\omega^3} \frac{Z_i^2 n_i e^6}{(4\pi\varepsilon_0)^3 m_e^2 v} \frac{\pi}{\sqrt{3}} G(v,\omega), \qquad (21.41)$$

in which the refractive index $n(\omega) = (1 - \omega_p^2/\omega^2)^{1/2}$ no longer appears explicitly.

The argument based on detailed balance relating the absorption coefficient to the probability given in §16.3 needs to be modified slightly here, due to the fact that an average over directions has been performed assuming an isotropic distribution of electrons. Only the changes in energy are considered, and then the transfer equation analogous to (20.21) reduces to

$$\frac{\mathrm{d}\bar{N}(k)}{\mathrm{d}t} = 4\pi \int_0^\infty \mathrm{d}p\, p^2\, \bar{w}(p,k) \left[f(p) + \bar{N}(k) \frac{\hbar\omega}{p} \frac{\mathrm{d}f(p)}{\mathrm{d}p} \right], \qquad (21.42)$$

where $\bar{N}(k)$ is the occupation number of the photons averaged over 4π steradians. The absorption coefficient is identified from the term on the right hand side of (21.42) that is proportional to $\bar{N}(k)$. One has

$$\gamma_c(\omega) = -4\pi \int_0^\infty \mathrm{d}p\, p^2\, \bar{w}(p,k) \frac{\hbar\omega}{p} \frac{\mathrm{d}f(p)}{\mathrm{d}p}. \qquad (21.43)$$

This form for the absorption coefficient is used to treat collisional damping by any non-relativistic isotropic distribution of electrons. It may be used to rederive (21.39) with (21.40) by inserting a Maxwellian distribution of electrons into (21.43) and carrying out the integral.

The absorption coefficient (21.40) cannot be negative. A necessary condition for negative absorption is that $\mathrm{d}f(p)/\mathrm{d}p$ be positive over some range of p, but this is not a sufficient condition. As follows by partially integrating (21.43), another necessary condition is that $\mathrm{d}\left[p^2\bar{w}(p,k)\right]/\mathrm{d}p$ be negative over some range of p. Inspection of (21.38) shows that this condition is not satisfied. Maser action due to bremsstrahlung in a plasma is not possible.

Exercise Set 21

21.1 The electron–ion collision frequency $\nu_e(\omega)$ has the following approximate numerical value

$$\nu_e(\omega) = \frac{\omega_p \ln \Lambda_c}{4\pi n_e \lambda_{De}^3} \approx 13.7 \times 10^{-6} \ln \Lambda_c \, n_e T_e^{-3/2} \, \text{s}^{-1}, \qquad (E21.1)$$

where n_e is per cubic meter and T_e is in kelvin. A numerical estimate of the Coulomb logarithm is given by (21.36).

(a) Estimate the collision frequency for the plasmas with properties listed below.

plasma	n_e	T_e	L
fusion plasma	$10^{20} \, \text{m}^{-3}$	$10^4 \, \text{eV}$	$1 \, \text{m}$
cool laboratory plasma	$10^{20} \, \text{m}^{-3}$	$1 \, \text{eV}$	$1 \, \text{m}$
ionosphere	$10^{13} \, \text{m}^{-3}$	$10^4 \, \text{K}$	$100 \, \text{km}$
solar corona	$10^{10} \, \text{cm}^{-3}$	$3 \times 10^6 \, \text{K}$	$10^{11} \, \text{cm}$
interplanetary plasma	$10^7 \, \text{m}^{-3}$	$10^5 \, \text{K}$	$10^{11} \, \text{m}$
surface of an X-ray pulsar	$10^{24} \, \text{cm}^{-3}$	$10^8 \, \text{K}$	$10^5 \, \text{cm}$

(b) A plasma of thickness L changes from being optically thin to optically thick as the parameter $\gamma_c(\omega)L/n(\omega)c$ passes through unity. Using the relation (21.39) between the absorption coefficient for bremsstrahlung and the estimated thickness L of each of the plasmas in part (a) to estimate the frequency at which the plasma becomes optically thin to bremsstrahlung.

(c) Illustrate the dependence of the specific intensity $I(\omega)$ on ω on a log–log plot over the range where the plasma passes from being optically thick to being optically thin.

21.2 For a radioastronomical source of bremsstrahlung that is optically thin and is spatially resolved (its angular size is greater than the beam size of the radiotelescope), the emission is characterized by the temperature and a quantity called the emission measure.

(a) Show that (21.34), (21.38) and (21.39) imply that the power radiated per unit frequency per thermal electron in bremsstrahlung is related to the collision frequency by

$$P(\omega) = \frac{4\pi n(\omega)\omega_p^2}{(2\pi c)^3 n_e m_e V_e^2} \, \nu_e(\omega). \qquad (E21.2)$$

(b) Show that on integrating the power radiated per unit volume per unit frequency and per unit solid angle $n_e P(\omega)/4\pi$, with $P(\omega)$

given by (21.34) through a source of thickness L along the line of sight (the z axis), the specific intensity of the bremsstrahlung implies a brightness temperature T_B, cf. (17.45) with Θ replaced by T_B,

$$\frac{T_B}{T_e} = EM \frac{32\pi^2}{3(2\pi)^{1/2}} \frac{r_0^3 c^2}{\omega^2} \frac{c}{V_e} \frac{3}{\sqrt{3}} \frac{\pi}{\sqrt{3}} \bar{G}(V_e, \omega), \qquad (E21.3)$$

with $r_0 = e^2/4\pi\varepsilon_0 mc^2$ the classical radius of the electron and where the *emission measure* for bremsstrahlung is defined by

$$EM = \int_0^L dz \sum_i Z_i^2 n_i n_e.$$

(c)   An HII region is a sphere of fully ionized plasma around a bright star. A model for an HII region implies $EM = 10^4\,\text{pc-cm}^{-6}$ ($1\,\text{pc} = 3.1 \times 10^{16}\,\text{m}$) and $T_e = 10^4\,\text{K}$. Estimate the expected radio brightness temperature T_B.

21.3 The following problem concerns the *bremsstrahlung emission of Langmuir waves*. The Langmuir waves are assumed to have phase speeds $\gg V_e$, and properties $\omega_L(k) = \omega_p$, $R_L(k) = \frac{1}{2}$. The power radiated in Langmuir waves per unit wavenumber $P(k)$ is defined by writing the total power radiated P in the form

$$P = \int_0^\infty dk\, P(k). \qquad (E21.4)$$

(a)   Evaluate $P(k)$ in the straight line approximation, deriving a result analogous to the expression (21.22) for $P(\omega)$ for transverse waves.

(b)   Show that the result obtained in part (a) reduces for $v \gg \omega_p/b_1$ to

$$P(k) = \sum_i \frac{8}{3} \frac{Z_i^2 n_i e^6}{(4\pi\varepsilon_0)^3 m_e^2 v} \frac{k^2}{\omega_p^2} \ln\left(\frac{2}{\Gamma} \frac{v}{b_1 \omega}\right). \qquad (E21.5)$$

(c)   Show that the power radiated per thermal electron is

$$P(k) = \sum_i \frac{8}{3} \frac{Z_i^2 n_i e^6}{(4\pi\varepsilon_0)^3 m_e^2} \frac{k^2}{\omega_p^2} \left(\frac{2}{\pi}\right)^{1/2} \frac{1}{V_e} \frac{\pi}{\sqrt{3}} \bar{G}(V_e, \omega). \qquad (E21.6)$$

21.4 Bremsstrahlung may be treated as emission in the wings of a collisionally broadened Cerenkov emission line. Let the collision frequency be ν. The model leading to (21.41) involves two assumptions. One is that the probability $p(\tau)d\tau$ that a particle experience a collision in the time interval $d\tau$ after an initial time $\tau = 0$ is

$$p(\tau)d\tau = \nu e^{-\nu\tau} d\tau. \qquad (E21.7)$$

The other assumption is that the wave train emitted by the particle has an abrupt change in phase whenever a collision occurs.

(a) The current associated with a particle in the straight line approximation in this model is separated into components for $t < 0$ and $t > 0$. Argue that the latter, denoted by a superscript $+$, is given by

$$\mathbf{J}^+(\omega, \mathbf{k}) = \int_0^\infty d\tau\, \nu\, e^{-\nu\tau} \int_0^\tau dt\, q\mathbf{v}\, e^{i(\omega - \mathbf{k}\cdot\mathbf{v})t}. \qquad (E21.8)$$

(b) Show that this current is

$$\mathbf{J}^+(\omega, \mathbf{k}) = q\mathbf{v} \left[\frac{i(\omega - \mathbf{k}\cdot\mathbf{v})}{(\omega - \mathbf{k}\cdot\mathbf{v})^2 + \nu^2} + \frac{\nu}{(\omega - \mathbf{k}\cdot\mathbf{v})^2 + \nu^2} \right]. \qquad (E21.9)$$

(c) Show that separation of the current associated with Cerenkov emission into parts for $t < 0$ and $t > 0$, cf. Exercise 4.5, gives for the latter:

$$\mathbf{J}^+(\omega, \mathbf{k}) = \frac{iq\mathbf{v}}{\omega - \mathbf{k}\cdot\mathbf{v} + i0}.$$

(d) Compare the real parts of these two currents to show that they imply replacement of the δ-function line profile by the Lorentzian line profile

$$\delta(\omega - \mathbf{k}\cdot\mathbf{v}) \to \frac{\nu}{\pi[(\omega - \mathbf{k}\cdot\mathbf{v})^2 + \nu^2]}. \qquad (E21.10)$$

(e) Show that if collisional broadening is included by replacing the δ-function profile in the probability (20.5) for Cerenkov emission by a Lorentzian line profile, the resulting probability describes bremsstrahlung. Further, show that for transverse waves in an isotropic plasma, after summing over the two states of polarization, this reproduces the form of the probability (21.41) in the limit $\omega \gg |\mathbf{k}\cdot\mathbf{v})|, \nu$.

(f) Show that the implied value of the collision frequency reproduces (21.40).

21.5 Bremsstrahlung due to electron–electron collisions has no electric dipole component. The quadrupolar component of the emission is treated in Exercise 17.8 and is extended in the following exercise. The power radiated in a single encounter between two electrons is given by (E17.5), viz.,

$$P = \frac{e^6}{480\pi^3 \varepsilon_0^3 m^2 c^5} \frac{v^2 + 11v_\phi^2}{r^4}. \qquad (E21.11)$$

(a) Assume that the speed at infinity is v_0 and that the impact parameter is b, and argue that the dependence of the total speed v and the azimuthal component v_ϕ of the velocity on radial distance r are determined by

$$v^2 = v_0^2 - \frac{e^2}{2\pi\varepsilon_0 mr}, \quad v_\phi = \frac{bv_0}{r}. \qquad (E21.12)$$

(b) Substitute $(E21.12)$ into $(E21.11)$ and derive an integral expression for the energy radiated in a single encounter as an integral over r, with

$$\dot{r} = (v^2 - v_\phi^2)^{1/2} = \tfrac{1}{2}\left(v_0^2 - \frac{e^2}{2\pi\varepsilon_0 mr} - \frac{b^2 v_0^2}{r^2}\right). \qquad (E21.13)$$

(c) Proceeding by analogy with (21.18), with v replaced by v_0, write down a formula for the power radiated by a single electron due to continuous encounters with surrounding electrons as an integral over impact parameter.

(d) Carry out the integrals over b and r, noting that the limits on the range are determined by $\dot{r} = 0, v$. Hence show that the mean power radiated per electron in electron–electron bremsstrahlung is given by

$$\bar{P} = \frac{n_e e^4 v_0^4}{36\pi\varepsilon_0^2 mc^5}. \qquad (E21.14)$$

Hints: Reduce the double integral to the form $\int_0^\infty dr (r - L)^{3/2}/r^{7/2}$, with $L = e^2/2\pi\varepsilon_0 m v_0^2$. Change variables by writing $r = L(1 + \tan^2 \phi)$. *Remark*: The emission due to electron–electron encounters is weaker than that due to electron–ion encounters by a numerical factor times v^2/c^2, where v is the speed of the emitting electron. It follows that electron–electron bremsstrahlung is unimportant in non-relativistic plasmas but becomes significant at sufficiently high temperatures.

21.6 Two standard properties of the Macdonald functions are

$$\frac{d}{dz}[z^\nu K_\nu(z)] = -z^\nu K_{\nu-1}(z), \quad K_{-\nu}(z) = K_\nu(z). \qquad (E21.15)$$

(a) Derive the first of these relations using the recursion formulas (21.11) and (21.12).

(b) Use the relations $(E21.15)$ to establish the identity (21.19), *viz.*,

$$\int_{z_1}^\infty dz\, z\left[K_0^2(z) + K_1^2(z)\right] = z_1 K_0(z_1) K_1(z_1).$$

(c) Derive the approximate forms (21.21) by substituting (21.17) into (21.18) and carrying out the integral over b.

21.7 An *impulsive model* for bremsstrahlung is developed by assuming that the velocity of a particle changes abruptly.

(a)   Let the velocity of a charge q change abruptly from $\mathbf{v}_1$ for $t < t_0$, when the particle is at $\mathbf{x} = \mathbf{x}_0$, to $\mathbf{v}_2$ for $t > 0$. Show that the current is

$$\mathbf{J}(\omega, \mathbf{k}) = iq e^{i(\omega t_0 - \mathbf{k}\cdot\mathbf{x}_0)} \left(\frac{\mathbf{v}_2}{\omega - \mathbf{k}\cdot\mathbf{v}_2 + i0} - \frac{\mathbf{v}_1}{\omega - \mathbf{k}\cdot\mathbf{v}_1 + i0} \right).$$
$$(E21.16)$$

(b)   Insert the current into the emission formula (16.10) and show that only the principal value terms contribute for emission of transverse waves in a medium with $n(\omega) \leq 1$.

(c)   Show that in the dipole approximation, which corresponds here to the lowest non-vanishing terms in an expansion in k, the energy radiated per unit frequency reduces to

$$U_{\text{rad}}(\omega) = \frac{q^2 n(\omega)}{4\pi\varepsilon_0} \frac{|\mathbf{v}_2 - \mathbf{v}_1|^2}{3\pi c^3}. \qquad (E21.17)$$

(d)   Show that if the force acting is $\mathbf{F}(t)$, then one has

$$\mathbf{v}_2 - \mathbf{v}_1 = \frac{1}{m_e} \int_{-\infty}^{\infty} dt\, \mathbf{F}(t). \qquad (E21.18)$$

(e)   Insert the expression (21.4) for the force into $(E21.17)$, compare the resulting expression with (21.16), and comment on the comparison.

21.8 The energy radiated per unit frequency *in vacuo* may be evaluated exactly for a relativistic particle in the impulsive model.

(a)   Show that $(E21.16)$ leads to an energy radiated

$$U_{\text{rad}}(\omega) = \frac{q^2}{4\pi\varepsilon_0} \frac{1}{(2\pi)^2 c^3} \int d^2\mathbf{\Omega}$$
$$\times \left| \boldsymbol{\kappa} \times \left(\frac{\mathbf{v}_2}{1 - \boldsymbol{\kappa}\cdot\mathbf{v}_2/c} - \frac{\mathbf{v}_1}{1 - \boldsymbol{\kappa}\cdot\mathbf{v}_1/c} \right) \right|^2. \qquad (E21.19)$$

(b)   Establish the identities $(\boldsymbol{\beta} = \mathbf{v}/c)$

$$\int d^2\mathbf{\Omega} \left| \frac{\boldsymbol{\kappa} \times \boldsymbol{\beta}}{1 - \boldsymbol{\kappa}\cdot\boldsymbol{\beta}} \right|^2 = 4\pi\gamma^2,$$

$$\int d^2\mathbf{\Omega} \frac{(\boldsymbol{\kappa} \times \boldsymbol{\beta}_1)\cdot(\boldsymbol{\kappa} \times \boldsymbol{\beta}_2)}{(1 - \boldsymbol{\kappa}\cdot\boldsymbol{\beta}_1)(1 - \boldsymbol{\kappa}\cdot\boldsymbol{\beta}_2)} = \frac{2\pi}{X} \ln \left| \frac{X + 1 - \boldsymbol{\beta}_1\cdot\boldsymbol{\beta}_2}{X - 1 + \boldsymbol{\beta}_1\cdot\boldsymbol{\beta}_2} \right|,$$

$$X = [(\boldsymbol{\beta}_1 - \boldsymbol{\beta}_2)^2 - (\boldsymbol{\beta}_1 \times \boldsymbol{\beta}_2)^2]^{1/2}. \qquad (E21.20)$$

Hint: In the second of these integrals use the *Feynman parametrization*

$$\frac{1}{AB} = \int_0^1 \frac{dx}{[1 - xA - (1 - x)B]^2},\qquad (E21.21)$$

and carry out the integral over solid angle using $(E21.20)$ before carrying out the integral over x.

(c)   Hence show

$$U_{\text{rad}}(\omega) = \frac{q^2}{4\pi^2\varepsilon_0} \left(\frac{u_1 u_2}{A_{12}} \ln \left| \frac{u_1 u_2 + A_{12}}{u_1 u_2 - A_{12}} \right| - 1 \right),\qquad (E21.22)$$

with $u_1 u_2 = \gamma_1 \gamma_2 (1 - \boldsymbol{\beta}_1 \cdot \boldsymbol{\beta}_2)$, $A_{12} = [(u_1 u_2)^2 - 1]^{1/2}$.

21.9 The following problem concerns the evaluation of the current in the straight line approximation including relativistic effects.

(a)   Show that the relativistic form of Newton's equation of motion $d\mathbf{p}/dt = \mathbf{F}(t)$ with $\mathbf{p} = \gamma m \mathbf{v}$, $\gamma = (1 - v^2 c^2)^{-1/2}$, implies an acceleration $\mathbf{a}(t)$ given by

$$a_i(t) = \frac{1}{m\gamma} \left(\delta_{ij} - \frac{v_i v_j}{c^2} \right) F_i(t).\qquad (E21.23)$$

(b)   The orbit in the straight line approximation with the first order term included is

$$\mathbf{X}(t) = \mathbf{x}_0 + \mathbf{v}t + \mathbf{X}^{(1)}(t).\qquad (E21.24)$$

Show

$$\ddot{X}_i^{(1)}(t) = \frac{1}{m\gamma} \left(\delta_{ij} - \frac{v_i v_j}{c^2} \right) F_i(t).\qquad (E21.25)$$

(c)   Show that the first order current, cf. (18.4), is

$$\mathbf{J}^{(1)}(\omega, \mathbf{k}) = q \, e^{i\mathbf{k}\cdot\mathbf{x}_0} \frac{(\omega - \mathbf{k}\cdot\mathbf{v})\delta_{ij} + k_j v_i}{(\omega - \mathbf{k}\cdot\mathbf{v})^2}$$

$$\times \int_{-\infty}^{\infty} dt \, \ddot{X}_i^{(1)}(t) \, e^{i(\omega - \mathbf{k}\cdot\mathbf{v})t}.\qquad (E21.26)$$

(d)   Using (21.15), show that $(E21.25)$ implies that the electron current is

$$\mathbf{J}^{(1)}(\omega, \mathbf{k}) = -\frac{2Ze^3 e^{i\mathbf{k}\cdot\mathbf{x}_0}}{4\pi\varepsilon_0 m_e \gamma v^2} \frac{(\omega - \mathbf{k}\cdot\mathbf{v})\delta_{ij} + k_j v_i}{(\omega - \mathbf{k}\cdot\mathbf{v})^2}$$

$$\times \left[i \frac{b_j}{b} K_1 \left(\frac{b}{v}(\omega - \mathbf{k}\cdot\mathbf{v}) \right) + \frac{1}{\gamma^2} \frac{v_j}{v} K_0 \left(\frac{b}{v}(\omega - \mathbf{k}\cdot\mathbf{v}) \right) \right].$$

$$(E21.27)$$

22

Formal Theory of
Gyromagnetic Emission

Preamble

Gyromagnetic emission is the generic name for the emission due to the accelerated motion of a charged particle in a magnetostatic field. The formal theory of gyromagnetic emission, or gyroemission for short, is developed here; its application to emission and absorption by mildly relativistic electrons in Chapter 23 and by ultrarelativistic electrons in Chapter 24.

22.1 The Current due to a Spiraling Charge

To treat gyromagnetic emission we first need to identify the relevant current, and this requires an explicit expression for the orbit in (18.4). The orbit is found by solving the equation of motion.

The equation of motion for a particle with charge q and mass m in a magnetic field $\mathbf{B}$ is

$$\frac{d\mathbf{p}}{dt} = q\mathbf{v} \times \mathbf{B}. \tag{22.1}$$

The following quantities are constants of the motion: the energy $\varepsilon = \gamma mc^2$, and the components $p_\perp = \gamma m v_\perp$ and $p_\parallel = \gamma m v_\parallel$ of the momentum perpendicular and parallel to the magnetic field, respectively. The motion of the particle is decomposed into a motion at constant velocity along the field lines plus a circular motion perpendicular to the field lines. This motion is referred to as a spiraling motion with the pitch of the spiral defining the *pitch angle* α:

$$v_\perp = v \sin \alpha, \quad v_\parallel = v \cos \alpha. \tag{22.2}$$

The frequency of the circular motion is called the *gyrofrequency* Ω, and the radius of the circular motion is called the *radius of gyration R*:

$$\Omega = \frac{\Omega_0}{\gamma}, \quad \Omega_0 = \frac{|q|B}{m}, \quad R = \frac{v_\perp}{\Omega} = \frac{p_\perp}{|q|B}. \tag{22.3}$$

The sense of gyration, which is the handedness of the circular motion, depends on the sign of the charge

$$\eta = q/|q|, \tag{22.4}$$

being right hand in a screw sense relative to the direction of $\mathbf{B}$ for negative charges ($\eta = -1$).

The orbit of the particle is written

$$\mathbf{X}(t) = \mathbf{x}_0 + \big(R\sin(\phi_0 + \Omega t), \eta R\cos(\phi_0 + \Omega t), v_\| t\big), \tag{22.5}$$

where ϕ_0 and $\mathbf{x}_0$ are determined by the position of the particle at $t = 0$, and where the 3 axis is chosen along the direction of $\mathbf{B}$. The instantaneous velocity of the particle is given by

$$\mathbf{v}(t) = \dot{\mathbf{X}}(t) = (v_\perp \cos(\Omega t), -\eta v_\perp \sin(\Omega t), v_\|). \tag{22.6}$$

Positively charged particles ($\eta = +1$) gyrate in a left hand screw sense relative to $\mathbf{B}$, and negatively charged particles ($\eta = -1$) gyrate in a right hand screw sense relative to $\mathbf{B}$.

The Fourier transform of the current, cf. (18.4), is

$$\mathbf{J}(\omega, \mathbf{k}) = q \int_{-\infty}^{\infty} dt\, \mathbf{v}(t)\, e^{i\omega t - i\mathbf{k}\cdot\mathbf{X}(t)}. \tag{22.7}$$

The wavevector is written

$$\mathbf{k} = (k_\perp \cos\psi, k_\perp \sin\psi, k_\|) = k(\sin\theta\cos\phi, \sin\theta\sin\phi, \cos\theta), \tag{22.8}$$

and then with (22.5) one has

$$i\omega t - i\mathbf{k}\cdot\mathbf{X}(t) = -i\mathbf{k}\cdot\mathbf{x}_0 + i(\omega - k_\| v_\|)t - ik_\perp \Omega\sin(\Omega t + \phi_0 + \eta\psi). \tag{22.9}$$

The integral in (22.7) may be evaluated explicitly after expanding in Bessel functions. The generating function for Bessel functions is

$$e^{iz\sin\phi} = \sum_{s=-\infty}^{\infty} e^{is\phi} J_s(z). \tag{22.10}$$

Then one has

$$e^{i\omega t - i\mathbf{k}\cdot\mathbf{X}(t)} = e^{-i\mathbf{k}\cdot\mathbf{x}_0} \sum_{s=-\infty}^{\infty} J_s(k_\perp R)e^{i(\omega - s\Omega - k_\| v_\|)t - is(\phi_0 + \eta\psi)}, \tag{22.11}$$

and hence

$$\mathbf{J}(\omega, \mathbf{k}) = q e^{-i\mathbf{k}\cdot\mathbf{x}_0} \sum_{s=-\infty}^{\infty} e^{-is(\phi_0 + \eta\psi)} \mathbf{V}(\mathbf{k}, \mathbf{p}; s)\, 2\pi\delta(\omega - s\Omega - k_\| v_\|), \tag{22.12}$$

with

$$\mathbf{V}(\mathbf{k}, \mathbf{p}; s) = \left(\tfrac{1}{2}v_\perp \left[e^{i\eta\psi} J_{s-1}(k_\perp R) + e^{-i\eta\psi} J_{s+1}(k_\perp R)\right],\right.$$
$$- i\eta\tfrac{1}{2}v_\perp \left[e^{i\eta\psi} J_{s-1}(k_\perp R) - e^{-i\eta\psi} J_{s+1}(k_\perp R)\right],$$
$$\left. v_\parallel J_s(k_\perp R)\right). \tag{22.13}$$

The expression (22.13) simplifies slightly if the coordinate axes are oriented such that $\mathbf{k}$ is in the 1–3 plane, corresponding to $\psi = 0$ in (22.8). Then use of the recurrence relations

$$J_{s-1}(z) + J_{s+1}(z) = \frac{2s}{z} J_s(z), \tag{22.14}$$

$$J_{s-1}(z) - J_{s+1}(z) = 2J_s'(z), \tag{22.15}$$

with $J_s'(z) = dJ_s(z)/dz$, implies that (22.13) reduces to

$$\mathbf{V}(\mathbf{k}, \mathbf{p}; s) = \left(v_\perp \frac{s}{k_\perp R} J_s(k_\perp R), -i\eta v_\perp J_s'(k_\perp R), v_\parallel J_s(k_\perp R)\right). \tag{22.16}$$

On inserting the current (22.12) into the emission formula (16.14), the power radiated into waves in the mode M in the elemental range $d^3\mathbf{k}/(2\pi)^3$ due to a spiraling charge reduces to

$$P_M(\mathbf{k}) = \sum_s \frac{2\pi q^2 R_M(\mathbf{k})}{\varepsilon_0} \left|\mathbf{e}_M^*(\mathbf{k}) \cdot \mathbf{V}(\mathbf{k}, \mathbf{p}; s)\right|^2 \delta\big(\omega_M(\mathbf{k}) - s\Omega - k_\parallel v_\parallel\big),$$
$$\tag{22.17}$$

By analogy with the definition (20.4), it is convenient to define a *probability of gyromagnetic emission* at the sth harmonic by writing

$$P_M(\mathbf{k}) = \sum_s \hbar\omega_M(\mathbf{k}) \, w_M(\mathbf{p}, \mathbf{k}, s). \tag{22.18}$$

Comparison with (22.17) then gives

$$w_M(\mathbf{p}, \mathbf{k}, s) = \frac{2\pi q^2 R_M(\mathbf{k})}{\varepsilon_0 \hbar\omega_M(\mathbf{k})} \left|\mathbf{e}_M^*(\mathbf{k}) \cdot \mathbf{V}(\mathbf{k}, \mathbf{p}; s)\right|^2 \delta\big(\omega_M(\mathbf{k}) - s\Omega - k_\parallel v_\parallel\big). \tag{22.19}$$

The probability (22.19) is the basis for the discussion of gyroemission and absorption here.

For transverse waves *in vacuo* the probability (22.19) is written as the polarization tensor

$$w^{\alpha\beta}(\mathbf{p}, \mathbf{k}, s) = \frac{\pi q^2}{\varepsilon_0 \hbar\omega} V^\alpha(\mathbf{k}, \mathbf{p}; s) V^{*\beta}(\mathbf{k}, \mathbf{p}; s) \delta\big(\omega - s\Omega - k_\parallel v_\parallel\big), \tag{22.20}$$

with $\omega = kc$ and with

$$V^\alpha(\mathbf{k}, \mathbf{p}; s) = \mathbf{e}^\alpha \cdot \mathbf{V}(\mathbf{k}, \mathbf{p}; s). \tag{22.21}$$

Gyroemission at each harmonic is completely polarized in the sense implied by (22.21).

22.2 The Resonance Condition

The resonance condition, also called the Doppler condition, for a particle in a magnetostatic field is implied by the δ-function in (22.17) or (22.19). For waves in the mode M, the resonance condition is

$$\omega_M(\mathbf{k}) - s\Omega - k_\parallel v_\parallel = 0. \qquad (22.22)$$

In terms of the refractive index (22.20) becomes

$$\omega = \frac{s\Omega_0}{\gamma[1 - n_M(\omega,\theta)\beta\cos\theta\cos\alpha]}, \qquad (22.23)$$

with $\beta = v/c$, $\gamma = (1 - \beta^2)^{-1/2}$, and where the angles α and θ are defined by (22.2) and (22.8), respectively. Resonances at $s > 0$ are said to be via the *normal Doppler effect*, and those at $s < 0$ are said to be via the *anomalous Doppler effect*. The resonance at $s = 0$ is sometimes said to be via the Cerenkov effect, although this is clearly open to confusion with the Cerenkov effect in the absence of a magnetic field. Resonances at $s \leq 0$ are possible only for $n_M(\omega,\theta)\beta > 1$.

As with the Cerenkov condition in the absence of a magnetostatic field, cf. §20.2, the resonance condition has quite different interpretations from classical and quantum mechanical viewpoints. From a classical viewpoint, consider the inertial frame in which the gyrocenter of the particle is at rest. On transforming to this frame, the resonance condition corresponds to the wave frequency being s times the particle gyrofrequency, as expected in a classical resonance.

For non-relativistic particles the resonance at $s = 1$ is dominant, and this corresponds to electric dipole emission. Emission at higher harmonics s becomes increasingly important with increasing energy of the particle, and emission at very high harmonics dominates in synchrotron emission. The resonances at negative harmonics $s < 0$ appear only when $\omega_M(\mathbf{k}) - k_\parallel v_\parallel$ is negative, and this corresponds to the wave frequency in the rest frame of the gyrocenter having the opposite sign to the wave frequency in the laboratory frame.

In a quantum mechanical treatment the energy eigenvalues of the particle play an important role. The solution of the relevant relativistic wave equation (the Klein–Gordon equation or the Dirac equation) for a particle in a magnetostatic field leads to energy eigenvalues

$$\varepsilon = \varepsilon_n(p_\parallel) = (m^2c^4 + p_\parallel^2 c^2 + 2n|q|B\hbar c^2)^{1/2}. \qquad (22.24)$$

In deriving (22.24) a spin contribution, which is different for the Dirac and the Klein–Gordon equations, is neglected. In both cases a factor of order unity is neglected in comparison with $2n$, which is justified

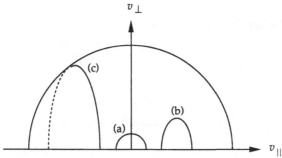

Fig. 22.1 Examples of resonance ellipses: (a) a semicircle centered on
the origin, (b) an ellipse inside $v = c$, (c) an ellipse touching $v = c$.

because the classical limit involves taking the limit $n \gg 1$. The quantum
numbers are the continuous parameter $p_\parallel$ and the discrete parameter
$n = 0, 1, 2, \ldots$. When a particle emits a wave quantum in a magnetic
field, momentum perpendicular to the magnetic field is not conserved
in general. Let $p'_\parallel$ and n' be the quantum numbers after emission of
the wave quantum. Conservation of parallel momentum and of energy
require

$$p'_\parallel = p_\parallel - \hbar k_\parallel, \quad \varepsilon' = \varepsilon_{n'}(p'_\parallel) = \varepsilon_n(p_\parallel) - \hbar \omega. \tag{22.25}$$

Now write $n' = n - s$ and assume that the changes in $p_\parallel$, ε, n are all
small. Then an expansion in the small quantities gives

$$\varepsilon - \hbar \omega = \left(1 - \hat{D}_s + \tfrac{1}{2}\hat{D}_s^2 + \cdots\right)\varepsilon, \quad \hat{D}_s = \hbar \left(\frac{s\Omega}{v_\perp} \frac{\partial}{\partial p_\perp} + k_\parallel \frac{\partial}{\partial p_\parallel}\right).$$
$$\tag{22.26}$$

To lowest order in $\hbar$ (22.26) with (22.25) reproduces (22.22). Thus, as
with the resonance condition for Cerenkov emission, cf. §20.2, from a
semiclassical viewpoint, the resonance condition reflects conservation of
energy and momentum on the microscopic scale.

The resonance condition (22.22) or (22.23) is amenable to a graphical
interpretation. It is convenient to plot the resonance curve in $v_\perp$–$v_\parallel$
space for given ω and $k_\parallel$. The resonance condition for each harmonic
defines a *resonance ellipse*. The resonance ellipse corresponds to all the
values of $v_\perp$ and $v_\parallel$ for which resonance with a wave at given ω, $k_\parallel$ and s
is possible. That is, a given wave resonates with all particles that lie on
the resonance ellipse that it defines. Similarly, a given particle resonates
with all waves that define resonance ellipses that pass the representative
point of the particle in $v_\perp$–$v_\parallel$ space.

The resonance ellipse is centered on the $v_\parallel$ axis at $v_\parallel = v_c$, with
semi-major axis v_R perpendicular to the $v_\parallel$ axis, and with eccentricity

e. These parameters are given by

$$\frac{v_c}{c} = \frac{\omega k_\| c}{s^2 \Omega_0^2 + k_\|^2 c^2}, \quad \left(\frac{v_R}{c}\right)^2 = \frac{s^2 \Omega_0^2 + k_\|^2 c^2 - \omega^2}{s^2 \Omega_0^2 + k_\|^2 c^2},$$

$$e^2 = \frac{k_\|^2 c^2}{s^2 \Omega_0^2 + k_\|^2 c^2}. \tag{22.27}$$

Some examples of resonance ellipses are illustrated in Figure 22.1.

22.3 The Power Radiated *in Vacuo*

The power in gyromagnetic emission *in vacuo* may be evaluated explicitly. This involves evaluating the sum and integrals in

$$P = \sum_{s=1}^{\infty} \int \frac{d^3 k}{(2\pi)^3} \, \hbar\omega \, w^{\alpha\alpha}(\mathbf{p}, \mathbf{k}, s). \tag{22.28}$$

For waves *in vacuo* one writes

$$\frac{d^3 k}{(2\pi)^3} = \frac{\omega^2}{(2\pi c)^3} \, d\omega d\cos\theta d\phi, \tag{22.29}$$

and carries out the trivial integral over ϕ. On introducing the variables used in (22.23), with the refractive index equal to unity here, the integral over ω is performed over the δ-function, giving

$$P = \sum_{s=1}^{\infty} \int_{-1}^{1} d\cos\theta \, \frac{q^2}{4\pi\varepsilon_0 c} \frac{s^2 \Omega_0^2 \beta_\perp^2}{\gamma^2 (1 - \beta_\| \cos\theta)^3}$$

$$\times \left\{ \left[\frac{z}{x} J_s(sx)\right]^2 + [J_s'(sx)]^2 \right\}, \tag{22.30}$$

with

$$z = \frac{\cos\theta - \beta_\|}{1 - \beta_\| \cos\theta}, \quad x = \frac{\beta_\perp \sin\theta}{1 - \beta_\| \cos\theta}. \tag{22.31}$$

The power emitted is separated in components in the two states of linear polarization. When the probability is written as a polarization tensor, the terms involving J_s^2 and $J_s'^2$ in (22.30) are the 11 and 22 terms, respectively, and these terms give the power emitted with polarization along the projection of $\mathbf{B}$ on the plane orthogonal to $\mathbf{k}$, and along the direction $\mathbf{B} \times \mathbf{k}$, respectively. Let the relevant powers be denoted $P^\|$ and $P^\perp$, with $P = P^\| + P^\perp$.

The $\cos\theta$-integral in (22.30) is performed after making a change of variable to $\cos\theta' = z$. Physically θ' corresponds to the angle of emission in the inertial frame in which the gyrocenter of the particle is at rest.

Denoting all quantities in this frame by primes, one has

$$\cos\theta' = \frac{\cos\theta - \beta_\parallel}{1 - \beta_\parallel \cos\theta}, \quad \beta'\sin\theta' = \frac{\beta_\perp \sin\theta}{1 - \beta_\parallel \cos\theta}, \quad \beta' = \frac{\beta_\perp}{(1 - \beta_\parallel^2)^{1/2}}.$$

(22.32)

Then with $d\cos\theta/d\cos\theta' = (1 - \beta_\parallel^2)/(1 - \beta_\parallel \cos\theta)^2$, (22.30) reduces to

$$P^{\parallel,\perp} = \sum_{s=1}^\infty \int_{-1}^1 d\cos\theta' \, \frac{q^2}{4\pi\varepsilon_0 c} s^2 \Omega_0^2 \beta'^2 (1 - \beta'^2)(1 + \beta_\parallel \cos\theta') F_s^{\parallel,\perp},$$

(22.33)

$$F_s^\parallel = \frac{\cos\theta'}{\beta'\sin\theta'} J_s(s\beta'\sin\theta'), \quad F_s^\perp = J_s'(s\beta'\sin\theta').$$

The integrals in (22.33) are performed as follows. First note that from symmetry the term proportional to $\beta_\parallel \cos\theta'$ does not contribute. One requires the following integral identities:

$$\int_{-1}^1 d\cos\theta' \, J_t^2(s\beta'\sin\theta') = \frac{2}{\beta'}\int_0^{\beta'} dy \, J_{2t}(2sy),$$

$$\int_{-1}^1 \frac{d\cos\theta'}{\sin^2\theta'} J_t^2(s\beta'\sin\theta') = 2\int_0^{\beta'} \frac{dy}{y} J_{2t}(2sy),$$

(22.34)

together with the identity, cf. (22.30),

$$\left[\frac{z}{x}J_s(sx)\right]^2 + [J_s'(sx)]^2 = \tfrac{1}{2}\left[J_{s-1}^2(z) + J_{s+1}^2(z)\right].$$

(22.35)

Using the recursion relations and the differential equation for the Bessel functions, cf. Exercise 22.6, one obtains

$$P^{\parallel,\perp} = \sum_{s=1} \frac{2q^2 s^2 \Omega_0^2}{4\pi\varepsilon_0 c} \beta'^2 (1 - \beta'^2) G_s^{\parallel,\perp},$$

(22.36)

with

$$G_s^\parallel = \frac{1}{\beta'^2}\int_0^{\beta'} dy \, \frac{J_{2s}(2sy)}{y} - \frac{1}{\beta'^3}\int_0^{\beta'} dy \, J_{2s}(2sy),$$

$$G_s^\perp = \frac{1}{s\beta'} J_{2s}'(2s\beta') - \frac{1}{\beta'^2}\int_0^{\beta'} \frac{dy}{y} J_{2s}(2sy) + \frac{1}{\beta'}\int_0^{\beta'} dy \, J_{2s}(2sy).$$

(22.37)

The final step is to perform the sum over s. The relevant sums are Kapteyn series, and the following series were first evaluated explicitly for the present purpose by Schott in 1912:

$$\sum_{s=1} s^2 J_{2s}(2sy) = \frac{y^2(1 + y^2)}{2(1 - y^2)^4},$$

(22.38)

$$\sum_{s=1} s J_{2s}'(2sy) = \frac{y}{2(1 - y^2)^2},$$

(22.39)

which apply for $0 \le y < 1$. The remaining integral is elementary, and the final result for the power emitted is

$$P = P^{\parallel} + P^{\perp} = \frac{2q^2\Omega_0^2 p_\perp^2}{12\pi\varepsilon_0 m^2 c^3} = \frac{2}{3}\frac{q^2\Omega_0^2\gamma^2\beta_\perp^2}{4\pi\varepsilon_0 c}, \qquad (22.40)$$

which is independent of $p_{\parallel}$. The result (22.40) for P reproduces (18.18) in the non-relativistic limit, and (22.40) generalizes (18.18) to include relativistic effects. The degree of linear polarization is determined by

$$\frac{P^{\parallel} - P^{\perp}}{P^{\parallel} + P^{\perp}} = -\frac{2 + \beta_\perp^2 - 2\beta_{\parallel}^2}{4(1 - \beta_{\parallel}^2)} = -\frac{2m^2c^2 + 3p_\perp^2}{4(m^2c^2 + p_\perp^2)}, \qquad (22.41)$$

which is also independent of $p_{\parallel}$.

An alternative way of deriving (22.40) and (22.41) is to perform the sum over s in (22.30) before carrying out the integral over θ. The relevant sums were also evaluated by Schott for this purpose. They are

$$\sum_{s=1} s^2 J_s^2(sx) = \frac{x^2(4 + x^2)}{16(1 - x^2)^{7/2}}, \qquad (22.42)$$

$$\sum_{s=1} s^2 J_s'^2(sx) = \frac{4 + 3x^2}{16(1 - x^2)^{5/2}}, \qquad (22.43)$$

for $0 \le x < 1$. Then one obtains

$$P = \frac{q^2\Omega_0^2}{16(4\pi\varepsilon_0)c}\beta_\perp^2(1 - \beta^2)$$

$$\times \int_{-1}^{1} \mathrm{d}\cos\theta \left\{ \frac{(\cos\theta - \beta_{\parallel})^2[4(1 - \beta_{\parallel}\cos\theta)^2 + \beta_\perp^2\sin^2\theta]}{[(1 - \beta_{\parallel}\cos\theta)^2 - \beta_\perp^2\sin^2\theta]^{7/2}} \right.$$

$$\left. + \frac{4(1 - \beta_{\parallel}\cos\theta)^2 + 3\beta_\perp^2\sin^2\theta}{[(1 - \beta_{\parallel}\cos\theta)^2 - \beta_\perp^2\sin^2\theta]^{5/2}} \right\}. \qquad (22.44)$$

The integral over $\cos\theta$ is then performed using standard integral identities.

22.4 Gyroemission and Absorption of Magnetoionic Waves

Gyromagnetic emission and absorption of magnetoionic waves is of major importance in plasmas. Here these processes are treated using the semiclassical formalism developed in §16.3.

The transfer equation for the waves is written down in semiclassical form in (16.33), viz.,

$$\frac{\mathrm{d}N_M(\mathbf{k})}{\mathrm{d}t} = \alpha_M(\mathbf{k}) - \gamma_M(\mathbf{k})N_M(\mathbf{k}). \qquad (22.45)$$

The emission and absorption coefficients are given by (16.34). In the present case, the quantum numbers q are n and $p_{\parallel}$, and the quantum numbers q' are $n - s$ and $p_{\parallel} - \hbar k_{\parallel}$. Noting from (22.24) that in the classical limit one has $n \to p_\perp^2/2|q|B\hbar$ so that s/n is of order $\hbar$, the expansion of the kinetic equations in powers of $\hbar$ gives the following leading terms:

$$\begin{bmatrix} \alpha_M(\mathbf{k}) \\ \gamma_M(\mathbf{k}) \end{bmatrix} = \sum_{s=-\infty}^{\infty} \int d^3p\, w_M(\mathbf{p}, \mathbf{k}, s) \begin{bmatrix} f(\mathbf{p}) \\ -\hat{D}_s f(\mathbf{p}) \end{bmatrix}, \qquad (22.46)$$

with $\hat{D}_s$ given by (22.26).

The semiclassical form for the absorption coefficient (22.46), with the probability given by (22.19), reproduces the corresponding result of a classical calculation. As shown in §17.5 for a quantum treatment of multipole emission and in §20.5 for Čerenkov absorption, the semiclassical expression reproduces the result obtained by inserting the antihermitian part of the relevant dielectric tensor into the expression (15.14) for the absorption coefficient. The relevant dielectric tensor for a magnetized plasma is

$$\begin{aligned}
K_{ij}(\omega, \mathbf{k}) = \delta_{ij} &+ \sum \frac{q^2}{4\pi\varepsilon_0} \int d^3p \left[\frac{v_\parallel}{v_\perp} \left(v_\perp \frac{\partial}{\partial p_\parallel} - v_\parallel \frac{\partial}{\partial p_\perp} \right) f(\mathbf{p}) b_i b_j \right. \\
&+ \sum_{s=-\infty}^{\infty} \frac{V_i(\mathbf{k}, \mathbf{p}; s) V_j^*(\mathbf{k}, \mathbf{p}; s)}{\omega - s\Omega - k_\parallel v_\parallel + i0} \\
&\left. \times \left(\frac{\omega - k_\parallel v_\parallel}{v_\perp} \frac{\partial}{\partial p_\perp} + k_\parallel \frac{\partial}{\partial p_\parallel} \right) f(\mathbf{p}) \right].
\end{aligned} \qquad (22.47)$$

Using (22.47) it is straightforward to repeat the steps outlined in §20.5 to prove that the two methods for treating absorption are equivalent in the present case.

For practical purposes it is often convenient to write the transfer equation in a form with ω and θ as the independent variables. The waves are described in terms of the specific intensity $I_M(\omega, \theta)$, which is the power crossing unit area per unit solid angle and per unit frequency range. Here we refer to unit solid angle about the wave normal direction. In terms of these variables, the emission by an individual particle is described by the *emissivity*, which is the power radiated per unit frequency and per unit solid angle. The emissivity $\eta_M(\omega, \theta)$ is identified directly from (22.18) using

$$\pi \int_0^\infty d\omega \int_{-1}^1 d\cos\theta\, \eta_M(\omega, \theta) = \int \frac{d^3k}{(2\pi)^3} P_M(\mathbf{k}).$$

Using the properties of the magnetoionic waves given in §13.4, together with the form (15.23) for $R_M(\mathbf{k})$, and the parameters introduced in

(22.23), the emissivity at the sth harmonic reduces to

$$\eta_M(s,\omega,\theta) = \frac{q^2 n_M \omega^2 \beta^2 \sin^2\alpha}{16\pi^2 \varepsilon_0 c(1 + T_M^2)}$$
$$\times \left| \frac{L_M \sin\theta + T_M(\cos\theta - n_M \beta \cos\alpha)}{n_M \beta \sin\alpha \sin\theta} J_s - \eta J_s' \right|^2$$
$$\times \delta[\omega(1 - n_M \beta \cos\theta \cos\alpha) - s\Omega], \qquad (22.48)$$

where the arguments (ω,θ) of n_M, T_M and L_M are omitted, and similarly the arguments $(\omega/\Omega)n_M \beta \sin\alpha \sin\theta$ of J_s and J_s' are omitted. The sign η of the charge is equal to -1 for electrons.

A form of the transfer equation in terms of these variables is

$$v_{gM} \frac{dI_M(\omega,\theta)}{ds} = J_M(\omega,\theta) - \gamma_M(\omega,\theta)I_M(\omega,\theta), \qquad (22.49)$$

where s denotes distance along the ray path. The emission coefficient $J_M(\omega,\theta)$ is identified as the volume emissivity, that is, the emissivity per unit volume. The emission coefficient is obtained by integrating the emissivity over the distribution of radiating particles:

$$J_M(\omega,\theta) = 2\pi \int_0^\infty dp\, p^2 \int_{-1}^1 d\cos\alpha\, \eta_M(\omega,\theta) f(p, \cos\alpha). \qquad (22.50)$$

with $\eta_M(\omega,\theta)$ identified as the sum of $\eta_M(s,\omega,\theta)$ over s. The absorption coefficient at the sth harmonic follows from (22.46), which reduces to

$$\gamma_M(s,\omega,\theta) = -2\pi \int_0^\infty dp\, p^2 \int_{-1}^1 d\cos\alpha\, \frac{(2\pi)^3 \eta_M(s,\omega,\theta)}{\beta n_M^2 \omega^2 \partial(\omega n_M)/\partial\omega}$$
$$\times \left(\frac{\partial}{\partial p} + \frac{\cos\alpha - n_M \beta \cos\theta}{p \sin\alpha} \frac{\partial}{\partial\alpha} \right) f(p, \cos\alpha). \quad (22.51)$$

Gyromagnetic emission and absorption by non-relativistic electrons may be treated approximately by noting that the power series expansions of the Bessel functions converge rapidly for $n_M \beta \sin\alpha \sin\theta \ll 1$. On making the power series expansion of the Bessel functions the lowest order terms are $J_s(z) = (\frac{1}{2}z)^s/s!$. Let us apply this approximation to the emissivity (22.48) for electrons ($q = -e$, $\eta = -1$). Then (22.48) gives

$$\eta_M(s,\omega,\theta) = \frac{e^2 \Omega_e^2}{8\pi^2 \varepsilon_0 n_M c} \frac{(s n_M \beta \sin\alpha)^{2s}}{2^{2s}(s!)^2(1 + T_M^2)}$$
$$\times \left[1 + L_M \sin\theta + T_M \cos\theta - n_M \beta \cos\alpha(T_M + \cos\theta) \right]^2$$
$$\times \delta[\omega(1 - n_M \beta \cos\theta \cos\alpha) - s\Omega_e(1 + \tfrac{1}{2}\beta^2)], \qquad (22.52)$$

It follows that emission and absorption decrease rapidly with increasing s for non-relativistic particles. The fundamental or cyclotron line usually dominates.

Gyromagnetic absorption may be negative under relatively mild conditions. The associated cyclotron maser emission is discussed in §23.2.

22.5 Kinetic Equation for Gyromagnetic Emission

The kinetic equation (16.35) for the particles due to the effect of gyromagnetic emission and absorption may be written down in terms of the probability (22.19) by taking the classical limit of the equations in a way similar to that done in §20.6 for Cerenkov emission. In the present case, the quantum numbers q are n and $p_\parallel$, the quantum numbers q' are $n - s$ and $p_\parallel - \hbar k_\parallel$ and the quantum numbers q'' are $n + s$ and $p_\parallel + \hbar k_\parallel$. Proceeding as in the derivation of (22.46), one obtains

$$\frac{df(\mathbf{p})}{dt} = \sum_{s=-\infty}^{\infty} \int \frac{d^3\mathbf{k}}{(2\pi)^3}\, \hat{D}_s w_M(\mathbf{p}, \mathbf{k}, s) \left[f(\mathbf{p}) + N_M(\mathbf{k}) \hat{D}_s f(\mathbf{p}) \right].$$

(22.53)

The term independent of $N_M(\mathbf{k})$ in (22.53) describes the back reaction to spontaneous emission. This term alone is of the form

$$\left[\frac{df(\mathbf{p})}{dt} \right]_{\text{spont}} = \frac{1}{p_\perp} \frac{\partial}{\partial p_\perp} \left[p_\perp A_\perp(\mathbf{p}) f(\mathbf{p}) \right] + \frac{\partial}{\partial p_\parallel} \left[A_\parallel(\mathbf{p}) f(\mathbf{p}) \right],$$

$$\begin{pmatrix} A_\perp(\mathbf{p}) \\ A_\parallel(\mathbf{p}) \end{pmatrix} = \sum_{s=-\infty}^{\infty} \int \frac{d^3\mathbf{k}}{(2\pi)^3}\, w_M(\mathbf{p}, \mathbf{k}, s) \begin{pmatrix} \hbar s \Omega / v_\perp \\ \hbar k_\parallel \end{pmatrix},$$

(22.54)

where subscript "spont" refers to spontaneous emission. As in the case of Cerenkov emission, one uses (22.54) to derive components of the radiation reaction force. Proceeding by analogy with the derivation of (20.36) one finds that any physical quantity $Q(p_\perp, p_\parallel)$ changes due to spontaneous emission according to

$$\left\langle \frac{dQ(p_\perp, p_\parallel)}{dt} \right\rangle = - \sum_{s=-\infty}^{\infty} \int \frac{d^3\mathbf{k}}{(2\pi)^3}\, w_M(\mathbf{p}, \mathbf{k}, s)\, \hat{D}_s Q(p_\perp, p_\parallel). \quad (22.55)$$

It is interesting to apply (22.55) to the case of radiation reaction to gyromagnetic emission *in vacuo* and to compare the result with that implied by the radiation reaction force (19.34), *viz.*, $\mathbf{F}_{\text{react}}(t) = (q^2/6\pi\varepsilon_0 c^3)\ddot{\mathbf{v}}(t)$. In making the comparison, it is important to note that (19.34) applies strictly only in the instantaneous rest frame of the particle and that one needs to average over a gyroperiod to make a meaningful comparison. Hence the only meaningful comparison is for the perpendicular components of momentum ($p_\parallel$ is zero and is unaffected by radiation reaction). In cases where they are both applicable, (22.55) and (19.34) lead to equivalent results.

As in the case of Cerenkov emission, the back reaction to the induced processes, described by the term proportional to $N_M(\mathbf{k})$ in (22.53), leads to diffusion in momentum space. Written in cylindrical polar coordinates in momentum space this has the form

$$\left[\frac{\mathrm{d}f(\mathbf{p})}{\mathrm{d}t}\right]_{\mathrm{ind}} = \frac{1}{p_\perp}\frac{\partial}{\partial p_\perp}\left\{p_\perp\left[D_{\perp\perp}(\mathbf{p})\frac{\partial}{\partial p_\perp} + D_{\perp\|}(\mathbf{p})\frac{\partial}{\partial p_\|}\right]f(\mathbf{p})\right\}$$
$$+ \frac{\partial}{\partial p_\|}\left\{\left[D_{\|\perp}(\mathbf{p})\frac{\partial}{\partial p_\perp} + D_{\|\|}(\mathbf{p})\frac{\partial}{\partial p_\|}\right]f(\mathbf{p})\right\},$$

$$(22.56)$$

where subscript "ind" refers to the induced processes. Alternatively, if (22.56) is written in spherical polar coordinates in momentum space then it has the form

$$\left[\frac{\mathrm{d}f(\mathbf{p})}{\mathrm{d}t}\right]_{\mathrm{ind}} = \frac{1}{\sin\alpha}\frac{\partial}{\partial\alpha}\left\{\sin\alpha\left[D_{\alpha\alpha}(\mathbf{p})\frac{\partial}{\partial\alpha} + D_{\alpha p}(\mathbf{p})\frac{\partial}{\partial p}\right]f(\mathbf{p})\right\}$$
$$+ \frac{1}{p^2}\frac{\partial}{\partial p}\left\{p^2\left[D_{p\alpha}(\mathbf{p})\frac{\partial}{\partial\alpha} + D_{pp}(\mathbf{p})\frac{\partial}{\partial p}\right]f(\mathbf{p})\right\}.$$

$$(22.57)$$

The diffusion coefficients in momentum space in (22.56) and (22.57) are evaluated using

$$D_{QQ'}(\mathbf{p}) = \sum_{s=-\infty}^{\infty}\int\frac{\mathrm{d}^3\mathbf{k}}{(2\pi)^3}\,w_M(\mathbf{p},\mathbf{k},s)\Delta Q\Delta Q', \qquad (22.58)$$

with $\Delta Q = \hat{D}_s Q$, for each quantity Q. Specifically, one has

$$\Delta p_\perp = \frac{s\Omega}{v_\perp}, \quad \Delta p_\| = \hbar k_\|, \quad \Delta\alpha = \frac{\hbar(\omega\cos\alpha - k_\| v)}{pv\sin\alpha}, \quad \Delta p = \frac{\hbar\omega}{v}.$$

$$(22.59)$$

Note that in deriving (22.59) it is implicit that the resonance condition (22.22) is satisfied.

The back reaction to maser emission is of particular interest. Specific calculations show that the back reaction tends to distort the distribution function in such a way as to reduce the growth rate of the maser. It is plausible that this should be the case in general.

22.6 Cyclotron Emission by Thermal Electrons

Cyclotron emission and absorption by non-thermal electrons occur as lines at the cyclotron frequency and its harmonics, with the intensity of the lines decreasingly rapidly with increasing harmonic number s. These lines have a finite width as a result of Doppler broadening due to the random thermal motions of the electrons. Cyclotron emission by a thermal distribution of electrons is treated here in the non-relativistic approximation in which case all the relevant integrals may be carried out explicitly. However, there is an intrinsic inconsistency in making the non-relativistic assumption in the present context: relativistic effects can be important for electrons of any energy when considering gyromagnetic emission.

With $\mathbf{p} = m_e \mathbf{v}$ for non-relativistic electrons, a Maxwellian distribution function of the electrons at temperature is

$$f(p) = \frac{n_e\, e^{-v^2/2V_e^2}}{(2\pi)^{3/2} m_e^3 V_e^3}, \qquad (22.60)$$

where $T_e = m_e V_e^2$ is the temperature in energy units. In the following it is usually convenient to use the parameter β_0 which is defined by writing the thermal speed of electrons as $V_e = \beta_0 c$. The emission coefficient $J_M(s, \omega, \theta)$ is given by (22.50) and its evaluation involves integrating the emissivity (22.48) over the distribution (22.60). Here we carry out the evaluation in the non-relativistic approximation, which involves replacing the relativistic gyrofrequency Ω by the non-relativistic electron cyclotron frequency Ω_e everywhere in the emissivity (22.48). It is convenient to use cylindrical polar coordinates in velocity space, that is, to use the variables $\beta_\perp$ and $\beta_\parallel$, rather than β and the pitch angle α. The non-relativistic approximation to the emissivity at the sth harmonic then reduces to

$$\eta_M(s, \omega, \theta) = \frac{e^2 n_M \omega^2 \beta_\perp^2}{16\pi^2 \varepsilon_0 c (1 + T_M^2)}$$
$$\times \left| \frac{L_M \sin\theta + T_M(\cos\theta - n_M\beta_\parallel)}{n_M \beta_\perp \sin\theta} J_s + J'_s \right|^2$$
$$\times \delta[\omega(1 - n_M\beta_\parallel \cos\theta) - s\Omega_e], \qquad (22.61)$$

with the argument of the Bessel functions being $z = (\omega/\Omega_e) n_M \beta_\perp \sin\theta$.
The evaluation of the integrals is outlined in Exercise 22.11. The resulting expression for the emission coefficient is

$$J_M(s, \omega, \theta) = \frac{4\pi \omega_p^2 n_M A_M(s, \omega, \theta) m_e \beta_0}{(2\pi)^{3/2} |\cos\theta|}$$
$$\times e^{-(\omega - s\Omega_e)^2 / 2\omega^2 n_M^2 \beta_0^2 \cos^2\theta}, \qquad (22.62)$$

and under the same assumptions the absorption coefficient is given by

$$\gamma_M(s, \omega, \theta) = \frac{\pi^{1/2} \omega_p^2 A_M(s, \omega, \theta)}{2\omega n_M \beta_0 |\cos\theta| \partial(\omega n_M)/\partial\omega}$$
$$\times e^{-(\omega - s\Omega_e)^2 / 2\omega^2 n_M^2 \beta_0^2 \cos^2\theta}, \qquad (22.63)$$

with

$$A_M(s,\omega,\theta)$$

$$= \frac{e^{-\lambda_M}}{1+T_M^2} \left\{ \left[\frac{\omega}{\Omega_e}(L_M\cos\theta - T_M\sin\theta)\tan\theta + sT_M\sec\theta \right]^2 \frac{I_s}{\lambda_M} \right.$$

$$+ 2 \left[\frac{\omega}{\Omega_e}(L_M\cos\theta - T_M\sin\theta)\tan\theta + sT_M\sec\theta \right]$$

$$\left. \times (I_s' - I_s) + \left(\frac{s^2}{\lambda_M} + 2\lambda_M \right) I_s - 2\lambda_M I_s' \right\}. \tag{22.64}$$

The formulas (22.62) and (22.63) apply to waves in any magnetoionic wave mode (and to other modes) in a magnetized plasma. Here we are concerned primarily with waves that can escape directly from the plasma. These are waves in the o mode at $\omega > \omega_p$ and in the x mode at $\omega > \omega_x$, cf. (13.27). These waves have refractive indices less than unity, and only emission via the normal Doppler effect is allowed for such waves. Only emission at $s \geq 1$ is considered in the following discussion.

The exponential functions in (22.62) and (22.63) determine the line profile. The line emission at the sth harmonic is centered on $\omega = s\Omega_e$ and it is Doppler broadened with a characteristic relative width

$$(\Delta\omega)_s/s\Omega_e = n_M\beta_0|\cos\theta|. \tag{22.65}$$

The Doppler broadening is due to the component of the random thermal motions along the magnetic field lines. That is, the Doppler broadening is due only to the random thermal motions of the gyrocenters of the electrons and does not involve their motion perpendicular to the field lines. As a consequence, the non-relativistic theory implies that the line width goes to zero for perpendicular propagation, that is, for $|\cos\theta| \to 0$. As discussed in §23.3, a relativistic effect ensures that the line width remains non-zero for $|\cos\theta| = 0$.

The detailed form (22.64) simplifies in the small gyroradius limit to

$$A_M(s,\omega,\theta) \approx \frac{(\frac{1}{2}\lambda_M)^{s-1}}{2s!(1+T_M^2)}$$

$$\times \left[\frac{\omega}{\Omega_e}(L_M\cos\theta - T_M\sin\theta)\tan\theta + sT_M\sec\theta + s \right]^2. \tag{22.66}$$

To obtain this result, the argument λ_M is rewritten as $k_\perp^2 R^2$, where $R = V_e/\Omega_e$ is the gyroradius of a thermal electron, and the small gyroradius limit then corresponds to $\lambda_M \ll 1$. The modified Bessel functions are approximated by the leading term in their power series expansions, specifically by $I_s(\lambda) \approx (\lambda/2)^{2s}/s!$, for $\lambda \ll 1$.

A further approximation applies to the magnetoionic waves, written $M = \sigma = \pm$ with $\sigma = +1$ for the o mode and $\sigma = -1$ for the x mode (for $X < 1$), in the limit (13.42). In this case (22.66) simplifies further to

$$A_\sigma(s, \omega, \theta) \approx \frac{s^2}{4} \frac{\left(\frac{1}{2}\lambda_M\right)^{s-1}}{s!} \left(1 - \sigma|\cos\theta|\right)^2. \qquad (22.67)$$

There is a rich variety of different analytic approximations for gyroemission and even in the case considered here of a non-relativistic thermal distribution several different analytic expressions are required to give a reasonable coverage of the various ranges of interest.

Thermal electron cyclotron emission is of importance as a diagnostic in laboratory plasma machines. The interpretation of the data is complicated by the fact that the plasma parameters, notably the temperature, density and the magnetic field strength, vary significantly along the line of sight, and the variations in these parameters must be deconvolved to obtain detailed information. Thermal electron cyclotron emission is also significant in astrophysics, notably in the slowly varying component of solar microwave emission, where is arises from hot plasma near sunspots ($B \approx 0.1$–0.3 T), and in optical cyclotron emission from white dwarf stars with strong magnetic fields $B \approx 10^3$ T. In these astrophysical applications the emission is optically thick. Gyromagnetic absorption by thermal electrons can also be important in preventing escape of radio emission generated at $\omega < 2\Omega_e$ or $\omega < 3\Omega_e$ from the solar corona, cf. Exercise 22.2.

Exercise Set 22

22.1 The parameters (22.27) of the resonance ellipse follow by writing the resonance condition (22.22) in the standard form of the equation for an ellipse.

(a) Show that by writing $\Omega = \Omega_0(1 - v_\perp^2/c^2 - v_\parallel^2/c^2)^{1/2}$ the resonance condition (22.22) reduces to the equation

$$(v_\parallel/c - v_c/c)^2/a^2 + v_\perp^2/b^2 = 1, \qquad (E22.1)$$

with $a = v_R/c$, $b^2 = (1 - e^2)a^2$, and where the parameters v_c, v_R, e are given by (22.27).

(b) Show that the ellipse touches the circle $v/c = 1$ for $\omega - k_\parallel v_\parallel < 0$ but does not cross it, cf. (c) in Figure 22.1, and find the values of $v_\parallel$ and $v_\perp$ for which this touching occurs.

(c) Show that part of the solution of $(E22.1)$ for $\omega - k_\parallel v_\parallel < 0$ is spurious, in that it does not satisfy the original equation (22.22).

22.2 Show that for waves in a mode M and for $k_\parallel^2 c^2 = n_M^2 \omega^2 \cos^2 \theta \ll s^2\Omega_e^2 \approx \omega^2$, the resonance ellipse is approximated by a circle whose radius is a fraction $2(s\Omega_0 - \omega)/\omega n_M \cos\theta$ of the displacement of its center from the origin of velocity space.

22.3 Show that the resonance equation (22.22) defines an ellipse in $p_\parallel$-$p_\perp$ space. Specifically, show that (22.22) implies

$$(\omega^2 - k_\parallel^2 c^2)\left(\frac{p_\parallel}{mc} - \frac{s\Omega_0 k_\parallel c}{\omega^2 - k_\parallel^2 c^2}\right)^2 + \omega^2\left(\frac{p_\perp}{mc}\right)^2$$

$$= \frac{s^2\Omega_0^2(\omega^2 + s^2\Omega_0^2 - k_\parallel^2 c^2)}{\omega^2 - k_\parallel^2 c^2}, \qquad (E22.2)$$

and hence identify the parameters of the ellipse.

22.4 Approximate forms for the quantity $\mathbf{V}(\mathbf{k}, \mathbf{p}; s)$ are obtained in the small gyroradius limit $k_\perp R \ll 1$ by retaining only the leading term in the power series expansion of the Bessel functions, as given by (22.52). Then approximate forms for the probability of gyromagnetic emission follow.

(a) Show, for $s = 0$ and $s = \pm 1$,

$$\mathbf{V}(\mathbf{k}, \mathbf{p}; 0) \approx (0, 0, 1), \quad \mathbf{V}(\mathbf{k}, \mathbf{p}; \pm 1) \approx \pm\tfrac{1}{2}v_\perp(1, \mp i\eta, 0). \quad (E22.3)$$

(b) Show, for $|s| > 1$,

$$\mathbf{V}(\mathbf{k}, \mathbf{p}; \pm|s|) \approx \left[\frac{(k_\perp R)^{|s|-1}}{2^{|s|+1}(|s| - 1)!}\right] \tfrac{1}{2}v_\perp(s/|s|, -i\eta, 0). \quad (E22.4)$$

(c)   Show that for $s = 0$, $|s| > 0$ the small gyroradius limits of the probability (22.19) are given by, respectively,

$$w_M(\mathbf{p}, \mathbf{k}, 0) = \frac{2\pi q^2 R_M(\mathbf{k})}{\varepsilon_0 \hbar \omega_M(\mathbf{k})} \left| \mathbf{e}_{Mz}(\mathbf{k}) \right|^2 \delta[\omega_M(\mathbf{k}) - k_\parallel v_\parallel],$$

$$w_M(\mathbf{p}, \mathbf{k}, s) = \frac{2\pi q^2 R_M(\mathbf{k})}{\varepsilon_0 \hbar \omega_M(\mathbf{k})} \left[\frac{(k_\perp R)^{|s|-1}}{2^{|s|+1}(|s|-1)!} \right] \left| \mathbf{e}_{Mx}(\mathbf{k}) \right.$$
$$\left. - \mathrm{i}(s/|s|)\eta \mathbf{e}_{My}(\mathbf{k}) \right|^2 \delta[\omega_M(\mathbf{k}) - s\Omega - k_\parallel v_\parallel],$$
$$(E22.5)$$

where the coordinates axes are defined by (22.8) with $\phi = 0$.

22.5 All information on the polarization of gyromagnetic emission of transverse waves *in vacuo* or in an isotropic medium is contained in the relevant polarization tensor.

(a)   Show that on choosing coordinates axes as defined by (22.8) and choosing

$$\mathbf{e}^1 = (\cos\theta\cos\phi, \cos\theta\sin\phi, -\sin\theta), \quad \mathbf{e}^2 = (-\sin\phi, \cos\phi, 0),$$
$$(E22.6)$$

the polarization tensor in (22.20) reduces to

$$V^\alpha(\mathbf{k}, \mathbf{p}; s) V^{*\beta}(\mathbf{k}, \mathbf{p}; s) =$$
$$v_\perp^2 \begin{pmatrix} |(z/x)J_s(sx)|^2 & \mathrm{i}\eta(z/x)J_s(sx)J_s'(sx) \\ -\mathrm{i}\eta(z/x)J_s(sx)J_s'(sx) & |J_s'(sx)|^2 \end{pmatrix},$$
$$(E22.7)$$

with z and x defined by (22.31).

(b)   Hence show that gyromagnetic emission at the sth harmonic *in vacuo* is completely polarized, and show that, with the polarization vector in the form $\mathbf{e} = (T\mathbf{e}^1 + \mathrm{i}\mathbf{e}^2)/(T^2 + 1)^{1/2}$, the axial ratio is given by

$$T = -\eta \frac{\cos\theta - \beta_\parallel}{\beta_\perp \sin\theta} \frac{J_s(sx)}{J_s'(sx)}, \quad x = \frac{\beta_\perp \sin\theta}{1 - \beta_\parallel \cos\theta}. \quad (E22.8)$$

22.6 Fill in the details of the derivation of (22.36). Specifically, derive and use the identity

$$J_{2s-2}(2sy) + J_{2s+2}(2sy) = 2J_{2s}(2sy) + 4J_{2s}''(2sy).$$

Hint: Use the recurrence relation (22.15) and the differential equation

$$J_\nu''(z) + J_\nu'(z)/z\,(1 - \nu^2/z^2)\,J_\nu(z) = 0. \quad (E22.9)$$

22.7 Compare the intensities of gyromagnetic emission at harmonics $s = 1$, $s = 2$ and $s > 2$ for a non-relativistic electron.

(a)   Derive (22.52) by retaining only the leading terms in the power series expansions of the Bessel functions

$$J_\nu(z) = \sum_n \frac{(-)^n}{n!\Gamma(\nu+n+1)} \left(\frac{z}{2}\right)^{2n+\nu}. \tag{E22.10}$$

(b)   Insert (22.52) in (22.36) with (22.37), and then carry out the appropriate integrals.

(c)   Hence show that the ratio of the power emitted at the second harmonic to power (22.40) emitted at the fundamental is

$$P_2/P_1 = 12\beta_\perp^2/5. \tag{E22.11}$$

(d)   Show that the ratio of the power emitted at the sth harmonic to that at $s = 1$ is given by

$$P_s/P_1 = 3s^{2s+1}(s+1)\beta_\perp^{2s}/(2s+1)!. \tag{E22.12}$$

22.8 The emissivity (22.48) simplifies for emission of transverse waves in an isotropic medium by non-relativistic electrons.

(a)   Show that for non-relativistic electrons in the small gyroradius limit, cf. Exercise 22.7, for $L_M = 0$, $T_M = T_\pm$ with $T_+T_- = -1$, corresponding to transverse waves with orthogonal polarizations, the term corresponding to the sth harmonic in (22.48) reduces to

$$\eta_M(\omega,\theta,s)$$
$$= \frac{e^2 n\omega^2 \beta^2 \sin^2\alpha \left(sn\beta\sin\alpha\sin\theta\right)^{2s-2}(1+\cos^2\theta)}{8\pi^2\varepsilon_0 c2^{2s}[(s-1)!]^2} \delta(\omega - s\Omega),$$
$$\tag{E22.13}$$

(b)   Show that evaluating the power emitted at the sth harmonic by integrating (E22.13) over solid angle and frequency gives

$$P_s = \frac{e^2\Omega_e^2}{2\pi c} \frac{s^{2s+1}(s+1)}{(2s+1)!} \left[n(s\Omega_e)\beta\sin\alpha\right]^{2s}. \tag{E22.14}$$

22.9 Show that (22.44) implies that the power radiated per unit solid angle in gyroemission is strongly peaked about the direction determined by $\beta\cos\theta = \cos\alpha$.

(a)   Specifically, use (22.44) to argue that maximum emission occurs where the denominator has a minimum, and determine this minimum as a function of $\cos\theta$.

(b)   The foregoing argument applies to the emission *in vacuo* summed over all s. How would you attempt to justify extending this argument (i) to all s, and (ii) to a medium where the maximum occurs at $n_M\beta\cos\theta = \cos\alpha$?

22.10 One uses (22.55) to find the change in any quantity as a result of spontaneous emission.

(a)   Determine the change of both $p_\perp$ and $p_\parallel$ for emission *in vacuo*.

(b)   Change variables to p and α and show that in the ultrarelativistic limit the change is predominantly in p.

22.11 Evaluate the emissivity (22.62) with (22.64) by making the non-relativistic approximation and inserting the thermal distribution (22.60).

(a)   Argue that the emission coefficient at the sth harmonic is given by

$$J_M(s,\omega,\theta) = \frac{n_e}{(2\pi)^{1/2}\beta_0^3} \int_{-\infty}^{\infty} \mathrm{d}\beta_\parallel \int_0^{\infty} \mathrm{d}\beta_\perp \beta_\perp\, e^{-\beta^2/2\beta_0^2}\eta_M(s,\omega,\theta).$$

$$(E22.15)$$

(b)   Perform the $\beta_\parallel$-integral over the δ-function.

(c)   The $\beta_\perp$-integral over the ordinary Bessel functions is evaluated in terms of modified Bessel functions I_s. The first of the following integrals is a standard integral; use it to evaluate the other two integrals

$$\int_0^{\infty} \mathrm{d}\beta_\perp \beta_\perp\, e^{-\beta_\perp^2/2\beta_0^2} \begin{bmatrix} J_s^2(z) \\ zJ_s(z)J_s'(z) \\ z^2 J_s'^2(z) \end{bmatrix}$$

$$= e^{-\lambda} \begin{bmatrix} I_s(\lambda) \\ \lambda[I_s'(\lambda) - I_s(\lambda)] \\ \{s^2 I_s(\lambda) - 2\lambda^2[I_s'(\lambda) - I_s(\lambda)]\} \end{bmatrix},$$

$$(E22.16)$$

with $z = k_\perp c\beta_\perp/\Omega_e$ and where the argument of the modified Bessel functions is $\lambda_M = (\omega/\Omega_e)^2 n_M^2 \beta_0^2 \sin^2 \theta$.

(c)   Show that Kirchhoff's law implies

$$J_M(s,\omega,\theta) = \frac{\omega^2 n_M^2 \partial(\omega n_M)\partial\omega}{(2\pi c)^3} \gamma_M(s,\omega,\theta) m_e \beta_0^2 c^2. \quad (E22.17)$$

and hence derive (22.63).

22.12 Consider thermal gyromagnetic absorption in the solar corona. The magnetic field is assumed to decrease with radial distance r from the center of the Sun as $\mathrm{d}B/\mathrm{d}r = -B/L_B$. Consider a radial ray and let **B** be at an angle θ to the radial direction. Assume that the refractive index is approximated by unity.

(a)   Assume that the absorption coefficient is equal to its maximum value γ_{Ms} at the center of the line at the sth harmonic, and that the line width is given by (23.9). Show that the intensity of the ray decreases by a factor $e^{-\tau_{Ms}}$ with the optical depth given by

$$\tau_{Ms} = \sqrt{2} n_e \beta_0 |\cos\theta| L_B \gamma_{Ms}/c. \qquad (E22.18)$$

(b) Estimate τ_{Ms} for both modes at $s = 2, 3$ assuming $n_e = 10^{15}\,\mathrm{m^{-3}}$, $T_e = 3 \times 10^6\,\mathrm{K}$, $B = 10^{-2}\,\mathrm{T}$, $L_B = 10^7\,\mathrm{m}$, $\theta = 45°$.

22.12 Considering gyroemission of transverse waves with refractive index $n(\omega) = (1 - \omega_p^2/\omega^2)^{1/2}$. Use (22.62) omit the subscript on n_M, set $L_M = 0$ and sum over the two orthogonal states of polarization.

(a) Sum over the two states of polarization in (22.62) by setting $T_M = 0$ and for $T_M = \infty$ and adding the results. Show that the same final result is obtained by summing over a pair of orthogonal circular polarizations $T_M = \pm 1$ and over arbitrary orthogonal elliptical polarizations $T_M, -1/T_M$.

(b) Show that the average emissivity per thermal electron in the small gyroradius limit is given by

$$\bar\eta(s,\omega,\theta) = \frac{e^2 s^2 \Omega_e (1 + \cos^2\theta)[sn(\omega)\beta_0 \sin\theta]^{2s-2}}{4\pi\varepsilon_0 (2\pi)^{3/2} \beta_0 c |\cos\theta| 2^s (s-1)!} \left(\tfrac{1}{2} s^2 \beta_0^2\right)^s$$
$$\times e^{-(\omega - s\Omega_e)^2/2[n(\omega)\omega\beta_0\cos\theta]^2}. \qquad (E22.19)$$

(c) Show that $(E22.19)$ implies that the power radiated at the sth harmonic per thermal electron is given by

$$P_s = \frac{e^2 \Omega^2}{8\pi^2\varepsilon_0 c} \frac{2^s s^{2s+1}(s+1)!\beta_0^{2s}}{(2s+1)!}. \qquad (E22.20)$$

(d) Show that the sum of $(E22.20)$ over s diverges, explain why this is a physically unacceptable result.

22.13 Show that in the approximation (22.67) gyroemission has a degree of polarization

$$r_x = 2|\cos\theta|/(1 + \cos^2\theta). \qquad (E22.21)$$

23

Gyrosynchroton Emission

Preamble

Cyclotron absorption can be negative under relatively mild conditions when the distribution function is an increasing function of $p_\perp$ at small $p_\perp$. It is essential to include a relativistic effect in the resonance condition to treat cyclotron maser emission correctly. Gyrosynchrotron emission, which applies for mildly relativistic electrons, is more cumbersome to treat than is gyroemission in either the cyclotron or synchrotron limits. Some useful analytic formulas are obtained using the Carlini approximation to Bessel functions.

23.1 Cyclotron Maser Emission

Electron cyclotron emission can be negative leading to electron cyclotron maser emission. The most favorable case for negative absorption is for x mode waves at $\theta \approx \pi/2$ at the fundamental $s = 1$ in a plasma with $\omega_p \ll \Omega_e$. The required source of free energy is a distribution of electrons with $\partial f/\partial p_\perp > 0$ at small $p_\perp$. The discussion of electron cyclotron maser emission here is centered around explaining why this case is the most favorable, and outlining some applications.

The case of nearly perpendicular propagation $\theta \approx \pi/2$ is of special significance because of the form of the Doppler condition (22.23), viz., $\omega - s\Omega_e/\gamma - \omega[1 - n_M(\omega, \theta)\beta_\parallel \cos\theta] = 0$. The non-relativistic approximation to (22.23) involves replacing the Lorentz factor γ by unity. This approximation is made here. In terms of the resonance ellipse this corresponds to assuming that the ellipse is approximated by a vertical line. In

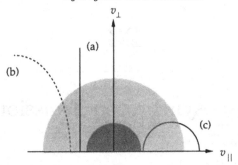

Fig. 23.1 Examples of resonance ellipses in the non-relativistic region
of velocity space. Thermal electrons occupy the darkly shaded region.
Curves (a), (b) and (c) are discussed in the text.

contrast, for $\theta = \pi/2$ the resonance ellipse reduces to a circle centered on
the origin of velocity space and with radius $\beta = (1 - \omega^2/s^2\Omega_e^2)^{1/2}$. A vertical line is a poor approximation to this circle when its radius is small.
The non-relativistic approximation is invalid for $\theta = \pi/2$ no matter how
small the energy of the electron. A "semirelativistic" approximation to
(22.23) is

$$\omega - s\Omega_e(1 - \tfrac{1}{2}\beta_\perp^2 - \tfrac{1}{2}\beta_\parallel^2) - \omega[1 - n_M(\omega,\theta)\beta_\parallel\cos\theta] = 0. \qquad (23.1)$$

This corresponds to a resonance ellipse that is a circle but with its center
displaced from the origin. Three cases are illustrated in Figure 23.1.
The resonance curve labeled (a) is a vertical line at $v_\parallel = (\omega - s\Omega_e)/k_\parallel$,
which corresponds to an ellipse with its center far to the left so that
the line is actually an arc of a very large ellipse. The resonance curve
labeled (b) illustrates an intermediate case where the center is closer to
the shaded regions. The resonance curve labeled (c) corresponds to the
semirelativistic case (23.1). The circle drawn is chosen so that it lies
just outside the region where the thermal electrons are located so that
thermal gyromagnetic absorption is absent.

The gyromagnetic absorption coefficient is given by (22.46), *viz.*,

$$\gamma_M(\mathbf{k}) = -\sum_{s=-\infty}^{\infty} 2\pi \int_{-\infty}^{\infty} dp_\parallel \int_0^{\infty} dp_\perp p_\perp\, w_M(\mathbf{p},\mathbf{k},s)$$

$$\times\, \hbar\left(\frac{s\Omega}{v_\perp}\frac{\partial}{\partial p_\perp} + k_\parallel\frac{\partial}{\partial p_\parallel}\right) f(p_\parallel, p_\perp). \qquad (23.2)$$

Here the integrals over $p_\perp$ and $p_\parallel$ are transformed into ones over $\beta_\perp$ and
$\beta_\parallel$. The variables $\beta_\perp$ and $\beta_\parallel$ are more appropriate when appealing to the
concept of a resonance ellipse in $\beta_\perp$–$\beta_\parallel$ space. One of these integrals is
performed over the δ-function. The remaining integral then has a simple
geometric interpretation: it is an integral around the resonance ellipse,

cf. Exercise 23.1. The following discussion is based on this geometric interpretation.

23.2 Loss-Cone Driven Maser Emission

In order to understand why the case of nearly perpendicular propagation $\theta \approx \pi/2$ is so favorable for negative absorption it is helpful to consider first the possibility of negative absorption when this condition is not satisfied. In the non-relativistic approximation the resonance ellipse is replaced by a vertical line, and the integral around it reduces to the integral over $0 < \beta_\perp < \infty$. In (23.2) there are two possible driving terms for negative absorption: $\partial f/\partial p_\perp > 0$ and $k_\parallel (\partial f/\partial p_\parallel) > 0$. However, these are necessary but not sufficient conditions for negative absorption. If one partially integrates over $p_\perp$ for the first of these terms one finds that it cannot lead to negative absorption. The argument is similar to that given at the end of §21.5, leading to the conclusion that collisional damping cannot be negative. It follows that in the non-relativistic approximation, in the sense used here (meaning the case where the resonance ellipse is replaced by a vertical line), negative absorption cannot be driven by a distribution with $\partial f/\partial p_\perp > 0$. It is possible for negative absorption to be driven by a distribution with $k_\parallel (\partial f/\partial p_\parallel) > 0$ but the conditions under which this might lead to maser action in practice are quite restrictive.

The situation is markedly different when the resonance ellipse is approximated by a circle close to the origin of velocity space. Suppose that the resonance ellipse lies entirely within a region where $\partial f/\partial p_\perp$ is positive. (For small $|\cos\theta|$ the other possible driving term $k_\parallel (\partial f/\partial p_\parallel)$ with $k_\parallel = k\cos\theta$ is unimportant.) The integral around the resonance circle is then all of one sign and can lead to a large negative absorption coefficient. This is believed to be the case relevant to electron cyclotron maser emission of practical interest.

There are two important requirements that need to be satisfied for electron cyclotron maser emission to be a possible source of escaping radiation. One is that there be a distribution of electrons with available free energy in the appropriate form, specifically with $\partial f/\partial p_\perp > 0$ at small $p_\perp$. One such distribution is the loss cone distribution illustrated in Figure 23.2. Loss cones form naturally in a magnetic trap where electrons are confined due to the magnetic mirror effect. Thus this first requirement is expected to be satisfied for magnetically trapped electrons under a wide variety of circumstances. The other requirement is that

Gyrosynchroton Emission

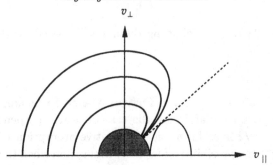

Fig. 23.2 A loss-cone distribution with loss-cone angle indicated by the dashed line. The resonance ellipse illustrated favors maser action.

there be waves with the dispersion properties needed (i) to define the most favorable resonant ellipse, as in the case illustrated in Figure 23.2, and (ii) to allow direct escape from the plasma. The most favorable case for emission is at $s = 1$ in the x mode above the cutoff frequency at $\omega = \omega_x$. According to (13.27) one has $\omega_x \approx \omega_p + \frac{1}{2}\Omega_e$ for $\omega_p \gg \frac{1}{2}\Omega_e$ and $\omega_x \approx \Omega_e + \omega_p^2/\Omega_e$ for $\omega_p \ll \frac{1}{2}\Omega_e$. Hence fundamental emission at $\omega \approx \Omega_e$ can be in the x mode only in a plasma with $\omega_p \ll \Omega_e$. Moreover, only emission that is above the line center at $\omega = \Omega_e$ by at least ω_p^2/Ω_e can be in the x mode.

This latter requirement restricts the emission to be at an angle θ significantly different from $\pi/2$. To see this, first recall from Figure 23.1 that $\theta - \pi/2$ must be small for the resonance ellipse to be approximated by a small circle rather than by a vertical line in the region of velocity space where the condition $\partial f/\partial p_\perp > 0$ is satisfied, and that it is only when the resonance ellipse closes within this region that effective growth can occur. The resonance condition (23.1) implies $\omega < \Omega_e$ for $\theta = \pi/2$, and it is only for $n_M \beta_\parallel \cos\theta > \frac{1}{2}\beta^2$ that resonance occurs at $\omega > \Omega_e$. The parameters (22.27) of the resonance ellipse imply that for small $|\cos\theta|$, the separation of the center of the resonance ellipse from the origin is proportional to $\cos\theta$, and the semimajor axis of the ellipse is determined by the difference $s\Omega_e - \omega$. Thus one infers that negative absorption occurs for emission at $|\theta - \pi/2|$ small but non-zero.

There is one further restriction on electron cyclotron maser emission that applies when the emission is attributed to a distribution of suprathermal electrons with a density much lower than that of thermal electrons. The resonance ellipse must avoid the region of velocity space occupied by the thermal electrons in order for gyromagnetic absorption by the thermal electrons swamping the maser emission. This point is also illustrated in Figures 23.1 and 23.2.

The evaluation of the absorption coefficient for electron cyclotron maser emission needs to be performed numerically in practice. By making appropriate simplifying assumptions one can estimate the maximum growth rate (the modulus of the maximum negative value of the absorption coefficient). For a loss-cone distribution of suprathermal electrons with mean speed $\langle\beta\rangle$ and loss-cone angle α_0, the maximum growth is estimated to be

$$|\gamma_x| \approx \frac{\pi\omega_p^2}{\Omega_e}\frac{n_1}{n_e}\frac{1}{\langle\beta\rangle^2 \sin\alpha_0}, \qquad (23.3)$$

where n_1 is the number density that would be required to fill the loss cone.

Applications of the theory of electron cyclotron maser emission outlined above are primarily of astrophysical and geophysical interest. Electron cyclotron maser emission is the accepted emission mechanism for auroral kilometric radiation from the Earth and analogous emissions from Jupiter and the other giant planets. It is also the favored mechanism for one particular class of solar radio burst (but not for most solar radio bursts) and for some analogous very bright radio emission from some flare stars. The auroral kilometric radiation (AKR) consists of bursts of cyclotron emission from regions above the auroral zones of the Earth. The electrons that generate AKR are accelerated downward by a potential drop of several kilovolts along the magnetic field lines. Below the acceleration region conservation of the adiabatic invariant $v^2 \sin^2\alpha/B$ implies that $\sin^2\alpha$ increases as B increases leading to the magnetic mirror effect. Electrons with sufficiently large initial pitch angle α reflect and electrons with smaller initial α precipitate into the atmosphere. One source of free energy that results from this is in the reflected electrons: there are no reflected electrons with small $\sin\alpha$ and hence the distribution is of the form illustrated in Figure 23.2. This is the favored source of free energy.

23.3 Mildly Relativisitic Electrons

There are two clearly defined limits of gyroemission: the non-relativistic or cyclotron limit and the ultrarelativistic or synchrotron limit. From a formal viewpoint it is convenient to define the cyclotron limit as (i) the strictly non-relativistic limit in sense that the resonance ellipse is approximated by a vertical line, cf. Figure 23.1, but (ii) allowing emission at a range of harmonics $s = 1, 2, \ldots$, with (iii) each Bessel function approximated by the leading term in its power series expansion. In the synchrotron limit the ultrarelativistic approximation is made, the harmonic number is regarded as a continuous variable, and the Bessel functions are approximated by their Airy integral forms. The cyclotron limit is the lowest order term in an expansion in β^2, and the synchrotron limit is the lowest order term in an expansion in $1/\gamma$. The intermediate gyrosynchrotron case, for example, for electrons $100\,\mathrm{keV}$ to $1\,\mathrm{MeV}$, is not covered adequately by these two limiting cases.

In the intermediate regime the procedure adopted is to assume that the gyroemission from a given electron is concentrated around a favored θ, specifically around $\beta \cos \theta = \cos \alpha$ *in vacuo*, cf. Exercise 22.9. An argument that generalizes this result is based on averaging the emissivity over the pitch angle distribution of the electrons, $\phi(\alpha)$. The integral over pitch angle is evaluated approximately using the method of steepest descents (Exercise 23.5). This method leads to the favored angle being determined by $\partial x / \partial \cos \alpha = 0$, where

$$x = \frac{n_M \beta \sin \alpha \sin \theta}{1 - n_M \beta \cos \alpha \cos \theta}, \tag{23.4}$$

is the argument of the Bessel functions in the emissivity (22.48). Thus the dominant angle of emission is given by the solution of

$$\cos \alpha = n_M \beta \cos \theta. \tag{23.5}$$

The fact that the emission is concentrated around a favored angle is used to develop a systematic method of approximation. The $\cos \alpha$-integral is performed as follows. Proceeding in a similar way to the evaluation of the $\cos \theta$-integral in (22.31), the integral is rewritten in terms of a variable $\cos \alpha'$ by changing variables from β and α to β' and α' with

$$\cos \alpha' = \frac{\cos \alpha - n_M \beta \cos \theta}{1 - n_M \beta \cos \alpha \cos \theta}, \quad \beta' = \frac{n_M \beta \sin \theta}{(1 - n_M^2 \beta^2 \cos^2 \theta)^{1/2}}. \tag{23.6}$$

This corresponds to making a Lorentz transformation such that a particle with $\cos \alpha = n_M \beta \cos \theta$ in the laboratory frame has $\cos \alpha' = 0$ in the new frame. The $\cos \alpha'$-integral is then approximated by expanding the integrand is powers of $\cos \alpha'$. The details are omitted here, but it is relevant to note that the resulting expression for the averaged emissivity involves Bessel functions $J_{2s}(2sx)$ similar to (22.37). Approximate forms for the averaged emissivity are obtained by making appropriate approximations to these Bessel functions.

One needs a different approximation to the Bessel functions in each of the three regimes defined above. The leading term in the power series is appropriate in the cyclotron limit. This gives

$$J_{2s}(2sx) \approx \frac{(sx)^{2s}}{(2s)!}, \quad J'_{2s}(2sx) \approx \frac{J_{2s}(2sx)}{x},$$

$$\int_0^x dy \, J_{2s}(2sy) \approx \frac{x J_{2s}(2sx)}{2s}. \tag{23.7}$$

The Carlini approximation (first introduced in this context by Trubnikov who developed the theory of gyroemission and absorption in detail in his doctoral dissertation in 1958)

$$J_{2s}(2sx) \approx \frac{[Z(x)]^{2s}}{(4\pi s)^{1/2}(1-x^2)^{1/2}}, \quad Z(x) = \frac{x e^{(1-x^2)^{1/2}}}{1+(1-x^2)^{1/2}}, \tag{23.8}$$

is appropriate in the intermediate regime. One finds

$$J'_{2s}(2sx) \approx \frac{(1-x^2)^{1/2} J_{2s}(2sx)}{x}, \quad \int_0^x dy \, J_{2s}(2sy) \approx \frac{x J_{2s}(2sx)}{2s(1-x^2)^{1/2}}. \tag{23.9}$$

The Airy integral approximation applies in the synchrotron limit. It gives

$$J_{2s}(2sx) \approx \frac{(1-x^2)^{1/2}}{\pi\sqrt{3}} K_{1/3}(R), \quad J'_{2s}(2sx) \approx \frac{(1-x^2)}{\pi\sqrt{3}} K_{2/3}(R),$$

$$\int_0^x dy \, J_{2s}(2sy) \approx \frac{1}{2s\pi\sqrt{3}} \int_R^\infty dt \, K_{1/3}(t), \tag{23.10}$$

with

$$R = 2s(1-x^2)^{3/2}/3. \tag{23.11}$$

Application of (23.10) to the ultrarelativistic limit is discussed in detail in §24.3.

The procedure outlined leads to the following expressions for the average emissivity for an isotropic distribution of electrons $\phi(\alpha) = 1$. Firstly, in the cyclotron limit the average emissivity is

$$\bar{\eta}(s,\omega,\theta) \approx \frac{e^2\omega^2(1 + L_M \sin\theta + T_M \cos\theta)^2}{8\pi^2\varepsilon_0 c\Omega_e n_N \sin^2\theta(1+T_M^2)} \frac{(s n_M \beta_\perp \sin\theta)^{2s}}{(2s)!}$$

$$\times \delta\left(s - \frac{\omega}{\Omega_e}(1 + \tfrac{1}{2}\beta^2 - n_M^2\beta^2\cos^2\theta)\right). \tag{23.12}$$

In the intermediate regime, the Carlini approximation allows one to regard the harmonic number s as a continuous variable. A counterpart of (23.12) in the Carlini approximation is

$$\bar{\eta}(s,\omega,\theta) \approx \frac{e^2\omega^2\gamma(1 - n_M^2\beta^2)}{16\pi^2\varepsilon_0 c\Omega_e n_M \sin^2\theta(1+T_M^2)} \Big\{[1/\xi + L_M \sin\theta$$

$$+ (1 - n_M^2\beta^2)T_M\cos\theta]^2 + n_M^2\beta^2\xi T_M^2\sin^4\theta/2s\Big\}$$

$$\times \frac{s\xi^{1/2}e^{2s/\xi}}{(4\pi s)^{1/2}} \left(\frac{\xi-1}{\xi+1}\right)^s \delta\left(s - \frac{\omega}{\Omega_e}\gamma\xi^2(1 - n_M^2\beta^2)\right). \quad (23.13)$$

with

$$\xi = (1 - \beta'^2)^{-1/2} = \left(1 + \frac{n^2\beta^2\sin^2\theta}{1 - n^2\beta^2}\right)^{1/2}. \quad (23.14)$$

In (23.13) the difference between the refractive indices of the two modes can usually be neglected, with n_M replaced by n, and further n may usually be replaced by unity except in the expression (23.14) for ξ. The cyclotron approximation (23.12) and the Carlini approximation (23.13) overlap for γ, $\xi \approx 1$ and for large s when $(2s)!$ is approximated by $(4\pi s)^{1/2}(2s/e)^{2s}$ using Stirling's formula.

More general but more cumbersome formulas may be written down based on the Carlini approximation and extensions of it. However, for most purposes (23.13), which overlaps in its range of validity with (23.12), suffices for non-relativistic and mildly relativistic particles. At its other limit of validity the Carlini approximation overlaps with the Airy integral approximation. The Airy integral approximation is treated in §24.3. In practice, when relativistic effects are significant the difference of the refractive indices of the two wave modes is not important, and neither is the longitudinal part of their polarization, and hence one replaces n_M by $n = 1 - \omega_p^2/2\omega^2$ with $L_M = 0$. The Airy integral approximation gives, after integrating over s,

$$\bar\eta(\omega,\theta) \approx \frac{\sqrt{3}e^2\Omega_e}{64\pi^3\varepsilon_0 c}\frac{\xi}{\gamma}\left[R\int_R^\infty dt\, K_{5/3}(t) + \frac{1-T_M^2}{1+T_M^2}RK_{2/3}(R)\right], \quad (23.15)$$

with $R = 2\omega\gamma\sin^2\theta/3\Omega_e\xi^3$, $\xi \approx \gamma\sin\theta/(1 + \gamma^2\omega_p^2/\omega^2)^{1/2}$. On expanding (23.14) in powers of $1/\gamma$ and of ω_p^2/ω^2 one obtains $\xi = \gamma\sin\theta/(1 + \gamma^2\omega_p^2/\omega^2)^{1/2}$. The Airy integral approximation and the Carlini approximation overlap in the limit of large R when one has $K_\nu(R) \approx (\pi/2R)^{1/2}e^{-R}$.

23.4 Gyrosynchrotron Formulas

The averaged emissivities that are derived in §23.3 are integrated over the momentum distribution of the electrons to obtain explicit expressions for the emission and absorption coefficients for gyrosynchrotron emission. The results of carrying out the integrals are quoted here only for a thermal distribution. A power-law distribution is more relevant for astrophysical application, but the analytic expressions for a power-law distribution are rather cumbersome, and usually need to be replaced by semiempirical approximations in detailed calculations.

When relativistic effects are included, the Maxwellian distribution (23.1) is replaced by

$$f(p) = \frac{n_1}{(2\pi)^{3/2}(m_e c \beta_0)^3} e^{-(\gamma-1)/\beta_0^2}, \qquad (23.16)$$

where the normalization coefficient is approximated for $\beta_0^2 \ll 1$. The integral of the average emissivity is performed by the method of steepest descents. The results for the emission coefficient or for the absorption coefficient simplify in two limiting cases. These cases correspond to the effective Lorentz factor $\xi = \xi_0$ at the optimum value determined by the steepest descent method being non-relativistic ($\xi_0 - 1 \ll 1$) and being highly relativistic ($\xi_0 \gg 1$), respectively. The optimum value for the effective Lorentz factor is $\xi_0 - 1 \ll 1$,

$$\xi_0 \approx \begin{cases} 1 + (\omega/\Omega_e)\beta_0^2 \sin^2\theta & \text{for } \xi_0 - 1 \ll 1, \\ \left[(4\omega/3\Omega_e)\beta_0^2 \sin^2\theta\right]^{1/3} & \text{for } \xi_0 - 1 \gg 1 \end{cases} \qquad (23.17)$$

The relevant approximation to the absorption coefficient is

$$\gamma(\omega, \theta) = \frac{\omega_p^2 n_1 (\pi \Omega_e/2\omega)^{1/2}}{\Omega_e \beta_0^2 n_e \sin^2\theta} \left[\frac{e\omega\beta_0^2 \sin^2\theta}{2\Omega_e(1 + T_M^2)}\right]^{\omega/\Omega_e}$$
$$\times \left[(1 + L_M \sin\theta + T_M \cos\theta)^2 + \beta_0^2 T_M^2 \sin^4\theta\right], \qquad (23.18a)$$
$$= \frac{3\pi^{1/2}\omega_p^2 n_1 \sin^2\theta}{2\kappa\Omega_e\beta_0 n_e} \frac{1 + \kappa^{-1/3}\beta_0^2 T_M^2 \sin\theta}{1 + T_M^2}$$
$$\times e^{(1 - \kappa^{1/3}/\sin\theta - 9/20\kappa^{1/3}\sin\theta)/\beta_0^2}, \qquad (23.18b)$$

in these two limits respectively, with $\kappa = (9\omega/2\Omega_e)\beta_0^2 \sin^2\theta$. For small β_0 (23.18a) overlaps with the corresponding result implied by (23.8) with (23.10), setting $\omega = s\Omega_e$ in the latter.

Qualitatively, gyrosynchrotron emission covers the range of harmonic numbers s from 2 or 3 to $\gtrsim 100$. The emission from a distribution of electrons leads to continuous emission over this frequency range, rather than to a sequence of lines. In the thermal case the overlapping of the lines occurs when the line width becomes comparable with the line separation. In the non-relativistic case, (23.9) implies that the lines begin to overlap for

$$s\beta_0 \gtrsim 1/n_M |\cos\theta|. \qquad (23.19)$$

This overlapping is due to the Doppler line widths. For sufficiently small $|\cos\theta|$ the relativistic correction is more important than the Doppler term in the resonance condition (23.1). The derivation of the line width in this case is carried out using the Carlini approximation by finding the mean frequency and the mean square frequency of the emitted radiation, for example, using the procedure outlined in Exercise 23.5. This leads to the line width at the sth harmonic for $\theta = \pi/2$ being estimated as $\Delta\omega_s = s^{3/2}\beta_0^2\Omega_e$. On combining this with (23.9) one obtains the following estimate for the relative line width:

$$(\Delta\omega)_s/s\Omega_e = (\beta_0^2\cos\theta + s\beta_0^4)^{1/2}. \tag{23.20}$$

It follows that for nearly perpendicular propagation the lines start to overlap for $s^{3/2}\beta_0^2 \gtrsim 1$.

In practice, overlapping of the harmonics occurs simply because real sources usually have an inhomogeneous magnetic field with a range of values of B and hence of Ω_e along any line of sight. However, the intrinsic line widths are of significance in connection with the possibility of maser action. As discussed in §23.2, cyclotron maser emission is possible at low harmonics. Maser emission becomes much less effective as the lines start to overlap, and for synchrotron maser emission is possible only under quite extreme conditions.

The effect of the refractive index being different from unity leads to the Razin effect. In the gyrosynchrotron case the Razin effect may be understood by noting that emission at the sth harmonic is 2^s electric multipole emission. According to the discussion in §17.2, the power emitted then varies with the refractive index to the power $2s - 1$. For large s one has

$$(1 - \omega_p^2/\omega^2)^{s-1/2} = e^{(s-1/2)\ln(1-\omega_p^2/\omega^2)} \approx e^{-s\omega_p^2/\omega^2}. \tag{23.21}$$

It follows that the emission is suppressed compared to emission *in vacuo* for $s\omega_p^2/\omega^2 \gtrsim 1$. With $s \approx \omega/\Omega_e$ this implies that Razin suppression occurs for

$$\omega \lesssim \omega_p^2/\Omega_e. \tag{23.22}$$

Note that this is a true suppression and not an absorption effect. Electrons emit less effectively in a medium with refractive index less than unity than they do *in vacuo*.

Exercise Set 23

23.1 The following exercise involves writing the integral over momentum space of any quantity involving the probability (22.20) as an integral around the resonance ellipse.

(a) Show that on transforming from momentum space to velocity space one has $d^3\mathbf{p} = d^3\mathbf{v}\, m^3\gamma^5$.

(b) The resonance ellipse has center at $v_\parallel = v_c$, $v_\perp = 0$ and has semimajor axis v_R and eccentricity e, given by (22.27). Show that if one transforms from the variables $v_\parallel$, $v_\perp$ to v', ϕ by writing $v_\parallel = v_c - v'(1 - e^2)^{1/2}\cos\phi$, $v_\perp = v'\sin\phi$, then the argument of the δ-function in the probability (22.20) is proportional to $v' - v_R$ with v_R given by (22.27).

(c) Show that the constant of proportionality in part (b) is the Jacobian

$$|J| = \frac{\gamma s\Omega_e}{v_R c^2}[v_\perp^2 + v_\parallel(v_\parallel - v_c)] - k_\parallel\frac{v_\parallel - v_c}{v_R}. \qquad (E23.1)$$

(d) Show that one has $dv_\parallel dv_\perp v_\perp = (1 - e^2)^{1/2}d\cos\phi\, dv_R v_R^2$.

(e) Hence show that the integral over the δ-function in the probability (22.19) reduces to the integral around the resonance ellipse in the form

$$\int d^3\mathbf{p}\,\delta(\omega - s\Omega_e - k_\parallel v_\parallel) = 2\pi m^3\gamma^5(1 - e^2)^{1/2}\int_{-1}^{1}\frac{d\cos\phi}{|J|}. \qquad (E23.2)$$

23.2 The following exercise concerns the treatment of gyroemission (at $s = 1$) in the case where the resonance ellipse is approximated by a circle displaced from the origin. The derivation is based on the form derived in Exercise 23.1 for integrating around the resonance ellipse.

(a) Show that for small eccentricity the Jacobian ($E23.2$) may be approximated by $J \approx \omega v_R/c^2$.

(b) Writing $\beta_R = v_R/c$, $\beta_c = v_c/c$, show that averaging the emissivity over a Maxwellian distribution gives

$$\bar{\eta}_M(s, \omega, \theta) =$$
$$\frac{e^2\Omega_e}{8\pi^2(2\pi)^{1/2}\varepsilon_0\beta_0^3 c^3}\frac{(sn_M\beta_R)^{2s-1}}{2^{2s}(s!)^2(1 + T_M^2)}\int_{-1}^{1}d\cos\phi\,\sin^{2s}\phi$$
$$\times\left[1 + L_M\sin\theta + T_M\cos\theta - n_M(\beta_c - \beta_R\cos\phi)\right.$$
$$\left.\times(T_M + \cos\theta)\right]^2 e^{-(\beta_c^2 - 2\beta_c\beta_R\cos\phi + \beta_R^2)/2\beta_0^2}. \qquad (E23.3)$$

(c) Show that when the resonance circle has $\beta_R \approx \beta_c \gg \beta_0$, so that an arc of it is approximated by the vertical line in Figure 23.1, (E23.3) involves an exponential factor with exponent $-(\beta_c - \beta_R)^2/2\beta_0^2$ and that the range of validity of this form overlaps with the results derived in §23.1, cf. (22.62).

(d) For $\beta_c \ll \beta_0$ compare the form of the exponential factor in part (c) with that appearing in (22.62) using an approximate form for β_R from (22.27) with $\omega^2 \approx s^2\Omega_e^2 \gg k_\parallel^2 c^2$.

23.3 Apply the method used in the previous problem to derive an analytic formula for the emission coefficient in the limit $\cos\theta \to 0$ where (22.62) incorrectly implies that there is no emission.

23.4 The emissivity for gyroemission simplifies for emission perpendicular to the field lines *in vacuo* when the average over an isotropic pitch angle distribution is performed exactly.

(a) Derive expressions for the emissivity from (22.48) for $\theta = \pi/2$ and for two orthogonal transverse linear polarization, $T_o = \infty$, $T_x = 0$.

(b) On averaging the emissivity obtained in part (a) over an isotropic pitch angle distribution, show that the resulting integrals are similar to (22.34) and that they give

$$\int_{-1}^{1} d\cos\alpha' \begin{bmatrix} \cos^2\alpha' J_s^2(s\beta\sin\alpha') \\ \sin^2\alpha' J_s'^2(s\beta\sin\alpha') \end{bmatrix} = \frac{1}{2s\beta} \begin{bmatrix} \chi_s^{(1)}(\beta) - \chi_s^{(2)}(\beta) \\ 3\chi_s^{(1)}(\beta) - \chi_s^{(2)}(\beta) \end{bmatrix},$$

$$\chi_s^{(1)}(\beta) = J_{2s}'(2s\beta) - \frac{1}{2s\beta}J_{2s}(2s\beta) + \frac{1}{2s\beta^2}\int_0^\beta dy\, J_{2s}(2sy),$$

$$\chi_s^{(2)}(\beta) = \frac{2s(1-\beta^2)}{\beta^2}\int_0^\beta dy\, J_{2s}(2sy).$$

(E23.4)

(c) Show that the ratio of the average emissivity in the o mode and the x mode simplifies in the following case:

$$\bar{\eta}_o(\omega, \pi/2)/\bar{\eta}_x(\omega, \pi/2) = \begin{cases} 1/3 & \text{for } s \ll 3\gamma^3/3, \\ \beta^2\gamma^3/2s & \text{for } s \gg 3\gamma^3/3. \end{cases}$$

(E23.5)

(d) Show that when the non-relativistic approximation (23.7) is made to the Bessel functions one obtains formulas first written down by Trubnikov

$$\begin{bmatrix} \bar{\eta}_o(\omega, \pi/2) \\ \bar{\eta}_x(\omega, \pi/2) \end{bmatrix} = \frac{e^2\Omega_e(s\beta)^{2s}}{8\pi^2\varepsilon_0 c(2s+1)!} \begin{bmatrix} \beta^2/(2s+3) \\ 1 \end{bmatrix}.$$

(E23.6)

(e) Show that when the Carlini approximation is made to the Bessel functions one obtains formulas also due to Trubnikov

$$\begin{bmatrix} \bar{\eta}_o(\omega, \pi/2) \\ \bar{\eta}_x(\omega, \pi/2) \end{bmatrix} = \frac{e^2 \Omega_e s \gamma^{1/2} e^{2s/\gamma}}{16\pi^2 \varepsilon_0 c \gamma^2 (4\pi s)^{1/2}} \left(\frac{\gamma - 1}{\gamma + 1} \right)^s \begin{bmatrix} \gamma(\gamma^2 - 1)/2s \\ 1 \end{bmatrix}.$$

$$(E23.7)$$

23.5 The following exercise concerns the estimation of the line width (23.20) for gyrosynchrotron emission perpendicular to the field lines ($\theta = \pi/2$) for a thermal distribution of electrons of the form (23.16). The procedure used in the original treatment by Trubnikov is based on the *method of steepest descents*. The integral around a contour C such that the end points contributions are negligible is approximated according to

$$\int_C dz\, e^{-f(z_0)} \approx e^{-f(z)} \int_{-\infty}^{\infty} dz\, e^{-(z-z_0)^2/2(\Delta z)^2}, \qquad (E23.8)$$

with $f'(z_0) = 0$ and $(\Delta z)^2 = 1/f''(z_0)$.

(a) Show that in the Carlini approximation the average (over pitch angle) emissivity for thermal electrons involves an integral of the form

$$\int_1^{\infty} d\gamma\, e^{-(\gamma-1)/\beta_0^2} Z^{2s}(\gamma) g(\gamma), \qquad (E23.9)$$

with $Z(\gamma) = [(\gamma - 1)/(\gamma + 1)]^{1/2} e^{1/\gamma}$ and where $g(\gamma)$ is a slowly varying function of γ.

(b) Change variable from γ to $x = \omega/s\Omega_e = 1/\gamma$ and apply the method of steepest descents to the integral ($E23.9$).

(c) Hence show that for small β_0 and large s the spread in frequency corresponds to the result $(\Delta\omega)_s/s\Omega_e = s^{1/2}\beta_0^2$ derived by Trubnikov.

24

Synchrotron Emission

Preamble

Synchrotron emission is gyromagnetic emission from ultrarelativistic particles. It is important in the laboratory as a source of radiation in synchrotrons and as an energy loss mechanism for relativistic electrons confined by a magnetic field. Synchrotron radiation is of particular importance in astrophysics because it is the dominant emission mechanism for the vast majority of radioastronomical sources.

24.1 Forward Emission by Relativistic Particles

Before discussing synchrotron emission in particular, it is appropriate to discuss emission by relativistic particles in general. Important features of the emission by relativistic particles may be determined by the special theory of relativity and are only weakly dependent on the specific emission mechanism involved. Emission by a particle which is highly relativistic in the laboratory frame K may be inferred from its emission pattern in its rest frame K_0. Let the instantaneous velocity $\mathbf{v}$ of the particle in K be along the z axis. The frame K_0, as viewed from K is moving along the z axis at velocity $\mathbf{v}$, as illustrated in Figure 24.1, and the frame K, as viewed from K_0, is moving along the z_0 axis at velocity $-\mathbf{v}$.

In K_0 the particle is momentarily at rest, and so its emission pattern is dipolar. For present purposes the only important point is that the emission pattern in K_0 is not highly anisotropic. Let ω_0, k_0, ψ_0 describe a plane wave in K_0, with ψ_0 the angle between $\mathbf{k}_0$ and the z_0 axis.

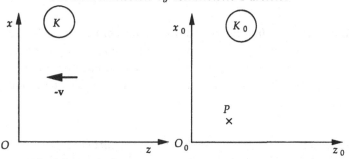

Fig. 24.1 The laboratory frame K moves at $-\mathbf{v}$ relative to the rest frame K_0. A particle at the point P at rest in K_0 has velocity $\mathbf{v}$ in K.

Applying a Lorentz transformation, the wave parameters ω, k, ψ in K are related to those in K_0 by

$$\omega = \gamma(\omega_0 + k_0 v \cos\psi_0), \qquad \omega_0 = \gamma(\omega - kv\cos\psi),$$

$$k\sin\psi = k_0\sin\psi_0, \qquad k_0\sin\psi_0 = k\sin\psi,$$

$$k\cos\psi = \gamma(k_0\cos\psi_0 + \omega_0 v/c^2), \quad k_0\cos\psi_0 = \gamma(k\cos\psi - \omega v/c^2),$$

$$(24.1)$$

with $\gamma = (1-\beta^2)^{-1/2}$, $\beta = v/c$. For emission *in vacuo* one has $k = \omega/c$, $k_0 = \omega_0/c$, and then (24.1) implies

$$\frac{\omega}{\omega_0} = \gamma(1+\beta\cos\psi_0), \quad \cos\psi = \frac{\cos\psi_0 + \beta_0}{1+\beta\cos\psi_0}, \quad \sin\psi = \frac{\sin\psi_0}{\gamma(1+\beta\cos\psi_0)}.$$

$$(24.2a,b,c)$$

For a highly relativistic particle one assumes $\gamma \gg 1$ and expands in powers of γ^{-1}. Thus one refers to terms of order unity, terms of order γ^{-1} and so on. In particular one has

$$\beta \approx 1 - 1/2\gamma^2, \qquad (24.3)$$

so that β differs from unity by a term of order γ^{-2}. Consider (24.2c); this implies, provided that $1+\beta\cos\psi_0$ and $\sin\psi_0$ are of order unity, that $\sin\psi$ is of order $1/\gamma$. The proviso is satisfied except for ψ_0 of order γ^{-1}. This leads to the following property: *All angles of emission in K_0, with the exception of a range of order γ^{-1} about the direction $-\mathbf{v}$, transform into a forward cone with half-angle of order γ^{-1} about the direction $\mathbf{v}$ in K.*

Inspection of (24.2a) shows that the frequency ω in K is of order γ times the frequency ω_0 in K_0, except for the small range of ψ_0 of order γ^{-1} around $\cos\psi_0 = -1$. In semiclassical language, it follows that nearly all the photons emitted in K are confined to the forward cone and have an energy of order γ times the energy of the corresponding photon in

K_0. Thus: *All but a fraction γ^{-1} of the power emitted in K is confined to a forward cone with half-angle of order γ^{-1} about the direction* **v**.

For gyromagnetic emission this implies that all but a fraction γ^{-1} of the power emitted by a particle with pitch angle α is confined to angles θ satisfying

$$\theta = \alpha + O(\gamma^{-1}).\qquad(24.4)$$

That is, all but a fraction γ^{-1} of the power is emitted on the surface of a cone with half-angle $\theta = \alpha$ and thickness of order γ^{-1}.

24.2 Semiquantitative Treatment of Synchrotron Emission

The properties of synchrotron emission are derived in a semiquantitative manner by combining some exact results for gyromagnetic emission and the foregoing arguments concerning emission by relativistic particles.

One exact result is for the power radiated: in the ultrarelativistic limit (22.40) gives

$$P \approx \frac{2q^2}{12\pi\varepsilon_0 c}\,\Omega_0^2\gamma^2\sin^2\alpha.\qquad(24.5)$$

The properties of emission by relativistic particles imply that this power is confined to a surface of angular thickness of order γ^{-1} about a hollow cone with half-angle $\theta = \alpha$ about the direction of the magnetic field. This argument gives us no information on the frequency spectrum of synchrotron radiation. The exact formula (22.23), which relates the frequency of emission to the harmonic number s gives, for emission *in vacuo* by a relativistic particle,

$$\omega \approx s\Omega_0/\gamma\sin^2\theta.\qquad(24.6)$$

However, we need an independent argument to estimate either ω or s.

The following argument enables one to estimate the typical frequency in synchrotron emission by considering the temporal distribution of the radiation received by an observer. First note that because of the highly anisotropic angular distribution of synchrotron emission, significant emission from a particle with pitch angle α is seen by an observer only in the narrow range of angles about $\theta = \alpha$. One pulse of radiation is received each revolution of the particle, with this pulse corresponding to the range of angles of order γ^{-1} during which the emission cone of the particle sweeps across the observer, as illustrated in Figure 24.2. Consider the case $\alpha = \pi/2$, which corresponds to the particle moving in a circle. The particle performs one gyration in a time $2\pi\gamma/\Omega_0$. The time

Fig. 24.2 The emission pattern of a gyrating particle with $\gamma \gg 1$. The beam sweeps around in a circle as the particle gyrates.

interval per gyration in which it is traveling towards the observer is a fraction of order γ^{-1} of this, that is, a time interval $\approx 2\pi/\Omega_0$. Consider the pulse received when a particle with speed βc moves directly towards the observer for a time Δt. Suppose that the particle is traveling in the x direction and that it starts radiating at $t = 0$ when it is at $x = 0$ and stops radiating at time $t = \Delta t$ when it is at $x = \beta c \Delta t$. (This starting and stopping of emission in the direction of the observer simulates the sweeping of the beam illustrated in Figure 24.2 across the direction to the observer.) The length of the wave train received by the observer is then $(1 - \beta)c\Delta t \approx c\Delta t/2\gamma^2$, and the duration of the pulse received per gyration by the observer is $\Delta t_{\mathrm{rec}} \approx \pi/\gamma^2\Omega_0$. A pulse of duration Δt contains Fourier components $\omega \lesssim 1/\Delta t$. Hence one expects synchrotron radiation to contain frequencies

$$\omega \lesssim \frac{\gamma^2 \Omega_0}{\pi}. \tag{24.7}$$

The estimate (24.7) applies only to emission by particles with pitch angle $\alpha = \pi/2$, but it is used to derive the corresponding result for arbitrary pitch angle by applying a Lorentz transformation. Consider a frame K in which a particle has pitch angle α and Lorentz factor γ, and another frame K' in which the particle has pitch angle $\alpha' = \pi/2$ and Lorentz factor γ'. The frame K' is moving relative to the frame K along the direction of the magnetic field with velocity $\beta c \cos \alpha$. The Lorentz factor of the transformation between the two frames is therefore $(1 - \beta^2 \cos^2)^{-1/2}\alpha \approx 1/\sin \alpha$. Hence we have

$$\omega \approx \omega'/\sin \alpha, \quad \gamma \approx \gamma'/\sin \alpha. \tag{24.8}$$

Then (24.7), now written $\omega' \lesssim \gamma'^2\Omega_0/\pi$, implies

$$\omega \lesssim \pi^{-1}\gamma^2\Omega_0 \sin \alpha. \tag{24.9}$$

It follows from (24.6) that the important harmonic numbers s involved in emission at $\theta \approx \alpha$ are

$$s \lesssim \pi^{-1}(\gamma \sin \theta)^3. \tag{24.10}$$

Thus, for $(\gamma \sin \theta)^3 \gg 1$, the emission is dominated by high harmonics and one is justified in treating s as a continuous variable.

24.3 The Emissivity for Synchrotron Emission

A detailed treatment of synchrotron emission is developed by assuming that s is a continuous variable, proportional to ω through (24.6), and using the Airy integral approximation (23.27) in evaluating the Bessel functions in the expression (22.36) for the power radiated.

The power radiated in gyromagnetic emission (22.36) is rewritten in the form

$$
P = \int_0^\infty ds \, \frac{2q^2 s^2 \Omega_0^2}{4\pi\varepsilon_0 c} \, \beta'^2 (1 - \beta'^2)
$$

$$
\times \left[\frac{1}{s\beta'} J'_{2s}(2s\beta') - \frac{(1 - \beta'^2)}{\beta'^3} \int_0^{\beta'} dy \, J_{2s}(2sy) \right]. \qquad (24.11)
$$

Making the relativistic approximation in (22.32) gives

$$
\beta' \approx 1 - 1/2\gamma^2 \sin^2 \theta. \qquad (24.12)
$$

One uses (24.11) to define an emissivity for synchrotron emission as follows. First, the integral over s is written in terms of an integral over ω using (24.6), thus allowing one to identify the power emitted per unit frequency. The emissivity is the power emitted per unit frequency and per unit solid angle, and to lowest order in the expansion in γ^{-1} the emission is at $\theta = \alpha$. Consequently, the power per unit solid angle is obtained simply by multiplying by $\delta(\cos\theta - \cos\alpha)/2\pi$. Thus the *emissivity for synchrotron emission* in the two states of linear polarization is given by

$$
\eta^{\parallel, \perp}(\omega, \theta) = \frac{q^2 s^2 \Omega_0 \gamma \sin\theta}{4\pi^2 \varepsilon_0 c} \, \delta(\theta - \alpha) \, \beta'^2 (1 - \beta'^2) \, G_s^{\parallel, \perp}, \qquad (24.13)
$$

with $G_s^{\parallel}$, $G_s^{\perp}$ given by (22.36), and with s determined by (24.6).

The Bessel functions involved have large order $2s$, and argument $2s\beta'$ which is very close to and slightly less than the order. An approximation that applies in this case is the *Airy integral approximation*, cf. (23.27),

$$
J_{2s}(2s\beta') \approx \frac{1}{\pi\sqrt{3}\gamma \sin\theta} \, K_{1/3}(R), \qquad (24.14)
$$

where $K_{1/3}(R)$ is a Macdonald function with argument

$$
R = 2s/3\gamma^3 \sin^3\theta = \omega/\omega_c, \quad \omega_c = \tfrac{3}{2}\Omega_0 \gamma^2 \sin\theta. \qquad (24.15)
$$

The Airy integral approximations (23.10) and also the following are required:

$$\int_0^{\beta'} dy \left(\frac{1}{y} - \frac{1}{\beta'} \right) J_{2s}(2sy)$$

$$\approx \frac{1}{4s\pi\sqrt{3}\gamma^2 \sin^2\theta} \left[\int_R^\infty dt \, K_{5/3}(t) - K_{2/3}(R) \right]. \qquad (24.16)$$

The emissivity (24.13) then reduces to

$$\eta^{\parallel,\perp}(\omega,\theta) = \frac{\sqrt{3}}{8\pi^2 c} \frac{q^2\Omega_0}{4\pi\varepsilon_0} \delta(\theta-\alpha) F^{\parallel,\perp}(\omega,\theta),$$

$$\qquad (24.17)$$

$$F^{\parallel,\perp}(\omega,\theta) = \frac{\omega}{\omega_c} \int_{\omega/\omega_c}^\infty dt \, K_{5/3}(t) \mp (\omega/\omega_c) K_{2/3}(\omega/\omega_c).$$

On integrating over solid angle and summing over the two states of polarization, the power emitted per unit frequency reduces to

$$P(\omega) = \frac{\sqrt{3}}{2\pi c} \frac{q^2\Omega_0 \sin\theta}{4\pi\varepsilon_0} F(\omega/\omega_c). \qquad (24.18)$$

with

$$F(R) = \tfrac{1}{2}[F^\parallel(R) + F^\perp(R)] = R \int_R^\infty dt \, K_{5/3}(t). \qquad (24.19)$$

Expansions of the function $F(R)$ for small and large arguments are given by

$$F(R) = \begin{cases} \dfrac{4\pi}{\sqrt{3}\Gamma(1/3)} \left(\dfrac{R}{2}\right)^{1/3} \left[1 - \tfrac{1}{2}\Gamma(1/3) \left(\dfrac{R}{2}\right)^{2/3} + \cdots \right] \\[4mm] \left(\dfrac{\pi R}{2}\right)^{1/2} e^{-R} \left[1 + \dfrac{55}{72R} + \cdots \right] \end{cases} \qquad (24.20)$$

for $R \ll 1$ and $R \gg 1$, respectively. In between these limiting cases, there is a maximum at $F(0.29) = 0.92$. The function $F(R)$ is plotted in Figure 24.3. A simple analytic approximation to it is $F(R) \approx 1.8R^{0.3}e^{-R}$.

24.4 Emission by a Power-Law Distribution

Synchrotron emission is the emission mechanism for a variety of radio-astronomical sources, including supernova remnants, quasars and radio galaxies. In these sources the frequency spectrum is usually a power of the form $I(\omega) \propto \omega^{-\alpha}$, where α is the *spectral index*. This type of spectrum is interpreted in terms of emission from relativistic electrons with a power-law energy distribution.

The emission due to a distribution $f(p,\alpha) = f(p)\phi(\alpha)$ of electrons

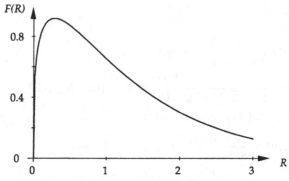

Fig. 24.3 The function $F(R)$ defined by (24.19).

is found by integrating the emissivity (24.13) over the distribution of electrons; this gives the emissivity per unit volume, referred to above as the emission coefficient $J(\omega, \theta)$. For a power-law distribution in energy it is convenient to introduce the *energy spectrum* $N(\varepsilon)$, with $\varepsilon = pc$ for relativistic particles, by writing

$$4\pi \int_0^\infty \mathrm{d}p\, p^2\, f(p) = \int_0^\infty \mathrm{d}\varepsilon\, N(\varepsilon), \quad N(\varepsilon) = \frac{4\pi}{c^3} \varepsilon^2 f(\varepsilon/c). \tag{24.21}$$

The energy spectrum is assumed to be a power-law of the form

$$N(\varepsilon) = \begin{cases} K\varepsilon^{-a} & \text{for } \varepsilon_1 \leq \varepsilon \leq \varepsilon_2, \\ 0 & \text{otherwise.} \end{cases} \tag{24.22}$$

In integrating the emissivity over this energy spectrum, the limits ε_1, ε_2 can often be ignored, but at least one limit is essential in normalizing the number of particles, specifically, one must have $\varepsilon_1 > 0$ for $a > 1$.

The volume emissivity follows by integrating (24.17) over the distribution of particles

$$J^{\parallel,\perp}(\omega, \theta) = \frac{\sqrt{3}}{8\pi^2 c} \frac{q^2 \Omega_0 \sin\theta \phi(\theta)}{4\pi\varepsilon_0} \int_0^\infty \mathrm{d}\varepsilon\, N(\varepsilon)\, F^{\parallel,\perp}(\omega/\omega_c). \tag{24.23}$$

The integral over ε is performed by setting $\varepsilon_1 = 0$, $\varepsilon_2 = \infty$, noting that the energy dependence of $F(\omega/\omega_c)$, as given by (24.19) with (24.15), involves $\omega/\omega_c \propto \varepsilon^{-2}$, and using the standard integral

$$\int_0^\infty \mathrm{d}x\, x^\mu K_\nu(ax) = 2^{\mu-1} a^{-\mu-1} \Gamma\left(\frac{1+\mu+\nu}{2}\right) \Gamma\left(\frac{1+\mu-\nu}{2}\right), \tag{24.24}$$

with

$$\Gamma(1+x) = x\Gamma(x), \quad \Gamma(1-x)\Gamma(x) = \frac{\pi}{\sin(\pi x)}. \tag{24.25}$$

The result is

$$J^{\|,\perp}(\omega,\theta) = A(\theta)\, j^{\|,\perp}(a) \left(\frac{2\omega}{3\Omega_0 \sin\theta} \right)^{-(a-1)/2}, \qquad (24.26)$$

with

$$A(\theta) = \frac{K(mc^2)^{-a+1}}{16\pi^2 c} \frac{\sqrt{3}q^2\Omega_0 \sin\theta\,\phi(\theta)}{4\pi\varepsilon_0},$$

$$j^{\|}(a) = \frac{2/3}{a+1} 2^{(a-3)/2}\, \Gamma\left(\frac{3a+7}{12} \right) \Gamma\left(\frac{3a-1}{12} \right), \qquad (24.27)$$

$$j^{\perp}(a) = \frac{a+5/3}{2/3}\, j^{\|}(a).$$

It follows that the spectral index a of the particle energy spectrum, and the spectral index α of the synchrotron emission are related by

$$\alpha = \tfrac{1}{2}(a-1), \quad a = 2\alpha + 1. \qquad (24.28)$$

A large fraction of radioastronomical sources are observed to have power-law frequency distributions over at least a decade in frequency. Such spectra are interpreted in terms of synchrotron emission from electrons with a power-law spectrum with spectral index a related to the spectral index α for the radio emission by (24.28). There is a variety of theories as to how these relativistic particles are accelerated. The favored theory, since it was proposed in the late 1970s, is that the acceleration occurs at shock fronts. Such acceleration leads naturally to power-law energy spectra of the form inferred from observation.

The degree of linear polarization of the synchrotron emission is given by

$$r_1 = -\frac{a+1}{a+7/3} = -\frac{\alpha+1}{\alpha+5/3}. \qquad (24.29)$$

The relatively high degree of polarization is a characteristic signature of synchrotron emission. For a value of $a = 3$, which is typical for many sources, (24.29) implies $r_1 = -3/4$, that is a degree of linear polarization of 75% in the direction orthogonal to the projection of the magnetic field $\mathbf{B}$ in the source on the plane of the sky. This prediction is based on the seemingly unrealistic assumption that the magnetic field in the source is uniform. In practice one expects the orientation of $\mathbf{B}$ to vary along the line of sight and across the source, and this would reduce the degree of polarization observed; the degree of polarization would be zero for a source in which the orientation of the magnetic field is random. Nevertheless many synchrotron sources show relatively high degrees of polarization, for example, $r_1 \gtrsim 30\%$ and a few have $r_1 \approx 70\%$. The predicted linear polarization was important in the identification (in the late 1940s) of the emission from radioastronomical sources as synchrotron emission.

24.5 Synchrotron Absorption

The absorption process corresponding to synchrotron emission is called synchrotron self-absorption, or more simply, *synchrotron absorption*. The absorption coefficient is written down by making the synchrotron approximation in (22.51). To lowest order in the expansion in γ^{-1}, the derivative with respect to α does not contribute. After changing variables from p to ε, and using (24.21) to introduce $N(\varepsilon)$, the absorption coefficients per unit length for the two states of linear polarization become

$$\mu^{\parallel,\perp}(\omega,\theta) = -\frac{(2\pi)^3 c^2}{2\omega^2} \int_{-1}^{1} d\cos\alpha\, \phi(\alpha) \int_0^{\infty} d\varepsilon\, \varepsilon^2\, \eta^{\parallel,\perp}(\omega,\theta)\, \frac{d}{d\varepsilon}\left[\frac{N(\varepsilon)}{\varepsilon^2}\right].$$
(24.30)

The absorption coefficient (24.30) cannot be negative, as may be shown by partially integrating, cf. Exercise 24.8.

For the power-law distribution (24.22), with the limits ε_1, ε_2 ignored, (24.30) gives

$$\mu^{\parallel,\perp}(\omega,\theta) = \frac{(2\pi)^3 c}{\omega^2} \frac{K(mc^2)^{-a}}{16\pi^2 c^2} \frac{\sqrt{3}q^2\Omega_0 \sin\theta\, \phi(\theta)}{4\pi\varepsilon_0}$$
$$\times (a+2)j^{\parallel,\perp}(a+1)\left(\frac{2\omega}{3\Omega_0 \sin\theta}\right)^{-a/2}, \qquad (24.31)$$

with $j^{\parallel,\perp}(a)$ given by (24.27).

The transfer equation for synchrotron radiation including the effects of the polarization is written in the Mueller calculus, cf. (14.45). For simplicity the circularly polarized component is ignored (it is of order γ^{-1} compared with the linearly polarized components), and the natural modes of the medium are assumed to be circularly polarized. The transfer equation, including spontaneous emission, is then of the form

$$\frac{d}{ds}\begin{pmatrix} I \\ Q \\ U \end{pmatrix} = \begin{pmatrix} \alpha_I \\ \alpha_Q \\ \alpha_U \end{pmatrix} + \begin{pmatrix} -\mu_I & -\mu_Q & 0 \\ -\mu_Q & -\mu_I & -\rho_V \\ 0 & \rho_V & -\mu_I \end{pmatrix}\begin{pmatrix} I \\ Q \\ U \end{pmatrix}. \qquad (24.32)$$

The emission coefficients are given by

$$\alpha_I = \frac{1}{c}\left[J^{\parallel}(\omega,\theta) + J^{\perp}(\omega,\theta)\right], \quad \alpha_Q = \frac{1}{c}\left[J^{\parallel}(\omega,\theta) - J^{\perp}(\omega,\theta)\right], \quad (24.33)$$

and the absorption coefficients by

$$\mu_I = \tfrac{1}{2}\left[\mu^{\parallel}(\omega,\theta) + \mu^{\perp}(\omega,\theta)\right], \quad \mu_Q = \tfrac{1}{2}\left[\mu^{\parallel}(\omega,\theta) - \mu^{\perp}(\omega,\theta)\right]. \quad (24.34)$$

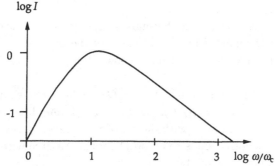

Fig. 24.4 Spectrum for a slab model of a synchrotron source with $a = 3$.
The asymptotic limits are $I \propto \omega^{5/2}$ and $I \propto \omega^{-(a-1)/2}$.

One integrates (24.32) through a source of thickness L in general, but let us consider simpler cases in which only one of the last two terms is retained. First, suppose that the absorption term is neglected. Then integration of (24.32) gives

$$I = \alpha_I L, \quad Q = \frac{\alpha_Q}{\rho_V} \sin(\rho_V L), \quad U = \frac{\alpha_Q}{\rho_V} [1 - \cos(\rho_V L)]. \quad (24.35)$$

This result describes the effect of spontaneous emission in a medium in which significant Faraday rotation occurs. It follows from (24.35) that in the limit of large Faraday rotation, that is, for $\rho_V L \gg 1$, Q and U do not exceed α_Q / ρ_V. This implies that the degree of polarization is of order $\alpha_Q / \alpha_I \rho_V L$, which is a factor $1/\rho_V L$ smaller than in the absence of Faraday rotation. This reduction in the degree of polarization due to Faraday rotation is called *Faraday depolarization*. It is due to the emission from different points along the line of sight having the plane of polarization rotated through different amounts due to the different path lengths through the plasma.

Next, suppose that Faraday rotation is neglected in (24.32) so that only emission and absorption are included. Integration of (24.32) then gives

$$I \pm Q = \frac{\alpha_I \pm \alpha_Q}{\mu_I \pm \mu_Q} \left(1 - e^{-(\mu_I \pm \mu_Q)L}\right), \quad U = 0. \quad (24.36)$$

The solution of (24.36) for I is plotted as a function of ω/ω_c for given L and given $N(\varepsilon)$ in Figure 24.4. The intensity has a maximum as a function of frequency. The *optically thin* region, where absorption is unimportant, corresponds to frequencies above the maximum, where one has $I \propto \omega^{-(a-1)/2}$, as in (24.26). The *optically thick* region, where absorption is important, corresponds to frequencies below the maximum, where one has $I \propto \omega^{5/2}$. The spectrum for an optically thick source is interpreted in terms of a thermal-like spectrum $I \propto \omega^2 \Theta$ with the temperature Θ replaced by the energy ε of the particle, and with this energy rewritten as $\varepsilon \propto \omega^{1/2}$ which follows from $\omega \approx \omega_c$, cf. (24.15) and Figure 24.3.

The degree of linear polarization $r_1 = Q/I$ reduces to a simple form for $(\mu_I \pm \mu_Q)L \ll 1$ and for $(\mu_I \pm \mu_Q)L \gg 1$. The former limit corresponds to an optically thin source with degree of polarization given by (24.29). The latter case corresponds to an optically thick source with the degree of polarization then given by

$$r_1 = \frac{Q}{I} = \frac{3}{6a + 13}, \quad \text{for} \quad \mu_I L \gg 1. \tag{24.37}$$

The polarization implied by (24.37) for $a = 3$ is about 10%. Comparison of (24.29) and (24.37) shows that as a source becomes optically thick the plane of polarization flips through 90° and decreases in magnitude.

24.6 The Razin Effect

Synchrotron emission in a plasma with plasma frequency ω_p, is suppressed at $\omega \lesssim \gamma\omega_p$, which is the *Razin effect*. This effect causes a turnover in the spectrum that is somewhat similar to synchrotron absorption, cf. Figure 24.4. However, the causes of the two types of turnover are quite different.

A treatment of synchrotron emission including the Razin effect involves repeating the derivation given in §24.3 retaining the refractive index $n(\omega) \neq 1$. The refractive index appears in such a way that most of its effects correspond to replacing β *in vacuo* by $n(\omega)\beta$ in a plasma. This is the case in the resonance condition and in the arguments of the Bessel functions, as noted in connection with (22.48). Suppose that if one writes $n(\omega)\beta = \beta_{\text{eff}}$ and $\gamma_{\text{eff}} = 1/(1 - \beta_{\text{eff}}^2)^{1/2}$, then for $n = (1 - \omega_p^2/\omega^2)^{1/2}$, $\gamma \gg 1$ implies $\gamma_{\text{eff}} \gg 1$ only for $\omega \gg \gamma\omega_p$. This suggests that emission at $\omega \lesssim \gamma\omega_p$ is more analogous to emission by a non-relativistic particle than to emission by a relativistic particle. Thus one expects a relativistic particle in a plasma to emit much less effectively at $\omega \lesssim \gamma\omega_p$ than a corresponding particle *in vacuo*.

An alternative argument based on a Lorentz transformation between the frames K and K_0 introduced in §24.1, cf. Figure 24.1, also suggests that emission in a plasma with refractive index $n = (1 - \omega_p^2/\omega^2)^{1/2}$ must be suppressed relative to emission *in vacuo* for $\omega \lesssim \gamma\omega_p$. First, note that $\omega^2 - k^2c^2$ is an invariant under Lorentz transformations, and is equal to ω_p^2. In K_0 one has $\omega_0^2 - k_0^2 c^2 = \omega_p^2$, so that the refractive index in K_0 is $n_0 = (1 - \omega_p^2/\omega_0^2)^{1/2}$. It follows that emission at $\omega_0 < \omega_p$ in K_0 cannot occur. Consequently, the corresponding emission in K also cannot occur. The first of (24.2) with $n_0 = 0$ implies that the frequency $\omega_0 = \omega_p$ in K_0 corresponds to $\omega = \gamma\omega_p$ in K. This leads one to expect that emission at $\omega < \gamma\omega_p$ is not possible in K, and while this is not strictly correct, emission at $\omega < \gamma\omega_p$ is strongly suppressed.

A semiquantitative treatment treatment of the Razin effect involves introducing the effective Lorentz factor, as defined above, in (24.15) by replacing R by

$$R = 2s/3\gamma_{\text{eff}}^3 \sin^3\theta = \omega/\omega_c, \quad \gamma_{\text{eff}} = \gamma(1 + \gamma^2\omega_p^2/\omega^2)^{-1/2}. \qquad (24.38)$$

The presence of the medium is unimportant for $\omega \gg \gamma\omega_p$ when one has $\gamma_{\text{eff}} \approx \gamma$. An important point in the following argument is that the emissivity is peaked around $R \approx 0.3$, that is, at $\omega \approx 0.3\omega_c$, and it decreases strongly ($\propto e^{-R}$) with $R \gg 1$. For $\omega \ll \gamma\omega_p$ inspection of (24.38) shows that $R \propto 1/\omega^3$ increases with decreasing frequency. Thus the effect of a medium is to cause R to have a maximum value as a function of ω at $\omega \approx \gamma\omega_p$. Provided that this maximum occurs at $\omega \gg \omega_c$ the medium has little effect on the synchrotron emission, but the emission is confined to the exponentially weak region if the maximum occurs at $\omega \ll \omega_c$ (with ω_c determined by ignoring the medium). The frequency at which this suppression effect sets in is found by solving $\omega/\omega_c = 1$ for the value of γ at which the turnover occurs, and substituting this into $\omega = \gamma\omega_p$. This gives the *Razin Tsytovich frequency*

$$\omega_{\text{RT}} = 2\omega_p^2/3\Omega_e \sin\theta. \qquad (24.39)$$

Below this turnover the spectrum rises much more steeply than for the self-absorbed case illustrated in Figure 24.4.

The Razin effect is of interest from a formal viewpoint. There is no known synchrotron source for which the effect is recognized as important.

Exercise Set 24

24.1 The temporal Fourier transform of a pulse of duration Δt varies slowly with frequency $\omega \leq 1/\Delta t$, and falls off rapidly for $\omega \gg 1/\Delta t$.

Confirm this statement by considering the Fourier transform of a square pulse, a sawtooth pulse, a gaussian pulse:

$$f_1(t) = \begin{cases} 0 & \text{for } t < 0 \text{ and } t > \Delta t, \\ 1 & \text{for } 0 < t < \Delta t, \end{cases}$$

$$f_2(t) = \begin{cases} 0 & \text{for } t < 0 \text{ and } t > \Delta t, \\ |t - \frac{1}{2}\Delta t| & \text{for } 0 < t < \Delta t, \end{cases}$$

$$f_3(t) = e^{-t^2/2(\Delta t)^2}.$$

24.2 The integral of the emissivity (24.17) over frequency involves the standard integral (24.24).

(a)   Carry out the integrals in (24.18) and compare the resulting expressions for the power radiated in the two linear polarizations with the exact results given by (22.39).

(b)   Show that in synchrotron emission by an electron with pitch angle α, photons are emitted at the rate

$$R = \frac{5}{2\sqrt{3}} \frac{e^2 \Omega_e \sin \alpha}{4\pi\varepsilon_0 \hbar c}. \tag{E24.1}$$

(c)   Show that the nth moment of the frequency distribution emitted by a single electron is given by

$$\langle \omega^n \rangle = \frac{\int_0^\infty d\omega\, \omega^{n-1} F(\omega/\omega_c)}{\int_0^\infty d\omega\, \omega^{-1} F(\omega/\omega_c)}. \tag{E24.2}$$

(d)   Use ($E24.2$) to show that the mean frequency and the variance in the frequency of the emission spectrum are determined by

$$\langle \omega \rangle = \frac{8}{15\sqrt{3}} \omega_c = 0.31\omega_c,$$

$$\langle (\omega - \langle \omega \rangle)^2 \rangle = \langle \omega^2 \rangle - \langle \omega \rangle^2 == \frac{211}{64} \langle \omega \rangle^2 = (1.8 \langle \omega \rangle)^2. \tag{E24.3}$$

24.3 Use the approximation (24.14) and the recursion formulas (21.11), (21.12) for the Macdonald functions to derive the results (24.16).

24.4 Fill in the details of the derivation of (24.26) with (24.27).

24.5 Due to synchrotron emission particles lose energy. This can affect the shape of the energy spectrum.

(a) Show that the average rate of energy loss by an electron in an isotropic distribution, that is, averaged over pitch angle, is given by

$$\dot{\varepsilon} = -b\varepsilon^2, \quad b = e^2\Omega_e^2/9\pi\varepsilon_0 m_e^2 c^5. \tag{E24.4}$$

(b) The evolution of the energy spectrum $N(\varepsilon, t)$ for electrons in a synchrotron source is modeled by an equation of the form

$$\frac{\partial N(\varepsilon, t)}{\partial t} = \frac{\partial}{\partial \varepsilon}\left[\dot{\varepsilon}N(\varepsilon, t)\right] - \nu_e N(\varepsilon, t) + Q(\varepsilon, t), \tag{E24.5}$$

where $Q(\varepsilon, t)$ is a source term and $-\nu_e N(\varepsilon, t)$ represents a loss term due to escape of electrons from the source. Solve $(E24.5)$ in the stationary case for $Q(\varepsilon) \propto \varepsilon^{-a}$

24.6 Show that the reversal of the sense of polarization as a synchrotron source becomes self-absorbed occurs at a frequency determined approximately by

$$e^{-\mu_\parallel L} \approx 2/(3a+5). \tag{E24.6}$$

24.7 Integrate (24.32) in two ways:

(a) Solve for the eigenvalues and eigenvectors of the matrix

$$[\rho - \mu] = \begin{pmatrix} -\mu_I & -\mu_Q & 0 \\ -\mu_Q & -\mu_I & -\rho_V \\ 0 & \rho_V & -\mu_I \end{pmatrix}. \tag{E24.7}$$

(b) Integrate the resulting equations for the three eigenvectors. Assume $I = I_0 = 0$ at $s = 0$ on the far side of the source.

(c) Integrate (23.32) formally as a matrix equation, obtaining, for the column matrix $[I] = (I, Q, U)$ in terms of $[\alpha] = (\alpha_I, \alpha_Q, \alpha_U)$,

$$[I] = e^{-[\rho - \mu]L} \int_0^L ds\, e^{[\rho - \mu]s}\, [\alpha], \tag{E24.8}$$

where $[r - \mu]$ denotes the 3×3 matrix $(E24.7)$.

(d) Show that these two procedures lead to the same results.

24.8 Show that synchrotron absorption *in vacuo* cannot be negative.

Specifically, partially integrate in (24.30) and show that negative absorption is possible only for $d(\varepsilon^2\eta^{\parallel,\perp})/d\varepsilon < 0$. Using the relation (24.15) between R and $\gamma = \varepsilon/mc^2$ and the recurrence relations (21.11) and (21.12), use the following and $K_\nu(R) > 0$ to complete the proof:

$$d(\varepsilon^2\eta^{\parallel,\perp})/d\varepsilon = 2\varepsilon R\left(\frac{\frac{3}{2}K_{5/3}(R) + \frac{1}{2}K_{1/3}(R)}{(2/3R)K_{2/3}(R)}\right). \tag{E24.9}$$

25

Scattering of Waves by Particles

Preamble

Any form of scattering of waves by particles may be treated as an emission process in which the scattered waves are identified as the emitted waves. The relevant extraneous current in a scattering process is that associated with the perturbed motion of the particle in the field of the unscattered wave. Examples of scattering processes discussed here include Thomson scattering, Rayleigh scattering and Raman scattering.

25.1 Thomson Scattering

Thomson scattering is scattering of waves by free electrons. The current associated with Thomson scattering is the first order current due to the perturbed motion of the scattering particle in the field of the unscattered wave. The following derivation of this current includes relativistic effects.

Let the unscattered waves be in the mode M'. The equation of motion is

$$d\mathbf{p}/dt = q[\mathbf{E}_{M'}(t, \mathbf{x}) + \mathbf{v} \times \mathbf{B}_{M'}(t, \mathbf{x})]. \qquad (25.1)$$

On expressing the fields in terms of the Fourier transform of the vector potential $\mathbf{A}_{M'}(\omega, \mathbf{k})$, (25.1) becomes

$$\frac{dp_i}{dt} = iq \int \frac{d\omega' d^3k'}{(2\pi)^4} \, g_{ij}(\omega', \mathbf{k}'; \mathbf{v}) A_{M'j}(\omega', \mathbf{k}') \, e^{-i(\omega' t - \mathbf{k}' \cdot \mathbf{x})}, \quad (25.2)$$

$$g_{ij}(\omega, \mathbf{k}; \mathbf{v}) = (\omega - \mathbf{k} \cdot \mathbf{v})\delta_{ij} + k_i v_j. \qquad (25.3)$$

The first order equation of motion is obtained from (25.2) by writing the orbit in the form

$$\mathbf{X}(t) = \mathbf{x}_0 + \mathbf{v}t + \mathbf{X}^{(1)}(t). \qquad (25.4)$$

In the fully relativistic case, the equation for the orbit is reexpressed as an equation for the velocity, and then (25.2) implies

$$\ddot{X}_i^{(1)}(t) = \frac{1}{m\gamma}\left(\delta_{ij} - \frac{v_i v_j}{c^2}\right) F_j(t), \tag{25.5}$$

where $F_i(t)$ is the quantity on the right hand side of (25.2) evaluated at $\mathbf{x} = \mathbf{x}_0 + \mathbf{v}t$. The first order current implied by (18.4) is

$$\mathbf{J}^{(1)}(\omega, \mathbf{k}) = q e^{-i\mathbf{k}\cdot\mathbf{x}_0} \int_{-\infty}^{\infty} dt \left[\dot{\mathbf{X}}^{(1)}(t) - i\mathbf{k}\cdot\mathbf{X}^{(1)}(t)\mathbf{v}\right] e^{i(\omega - \mathbf{k}\cdot\mathbf{v})t}. \tag{25.6}$$

On inserting the solution for the first order perturbation in the orbit from (25.5) with (25.2) into (25.6), one obtains the current associated with Thomson scattering:

$$J_i^{(\text{ts})}(\omega, \mathbf{k}) = -\frac{q^2}{m} \int \frac{d\omega' d^3\mathbf{k}'}{(2\pi)^4} e^{-i(\mathbf{k}-\mathbf{k}')\cdot\mathbf{x}_0} a_{ij}(\omega, \mathbf{k}; \omega', \mathbf{k}'; \mathbf{v})$$
$$\times A_{M'j}(\omega', \mathbf{k}')\, 2\pi\delta\big((\omega - \mathbf{k}\cdot\mathbf{v}) - (\omega' - \mathbf{k}'\cdot\mathbf{v})\big), \tag{25.7}$$

$$a_{ij}(\omega, \mathbf{k}; \omega', \mathbf{k}'; \mathbf{v}) = \frac{1}{\gamma}\Bigg[\delta_{ij} + \frac{k_i' v_j}{\omega' - \mathbf{k}'\cdot\mathbf{v}} + \frac{k_j v_i}{\omega - \mathbf{k}\cdot\mathbf{v}}$$
$$+ \frac{(\mathbf{k}\cdot\mathbf{k}' - \omega\omega'/c^2)v_i v_j}{(\omega - \mathbf{k}\cdot\mathbf{v})(\omega' - \mathbf{k}'\cdot\mathbf{v})}\Bigg], \tag{25.8}$$

In the non-relativistic case, (25.8) simplifies to $a_{ij} = \delta_{ij}$, and (25.7) then implies that the current is along the direction of the electric field in the unscattered waves.

On inserting the current (25.7) into the emission formula (16.10) and using (16.13), one obtains the following expression for the power emitted in scattered waves in the mode M:

$$P_M(\mathbf{k}) = \lim_{T\to\infty} \frac{1}{T}\frac{q^4 R_M(\mathbf{k})}{m^2\varepsilon_0} e_{Mi}^*(\mathbf{k}) e_{Mj}(\mathbf{k}) \int \frac{d\omega' d^3\mathbf{k}'}{(2\pi)^4} \int \frac{d\omega'' d^3\mathbf{k}''}{(2\pi)^4}$$
$$\times e^{i(\mathbf{k}'-\mathbf{k}'')\cdot\mathbf{x}_0} a_{ir}(\omega, \mathbf{k}; \omega', \mathbf{k}'; \mathbf{v}) a_{js}(\omega, \mathbf{k}; \omega'', \mathbf{k}''; \mathbf{v})$$
$$\times A_{M'r}(\omega', \mathbf{k}') A_{M's}^*(\omega'', \mathbf{k}'')\, 2\pi\delta\big((\omega - \mathbf{k}\cdot\mathbf{v}) - (\omega' - \mathbf{k}'\cdot\mathbf{v})\big)$$
$$\times 2\pi\delta\big((\omega - \mathbf{k}\cdot\mathbf{v}) - (\omega'' - \mathbf{k}''\cdot\mathbf{v})\big). \tag{25.9}$$

with $\omega = \omega_M(\mathbf{k})$. To proceed further some assumption needs to be made concerning the phase of the unscattered waves. Here we make the random phase approximation. In practice this assumption is nearly always made in treating scattering processes; if there is some intrinsic dependence on the phase of the waves then it is usually appropriate to use a different theory that is developed in terms of wave amplitudes.

An average over random phases is performed as follows. Suppose that the field $\mathbf{A}_M(\omega, \mathbf{k})$ for waves in the mode M is written in the form (15.1) with the amplitude $a_M(\mathbf{k})$ having an explicit phase factor

$e^{i\psi_M(\mathbf{k})}$. The outer product of this with with its complex conjugate at ω', $\mathbf{k}'$ has a combined phase factor $\psi_M(\mathbf{k}) - \psi_M(\mathbf{k}')$. If the phase is a random function, as implied by the random phase approximation, then averaging over the combined phase factor gives zero except for $\mathbf{k}' = \mathbf{k}$. The dispersion relation, contained in a δ-function in (15.1), then implies $\omega' = \omega$. Denoting the average by angular brackets, use of (15.1), (15.5), (15.13) and (4.37) gives

$$\left\langle e^{i\phi_{M'}(\mathbf{k}') - i\phi_{M'}(\mathbf{k}'')} \right\rangle = \frac{1}{V}(2\pi)^3 \delta^3(\mathbf{k}' - \mathbf{k}''), \qquad (25.10)$$

$$\left\langle A_{M'i}(\omega',\mathbf{k}') A^*_{M'j}(\omega'',\mathbf{k}'') \right\rangle = \frac{1}{TV} A_{M'i}(\omega',\mathbf{k}') A^*_{M'j}(\omega',\mathbf{k}')$$
$$\times (2\pi)^4 \delta(\omega' - \omega'') \delta^3(\mathbf{k}' - \mathbf{k}''), \qquad (25.11)$$

where the limits $T \to \infty$, $V \to \infty$ are implicit. The amplitude for the waves in the mode M' is rewritten using (15.1), with (15.5) and (15.13) used to reexpress the result in terms of the total energy $W_{M'}(\mathbf{k}')$ in the unscattered waves:

$$A_{M'i}(\omega',\mathbf{k}') A^*_{M'j}(\omega',\mathbf{k}') = T\frac{R_{M'}(\mathbf{k}') W_{M'}(\mathbf{k}')}{\varepsilon_0 [\omega_{M'}(\mathbf{k}')]^2}$$
$$\times e_{M'i}(\mathbf{k}') e^*_{M'j}(\mathbf{k}') \, 2\pi\delta(\omega' - \omega_{M'}(\mathbf{k}')). \qquad (25.12)$$

The resulting expression for the power emitted in scattered waves is

$$P_M(\mathbf{k}) = \frac{4(2\pi)^3 q^4}{(4\pi\varepsilon_0)^2 m^2} R_M(\mathbf{k}) \int \frac{d^3k'}{(2\pi)^3} \frac{|a_{MM'}(\mathbf{k},\mathbf{k}';\mathbf{v})|^2}{[\omega_{M'}(\mathbf{k}')]^2} R_{M'}(\mathbf{k}')$$
$$\times W_{M'}(\mathbf{k}') \delta([\omega_M(\mathbf{k}) - \mathbf{k} \cdot \mathbf{v}] - [\omega_{M'}(\mathbf{k}') - \mathbf{k}' \cdot \mathbf{v}]), \qquad (25.13)$$

$$a_{MM'}(\mathbf{k},\mathbf{k}';\mathbf{v}) = e^*_{Mi}(\mathbf{k}) e_{M'j}(\mathbf{k}')$$
$$\times a_{ij}(\omega_M(\mathbf{k}),\mathbf{k};\omega_{M'}(\mathbf{k}'),\mathbf{k}';\mathbf{v}). \qquad (25.14)$$

It is convenient to define a probability of scattering, analogous to the probability of emission, cf. (20.4). Let the probability per unit time that a particle with momentum $\mathbf{p}$ scatters a wave quantum in the mode M in the range $d^3k/(2\pi)^3$ into a wave quantum in the mode M' in the range $d^3k'/(2\pi)^3$ be $w_{MM'}(\mathbf{k},\mathbf{k}',\mathbf{p})$. In (25.13) one uses (15.35) to rewrite $W_{M'}(\mathbf{k}')$ as $\hbar\omega_{M'}(\mathbf{k}') N_{M'}(\mathbf{k}')$. One sets $N_{M'}(\mathbf{k}') = 1$ to correspond to one initial wave quantum in identifying the probability. Thus one identifies the *probability for Thomson scattering*:

$$w_{MM'}(\mathbf{k},\mathbf{k}',\mathbf{p}) = \frac{4(2\pi)^3 q^4}{(4\pi\varepsilon_0)^2 m^2} \frac{R_M(\mathbf{k}) R_{M'}(\mathbf{k}')}{\omega_M(\mathbf{k}) \omega_{M'}(\mathbf{k}')} |a_{MM'}(\mathbf{k},\mathbf{k}';\mathbf{v})|^2$$
$$\times \delta([\omega_M(\mathbf{k}) - \mathbf{k} \cdot \mathbf{v}] - [\omega_{M'}(\mathbf{k}') - \mathbf{k}' \cdot \mathbf{v}]). \qquad (25.15)$$

The probability (25.15) is symmetric under the interchange of the roles of scattered and unscattered waves:

$$w_{MM'}(\mathbf{k}, \mathbf{k}', \mathbf{p}) = w_{M'M}(\mathbf{k}', \mathbf{k}, \mathbf{p}). \qquad (25.16)$$

This implies that the same probability is to be used in treating scatterings $M' \to M$ and $M \to M'$. The symmetry (25.16) is required when applying detailed balance to a scattering process.

25.2 Inverse Compton Scattering

The terminology associated with the scattering of waves by free electrons is somewhat complicated by the historical development of the topic. "Thomson scattering" is the classical theory. "Compton scattering" is used both in the specific historical sense of the scattering of electrons by photons, and also in the generic sense of the quantum mechanical treatment of Thomson scattering. The scattering of radiation by relativistic particles is usually called "inverse Compton scattering" irrespective of whether it is treated classically or quantum mechanically.

Here we present a classical treatment of inverse Compton scattering, in which relevant integrals are performed exactly for transverse waves *in vacuo*. The probability of Thomson scattering of unpolarized transverse (T) waves *in vacuo* follows from (25.15) by inserting $M = M' = \mathrm{T}$, $R_\mathrm{T} = \frac{1}{2}$, $k = \omega/c$, $k' = \omega'/c$, and averaging over the initial states of polarization and summing over the final states of polarization. A lengthy calculation gives, for electrons,

$$w(\mathbf{k}, \mathbf{k}', \mathbf{p}) = \frac{(2\pi)^3 e^4}{(4\pi\varepsilon_0)^2 m_\mathrm{e}^2} \frac{\bar{X}}{\omega\omega'} \delta\big(\omega(1 - \boldsymbol{\kappa} \cdot \boldsymbol{\beta}) - \omega'(1 - \boldsymbol{\kappa}' \cdot \boldsymbol{\beta})\big), (25.17)$$

$$\bar{X} = \tfrac{1}{2}\left\{1 + \left[1 - \frac{1 - \boldsymbol{\kappa} \cdot \boldsymbol{\kappa}'}{\gamma^2(1 - \boldsymbol{\kappa} \cdot \boldsymbol{\beta})(1 - \boldsymbol{\kappa}' \cdot \boldsymbol{\beta})}\right]^2\right\}, \qquad (25.18)$$

where the following notation is used:

$$\mathbf{k} = k\boldsymbol{\kappa}, \quad \mathbf{k}' = k'\boldsymbol{\kappa}', \quad \mathbf{v} = \boldsymbol{\beta}c. \qquad (25.19)$$

Integrals of the following forms may be evaluated exactly in the case of isotropic distribution of unscattered photons:

$$\begin{pmatrix} I^{(n)} \\ I'^{(m)} \end{pmatrix} = \int \frac{d^3k}{(2\pi)^3} \int \frac{d^3k'}{(2\pi)^3} \, w(\mathbf{k}, \mathbf{k}', \mathbf{p}) \begin{pmatrix} \omega^n \\ \omega'^m \end{pmatrix}. \qquad (25.20)$$

The following quantities have simple physical interpretations,

$$R = I^{(0)}, \quad P = \hbar I^{(1)}, \quad P' = -\hbar I'^{(1)}, \qquad (25.21)$$

as the rate photons are scattered, the power gained by the scattered waves, and the power gained (a negative quantity) by the unscattered waves. Let N'_{ph} be the number density of the unscattered photons, and W'_{ph} be their energy density:

$$\begin{pmatrix} N'_{\text{ph}} \\ W'_{\text{ph}} \end{pmatrix} = \int \frac{d^3\mathbf{k}'}{(2\pi)^3} N(\mathbf{k}') \begin{pmatrix} 1 \\ \hbar\omega' \end{pmatrix}. \tag{25.22}$$

Then one finds

$$R = \sigma_{\text{T}} c N'_{\text{ph}}, \quad P = \sigma_{\text{T}} c W'_{\text{ph}} \frac{3+\beta^2}{3(1-\beta^2)},$$

$$P' = -\sigma_{\text{T}} c W'_{\text{ph}}, \quad \sigma_{\text{T}} = \frac{8\pi r_0^2}{3}, \tag{25.23}$$

where σ_{T} is the Thomson cross-section and $r_0 = e^2/4\pi\varepsilon_0 m_e c^2$ is the classical radius of the electron. The ratio of the mean frequency $\langle\omega\rangle$ of the scattered photons to the mean frequency $\langle\omega'\rangle$ of the unscattered photons is found by writing $P = \hbar\langle\omega\rangle R$ and $W'_{\text{ph}} = \hbar\langle\omega'\rangle N'_{\text{ph}}$. This gives

$$\frac{\langle\omega\rangle}{\langle\omega'\rangle} = \frac{3+\beta^2}{3(1-\beta^2)} \approx \begin{cases} 1 & \text{for } \beta^2 \ll 1, \\ \frac{4}{3}\gamma^2 & \text{for } \gamma \gg 1. \end{cases} \tag{25.24}$$

In the highly relativistic case the fact that the mean frequency of the scattered photons exceeds that of the unscattered photons by a factor of order γ^2 is attributed to the relativistic boost associated with emission in a forward cone with half-angle $\approx \gamma^{-1}$ about the direction of motion of the particle.

The rate of change $\dot{\varepsilon}$ of the energy of the scattering electron follows from conservation of energy:

$$P + P' + \dot{\varepsilon} = 0. \tag{25.25}$$

Then (25.23) implies

$$\dot{\varepsilon} = -\frac{4}{3} \sigma_{\text{T}} c W'_{\text{ph}} \frac{\beta^2}{1-\beta^2}, \tag{25.26}$$

for scattering of an isotropic distribution of photons.

25.3 Scattering by a Classical Oscillator

Another important class of scattering processes involves scattering by electrons in bound states of atoms or molecules. Scattering by a bound electron should be treated quantum mechanically, but some important features are contained in a simplified classical model. The model used here is the same as that used in §9.3, that is, a damped oscillator with natural frequency ω_0 and decay constant Γ, as described by (9.12).

The temporal Fourier transform of the dipole moment $q\mathbf{X}$ of the oscillator follows directly from (9.13):

$$\mathbf{d}(\omega) = \frac{q^2 \mathbf{E}(\omega)}{m(\omega_0^2 - \omega^2 - i\omega\Gamma)}. \tag{25.27}$$

No spatial dependence is included in (25.27) because of the use of the dipole approximation. However, we may make the implicit dependence of $\mathbf{E}(\omega)$ on $\mathbf{x}$ explicit by writing, cf. (5.9),

$$\mathbf{E}(\omega, \mathbf{x}) = \int \frac{d\omega' d^3 k'}{(2\pi)^4} e^{i\mathbf{k}' \cdot \mathbf{x}} i\omega' \mathbf{A}(\omega', \mathbf{k}') 2\pi\delta(\omega - \omega'). \tag{25.28}$$

Then $\mathbf{A}(\omega', \mathbf{k}')$ is identified as the amplitude of the unscattered waves in the mode M', that is, $\mathbf{A}(\omega', \mathbf{k}')$ is assumed to be of the form (15.1) with M, ω, $\mathbf{k}$ replaced by M', ω', $\mathbf{k}'$.

On inserting (25.27) with (25.28) into the emission formula (16.10) in the dipole approximation, the square of the electric vector (25.28) appears. Assuming random phases, the average over the phase is replaced by an average over all of space, denoted by angular brackets, and then (25.28) gives

$$\langle E_i(\omega) E_j^*(\omega) \rangle = \frac{T}{V} \int \frac{d^3 k'}{(2\pi)^3} \omega^2 |a_{M'}(\mathbf{k}')|^2$$
$$\times e_{M'i}(\mathbf{k}') e_{M'j}^*(\mathbf{k}') 2\pi\delta(\omega - \omega_{M'}(\mathbf{k}')). \tag{25.29}$$

A power of the normalization time T appears from using (4.37) to rewrite the square of the δ-function in (15.1), and this factor of T cancels that in (16.10). After using (15.5) and (15.13) and (15.35) for the mode M', one finds that the power emitted in scattered waves in the mode M is written in the form:

$$P_M(\mathbf{k}) = \int \frac{d^3 k'}{(2\pi)^3} w_{MM'}(\mathbf{k}, \mathbf{k}') \hbar\omega_{M'}(\mathbf{k}') N_{M'}(\mathbf{k}'), \tag{25.30}$$

with

$$w_{MM'}(\mathbf{k}, \mathbf{k}') = \left(\frac{q^2}{4\pi\varepsilon_0 m}\right)^2 \frac{\omega_M^2(\mathbf{k}) R_M(\mathbf{k}) R_{M'}(\mathbf{k}')}{\left[\omega_M^2(\mathbf{k}) - \omega_0^2\right]^2 + \omega_M^2(\mathbf{k})\Gamma^2}$$
$$\times |e_M^*(\mathbf{k}) \cdot e_{M'}(\mathbf{k}')|^2 2\pi\delta([\omega_M(\mathbf{k}) - \mathbf{k} \cdot \mathbf{v}] - [\omega_{M'}(\mathbf{k}') - \mathbf{k}' \cdot \mathbf{v}]), \tag{25.31}$$

where in the δ-function allowance has been made for motion of the oscillator at constant velocity $\mathbf{v}$, with $|\mathbf{k} \cdot \mathbf{v}| \ll \omega_M(\mathbf{k})$ and $|\mathbf{k}' \cdot \mathbf{v}| \ll \omega_{M'}(\mathbf{k}')$ ignored elsewhere in accord with the dipole approximation.

The inclusion of the Doppler shift of the frequencies in the δ-function in (25.31) is important when averaging over a distribution of oscillators. Thermal motions cause a small spread in the frequency of the scattered waves, in the sense that even in the idealized case of strictly monochromatic unscattered waves, the scattered waves have a frequency spread. Averaging (25.31) over a Maxwellian distribution $\propto e^{-v^2/2V_0^2}$ with thermal speed V_0 gives

$$
\langle w_{MM'}(\mathbf{k}, \mathbf{k}') \rangle = \left(\frac{q^2}{4\pi\varepsilon_0 m} \right)^2 \frac{\omega_M^2(\mathbf{k}) R_M(\mathbf{k}) R_{M'}(\mathbf{k}')}{\left[\omega_M^2(\mathbf{k}) - \omega_0^2\right]^2 + \omega_M^2(\mathbf{k})\Gamma^2}
$$
$$
\times |\mathbf{e}_M^*(\mathbf{k}) \cdot \mathbf{e}_{M'}(\mathbf{k}')|^2 \frac{(2\pi)^{1/2}}{|\mathbf{k} - \mathbf{k}'|V_0}
$$
$$
\times e^{[\omega_M(\mathbf{k}) - \omega_{M'}(\mathbf{k}')]^2 / 2|\mathbf{k} - \mathbf{k}'|^2 V_0^2}, \tag{25.32}
$$

where the angular brackets denote the average. Thus there is a sread of order $|\mathbf{k} - \mathbf{k}'|V_0$ about the frequency of the unscattered waves.

The probability (25.31) reduces to the probability for Thomson scattering for $\omega \gg \omega_0$; in this case the natural oscillation of the particle is unimportant, and the particle scatters like a free charge. In the opposite limit $\omega \ll \omega_0$ the probability (25.31) implies a Rayleigh type scattering law. To see this consider the cross-section for scattering of transverse waves *in vacuo*. The differential cross-section is defined as the ratio of the specific intensity of the scattered radiation to the energy flux per unit frequency of the unscattered radiation. The differential cross-section, per unit solid angle $d^2\Omega$ about the direction $\boldsymbol{\kappa}$, follows from (25.30) with (25.31):

$$
\frac{d\sigma(\omega, \boldsymbol{\kappa})}{d^2\Omega} = \left(\frac{q^2}{4\pi\varepsilon_0 m} \right)^2 \frac{\omega^4}{(\omega^2 - \omega_0^2)^2 + \omega^2\Gamma^2} |\mathbf{e}^* \cdot \mathbf{e}'|^2. \tag{25.33}
$$

At low frequencies the cross-section is proportional to the fourth power of the frequency, which is the Rayleigh scattering law.

25.4 Rayleigh Scattering and Raman Scattering

Scattering in a medium such as a gas, a liquid or a solid may be due to scattering off the particles of the medium itself, due to scattering off irregularities or other inhomogeneities in the medium, or due to scattering off impurities or other inclusions. Here a classical model is developed for aspects of the scattering off the particles of the medium itself. This scattering includes both Rayleigh and Raman scattering.

A classical model that may be applied to Rayleigh and Raman scatterings is developed as follows. Consider a non-dispersive dielectric in which the dielectric tensor K_{ij} has local spatial variations. Let $\delta K_{ij}(\mathbf{x})$ denote the part of the dielectric tensor which depends on the local spatial variations. Then in the dipole approximation the polarization $\mathbf{P}$ contains a part $\delta\mathbf{P}(t,\mathbf{x})$ that varies temporally with the electric field $\mathbf{E}(t)$ and that varies spatially with $\delta K_{ij}(\mathbf{x})$:

$$\delta P_i(t,\mathbf{x}) = \varepsilon_0 \, \delta K_{ij}(\mathbf{x}) E_j(t). \qquad (25.34)$$

The extraneous current, to be identified as the source term in the emission formula, follows from the Fourier transform of (25.34) with (6.14):

$$J_i(\omega,\mathbf{k}) = \varepsilon_0 \, \omega^2 \delta K_{ij}(\mathbf{k}) A_j(\omega), \qquad (25.35)$$

where the electric vector is written in terms of the vector potential in the temporal gauge using (5.9). The field $\mathbf{A}$ is now identified as the field of the unscattered waves, in the mode M' say, and the current (25.35) is inserted into the emission formula to find the power emitted in the scattered waves in the mode M, as is done above in the treatment of Thomson scattering. This model is used to treat a variety of scattering processes by identifying different specific forms for the spatial variations in $\delta K_{ij}(\mathbf{x})$.

An important simplifying feature of the foregoing model is that the spatial and temporal dependences on the right hand side of (25.34) are separated into the functions $\delta K_{ij}(\mathbf{x})$ and $\mathbf{E}(t)$, respectively. The assumption that the medium is non-dispersive is made so that the temporal (or frequency) dependence of $\delta K_{ij}(\mathbf{x})$ is neglected, and the dipole approximation is made so that the spatial dependence of $\mathbf{E}(t)$ is neglected.

One application of this model is to scattering by a random distribution of individual particles each with polarizability a_{ij}, cf. §9.4. This case is treated by making the identification

$$\delta K_{ij}(\mathbf{x}) = \sum_R \frac{a_{ij}}{\varepsilon_0} \, \delta^3(\mathbf{x} - \mathbf{x}_R), \qquad (25.36)$$

where R labels a particle in the medium at position $\mathbf{x} = \mathbf{x}_R$, and where the sum is over all particles. The Fourier transform of (25.36) is

$$\delta K_{ij}(\mathbf{k}) = \sum_R \frac{a_{ij}}{\varepsilon_0} \, e^{-i\mathbf{k}\cdot\mathbf{x}_R}. \qquad (25.37)$$

On inserting (25.37) into (25.35) and then repeating the derivation of the scattering probability, the factor

$$F(\mathbf{k}) = \left| \sum_R e^{-i\mathbf{k}\cdot\mathbf{x}_R} \right|^2 = \sum_{R,S} e^{-i\mathbf{k}\cdot(\mathbf{x}_R - \mathbf{x}_S)} \qquad (25.38)$$

appears. It is appropriate to average over the positions of the particle so that the scattering by each individual particle in the medium is independent of the other particles. This gives $F(\mathbf{k}) = n_0 V$, where $n_0 V$ is the total number of scatterers, with n_0 their number density and V the volume of the system. The total scattering probability is the scattering probability per particle times the number of particles. The resulting expression for the scattering probability is

$$w_{MM'}(\mathbf{k}, \mathbf{k}') = \frac{R_M(\mathbf{k}) R_{M'}(\mathbf{k}')}{\varepsilon_0^2} \, \omega_M^2(\mathbf{k}) \, n_0 V \, |e_{Mi}^*(\mathbf{k}) e_{M'j}(\mathbf{k}') a_{ij}|^2$$
$$\times 2\pi\delta\big(\omega_M(\mathbf{k}) - \omega_{M'}(\mathbf{k}')\big). \tag{25.39}$$

Simplification of (25.39) follows by assuming that the scattering by individual particles is isotropic, by assuming that the waves are transverse waves in the isotropic dielectric, and by approximating the dielectric constant K by unity except where the difference $K - 1$ appears explicitly. One writes $a_{ij} = a\delta_{ij}$, with the polarizability related to the dielectric constant by $a = \varepsilon_0(K - 1)/n_0$. Then the *differential cross-section for Rayleigh scattering* follows from (25.39) by integrating over $dk\, k^2/(2\pi)^3$ with $k \approx \omega/c$ and dividing by the group speed $v_g \approx c$. Thus one finds

$$\frac{d\sigma(\omega, \boldsymbol{\kappa})}{d^2\Omega} = \frac{\omega^4}{16\pi^2 c^4} \frac{(K-1)^2 V}{n_0} \, |\mathbf{e}^* \cdot \mathbf{e}'|^2. \tag{25.40}$$

The factor of V in (25.40) is not present if one considers the cross-section per particle, which is given by dividing by $n_0 V$. On averaging over the states of polarization and integrating over solid angle, one obtains

$$\sigma_0 = \frac{\omega^4}{6\pi n_0^2 c^4} \, (K - 1)^2. \tag{25.41}$$

The *extinction coefficient* α_0, which is defined such that the intensity of the incident light varies with distance s along the ray path as $e^{-\alpha_0 s}$, is then identified as

$$\alpha_0 = n_0 \sigma_0. \tag{25.42}$$

A second application of the model outlined above is to another form of Rayleigh scattering in which the spatial variations in $\delta K_{ij}(\mathbf{x})$ are attributed to thermal fluctuations in the number density of scatterers. In this case one writes

$$\delta K_{ij}(\mathbf{x}) = \delta n_0(\mathbf{x}) \frac{\partial K_{ij}}{\partial n_0}, \tag{25.43}$$

where $\delta n_0(\mathbf{x})$ describes the fluctuations in the number density. After insertion in the emission formula the density fluctuation appears squared and an ensemble average is taken. The resulting correlation function for the density fluctuations is identified as the level of thermal fluctuations in a medium, and is evaluated using thermodynamic arguments.

The final application of the model that we consider is to a classical version of Raman scattering. In this case the scattering by individual particles is modified from that assumed in (25.37) by including a spatial dependence in the polarizability:

$$a_{ij}(\mathbf{x}) = (a_{ij})_0 + \mathbf{X} \cdot \left(\frac{\partial}{\partial \mathbf{x}} a_{ij}\right)_0 + \cdots, \qquad (25.44)$$

where subscript 0 indicates the value in the absence of the spatial variations and where $\mathbf{X}$ is a local perturbation in position. The variable $\mathbf{X}$ is assumed to be a classical oscillator of the form discussed in §25.3. The free oscillations are assumed of the form, cf. (9.12) with the right hand side replaced by zero,

$$\mathbf{X}(t) = \mathbf{X}_0 \, e^{-\Gamma t/2} \cos(\bar{\omega}_0 t + \phi), \qquad (25.45)$$

with $\bar{\omega}_0 = (\omega_0^2 - \Gamma^2/4)^{1/2}$, and where $\mathbf{X}_0$, ϕ are arbitrary parameters. A calculation similar to that leading to (25.39) gives the probability for Raman scattering:

$$w_{MM'}(\mathbf{k}, \mathbf{k}') = \sum_{\pm} \frac{R_M(\mathbf{k}) R_{M'}(\mathbf{k}')}{\varepsilon_0^2} \, \omega_M^2(\mathbf{k}) \, n_0 V \, |e_{Mi}^*(\mathbf{k}) e_{M'j}(\mathbf{k}') c_{ij}|^2$$

$$\times 2\pi\delta\big(\omega_M(\mathbf{k}) - \omega_{M'}(\mathbf{k}') \pm \omega_0\big). \qquad (25.46)$$

Here the scattering tensor c_{ij} is identified as $\frac{1}{4}\mathbf{X}_0 \cdot \partial a_{ij}/\partial \mathbf{x}$. There are two lines corresponding to frequencies $\omega_M = \omega_{M'} \pm \bar{\omega}_0$, with the frequency of the scattered radiation shifted either down or up by the natural frequency of the oscillator. This frequency shift is characteristic of Raman scattering. The lower frequency $\omega_M = \omega_{M'} - \bar{\omega}_0$ is called the *Stokes line*, and the higher frequency $\omega_M = \omega_{M'} + \bar{\omega}_0$ is called the *anti-Stokes line*.

Scattering by an atom or molecule is attributed to an atom or molecule absorbing the incident photon and emitting the scattered photon. The Stokes and anti-Stokes lines have an interpretation in terms of conservation of energy in a quantum mechanical treatment. In Raman scattering the energy of the atom or molecule changes during the scattering process, say from the state n before to the state n' after, with the energy difference given by $E_n - E_{n'} = \hbar\omega_{nn'}$, cf. (9.25). This energy difference appears as the difference $\omega_{nn'}$ in frequency between the scattered and unscattered waves.

25.5 Stimulated Scattering

As with the relation between an emission process and the corresponding absorption process, there is an absorption-like counterpart to each specific scattering process. This absorption-like process is called *stimulated scattering* or *induced scattering*. Stimulated scattering is related to the spontaneous scattering process by appealing to Einstein coefficients for scattering.

As in §16.4, consider a transition between two states labeled $\{q\}$ and $\{q'\}$. Let us denote a scattering process in which the initial wave quantum is in the mode M' and the final wave quantum is in the mode M by $q + M' \to q' + M$, and write its probability as $w_{MM'qq'}(\mathbf{k}, \mathbf{k}')$. Conservation of energy on a microscopic scale requires $\varepsilon_q + \hbar\omega_{M'} = \varepsilon_{q'} + \hbar\omega_M$. Appealing to the Einstein coefficients, the rate of transitions $q + M' \to q' + M$ is proportional to $[1 + N_M(\mathbf{k})]N_{M'}(\mathbf{k}')n_q$, and the rate of transitions $q' + M \to q + M'$ is proportional to $[1 + N_{M'}(\mathbf{k}')]n_{q'}$. The net rate of transitions is determined by the difference between these two rates, and the implied kinetic equation for the waves in the mode M is

$$\frac{dN_M(\mathbf{k})}{dt} = \sum_{q,q'} \int \frac{d^3k'}{(2\pi)^3} \, w_{MM'qq'}(\mathbf{k}, \mathbf{k}') \left[N_{M'}(\mathbf{k}')n_q \right.$$

$$\left. - N_M(\mathbf{k})n_{q'} + N_{M'}(\mathbf{k}')N_M(\mathbf{k})(n_q - n_{q'}) \right], \quad (25.47)$$

$$\frac{dN_{M'}(\mathbf{k}')}{dt} = -\sum_{q,q'} \int \frac{d^3k}{(2\pi)^3} \, w_{MM'qq'}(\mathbf{k}, \mathbf{k}') \left[N_{M'}(\mathbf{k}')n_q \right.$$

$$\left. - N_M(\mathbf{k})n_{q'} + N_{M'}(\mathbf{k}')N_M(\mathbf{k})(n_q - n_{q'}) \right]. \quad (25.48)$$

The three terms inside the square brackets describe spontaneous scattering $q + M' \to q' + M$, spontaneous scattering $q' + M \to q + M'$, and stimulated scattering, respectively.

Stimulated scattering by a thermal distribution of particles causes an exponential rate of transfer of wave quanta from higher to lower frequencies. For Raman scattering and other forms of combination scattering this implies that the Stokes line grows and the anti-Stokes line is damped. Stimulated Raman scattering of laser light can cause a transfer of radiation from the laser line to the Stokes line with high efficiency under favorable conditions. In a plasma, stimulated (or induced) scattering of Langmuir waves can transfer energy from higher to lower k values efficiently. This process is also called non-linear Landau damping because the beat between the scattered and unscattered waves is Landau damped by thermal particles. Thus the Langmuir waves tend to form a so-called condensate at small k. This condensate can be unstable to modulational instabilities including Langmuir collapse, but such strong turbulence effects are outside the scope of the present discussion.

Exercise Set 25

25.1 Fill in the steps between (25.2) and (25.8), including the following specific steps.

(a)   By a partial integration and use of (25.3), show that (25.6) may be written in the form

$$J_i^{(1)}(\omega, \mathbf{k}) = -iqe^{-i\mathbf{k}\cdot\mathbf{x}_0} \int_{-\infty}^{\infty} dt\, g_{ji}(\omega, \mathbf{k}; \mathbf{v}) X_j^{(1)}(t)\, e^{i(\omega - \mathbf{k}\cdot\mathbf{v})t}.$$

(b)   Show that the tensor

$$a_{ij}(\omega, \mathbf{k}; \omega', \mathbf{k}'; \mathbf{v}) = \gamma^{-1}(\omega - \mathbf{k}\cdot\mathbf{v})^{-2}$$
$$\times\, g_{ri}(\omega, \mathbf{k}; \mathbf{v}) \left(\delta_{rs} - \frac{v_r v_s}{c^2}\right) g_{sj}(\omega', \mathbf{k}'; \mathbf{v}).$$
$$(E25.1)$$

may be written in the form (25.8).

Hint: You need to use the relation $\omega - \mathbf{k}\cdot\mathbf{v} = \omega' - \mathbf{k}'\cdot\mathbf{v}$ implied by the δ-function in (25.7).

25.2 According to (25.15) with (25.14) the polarization of the scattered and unscattered waves in Thomson scattering appears in the factor $a_{MM'}(\mathbf{k}, \mathbf{k}'; \mathbf{v})$.

(a)   Show that for non-relativistic electrons one has

$$|a_{MM'}(\mathbf{k}, \mathbf{k}'; \mathbf{v})|^2 \approx |\mathbf{e}_M^*(\mathbf{k}) \cdot \mathbf{e}_{M'}(\mathbf{k}')|^2.$$

(b)   For unpolarized transverse waves one is to average over the initial states of polarization and sum over the final states of polarization. Summing over the two states of polarization, cf. (16.13), show that these sums and averages, denoted by an overline, reduce to

$$\overline{|\mathbf{e}^* \cdot \mathbf{e}'|^2} = \tfrac{1}{2}\left(1 + |\boldsymbol{\kappa} \cdot \boldsymbol{\kappa}'|^2\right). \qquad (E25.2)$$

(c)   Show that the result corresponding to $(E25.2)$ for scattering of Langmuir waves into transverse waves is

$$\overline{|\mathbf{e}^* \cdot \mathbf{e}'|^2} = |\boldsymbol{\kappa} \times \boldsymbol{\kappa}'|^2. \qquad (E25.3)$$

(d)   Using the transformation formulas of §18.4, show that the generalization of $(E25.2)$ for scattering of transverse waves *in vacuo* by a particle of any energy is ($\boldsymbol{\beta} = \mathbf{v}/c$)

$$\overline{|a_{\mathrm{TT}}(\mathbf{k}, \mathbf{k}'; \mathbf{v})|^2} = \tfrac{1}{2}\left\{1 + \left[1 - \frac{1 - \boldsymbol{\kappa}\cdot\boldsymbol{\kappa}'}{\gamma^2(1 - \boldsymbol{\kappa}\cdot\boldsymbol{\beta})(1 - \boldsymbol{\kappa}'\cdot\boldsymbol{\beta})}\right]^2\right\}.$$
$$(E25.4)$$

25.3 The probability for Thomson scattering of unpolarized transverse waves *in vacuo* is given by (25.17) with (25.18).

(a) Show that the mean square frequencies of the scattered and unscattered waves are in the ratio

$$\frac{\langle \omega^2 \rangle}{\langle \omega'^2 \rangle} = \frac{1 + \frac{4}{3}\beta^2 + \frac{7}{15}\beta^4}{(1 - \beta^2)^2} \approx \begin{cases} 1 & \text{for } \beta^2 \ll 1, \\ \frac{14}{5}\gamma^4 & \text{for } \gamma \gg 1. \end{cases} \qquad (E25.5)$$

(b) Hence, using (25.24), show that for $\gamma \gg 1$ and for a monochromatic spectrum of incident waves one has

$$\left\langle (\omega - \langle \omega \rangle)^2 \right\rangle = \frac{23}{40} \langle \omega \rangle^2. \qquad (E25.6)$$

25.4 Consider Thomson scattering of collimated unpolarized transverse waves *in vacuo*. Let the unscattered waves be along the direction **n** and let χ be the angle between **v** and **n**.

(a) Show that the results (25.23) and (25.26) for isotropic unscattered waves are replaced by ($\sigma_{\mathrm{T}} = 8\pi r_0^2/3$)

$$P = \sigma_{\mathrm{T}} c W'_{\mathrm{ph}} \frac{(1 - \beta \cos \chi)^2}{1 - \beta^2},$$

$$P' = -\sigma_{\mathrm{T}} c W'_{\mathrm{ph}} (1 - \beta \cos \chi), \qquad (E25.7)$$

$$\dot{\varepsilon} = -\sigma_{\mathrm{T}} c W'_{\mathrm{ph}} \frac{\beta(1 - \beta \cos \chi)(\beta - \cos \chi)}{1 - \beta^2}.$$

(b) Comment on the fact that for $\beta < \cos \chi$ the energy of the scattering particle increases rather than decreases.

25.5 Inverse Compton scattering is treated by making the ultrarelativistic approximation in the probability rather than carrying out all the integrals for arbitrary β. In the following let θ, θ' be the angles between $\boldsymbol{\beta}$ and $\boldsymbol{\kappa}$, $\boldsymbol{\kappa}'$, respectively.

(a) Use the fact that all but a fraction γ^{-1} (which fraction is neglected) of the power emitted by a highly relativistic particle is confined to a forward cone with half-angle $\theta \approx \gamma^{-1}$ to show that the δ-function in (25.17) implies

$$\omega \approx \frac{4\gamma^2 \omega' \sin^2(\theta'/2)}{1 + \gamma^2 \theta^2}. \qquad (E25.8)$$

(b) Show that the quantity $\bar{X}$ defined by (25.18) reduces, in the ultrarelativistic limit, to

$$\bar{X} \approx (1 + \gamma^4 \theta^4)/(1 + \gamma^2 \theta^2)^2. \qquad (E25.9)$$

(c)   Hence show that on evaluating

$$P = \int \frac{d^3k}{(2\pi)^3} \int \frac{d^3k'}{(2\pi)^3}\, \hbar\omega\, w(\mathbf{k}, \mathbf{k}', \mathbf{p}) \qquad (E25.10)$$

the result given by (25.23) is reproduced.

Hint: Set $y = \gamma^2\theta^2$ and use

$$\int_0^\infty dy\, \frac{1+y^2}{(1+y)^5} = \frac{2}{3}.$$

25.6 Consider Thomson scattering of Langmuir waves into transverse waves. Assume that the Langmuir waves have frequency $\omega_L(\mathbf{k}) \approx \omega_p$, $R_L(\mathbf{k}) \approx \frac{1}{2}$ and wavevector $\mathbf{k}' = k'\boldsymbol{\kappa}'$.

(a)   Show that the probability for the scattering may be written as the polarization tensor, cf. (25.14) and (25.15),

$$w_{LT}^{\alpha\beta}(\mathbf{k}, \mathbf{k}', \mathbf{p}) = \frac{4(2\pi)^3 q^4 e_i^\alpha e_j^{*\beta} \kappa_r' \kappa_s' a_{ir} a_{js}}{(4\pi\varepsilon_0)^2 m^2 \omega\omega_p}$$

$$\times\, \delta\big(\omega(1 - \boldsymbol{\kappa}\cdot\boldsymbol{\beta}) - \omega_p + \mathbf{k}'\cdot\mathbf{v})\big), \quad (E25.11)$$

where a_{ij} denotes $a_{ij}(\omega, \mathbf{k}; \omega_p, \mathbf{k}'; \mathbf{v})$.

(b)   Show that after summing over the states of polarization for the transverse waves, the factor $e_i^\alpha e_j^{*\beta} \kappa_r' \kappa_s' a_{ir} a_{js}$ in $(E25.11)$ is replaced by

$$(\delta_{ij} - \kappa_i\kappa_j)\kappa_r'\kappa_s' a_{ir} a_{js} =$$

$$\frac{\omega_p^2}{\omega^2}\, \frac{\gamma^2(1 - \boldsymbol{\kappa}\cdot\boldsymbol{\beta})^2[1 - (\boldsymbol{\kappa}\cdot\boldsymbol{\beta})^2] - (\boldsymbol{\kappa}\cdot\boldsymbol{\kappa}' - \boldsymbol{\kappa}'\cdot\boldsymbol{\beta})^2}{\gamma^4(1 - \boldsymbol{\kappa}\cdot\boldsymbol{\beta})^4},$$

$$(E25.12)$$

and show that this reduces to the factor $(E25.3)$ for $\beta = 0$.

(c)   By making the ultrarelativistic approximation, show that the characteristic frequency of the scattered waves is of order $\omega \approx \gamma^2 k'c$ for $k'c \gg \omega_p$, and that the power in the scattered waves is related to the energy density W^L in the Langmuir waves approximately by

$$P \approx \frac{8\pi r_0^2}{9}\, cW^L\gamma^2. \qquad (E25.13)$$

25.7 The polarization of transverse waves is included in the scattering probability by writing it as a polarization tensor for both the unscattered and the scattered waves.

(a)   Argue that the scattering may be described by an equation of the form

$$\frac{dN^{\alpha\beta}(\mathbf{k})}{dt} = \int \frac{d^3k'}{(2\pi)^3}\, w^{\alpha\beta\alpha'\beta'}(\mathbf{k}, \mathbf{k}', \mathbf{p})\, N^{\alpha'\beta'}(\mathbf{k}'), \qquad (E25.14)$$

where $N^{\alpha\beta}(\mathbf{k})$ and $N^{\alpha'\beta'}(\mathbf{k}')$ are polarization tensors corresponding to the occupation numbers for the scattered and unscattered photons.

(b)   Show that for the probability in $(E25.14)$ is given by evaluating (25.31) or (25.39) for transverse waves and replacing the scalar product of the polarization vectors according to

$$|\mathbf{e}_M^*(\mathbf{k}) \cdot \mathbf{e}_{M'}(\mathbf{k}')|^2 \rightarrow e_i^\alpha e_j^{*\beta} \, e_i'^{\alpha'} e_j'^{*\beta'}. \qquad (E25.15)$$

(c)   Assume that the unscattered radiation is unpolarized and is collimated along $\mathbf{n}$. Show that the scattered radiation in the direction $\boldsymbol{\kappa}$ (i) is partially linearly polarized along the direction $\boldsymbol{\kappa} \times \mathbf{n}$, and (ii) has degree of polarization $\cos^2 \theta/(1+\cos^2 \theta)$, with $\cos\theta = \boldsymbol{\kappa}\cdot\mathbf{n}$.

25.8 In general Rayleigh and Raman scatterings by particles involve a scattering tensor c_{ij}. The individual particles are assumed oriented at random, and an average is taken over the orientations in defining an effective scattering probability. The scattering tensor is separated into a scalar part, a traceless symmetric part and an antisymmetric part by writing

$$c_{ij} = c^{(0)}\delta_{ij} + c_{ij}^{(s)} + c_{ij}^{(a)}. \qquad (E25.16)$$

Let $\langle c_{ij} c_{rs}^* \rangle$ be the average of the outer product of this tensor with itself.

(a)   Argue that $\langle c_{ij} c_{rs}^* \rangle$ can have only three independent components, along say $\delta_{ij}\delta_{rs}$, $\delta_{ir}\delta_{js}$, $\delta_{is}\delta_{jr}$.

(b)   Write the average in the form

$$\langle c_{ij} c_{rs}^* \rangle = \delta_{ij}\delta_{rs} C^{(0)} + (\delta_{ir}\delta_{js} + \delta_{is}\delta_{jr} - \tfrac{2}{3}\delta_{ij}\delta_{rs})C^{(s)}$$
$$+ (\delta_{ir}\delta_{js} - \delta_{is}\delta_{jr})C^{(a)}, \qquad (E25.17)$$

and show that one has

$$C^{(0)} = \tfrac{1}{9}\, \delta_{ij}\delta_{rs} \, \langle c_{ij} c_{rs}^* \rangle,$$
$$C^{(s)} = \tfrac{1}{20}\, (\delta_{ir}\delta_{js} + \delta_{is}\delta_{jr} - \tfrac{2}{3}\delta_{ij}\delta_{rs}) \, \langle c_{ij} c_{rs}^* \rangle, \qquad (E25.18)$$
$$C^{(a)} = \tfrac{1}{12}\, (\delta_{ir}\delta_{js} - \delta_{is}\delta_{jr}) \, \langle c_{ij} c_{rs}^* \rangle.$$

(c)   By inserting $(E25.18)$ into $(E25.17)$ show that the averages imply

$$C^{(0)} = \langle |c^{(0)}|^2 \rangle, \quad C^{(s)} = \tfrac{1}{10}\, \langle c_{ij}^{(s)} c_{ij}^{*(s)} \rangle, \quad C^{(a)} = \tfrac{1}{6}\, \langle c_{ij}^{(a)} c_{ij}^{*(a)} \rangle. \qquad (E25.19)$$

Remark: General scattering is regarded as a superposition of scalar, symmetric and antisymmetric scatterings.

25.9 The polarization characteristics of scalar, symmetric and antisymmetric scatterings, cf. Exercise 25.8, are treated independently.

(a) Show that on averaging over the initial and summing over the final states of polarization, one has

$$\tfrac{1}{2}(\delta_{ir} - \kappa_i\kappa_r)(\delta_{js} - \kappa'_j\kappa'_s)\,\langle c_{ij}c^*_{rs}\rangle = \tfrac{1}{2}[1 + (\boldsymbol{\kappa}\cdot\boldsymbol{\kappa}')^2]\,C^{(0)}$$
$$+ \tfrac{1}{6}[13 + (\boldsymbol{\kappa}\cdot\boldsymbol{\kappa}')^2]\,C^{(s)} + \tfrac{1}{2}[3 - (\boldsymbol{\kappa}\cdot\boldsymbol{\kappa}')^2]\,C^{(a)}.$$

$$(E25.20)$$

(b) Show that the degrees of polarization for scalar, symmetric and antisymmetric scatterings of unpolarized radiation are $\cos^2\theta/(1 + \cos^2\theta)$, $(6 + \cos^2\theta)/(13 + \cos^2\theta)$, and $(2 - \cos^2)\theta/(3 - \cos^2\theta)$, respectively, with $\cos\theta = \boldsymbol{\kappa}\cdot\boldsymbol{\kappa}'$.

(c) Suppose that the unscattered radiation is linearly polarized along the direction $\mathbf{e}'_0$. Let ζ be the angle between $\mathbf{e}'_0$ and $\boldsymbol{\kappa}$. Show that for scalar, symmetric and antisymmetric scatterings the intensity of the scattered radiation varies with ζ as $\sin^2\zeta$, $1 + \tfrac{1}{3}\sin^2\zeta$ and $\cos^2\zeta$, respectively.

26

Non-linear Emission Processes

Preamble

Non-linear emission processes are important in both non-linear optics and non-linear plasma physics. There is a close analogy between the physical processes involved in these two branches of electromagnetic theory, but the details are qualitatively different. Processes in non-linear optics include second harmonic generation, Brillouin scattering and parametric three-wave interactions, which involve the quadratic non-linearity, together with four-wave mixing, self-focusing of light and multi-photon absorption, which also involve the cubic non-linearity. In plasma physics the main emphasis is on the evolution of plasma turbulence, due to wave–wave interactions and non-linear wave–particle interactions, which involve the quadratic non-linearity, and modulational instabilities, which also involve the cubic non-linearity. The detailed discussion here is restricted to processes that involve only the quadratic non-linear response. The specific processes discussed are chosen as illustrative examples of non-linear emission processes.

26.1 Three-Wave Interactions

The formal theory of emission processes developed in §16.2 may be applied to treat three-wave interactions by identifying the extraneous current as that due to the beating of two waves. The idea is that the quadratic non-linear response of the medium, as discussed briefly in §6.4, implies a response of the medium simultaneously to two different fields (or to the same field appearing twice), and this current acts as a source of a third field. Let the two initial fields be at $\omega', \mathbf{k}'$ and $\omega'', \mathbf{k}''$, and the third field be at $\omega, \mathbf{k}$. The δ-functions in the response, cf. (6.26), require that the beat conditions

$$\omega' + \omega'' = \omega, \quad \mathbf{k}' + \mathbf{k}'' = \mathbf{k}, \tag{26.1}$$

be satisfied. The kinematic conditions for a three-wave interaction to occur are that (26.1) be satisfied for the three waves and that each wave satisfies the appropriate dispersion relation. Let us suppose that the three waves are in modes P, Q and M, respectively. A *coalescence* process $P + Q \to M$ involves two waves beating to generate the third, and a *decay* process $M \to P + Q$ is the inverse of a coalescence process. Here we treat the coalescence process using the emission formula and relate the decay process to it by appealing to detailed balance. The method also leads to kinetic equations for three-wave interactions.

The extraneous current is identified as the quadratic non-linear current in (6.25). This is

$$J_i^{(2)}(\omega, \mathbf{k}) = \int \frac{d\omega d^3 \mathbf{k}}{(2\pi)^3} (2\pi)^4 \delta(\omega - \omega' - \omega'') \delta^3(\mathbf{k} - \mathbf{k}' - \mathbf{k}'')$$
$$\times \alpha_{ijl}^{(2)}(\omega, \mathbf{k}; \omega', \mathbf{k}'; \omega'', \mathbf{k}'') A_j(\omega', \mathbf{k}') A_l(\omega'', \mathbf{k}''). \tag{26.2}$$

The δ-functions in (26.2) imply the beat conditions (26.1), as anticipated.

In the definition (26.2) of the non-linear response tensor, it is evident that one is free to symmetrize over the two fields in the integrand without loss of generality. This leads to the first of the following two symmetry properties of the non-linear response tensor:

$$\alpha_{ijl}^{(2)}(\omega, \mathbf{k}; \omega', \mathbf{k}'; \omega'', \mathbf{k}'') = \alpha_{ilj}^{(2)}(\omega, \mathbf{k}; \omega'', \mathbf{k}''; \omega', \mathbf{k}')$$
$$= \alpha_{jil}^{(2)}(-\omega', -\mathbf{k}'; -\omega, -\mathbf{k}; \omega'', \mathbf{k}''). \tag{26.3}$$

The second symmetry property applies only to the non-resonant part of the response tensor (in which the principal values are taken of all resonant denominators) and is derived by generalizing the Onsager relations to the non-linear response. The second symmetry is important in establishing crossing symmetries, as discussed below.

On inserting (26.2) into the emission formula (16.10) the square of $\mathbf{e}_M^*(\mathbf{k}) \cdot \mathbf{J}^{(2)}(\omega_M(\mathbf{k}), \mathbf{k})$ appears. The current (26.2) depends on the phases of the beating fields and an assumption needs to be made concerning the phase. Here we assume the random phase approximation. The average over random phases is performed in (25.10) with (25.11). Denoting the average over phase by angular brackets, use of (15.1), (15.5),

(15.13), (4.37) and (15.35) gives

$$\langle A_{Mi}(\omega, \mathbf{k}) A_{Mj}^*(\omega', \mathbf{k}') \rangle = (2\pi)^4 \delta(\omega - \omega') \delta^3(\mathbf{k} - \mathbf{k}') \frac{R_M(\mathbf{k}) \hbar N_M(\mathbf{k})}{\varepsilon_0 \omega_M(\mathbf{k})}$$

$$\times e_{Mi}(\mathbf{k}) e_{Mj}^*(\mathbf{k}) \, 2\pi \delta(\omega - \omega_M(\mathbf{k})). \qquad (26.4)$$

Let us assume that two waves from different wave distributions coalesce. Then in (26.2) one makes the replacements $A_j \to A_{Pj} + A_{Qj}$ and $A_l \to A_{Pl} + A_{Ql}$ and retains only the two cross terms in the current. In view of the symmetry property (26.3), these two terms are identical so that the relevant current is of the form (26.2) with $A_j A_l$ replaced by $2A_{Pj} A_{Ql}$. On inserting the resulting current into the emission formula (16.10) the average over phases is performed for both waves using (26.4). The δ-functions expressing the conditions (26.1) appear squared and are evaluated using (4.37) and (4.41). The resulting power radiated per unit volume in the mode M may be written in the semiclassical form

$$\frac{dN_M(\mathbf{k})}{dt} = \int \frac{d^3k'}{(2\pi)^3} \frac{d^3k''}{(2\pi)^3} \, u_{MPQ}(\mathbf{k}, \mathbf{k}', \mathbf{k}'') \, N_P(\mathbf{k}') N_Q(\mathbf{k}''), \qquad (26.5)$$

with the probability for the three-wave interaction identified as

$$u_{MPQ}(\mathbf{k}, \mathbf{k}', \mathbf{k}'') = \frac{4\hbar}{\varepsilon_0^3} \frac{R_M(\mathbf{k}) R_P(\mathbf{k}') R_Q(\mathbf{k}'')}{\omega_M(\mathbf{k}) \omega_P(\mathbf{k}') \omega_Q(\mathbf{k}'')} \, |\alpha_{MPQ}(\mathbf{k}, \mathbf{k}', \mathbf{k}'')|^2$$

$$\times (2\pi)^4 \delta^3(\mathbf{k} - \mathbf{k}' - \mathbf{k}'') \, \delta[\omega_M(\mathbf{k}) - \omega_P(\mathbf{k}') - \omega_Q(\mathbf{k}'')], \quad (26.6a)$$

$$\alpha_{MPQ}(\mathbf{k}, \mathbf{k}', \mathbf{k}'') = e_{Mi}^*(\mathbf{k}) e_{Pj}(\mathbf{k}') e_{Ql}(\mathbf{k}'')$$

$$\times \alpha_{ijl}^{(2)}(\omega_M(\mathbf{k}), \mathbf{k}; \omega_P(\mathbf{k}'), \mathbf{k}'; \omega_Q(\mathbf{k}''), \mathbf{k}''). \quad (26.6b)$$

The kinetic equation (26.5) is generalized by appealing to the Einstein coefficients. The rate of transitions for the coalescence process $P + Q \to M$ is proportional to $N_P(\mathbf{k}') N_Q(\mathbf{k}'') [1 + N_M(\mathbf{k})]$, and the rate of decay transitions $M \to P + Q$ is proportional to $[1 + N_P(\mathbf{k}')] [1 + N_Q(\mathbf{k}'')] N_M(\mathbf{k})$. The kinetic equations that result from subtracting these two rates are

$$\frac{dN_M(\mathbf{k})}{dt} = \int \frac{d^3k'}{(2\pi)^3} \frac{d^3k''}{(2\pi)^3} \, u_{MPQ}(\mathbf{k}, \mathbf{k}', \mathbf{k}'') \, \{N_P(\mathbf{k}') N_Q(\mathbf{k}'')$$

$$- N_M(\mathbf{k}) [1 + N_P(\mathbf{k}') + N_Q(\mathbf{k}'')]\}, \quad (26.7)$$

$$\frac{dN_P(\mathbf{k}')}{dt} = -\int \frac{d^3k}{(2\pi)^3} \frac{d^3k''}{(2\pi)^3} \, u_{MPQ}(\mathbf{k}, \mathbf{k}', \mathbf{k}'') \, \{N_P(\mathbf{k}') N_Q(\mathbf{k}'')$$

$$- N_M(\mathbf{k}) [1 + N_P(\mathbf{k}') + N_Q(\mathbf{k}'')]\}, \quad (26.8)$$

$$\frac{dN_Q(\mathbf{k}'')}{dt} = -\int \frac{d^3k}{(2\pi)^3} \frac{d^3k'}{(2\pi)^3} \, u_{MPQ}(\mathbf{k}, \mathbf{k}', \mathbf{k}'') \, \{N_P(\mathbf{k}') N_Q(\mathbf{k}'')$$

$$- N_M(\mathbf{k}) [1 + N_P(\mathbf{k}') + N_Q(\mathbf{k}'')]\}. \quad (26.9)$$

The term (26.5) is reproduced by (26.7) only in the limit in which the terms involving $N_M(\mathbf{k})$ are negligible on the right hand side, that is, only in the limit in which the level of the waves in the mode M is negligible. The significance of the other terms on the right hand sides of (26.7)–(26.9) is discussed in several different specific contexts below.

Implicit in the derivation of the kinetic equations (26.7)–(26.9) is that the probabilities for the coalescence $P+Q \to M$ and of the inverse decay process $M \to P + Q$ are identical. There are also crossing relations. An example is for Brillouin scattering in a uniaxial crystal. Specifically consider coalescence of an ordinary wave ($P = $ o) and a sound wave ($Q = $ s) to form an extraordinary wave ($M = $ x). The process described is o + s $\to$ x and the inverse process is x $\to$ o+s. However, the *crossed* process o $\to$ x + s and its inverse x + s $\to$ o are also allowed. The probability for the crossed process is obtained from the probability for the initial process by changing the sign of the wavevector $\mathbf{k}_s \to -\mathbf{k}_s$ of the sound wave and using the relations (11.13) and (11.19). Specifically, if the crossing symmetry is applied to the wave mode labeled P in (26.6a, b), then one uses

$$\omega_P(-\mathbf{k}') = -\omega_P(\mathbf{k}'), \quad \mathbf{e}_P(-\mathbf{k}') = \mathbf{e}_P^*(\mathbf{k}'), \quad R_P(-\mathbf{k}') = R_P(\mathbf{k}'),$$
(26.10)

and the modified probability then describes the crossed processes $M + P \leftrightarrow Q$. Formally, the crossing symmetry depends on the second of the symmetry properties (26.3).

Four-wave interactions may be treated in a similar way to three-wave interactions. The relevant current includes the cubic response in (6.25), and it also includes currents arising from the quadratic response acting twice. For example, waves at k_1, k_2 and k_3, where k denotes ω and $\mathbf{k}$ together, can coalesce together by two of them beating to form a current, at $k' = k_1 + k_2$, this current generates a field at k' and this field can coalesce with the third field to produce a current at $k = k' + k_3 = k_1 + k_2 + k_3$. There are obviously three ways of combining two quadratic responses to produce an effective cubic response, and all three must be added to find the total effective cubic response in general. Different terms in the effective cubic response contribute in different applications, and appropriate simplifying approximations need to be considered separately in each case. Four-wave interactions include coalescence of three waves to produce the fourth, and the crossed process of wave–wave scattering. These are weak turbulence processes for which the random phase approximation is appropriate. The cubic response also allows strong turbulence processes such as self-focusing of laser light and Langmuir collapse; such processes cannot be treated using the random phase approximation.

26.2 Frequency Doubling in a Crystal

One application of the theory of wave–wave interactions is to frequency doubling of laser light in a crystal. In this case the two initial waves are in the same mode and have the same frequency so that the final wave has double this frequency.

Two modifications are made to the probability (26.6a) for this case. First, (26.6a) is derived assuming that the two initial waves are different and if the two waves are the same one multiplies the probability (26.6a) by 2 to obtain the correct probability. (This factor of 2 arises as follows. First, in identifying the current there is a factor of 2 that arises in (26.2) from the two cross terms when the total field $\mathbf{A}(\omega, \mathbf{k})$ is a sum of two independent fields, and this factor of 2 is absent when there is only one field, as is the case here. This factor of 2 appears squared in the probability. Second, an additional factor of 2 arises because the phase average of the outer product of two fields with their complex conjugates, each of which average is of the form (26.4), can be taken in two equivalent ways if the two fields are the same field.) The other change is to rewrite the non-linear response tensor in (26.4) in terms of the corresponding non-linear susceptibility tensor introduced in (6.23). Spatial dispersion is neglected and then one has

$$\alpha_{ijl}^{(2)}(\omega, \omega', \omega'') = i\omega\omega'\omega''\chi_{ijl}^{(2)}(\omega, \omega', \omega''). \qquad (26.11)$$

The quantum mechanical procedure applied in §9.4 to an atom or molecule allows one to calculate the non-linear polarizability, and the non-linear susceptibility tensor follows directly from the non-linear polarizability. The counterpart of (9.30) for the non-linear polarizability is

$$a_{ijl}(\omega, \omega', \omega'') = \sum_{q, q', q''} \frac{\rho_{qq}^{(0)}}{\hbar^2} \left[\frac{(d_i)_{qq'}(d_j)_{q'q''}(d_i)_{q''q}}{(\omega - \omega_{q''q})(\omega' - \omega_{q'q})} + \text{perm} \right], \qquad (26.12)$$

where "perm" denotes five other terms that are constructed from the given term by imposing the symmetry properties (26.3), cf. Exercise 26.2.

There are two notable limitations on frequency doubling in a crystal. First, the non-linear susceptibility must be non-zero and, as argued in §6.4, the tensor is identically zero for an isotropic dielectric when spatial dispersion is ignored. Frequency doubling can occur only in an anisotropic crystal in which the quadratic non-linear susceptibility is non-zero. There is no corresponding restriction in principle for frequency trebling because all media, including the vacuum, have non-zero cubic non-linear response tensors.

The other limitation on frequency doubling is that the beat conditions (26.1) need to be satisfied. These require $\omega = 2\omega'$ and $\mathbf{k} = 2\mathbf{k}'$. It follows that the three waves, that is, the two identical coalescing waves and the wave at double the frequency, are collinear, and hence they have the same angles of propagation, $\theta = \theta'$ and $\phi = \phi'$. Thus they must satisfy the *refractive index matching* condition

$$n_M(2\omega, \theta, \phi) = n_P(\omega, \theta, \phi). \tag{26.13}$$

Although frequency doubling is not possible for another reason in an isotropic dielectric, let us consider whether (26.13) can be satisfied in this case. The labels M and P are redundant (the only waves are transverse waves), and there is no dependence on angle. Then (26.13) can be satisfied if the dielectric is strictly non-dispersive. Weak dispersion implies a term in the refractive index proportional to ω^2, and then (26.13) cannot be satisfied. An analogous argument leads to the conclusion that the condition (26.13) cannot be satisfied for waves in one specific mode of an anisotropic crystal which has any dispersion. To satisfy (26.13) the initial wave and the wave at double the frequency must be in different wave modes. That is, one must have $M \neq P$. For example, in a uniaxial crystal there are only two possibilities: $n_o(2\omega) = n_x(\omega, \theta)$, or $n_x(2\omega, \theta) = n_o(\omega)$. In practice one or other of these conditions can be satisfied in a given crystal only for a specific angle θ of propagation.

The kinetic equations (26.7) to (26.9) are modified to treat frequency doubling by the following relabeling: one writes $M = 2$ for the radiation at 2ω and $P = Q = 1$ for the initial radiation at ω. Then (26.8) and (26.9) are identical. Let us ignore the unit term inside the square brackets for the present; this term describes the intrinsically quantum mechanical effect of photon splitting. Then (26.7) implies that the occupation number N_2 of the 2ω-emission builds up until the quantity inside the curly brackets in (26.7) approaches zero, which defines the saturation level for the frequency doubling process. It follows that the saturation level is $2N_2(2\mathbf{k}) = N_1(\mathbf{k})$. On multiplying by $\hbar\omega$, the definition (17.42) implies that saturation occurs when the effective temperature $T_2(2\mathbf{k})$ of the 2ω-emission is equal to the effective temperature $T_1(\mathbf{k})$ of the ω-radiation. Thus one has

$$T_2(2\mathbf{k}) \leq T_1(\mathbf{k}), \tag{26.14}$$

with the equality applying at the saturation level. It is tempting to interpret (26.14) in terms of the second law of thermodynamics. However, although a thermodynamic argument leads to the correct saturation level in this case, in other cases the saturation level implied by (26.7) is not so readily interpreted.

Frequency doubling is one example of a much wider class of non-linear processes of interest in non-linear optics. Another example is parametric generation of frequencies $\omega_1 \pm \omega_2$ when light at ω_1 and ω_2 from two different lasers is focused on an appropriate crystal. Frequency trebling and higher harmonic generation can also occur in principle, and they require that index matching conditions analogous to (26.13) be satisfied. A further example is for light from a single laser $(\omega, \mathbf{k})$ that generates a Stokes line $(\omega_S, \mathbf{k}_S)$ through stimulated scattering, cf. §25.5. The anti-Stokes line $(\omega_A, \mathbf{k}_A)$ automatically satisfies the four-wave frequency matching condition $2\omega = \omega_S + \omega_A$ and it may be generated through a wave–wave interaction provided that the index matching allows the relation $2\mathbf{k} = \mathbf{k}_S + \mathbf{k}_A$ to be satisfied. This can produce an otherwise unexpected signal at the anti-Stokes line.

A three-wave process in a different context is photon splitting in a superstrong magnetic field. A vacuum with a static magnetic field, referred to here as a magnetized vacuum, has properties like those of a material medium. The magnetized vacuum is birefringent, cf. Exercise 12.2, and it also has a non-zero quadratic response tensor. In a magnetized vacuum the splitting of one photon into two photons is allowed. As for frequency doubling in a crystal, the initial photon and the daughter photons cannot all be in the same mode. Photon splitting is an intrinsically quantum mechanical effect in the sense that it cannot be described quantitatively without introducing $\hbar$. However, it is possible to describe the process semiclassically. The relevant non-linear response tensor follows from the Heisenberg–Euler Lagrangian, cf. Exercise 26.4, and this allows one to write down the relevant probability using (26.6a). The kinetic equation for photon splitting is dominated by the term N_M, that arises from the unit term inside the square brackets in (26.7)–(26.9), which term is intrinsically quantum mechanical. It is the dominant term in photon splitting because the process is of interest at high frequencies (for γ-rays) in which case the occupation number is less than unity in most applications; for example it is approximately $e^{-\hbar\omega/\Theta}$ for $\hbar\omega \gg \Theta$, where Θ is the temperature in energy units. In contrast, in non-linear optics and in plasma physics the occupation numbers tend to be much greater than unity, and then this intrinsically quantum mechanical term is negligible.

26.3 The Non-linear Response of a Plasma

The quadratic non-linear response tensor of a plasma is non-zero even in the simplest case of a cold isotropic plasma. For many purposes, it suffices to have two different approximate forms for the non-linear response tensor. One of these is derived using the cold plasma model and the other is derived using the electrostatic approximation in a thermal plasma.

The cold plasma model is used in §10.2 to derive the response of a cold magnetized plasma. Here we apply the model to a cold unmagnetized electron gas and derive the linear and quadratic response tensors. The basic equations are

$$\partial \mathbf{v}/\partial t + \mathbf{v} \cdot \operatorname{grad} \mathbf{v} = -(e/m_e)[\mathbf{E} + \mathbf{v} \times \mathbf{B}], \qquad (26.15)$$

$$\partial n/\partial t + \operatorname{div}(n\mathbf{v}) = 0, \quad \mathbf{J} = -en\mathbf{v}, \qquad (26.16)$$

where all quantities are functions of t and $\mathbf{x}$. These are the equation of fluid motion, the continuity equation for electrons and the fluid model for the current, respectively. The quantities $\mathbf{E}$ and $\mathbf{B}$ are of first order in the wave amplitude and the other quantities are expanded in powers of the wave amplitude:

$$\mathbf{v} = \mathbf{v}^{(1)} + \mathbf{v}^{(2)} + \cdots,$$
$$n = n^{(0)} + n^{(1)} + \cdots, \qquad (26.17)$$
$$\mathbf{J} = \mathbf{J}^{(1)} + \mathbf{J}^{(2)} + \cdots.$$

Equations (26.15) and (26.16) are solved using a perturbation approach to derive sets of equations at successive order. The electromagnetic fields are expressed in terms of the vector potential, and each set of equations is Fourier transformed.

The resulting first order equations give

$$\mathbf{v}^{(1)}(\omega, \mathbf{k}) = -(e/m_e)\mathbf{A}(\omega, \mathbf{k}), \quad n^{(1)}(\omega, \mathbf{k}) = n^{(0)}\mathbf{k} \cdot \mathbf{v}^{(1)}(\omega, \mathbf{k})/\omega. \qquad (26.18)$$

The second order equation of motion reduces to

$$-im_e\omega\mathbf{v}^{(2)}(\omega, \mathbf{k}) = \int d\lambda^{(2)} \left\{ -i\mathbf{k}_2 \cdot \mathbf{v}^{(1)}(\omega_1, \mathbf{k}_1)\mathbf{v}^{(1)}(\omega_2, \mathbf{k}_2) \right.$$
$$\left. - ie\mathbf{v}^{(1)}(\omega_1, \mathbf{k}_1) \times \left[\mathbf{k}_2 \times \mathbf{E}(\omega_2, \mathbf{k}_2)\right]/\omega_2 \right\}, \qquad (26.19)$$

where the convolution integral $d\lambda^{(2)}$ is defined by (6.26). The second order current is

$$\mathbf{J}^{(2)}(\omega, \mathbf{k}) = -en^{(0)}\mathbf{v}^{(2)}(\omega, \mathbf{k}) - e \int d\lambda^{(2)} n^{(1)}(\omega_1, \mathbf{k}_1)\mathbf{v}^{(1)}(\omega_2, \mathbf{k}_2). \qquad (26.20)$$

After substituting (26.18) and (26.19) into (26.20), the non-linear response tensor is identified by writing the expression that results from (26.20) in the form (26.2). The first of the symmetry properties (26.3) must be imposed to obtain a result that is independent of the details of the calculation. This calculation leads to the following expression for the quadratic non-linear response tensor for a cold isotropic electron gas:

$$\alpha_{ijl}^{(2)}(\omega, \mathbf{k}; \omega', \mathbf{k}'; \omega'', \mathbf{k}'') = -\frac{e^3 n_e}{2m_e^2} \left(\delta_{jl}\frac{k_i}{\omega} + \delta_{il}\frac{k_i'}{\omega'} + \delta_{ij}\frac{k_i''}{\omega''} \right), \quad (26.21)$$

where the unperturbed electron number density is rewritten as n_e.

The expression (26.21) applies provided that all disturbances are fast, in the sense that they have phase speeds much greater than the thermal speeds of the particles. Thus, (26.21) applies for $|\omega| \gg kV_e$, $|\omega'| \gg k'V_e$, $|\omega''| \gg k''V_e$, and only when all three of these inequalities are satisfied is (26.21) to be used in the probability (26.6b). In an isotropic plasma only Langmuir (L) waves and transverse (T) waves have phase speeds $\gg V_e$, and the only allowed three-wave processes involving fast waves are $L + L \leftrightarrow T$ and $L + T \leftrightarrow T$. Alternative approximate forms for the non-linear response tensor of a thermal plasma are required for cases where one or more of the three waves or disturbances is slow in the sense that it has a phase speed $\lesssim V_e$.

A general expression for the quadratic response tensor of an unmagnetized plasma is obtained using the Vlasov theory, that is, by extending the calculation of the linear response tensor given in §10.4 to the non-linear case. The general expression that results may be separated into contributions from each species of particles in the plasma. The contribution of one (unlabeled) species is

$$\alpha_{ijl}^{(2)}(\omega, \mathbf{k}; \omega', \mathbf{k}'; \omega'', \mathbf{k}'') = q^3 \int d^3\mathbf{p} \, \frac{v_i g_{rj}(\omega', \mathbf{k}'; \mathbf{v})}{\omega - \mathbf{k} \cdot \mathbf{v}}$$
$$\times \frac{\partial}{\partial p_r} \left[\frac{g_{sl}(\omega'', \mathbf{k}''; \mathbf{v})}{\omega'' - \mathbf{k}'' \cdot \mathbf{v}} \frac{\partial f(\mathbf{p})}{\partial p_s} \right], \quad (26.22a)$$

$$g_{ij}(\omega, \mathbf{k}; \mathbf{v}) = (\omega - \mathbf{k} \cdot \mathbf{v})\delta_{ij} + k_i v_j. \quad (26.22b)$$

The form (26.22a) is unsymmetrized, and the first of the symmetry properties (26.3) needs to be imposed on it to obtain a general expression. However, if we assume that one disturbance, say that at ω'', $\mathbf{k}''$, is different from the other two in that it is slow, then it is appropriate to use the unsymmetrized form (26.22a). Three simplifying assumptions (in addition to the neglect of relativistic effects) are made to derive a suitable approximation. First, it is assumed that for the slow disturbance at ω'', $\mathbf{k}''$ the main contribution arises from where the denominator $\omega'' - \mathbf{k}'' \cdot \mathbf{v}$

is small, and only the leading term in an expansion in $\omega'' - \mathbf{k}'' \cdot \mathbf{v}$ is retained. Second, the other two disturbances are assumed to be fast, the approximation $v_i g_{rj}(\omega', \mathbf{k}'; \mathbf{v})/\omega - \mathbf{k} \cdot \mathbf{v} \approx v_i \delta_{rj}$ is made, and a partial integration is performed over p_r. Third, the disturbances at ω'', $\mathbf{k}''$ are assumed to be longitudinal, specifically due only to fluctuations in the electron density. Then (26.22a) gives

$$\alpha_{ijl}^{(2)}(\omega, \mathbf{k}; \omega', \mathbf{k}'; \omega'', \mathbf{k}'') \approx \frac{e\varepsilon_0}{m_e}\omega'' \chi^{\mathrm{Le}}(\omega'', \mathbf{k}'')\delta_{ij}k_l'', \qquad (26.23)$$

where $\chi^{\mathrm{Le}}(\omega, \mathbf{k}) = \left[1 - \phi(y_e)\right]/k^2\lambda_{\mathrm{De}}^2$ is the real part of the electronic contribution to $K^L(\omega, \mathbf{k})$ in (10.23). For slow disturbances the further approximation $\phi(y_e) \ll 1$ is usually justified, cf. (10.29).

The approximate form (26.23) is appropriate for treating three-wave processes in an isotropic plasma that involve an ion sound (s) wave and two fast waves. These processes are $L + s \leftrightarrow T$, $L + s \leftrightarrow L$, $T + s \leftrightarrow T$. The form (26.23) is also appropriate when the low-frequency field is not a wave but is associated with fluctuations due to the thermal motions of the plasma particles. These fluctuations lead to an important modification to Thomson scattering for long wavelength ($k\lambda_{\mathrm{De}} \lesssim 1$) waves in a plasma, as discussed below.

26.4 Radiation from a Turbulent Plasma

A plasma is said to be *turbulent* when waves in one or more wave modes are excited well above the thermal level. For waves in the mode M this corresponds to their effective temperature satisfying $T_M(\mathbf{k}) = \hbar\omega_M(\mathbf{k})N_M(\mathbf{k}) \gg T_e$. Here we discuss radiation that can occur in an isotropic plasma due to the presence of turbulence in the Langmuir and ion sound modes. Three processes are of interest: $L + s \leftrightarrow T$, which results in *fundamental plasma emission*, $L + s \leftrightarrow L$, which causes the distribution of Langmuir waves to evolve, and $L + L \leftrightarrow T$, which results in *second harmonic plasma emission*. Other relevant processes are $T + s \leftrightarrow T$ and $T + L \leftrightarrow T$, which are counterparts of Brillouin and Raman scatterings in a turbulent plasma, and which can lead to a relatively large cross-section for the scattering of transverse waves. A related turbulent process mentioned below is non-linear scattering, in which a fluctuation associated with the thermal motion of an ion replaces the ion sound wave.

Plasma emission is the dominant emission process for solar radio bursts, which were first observed and studied in the 1940s. The earliest theory was presented by Ginzburg and Zheleznyakov in 1958. There are three ingredients in their theory, and although all the quantitative details are no longer acceptable in view of subsequent developments in plasma theory, these three ingredients remain in most current versions of the theory of plasma emission. The ingredients are:

(1) Langmuir turbulence is generated through a streaming or other instability. In particular, the exciting agency of so-called type III solar radio bursts is a stream of electrons with the properties required to cause Langmuir waves to grow, cf. §20.5.

(2) Scattering of Langmuir waves off low-frequency fluctuations (ion sound waves or other fluctuations) produces both scattered Langmuir waves and scattered transverse waves, and the latter escape to be observed as fundamental ($\omega \approx \omega_{\mathrm{p}}$) plasma emission.

(3) Coalescence of two Langmuir waves produces second harmonic ($\omega \approx 2\omega_{\mathrm{p}}$) plasma emission.

The source of the low-frequency turbulence is not known, although there is evidence from *in situ* observations in the solar wind that the plasma contains an adequate non-thermal level of low-frequency fluctuations.

Scattering by Ion Sound Waves

The scattering of one fast wave in the mode P into a fast wave in the mode M by an ion sound wave can be due either to the coalescence process $P + s \to M$ or due to the decay process $P \to M + s$. The probability for either of these processes follows by inserting (26.23) into (26.6a, b), taking a factor of 2 in account due to the lack of symmetrization in (26.23), and inserting the expression (13.12) for $R_{\mathrm{s}}(\mathbf{k}'')$:

$$u_{MPs}(\mathbf{k}, \mathbf{k}', \mathbf{k}'') = \frac{e^2 \hbar \omega_{\mathrm{p}}^2}{2\varepsilon_0 m_{\mathrm{e}}^2 V_{\mathrm{e}}^2} \frac{\omega_{\mathrm{s}}(\mathbf{k}'') R_M(\mathbf{k}) R_P(\mathbf{k}')}{\omega_M(\mathbf{k}) \omega_P(\mathbf{k}')} |\mathbf{e}_M^*(\mathbf{k}) \cdot \mathbf{e}_P(\mathbf{k}')|^2$$
$$\times (2\pi)^4 \delta^3(\mathbf{k} - \mathbf{k}' \mp \mathbf{k}'') \, \delta\big(\omega_M(\mathbf{k}) - \omega_P(\mathbf{k}') \mp \omega_{\mathrm{s}}(\mathbf{k}'')\big).$$
$$(26.24)$$

For transverse waves the factor involving the polarization vectors is summed over the two final states of polarization or averaged over the initial states of polarization. The various possible cases then give

$$|\mathbf{e}_M^*\mathbf{k}) \cdot \mathbf{e}_P(\mathbf{k}')|^2 = \begin{cases} |\boldsymbol{\kappa}_{\mathrm{L}} \cdot \boldsymbol{\kappa}_{\mathrm{L}'}|^2 & M = \mathrm{L}, \, P = \mathrm{L}', \\ |\boldsymbol{\kappa}_{\mathrm{L}} \times \boldsymbol{\kappa}_{\mathrm{T}}|^2 & M = \mathrm{L}, \, P = \mathrm{T}, \\ \frac{1}{2} |\boldsymbol{\kappa}_{\mathrm{T}} \times \boldsymbol{\kappa}_{\mathrm{L}}|^2 & M = \mathrm{T}, \, P = \mathrm{L}, \\ \frac{1}{2}(1 + |\boldsymbol{\kappa}_{\mathrm{L}} \cdot \boldsymbol{\kappa}_{\mathrm{T}}|^2) & M = \mathrm{T}, \, P = \mathrm{T}'. \end{cases} \quad (26.25)$$

For the three-wave interaction to occur the beat conditions (26.1)

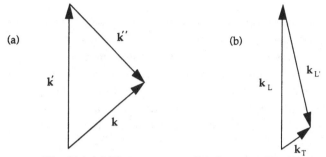

Fig. 26.1 (a) The vector sum (26.1) in an arbitrary case.
(b) Coalescence of two Langmuir waves in the head-on case.

must be satisfied. The vector sum of the wavevectors is illustrated in Figure 26.1(a). The frequency of Langmuir waves or of transverse waves is much greater than that of the ion sound waves, so that the initial and final fast waves must have approximately the same frequency. The condition $\mathbf{k} = \mathbf{k}' \pm \mathbf{k}''$ has different implications for the specific processes discussed in (26.25). For $L \pm s \rightarrow L'$ it imposes a somewhat subtle kinematic restriction that is not discussed here, cf. Exercise 26.6 however. For $L \pm s \rightarrow T$ the transverse wave must have frequency close to that of the Langmuir wave (fundamental plasma emission) and then the transverse waves have wavenumber k very much smaller than the k' of Langmuir waves of relevance; one then requires that the ion sound wave has $\mathbf{k}'' \approx \mp \mathbf{k}'$. For $T \pm s \rightarrow T'$ the ion sound wave must have very small k'' comparable to the wavenumber of the transverse waves.

In practice there is little interest in the detailed rate of the scattering process. If the scattering is too weak to lead to saturation of the three-wave interaction then it is of little interest, and if saturation does occur then one is interested primarily in the saturation level. The simplest case is when the level of the ion sound waves is arbitrarily high when the dominant terms inside the curly brackets in (26.7)–(26.9) are $N_s(\mathbf{k}'')\big[N_P(\mathbf{k}') - N_M(\mathbf{k})\big]$. It follows that saturation occurs for $N_M(\mathbf{k}) = N_P(\mathbf{k}')$. As the two fast waves have nearly equal frequencies, this corresponds to saturation when the effective temperature of the scattered waves approaches that of the initial waves. This saturation level is what one would expect on the basis of a thermodynamic argument, as for the case of frequency doubling in a crystal, cf. (26.14). However, this simple result applies only for $N_s(\mathbf{k}'') \gg N_P(\mathbf{k}')$, and when this inequality is relaxed the predicted saturation level is not so simple to interpret, cf. Exercise 26.1.

Provided that saturation of the scattering by low-frequency waves is a reasonable assumption, the foregoing discussion implies that the effective temperature of escaping fundamental plasma emission should never exceed the effective temperature of the Langmuir waves. The latter may be estimated from the theory of the streaming instability. The predicted maximum effective temperature is compatible with an observed typical maximum value of the brightness temperature of type III bursts.

Second Harmonic Plasma Emission

The probability for the process $L + L' \to T$ of second harmonic plasma emission follows by assuming the approximate form (26.21) for the non-linear response tensor and inserting it and the properties of Langmuir waves, cf. §13.1, into (26.6a, b). After summing over the two states of polarization of the transverse waves, one has

$$u_{\mathrm{TLL}}(\mathbf{k}_T, \mathbf{k}_L, \mathbf{k}_{L'}) = \frac{e^2 \hbar}{32\varepsilon_0 m_e^2 \omega_p} \frac{(k_L^2 - k_{L'}^2)^2}{k_T^2} |\boldsymbol{\kappa}_L \times \boldsymbol{\kappa}_{L'}|^2$$
$$\times (2\pi)^4 \delta^3(\mathbf{k}_T - \mathbf{k}_L - \mathbf{k}_{L'}) \, \delta(\omega_T(\mathbf{k}_T) - \omega_L(\mathbf{k}_L) - \omega_L(\mathbf{k}_{L'})). \tag{26.26}$$

The probability simplifies in the 'head-on' approximation, as illustrated in Figure 26.1(b), which applies when the wavenumber of the transverse waves is neglected in comparison with that of the Langmuir waves. With $\omega \approx 2\omega_p$, the dispersion relation $\omega^2 = \omega_p^2 + k_T^2 c^2$ implies $k_T^2 = 3\omega_p^2/c^2$. Let the phase speed of the Langmuir wave be $v_\phi \approx \omega_p/k_L$, which is assumed to be approximately the same as the speed of the streaming electrons that generate the Langmuir waves. Then one has $k_T^2/k_L^2 \approx 3v_\phi^2/c^2$. Provided this ratio is small, the beat condition in (26.1) requires $\mathbf{k}_{L'} \approx -\mathbf{k}_L$, which corresponds to the two Langmuir waves meeting head on.

The head-on approximation for waves generated by a streaming instability requires Langmuir waves propagating oppositely to the streaming direction to coalesce with the Langmuir waves in the streaming direction. Scattering $L \pm s \to L'$ is a possible mechanism for the generation of the required backward propagating Langmuir waves.

In the head-on approximation the following apply in (26.26):

$$\delta[\omega_T(\mathbf{k}_T) - \omega_L(\mathbf{k}_L)] \approx (1/\sqrt{3}c)\delta(k_T - \sqrt{3}\omega_p/c),$$
$$\frac{(k_L^2 - k_{L'}^2)^2}{k_T^2} |\boldsymbol{\kappa}_L \times \boldsymbol{\kappa}_{L'}|^2 \approx (3\omega_p^2/c^2)|\boldsymbol{\kappa}_T \cdot \boldsymbol{\kappa}_L|^2 |\boldsymbol{\kappa}_T \times \boldsymbol{\kappa}_L|^2. \tag{26.27}$$

The angular dependence implied by (26.27) is quadrupolar, as is seen by writing the angle between $\boldsymbol{\kappa}_T$ and $\boldsymbol{\kappa}_L$ as θ and noting that the angular dependence is then as $\sin^2\theta\cos^2\theta$.

Saturation of second harmonic emission occurs where the factor in the curly brackets in (26.7)–(26.9) vanishes. Ignoring the intrinsically quantum mechanical term (the unit term inside the square brackets), writing $M = $ T, $P = $ L and $Q = $ L$'$, solving for N_T and expressing the result in terms of effective temperatures, one finds

$$T_T(\mathbf{k}_T) \leq \frac{2T_L(\mathbf{k}_L)T_{L'}(\mathbf{k}_{L'})}{T_L(\mathbf{k}_L) + T_{L'}(\mathbf{k}_{L'})}, \qquad (26.28)$$

where saturation corresponds to the equality. For $T_{L'} = T_L$ saturation occurs at $T_T = T_L$, and for $T_{L'} \ll T_L$ saturation occurs at $T_T = 2T_{L'}$. It follows that if all stages saturate, then the effective temperature of the second harmonic is the same as that of the fundamental. That is, provided that the production of the backward propagating Langmuir waves by scattering $L\pm s \rightarrow L'$ saturates ($T_{L'} \approx T_L$), then the saturation level of the second harmonic is the same as that of the fundamental and is equal to T_L. However, if the production of the backward propagating Langmuir waves does not saturate ($T_{L'} \ll T_L$), then the saturation level of the second harmonic is much less than that of the fundamental ($T_T \approx 2T_{L'} \ll T_L$).

In type III solar radio bursts there is a wide variation in the intensity of the observed radio emission. Much of this variation may be attributed to the various stages in the plasma emission processes not reaching saturation. If one concentrates on the maximum brightness temperatures observed then there is reasonable agreement between theory and observation. The brightest second harmonic plasma emission and the brightest fundamental emission have about the same brightness temperature (up to about 10^{15} K) and this is in reasonable agreement with what one expects for the effective temperature of the Langmuir waves from the saturation of the streaming instability.

Non-linear Scattering by Thermal Ions

The existence of the quadratic non-linear response of a plasma has an important qualitative effect on scattering of waves by particles at long wavelengths. When the plasma response is neglected the scattering is Thomson scattering by electrons. Thomson scattering by ions is possible, but its cross-section is smaller than that for electrons by the square of the ratio of the masses, and so it is negligible in practice. Note for a later comparison that if a monochromatic beam of radiation is Thomson scattered by thermal electrons, then the scattered radiation has a Doppler width $\Delta\omega$ characteristic of the thermal motion of electrons, that is, $\Delta\omega/\omega \approx V_e/c$.

The inclusion of the response of the plasma allows a different form of scattering, called *non-linear scattering*, that interferes with Thomson scattering. This is associated with the self-consistent or shielding field that a charge q generates. To see this, consider the Fourier transform of the current of the charge, as given by (20.2). When this is inserted as the source term in the wave equation, the resulting field, cf. (16.3) with (16.5), is the shielding field associated with the charge q. This field is

$$A_i^{(q)}(\omega, \mathbf{k}) = -\frac{2\pi q \lambda_{ij}(\omega, \mathbf{k}) v_j}{\varepsilon_0 \omega^2 \Lambda(\omega, \mathbf{k})} \, e^{-i\mathbf{k}\cdot\mathbf{x}_0} \delta(\omega - \mathbf{k}\cdot\mathbf{v}). \tag{26.29}$$

Formally, non-linear scattering is due to the shielding field $\mathbf{A}^{(q)}$ associated with the scattering particle beating with the field $\mathbf{A}_{M'}$ of the initial wave to produce an extraneous current through the response (26.2); this current is added to the current (25.7) for Thomson scattering to produce the total current for scattering. The total scattering probability for a non-relativistic particle with charge q and mass m is of the form (25.15) with the factor $a_{MM'}(\mathbf{k}, \mathbf{k}'; \mathbf{v})$ defined in (25.14) replaced by

$$a_{MM'}(\mathbf{k}, \mathbf{k}'; \mathbf{v}) \approx \mathbf{e}_M^*(\mathbf{k}) \cdot \mathbf{e}_{M'}(\mathbf{k}') \left[1 + \frac{em}{qm_e} \frac{\chi^{\mathrm{Le}}(\omega - \omega', \mathbf{k} - \mathbf{k}')}{K^{\mathrm{L}}(\omega - \omega', \mathbf{k} - \mathbf{k}')} \right],$$

$$\tag{26.30}$$

where the unit term inside the square brackets arises from Thomson scattering and where the approximate form (26.23) is used to include non-linear scattering.

There are two limiting cases in which non-linear scattering, described by the final term inside the square brackets in (26.30), simplifies. Note first that the denominator is $K^{\mathrm{L}} = 1 + \chi^{\mathrm{Le}} + \chi^{\mathrm{Li}}$, where χ^{Li} denotes the contribution of the ions. The two limiting cases are $K^{\mathrm{L}} \approx 1 \gg \chi^{\mathrm{Le}}$ and $K^{\mathrm{L}} \approx \chi^{\mathrm{Le}}$. The former applies for $|\omega - \omega'| \gg |\mathbf{k} - \mathbf{k}'| V_e$, that is, when the beat between the scattered and unscattered waves is a fast disturbance. The latter approximation applies for $|\omega - \omega'| \ll |\mathbf{k} - \mathbf{k}'| V_e$, when in (26.30) one has $\chi^{\mathrm{Le}} \approx 1/|\mathbf{k} - \mathbf{k}'|^2 \lambda_{\mathrm{De}}^2$, which is assumed $\gg 1$. Let us refer to these two limiting cases as the short wavelength limit and the long wavelength limit, respectively.

Consider scattering by electrons, that is, $q = -e$ and $m = m_e$ in (26.30). In the short wavelength limit the correction due to non-linear scattering is small, and the scattering is essentially Thomson scattering. However, in the long wavelength limit the final term inside the square brackets is approximately equal to -1, and there is strong destructive interference between Thomson scattering and non-linear scattering. This may be attributed to the Debye shielding of an electron. The Debye shielding is associated with a local deficiency in the electron density comoving with the scattering electron. A wave with wavelength $\gg \lambda_{De}$ sees the Debye shielding effectively as a positively charged quasi-particle with the same mass as the electron, smeared out over a radius $\gtrsim \lambda_{De}$. Thomson scattering by the negatively charged electron and that by this positively charged quasi-particle interfere destructively in the long wavelength limit, so that the net scattering by electrons is greatly reduced.

Now consider scattering by ions. Thomson scattering alone is ineffective due to the probability being inversely proportional to the square of the mass, specifically to q^2/m^2. However, in the long wavelength limit the final term in (26.30) may be approximated by em/qm_e, and because this factor appears squared in the probability (25.15), the probability becomes proportional to e^2/m_e^2 rather than q^2/m^2. Thus in the long wavelength limit scattering by thermal ions occurs with essentially the same cross-section as Thomson scattering by electrons *in vacuo*, and the scattering in a plasma is dominated by the ions rather than the electrons. This implies that the line width of scattered radiation should be characteristic of the Doppler spread associated with the thermal motion of the ions, which is much less than the Doppler spread associated with thermal electrons. Historically, this is of interest because the confirmation that backscattered radiation from the ionosphere has this narrower spread was one of the first notable successful predictions of non-linear plasma theory.

Exercise Set 26

26.1 The saturation level for a three-wave interaction $P+Q \leftrightarrow M$ occurs when the quantity inside the curly brackets in (26.7)–(26.9) is zero. This condition corresponds to

$$N_M = \frac{N_P N_Q}{1 + N_P + N_Q}, \quad N_P = N_M \frac{1 + N_Q}{N_Q - N_M}, \quad N_Q = N_M \frac{1 + N_P}{N_P - N_M},$$
$$(E26.1)$$

where arguments are omitted. These conditions may be expressed in terms of the effective temperatures by writing $N_M = T_M/\hbar\omega_M$, etc.

(a) Apply the first of the relations ($E26.1$) to second harmonic generation ($T_M \to T_2$, $T_P = T_Q \to T_1$) of laser light with frequency ω_0 and show that it reproduces the equality in (26.14) for $T_1 \gg T_0$, with $T_0 = \hbar\omega_0$.

(b) Estimate to within a factor of 2 the values of T_0 and T_1 for a pulse of ruby laser light in a range $\Delta\lambda = 0.1\,\mathrm{nm}$ about the wavelength $\lambda = 694.3\,\mathrm{nm}$, with a power of $1\,\mathrm{MW}$ confined to a solid angle $\Delta\Omega = \pi\theta^2$ with $\theta = 2\,\mathrm{mrad}$. Hence confirm that the neglect of the unit term in ($E26.1$) is well justified in this case. (Boltzmann's constant $= 1.38 \times 10^{-23}\,\mathrm{J\,K^{-1}}$, $\hbar = 1.05^{-34}\,\mathrm{J\,s}$.)

(c) Suppose $N_M, N_Q \gg N_P$ initially, so that the second of ($E26.1$) determines the saturation level for N_P for $N_Q > N_M$. For $N_Q < N_M$ the saturation level is negative, which requires interpretation. Explain what occurs in this case by discussing the evolution of the three occupation numbers in a qualitative way.

(d) Consider the case of photon splitting with $N_M \ll 1$ and $N_P = N_Q = 0$ initially. Does saturation occur in this case? By discussing the evolution qualitatively, either interpret the saturation level or explain why none occurs.

(e) Discuss the saturation level for the waves M in the processes $P \pm s \leftrightarrow M$ when the level of the ion sound waves corresponds to $N_M \ll N_s \ll N_P$ initially.

26.2 Derive the non-linear polarizability (26.12) by extending the calculation of the linear polarizability given in §9.4 as follows.

(a) Show that (9.26) implies that the second order correction to the density matrix satisfies

$$i\hbar \frac{\mathrm{d}}{\mathrm{d}t}\left[\rho_{qq'}^{(2)}(t)e^{i\omega_{qq'}t}\right] = -\sum_{q''}\left[\rho_{qq''}^{(1)}(t)H_{q''q'}^{(1)}(t)\right.$$
$$\left. - H_{qq''}^{(1)}(t)\rho_{q''q'}^{(1)}(t)\right]e^{i\omega_{qq'}t}. \quad (E26.2)$$

(b) Show that after inserting (9.23) with (9.25), *viz.*, $H_{qq'}^{(1)}(t) = -\mathbf{d}_{qq'} \cdot \mathbf{E}(t)e^{i\omega_{qq'}t}$, then Fourier transforming and writing the convolution integral in the form (6.24) that (E26.2) gives

$$\hbar(\omega - \omega_{qq'})\rho_{qq'}^{(2)}(\omega) = \sum_{q''} \int d\omega^{(2)} \left[\rho_{qq''}^{(1)}(\omega_1)\mathbf{d}_{q''q'} \right.$$
$$\left. - \mathbf{d}_{qq''}\rho_{q''q'}^{(1)}(\omega_1) \right] \cdot \mathbf{E}(\omega_2). \quad (E26.3)$$

(c) Using (9.28) and (E26.3), identify the non-linear polarizability by writing

$$\langle d_i^{(2)} \rangle = \sum_{qq'} (d_i)_{qq'}\rho_{q'q}^{(2)}(\omega) = \int d\omega^{(2)} a_{ijl}E_j(\omega_1)E_l(\omega_2). \quad (E26.4)$$

(d) Arrange the result in three terms proportional to $\rho_{qq}^{(0)}$, $\rho_{q'q'}^{(0)}$, $\rho_{q''q''}^{(0)}$, relabel the states q, q' and q'' so that the density matrix appears only for the state q, and then impose the first of the symmetry properties (26.3) explicitly.

(e) Show that the resulting six terms include the term shown explicitly in (26.12) and that the other five terms are such that they may be generated from the one written using the all symmetries implied by (26.3).

26.3 Frequency doubling of ruby laser light occurs in the particular uniaxial crystal known as KDP (chemically KH_2PO_4) for $n_x(2\omega, \theta) = n_o(\omega)$ at $\theta \approx 50°$. What can you deduce about the values of $K_\perp$ and $K_\parallel$ for KDP given this information?

26.4 The linear non-linear response tensors for a magnetized vacuum may be derived from the Heisenberg–Euler Lagrangian,

$$L = \tfrac{1}{2}\varepsilon_0 \left[A + \frac{\alpha}{45\pi B_c^2}(A^2 + 7C^2) + \frac{2\alpha}{315\pi B_c^4}(2A^3 + 13AC^2) + \cdots \right],$$

$$A = |\mathbf{E}|^2 - |\mathbf{B}|^2, \quad C = \mathbf{E}\cdot\mathbf{B}, \quad (E26.5)$$

where $B_c = m_e^2 c^2/\hbar e$ is the critical magnetic field, by identifying $\mathbf{E}$ as the electric field in a wave, separating $\mathbf{B}$ into a background part $\mathbf{B}_0$ plus the magnetic field $\delta\mathbf{B}$ in the wave, and using

$$\mathbf{P} = -\partial L/\partial\mathbf{E}, \quad \mathbf{M} = -\partial L/\partial\delta\mathbf{B}. \quad (E26.6)$$

(a) Show that this procedure reproduces the result quoted in Exercise 12.2 for the linear response tensor for a magnetized vacuum.

(b) Explain in detail how the quadratic non-linear response tensor $\alpha_{ijl}^{(2)}$ is to be calculated from the non-linear correction terms in (E26.6).

26.5 Show that in an isotropic plasma a three-wave interaction involving three transverse waves does not occur for two independent reasons:

(a)   Show that the beat conditions (26.1) cannot be satisfied for three waves with dispersion relation $\omega^2 = \omega_{\rm p}^2 + k^2 c^2$.

(b)   Show that on substituting the form (26.21) for the non-linear response tensor into the probability (26.6b) that when all waves have transverse polarization the quantity α_{MPQ} with $M = P = Q = {\rm T}$ is identically zero.

26.6 Consider the kinematic restrictions implied by the δ-functions in (26.24) for the processes ${\rm L} \pm {\rm s} \to {\rm L}'$.

(a)   Using $\omega_{\rm L}(k) \approx \omega_{\rm p} + 3k^2 V_{\rm e}^2/2\omega_{\rm p}$ and $\omega_{\rm s}(k) \approx kv_{\rm s}$, show that frequency matching condition requires

$$k_{\rm L}^2 \pm k_0 k_{\rm s} = k_{{\rm L}'}^2, \quad k_0 = 2v_{\rm s}\omega_{\rm p}/3V_{\rm e}^2. \qquad (E26.7)$$

(b)   Show that the wavevector matching requires

$$\boldsymbol{\kappa}_{\rm L} \cdot \boldsymbol{\kappa}_{\rm s} = (k_0 \mp k_{\rm s})/2k_{\rm L}. \qquad (E26.8)$$

(c)   Determine the limitations on $k_{\rm s}$ and $k_{\rm L}$ implied by the requirement $-1 \le \boldsymbol{\kappa}_{\rm L} \cdot \boldsymbol{\kappa}_{\rm s} \le 1$.

(d)   Show that for $k_{\rm s} \ll k_{\rm L}$ the three-wave interaction is possible only for Langmuir waves with $k_{\rm L} > k_0$.

26.7 The non-linear response tensor for three static slow disturbances in a thermal plasma may be evaluated using a model in which the response is described by the induced charged density responding to an electrostatic field in the Coulomb gauge. The response of one species with charge q, mass m, mean number density n and thermal speed V is described by

$$\rho(\mathbf{x}) = qne^{-q\phi(\mathbf{x})/mV^2}. \qquad (E26.9)$$

(a)   Show that the second term in the expansion of the charge density is $\rho^{(2)} = q^3 n|\phi(\mathbf{x})|^2/2m^2 V^4$.

(b)   Fourier transform the response in part (a) in space.

(c)   Show that the result implies that in the low-frequency (for all frequencies) limit the non-linear response tensor is

$$\alpha_{ijl}^{(2)}(\omega, \mathbf{k}; \omega', \mathbf{k}'; \omega'', \mathbf{k}'') = \frac{q^2 n}{2m^2 V^4} \frac{\omega\omega'\omega''}{|\mathbf{k}|^2|\mathbf{k}'|^2|\mathbf{k}''|^2} k_i k_j' k_l''. \qquad (E26.10)$$

Bibliographic Notes

No explicit references to the published literature are made in the text. The following list of textbooks and monographs (plus several review papers) is intended to provide an avenue into the literature, as well as indicating sources for further background reading. First we list some general references on electromagnetic theory and on mathematical methods, and then give more specific references for each of the five Parts.

General references

Abramowitz, M. & Stegun, I.A. (1965). *Handbook of Mathematical Functions*. Dover Publications: New York.

Gradshteyn, I.S. & Ryzhik, I.M. (1980). *Table of Integrals, Series, and Products*. Academic Press: New York.

Jackson, W.D. (1975), *Classical Electrodynamics*. John Wiley & Sons: New York.

Jones, D.S. (1964). *The Theory of Electromagnetism*. Pergamon Press: Oxford.

Landau, L.D. & Lifshitz, E.M. (1971). *The Classical Theory of Fields*. Pergamon Press: Oxford.

Morse, P.M. & Feshbach, H. (1953). *Methods of Mathematical Physics*. McGraw-Hill: New York.

Panofsky, W.K.H. & Phillips, M. (1962). *Classical Electricity and Magnetism*. Addison-Wesley: Reading Mass.

Stratton, J.A. (1941). *Electromagnetic Theory*. McGraw-Hill: New York.

Watson, G.N. (1944). *A Treatise on the Theory of Bessel Functions*. Cambridge University Press: Cambridge.

Whittaker, E.T. & Watson, G.N. (1965). *A Course of Modern Analysis*. Cambridge University Press: Cambridge.

Van Bladel, J. (1985). *Electromagnetic Fields*. Springer-Verlag: Berlin.

Part One: Electromagnetic fields in vacuo
Abram, J. (1965). *Tensor Calculus Through Differential Geometry*. Butterworths: London.
Bracewell, R. (1965). *The Fourier Transform and its Applications*. McGraw-Hill: New York.
Brillouin, L. (1965).*Tensors in Mechanics and Elasticity*. Academic Press: New York.
Champeney, D.C (1973). *Fourier Transforms and their Physical Applications*. Academic Press: New York.
Lighthill, M.J. (1960). *Introduction to Fourier Analysis and Generalized Functions*. Cambridge University Press: Cambridge.

Part Two: Electromagnetic responses of media
Fried, B.D. & Conte, S.D. (1961). *The Plasma Dispersion Function*. Academic Press, New York.
Landau, L.D. & Lifshitz, E.M. (1960). *Electrodynamics of Continuous Media*. Pergamon Press: Oxford.
Lifshitz, E.M. & Pitaevskii, L.P. (1981). *Physical Kinetics*. Pergamon Press: Oxford.
O'Dell, T.H. (1970). *Electrodynamics of Magneto-electric Media*. North-Holland: Amsterdam
Sitenko, A.G. (1967). *Electromagnetic Fluctuations in Plasma*. Academic Press, New York.

Part Three: Wave properties
Akhiezer, A.I., Akhiezer, I.A., Polovin, R.V., Sitenko, A.G. & Stepanov, K.N. (1967). *Collective Oscillations in a Plasma*. M.I.T. Press: Cambridge, Mass.
Born, M. & Wolf, E. (1966). *Principles of Optics*. Pergamon Press: Oxford.
Budden, K.G. (1961). *Radio Waves in the Ionosphere*. Cambridge University Press: Cambridge.
Chen, H.C. (1985). *Theory of Electromagnetic Waves*. McGraw-Hill: New York.
Clarke, D. & Grainger, J.F. (1971). *Polarized Light and Optical Measurement*. Pergamon Press: Oxford.
Ginzburg, V.L. (1964). *The Propagation of Electromagnetic Waves in Plasmas*. Pergamon Press: Oxford.
Shurcliff, W.A. (1962). *Polarized Light*. Oxford University Press: Oxford.
Stix, T.H. (1962). *The Theory of Plasma Waves*. McGraw-Hill: New York.

Part Four: Theory of emission processes
Batygin, V.V. & Toptygin, I.N. (1964). *Problems in Electrodynamics*. Academic Press: New York.
Bloembergen, N. (1965). *Nonlinear Optics*. W.A. Benjamin Inc: New York.
Kaplan, S.A. & Tsytovich, V.N. (1973). *Plasma Astrophysics*. Pergamon Press: Oxford.
Melrose, D.B. (1980). *Plasma Astrophysics Volume I Emission, Absorption and Transfer of Waves in Plasmas*. Gordon & Breach: New York.

Rohrlich, F. (1965). *Classical Charged Particles*. Addison-Wesley: Reading, Mass.

Tsytovich, V.N. (1972). *An Introduction to the Theory of Plasma Turbulence*. Pergamon Press: Oxford.

Part Five: Specific emission processes

Baldwin, G.C. (1969). *An Introduction to Nonlinear Optics*. Plenum Press: New York.

Bekefi, G. (1966). *Radiation Processes in Plasmas*. John Wiley & Sons: New York.

Blumenthal, G.B. & Gould, R.J. (1970). Bremsstrahlung, synchrotron radiation and Compton scattering of high-energy electrons traversing dilute gases. *Rev. Mod. Phys.* **42**, 237.

Dulk, G.A. (1985). Radio emission from the Sun and stars. *Ann. Rev. Astron. Astrophys.* **23**, 169.

Ginzburg, V.L. (1979). *Theoretical Physics and Astrophysics*. Pergamon Press: Oxford.

Ginzburg, V.L. & Syrovatskii, S.I. (1965). Cosmic magnetobremsstrahlung (synchrotron radiation). *Ann. Rev. Astron. Astrophys.* **3**, 297.

Jelley, J.V. (1958). *Cerenkov Radiation and its Application*. Pergamon Press: Oxford.

McLean, D.J. & Labrum, N.R. (1985). *Solar Radiophysics*. Cambridge University Press: Cambridge.

Melrose, D.B. (1980). *Plasma Astrophysics Volume 2 Astrophysical Applications*. Gordon & Breach: New York.

Melrose, D.B. (1986). *Instabilities in Space and Laboratory Plasmas*. Cambridge University Press: Cambridge.

Schott, G.A. (1912). *Electromagnetic Radiation*. Cambridge University Press: Cambridge.

Sokolov, A.A. & Ternov, I.M. (1968). *Synchrotron Radiation*. Pergamon Press: Oxford.

Trubnikov, B.A. (1985). Particle interactions in fully ionized plasma. *Rev. Plasma Phys.* **1**, 105.

Index